S0-CAS-676

Carbocyclic Cage Compounds: Chemistry and Applications

Methods in Stereochemical Analysis

Series Editor: Alan P. Marchand
Department of Chemistry
University of North Texas
NT Station, Box 5068
Denton, TX 76203-5068

Carbocyclic Cage Compounds: Chemistry and Applications

Edited by
Eiji Osawa
and
Osamu Yonemitsu

Eiji Osawa
Department of Knowledge-Based
Information Engineering
Toyohashi University of Technology
Tempaku-cho, Toyohashi 441
Japan

Osamu Yonemitsu
Department of Chemistry
Faculty of Science
Hokkaido University
Sapporo 060
Japan

Library of Congress Cataloging-in-Publication Data

Carbocyclic cage compounds : chemistry and applications / editors,
Eiji Osawa, Osamu Yonemitsu.
p. cm. — (Methods in stereochemical analysis)
Includes bibliographical references and index.
ISBN 0-89573-728-0
1. Cage hydrocarbons. I. Osawa, Eiji, 1936– . II. Yonemitsu,
Osamu, 1930– . III. Series.
QD305.H9C37 1992
547'.5 – dc20 92-9636
CIP

Cover illustration by Harry Kroto and Christer Aakeröy

Printed in the United States of America

ISBN 0-89573-728-0 VCH Publishers
ISBN 3-527-27933-4 VCH Verlagsgesellschaft

Printing History:
10 9 8 7 6 5 4 3 2 1

Published jointly by:

VCH Publishers, Inc.
220 East 23rd Street
Suite 909
New York, New York 10010

VCH Verlagsgesellschaft mbH
P.O. Box 10 11 61
D-6940 Weinheim
Federal Republic of Germany

VCH Publishers (UK) Ltd.
8 Wellington Court
Cambridge CB1 1HZ
United Kingdom

Preface

About 10 years ago, the present editors and Professor T. Mukai, now retired from Tohoku University, were granted a three-year fund from the Ministry of Education of Japan, which allowed us to organize several symposia for a group of about two dozen organic chemists in Japan, who were engaged in the research of *three-dimensional polycyclic molecules*. We convened for two days on an annual basis to discuss various aspects of the chemistry of cage compounds. After three years, many of us thought it would be good if the grant could be extended. To our pleasure, the grant was renewed again and again, and now the traditional meeting is still held almost every year.

Then came a timely suggestion from Professor Y. Takeuchi that we should write up the results of our research activities. After several discussions, we finally agreed to publish a collection of review articles that summarized the topics of cage compounds. In order to cover all of the pertinent topics, we invited several people who are active in this area from outside our organization to contribute articles. We intentionally omitted coverage of most of the highly strained saturated molecules and the cyclodextrins, which recently had been reviewed extensively elsewhere.

"Cage molecules" have long been chemists' sirens: Many of them possess incredibly beautiful structures with high symmetry. Nonchemists might wonder why it is possible in the chemical world that people are allowed to spend enormous amounts of time and resources to synthesize compounds that apparently have no understandably useful ends. There are, however, many reasons that justify chemists' addiction to their favorite molecules. Syntheses of these molecules are intellectually challenging. One would expect nonnatural compounds with unusual structures to exhibit hitherto unknown properties. Cage molecules have fixed geometries and hence are free from complications that might arise from conformational flexibility. Because of these advantages, cage molecules will never cease to attract chemists.

Since the time that the preparaton of this book began in earnest, at least three notable incidents have occurred that relate crucially to this book. First, the 1987 Nobel Prize in chemistry was awarded to three cage chemists (Cram, Lehn, and Pedersen). This provided invaluable encouragement to us. Second, a book of very similar title, *Cage Hydrocarbons* (G. A. Olah, Ed., John Wiley & Sons, Inc., New York, 1990) was published, which is dedicated to Professor Paul von R. Schleyer to honor his career, which spans 30 years of excellence in cage hydrocarbon chemistry. Fortunately, as Professor Schleyer himself told me, there was virtually no overlap between Professor Olah's book and ours, with the possible exception of Dr. Camille Ganter's chapter on adamantane rearrangements. I believe that Dr. Ganter's chapter nicely complements the chapters in the other book by supplying a highly microscopic view of "adamantane-land."

Finally, a fantastic superstar of cage molecules, C_{60}, came to the foreground! Though the molecule came into being in 1985, this nascent research area virtually exploded (in the words of Professor Harry Kroto) when a convenient method of preparation (extraction of carbon soot with solvent) was discovered in 1990. Subsequently, early in 1991, high-temperature superconductivity was observed in its doped film. It is hoped that this cage molecule will really have significant industrial applications so that it no longer will be necessary to apologize for the "fundamental" nature of research on cage molecules. We have been fortunate to obtain a chapter by Kroto and Walton on the chemistry of C_{60}, which also contains suggestions that are likely to prove to be highly stimulating to cage synthetic chemists.

In Chapter 1, Professor Alan Marchand summarizes his well-contemplated view on the perspective of cage chemistry, an area in which he is now one of the most productive researchers. Alan is the man behind the scenes who helped to develop this book, not only as editor of the series *Methods in Stereochemical Analysis*, but also by contributing his readable chapter.

Chapter 2 deals with a unique topic of chiral cage molecules, the fruitful research area long explored by Professor K. Naemura, with his mentor, Professor K. Nakazaki, now retired. I have a small personal recollection about this topic. When I was a postdoctoral fellow under Professor Schleyer, I proposed a way to depict D_3-trishomocubane that, I thought, best conveyed its unique chirality (see the right-hand drawing of **115** in Scheme 2.12 of Chapter 2). However, for a reason that I never understood, he did not like that and preferred instead to use a conventional drawing (like the left-hand one of **115**) in our publication on the first synthesis of this molecule. Soon after that, Professor Nakazaki's group launched their famous series of research on D_3-trishomocubane, probably stimulated by our synthesis, and used exactly the same drawing as my unlucky one!

I thought it my duty to almost force Professor H. Shirahama to write Chapter 3 for this book. The reason is that he is virtually the first person to utilize the concept of conformational control in the synthesis of three-dimensional natural products. His view originates from his keen perspective on natural products, which have been taught in terms of the traditional two-dimensional representation in this old field of chemistry. (Think how morphine is depicted in textbooks!) He is also one of the first natural product chemists to have utilized computational chemistry (he still maintains that MMI is the best) in his synthetic design. Dr. F. Matsuda helped him to include this historical story in their chapter.

One of the most promising young chemists, Dr. Y. Tobe, has turned the chemistry of propellanes into an exciting area of research by incorporating strain into these systems in much the same way as has been done by other pioneers in this field like Professors K. B. Wiberg and P. E. Eaton. Dr. Tobe's latest results and views are presented in Chapter 5.

Another promising man, Professor J. Nishimura, has contributed an important chapter on cyclophane chemistry. Upon reading his contribution (Chapter 6), we realized that there is hardly any limit to his imagination in expanding the scope of cyclophane chemistry.

I met Professor G. Mehta for the first time in Germany in 1990 during a NATO symposium on strain in organic chemistry. His talk on the projected (and still ongoing) synthesis of dodecahedrane impressed the audience very deeply. One other area in which Professor Mehta and his student S. Padma work extremely hard is the synthesis of unknown prismanes. Their Chapter 7 on this topic provides concise descriptions of experiments that have been performed in this fascinating area of synthetic organic chemistry. Actually I have been collaborating with them on prismanes for some years, and this connection is the reason for my success in obtaining this chapter.

Chapter 8 is written by Dr. H. Higuchi, who recently returned from a postdoctoral stint in Professor Eaton's laboratory in Chicago. With the help of Dr. I. Ueda, he relates the latest developments in the ever-changing world of cubane chemistry, which all chemists have found to be fascinating for nearly 30 years.

Dr. W. L. Dilling is another name familiar to anyone who has ever looked into the complicated array of polycyclic structures related to Aldrin and other polyhalogenated insecticides. The only author who has provided a private address for correspondence, his chapter systematically presents further developments in the chemistry of this old but interesting class of cage molecules.

Professors Waegell and Fournier contributed a substantial chapter dedicated to the history and developments of reflex and anti reflex effects. To our knowledge, this chapter provides the first detailed description of this important concept. In my personal opinion, the reflex effect represents one of the earlier recognitions of the importance of long-range nonbonded interactions, beyond the gauche (1,4) effect. In retrospect, conformational analysis has primarily been dominated by 1,4-interactions for a long time, and only recently has perspective gradually shifted to recognize the importance of 1,5-, 1,6-, and longer interactions.

Chapter 12 is written by the group headed by Professor T. Mukai, one of the founders of our cage chemists group. Professor Miyashi succeeded him at Tohoku University, while Dr. Yamashita moved to the Institute of Molecular Science shortly after completing this chapter. Their experience in photochemistry dates back to the time of an early study of the irradiation of tropolone, which Professor Mukai pursued when he was working with Professor T. Nozoe.

The final chapter is written by the Hirao–Yamashita–Yonemitsu team. The photochemical interconversion of norbornadiene to quadricyclane became a hot research project when the energy shortage was acute. This group played a leading role among the many groups in Japan who studied the norbornadiene–quadricyclane interconversion. Professor Hirao received a prize from the Pharmaceutical Society of Japan for his achievements in this area, many of which are detailed in this chapter.

Last but not least, we heartily express our gratitude to the officials of the Ministry of Education of Japan for their continued understanding of the significance of basic scientific research.

Because of our long-lasting affection for cage molecules, we believe that this volume will also be of interest to a broad segment of the chemical community, ranging from graduate organic students to professional chemists.

Eiji Osawa, Toyohashi
Osamu Yonemitsu, Sapporo
August 1991

Contents

Chapter 1. Polycyclic Cage Molecules: Useful Intermediates in Organic Synthesis and an Emerging Class of Substrates for Mechanistic Studies 1
Alan P. Marchand

1.1. Introduction 2
1.2. Synthetically Useful Cage Fragmentation Processes 3
1.3. Synthetic Applications of Cage Fragmentation Products 30
1.4. Caged Reactive Intermediates 39
1.5. Cage Compounds as Substrates for Mechanistic, Spectral, and Physical–Organic Studies 50
1.6. Epilog 53
Acknowledgments 53
References and Footnotes 54

Chapter 2. High-Symmetry Chiral Cage-Shaped Molecules 61
Koichiro Naemura

2.1. Introduction 62
2.2. Chiral Tricyclic Cage-Shaped Molecules 63
2.3. Chiral Tetracyclic Cage-Shaped Molecules 69
2.4. Chiral Pentacyclic Cage-Shaped Molecules 73
2.5. Molecules with Chiral Polyhedral Symmetry 78
2.6. Chiral Cage-Shaped Molecules of Adamantane Homolog 80
2.7. Chiroptical Properties of Cage-Shaped Molecules 83
2.8. Biocatalysts in Syntheses of Optically Active Cage-Shaped Molecules 86
References 87

Chapter 3. Postfullerene Organic Chemisty 91
H. W. Kroto and D. R. M. Walton

3.1. Introduction 91
3.2. Spontaneous Fullerene Formation 92
3.3. C_{60} Formation in Flames 94
3.4. Pentagons among the Hexagons 94
3.5. The Significance of Barth and Lawton's Corannulene Synthesis 94
3.6. Routes to Nonplanar sp^2 Networks 95
3.7. Discussion 98
Acknowledgments 99
References 99

Chapter 4. A New Look at Natural Products Chemistry in Three Dimensions 101
Fuyuhiko Matsuda and Haruhisa Shirahama

4.1. Introduction 101
4.2. Stereocontrolled Transannular Cyclization of Macrocyclic Compounds 102
4.3. Stereoselective Intermolecular Reactions of Macrocycles 111
References 121

Chapter 5. Propellanes 125
Yoshito Tobe

5.1. Introduction 125
5.2. Small-Ring Propellanes 126
5.3. Propellanes as Stereochemical Models 137
5.4. Synthetic Application of Propellanes 140
5.5. Naturally Occurring Propellanes 149
References 150

Chapter 6. Cyclophanes from Vinylarenes 155
Jun Nishimura

6.1. Introduction 155
6.2. Intramolecular [2 +2] Photocycloaddition of Styrene Derivatives toward a New Class of Cyclophanes 159
6.3. Cationic Cyclocodimerization 172
Acknowledgments 179
References 179

Chapter 7. Syntheses of Prismanes 183
Goverdhan Mehta and S. Padma

7.1. Introduction 183
7.2. Synthetic Challenge and Tactics 187
7.3. [3]-Prismane 189
7.4. [4]-Prismane 189
7.5. [5]-Prismane 196
7.6. Toward [6]-Prismane 199
7.7. Toward [7]-Prismane 209
7.8. Outlook 211
Acknowledgments 212
References 212

Chapter 8. Recent Developments in the Chemistry of Cubane 217
Hiroyuki Higuchi and Ikuo Ueda

8.1. Introduction 218
8.2. Methodology for Direct Functionalization of Cubane 219
8.3. Reactions of Cubanes with Electron-Deficient Species 227
8.4. Highly Pyramidalized Olefins in the Cubane System—Homocubene and Cubene 232
8.5. New Type of Cubane Derivatives—Combination of Cubane with π-Electronic Ring System 234
8.6. Developments of the Related Chemistry of Cubane 237
8.7. Future Chemistry of Cubane and Its Perspectives 240
8.8. Conclusion 243
Acknowledgments 244
References 244

Chapter 9. Recent Advances in Selected Aspects of Bishomocubane Chemistry 249
Wendell L. Dilling

9.1. Introduction 249
9.2. Theoretical Calculations 250
9.3. Formation Reactions 254
9.4. Reactions 268
References 290

Chapter 10. A New Approach to the Adamantane Rearrangements 293
Camille Ganter

10.1. Introduction 294
10.2. Aluminum Halide Catalyzed Rearrangements 295
10.3. Rearrangements of Regioselectively Generated Carbocations under Ionic Hydrogenation Conditions 300

10.4. Product Analyses 315
10.5. Concluding Remarks 315
Acknowledgments 315
References 316

Chapter 11. Reflex and Anti Reflex Effects: Discovery and Developments 319
Josette Fournier and Bernard Waegell

11.1. Introduction 320
11.2. Experimental Data 322
11.3. Theoretical Approach to the Reflex and Anti Reflex Effects 346
11.4. Conclusion 360
Acknowledgments 362
References 362

Chapter 12. Cage Molecules in Photochemistry 365
Tsutomu Miyashi, Yoshiro Yamashita, and Toshio Mukai

12.1. Introduction 365
12.2. Photochemical Syntheses of Cage Molecules 366
12.3. Photochemical Reactions of Cage Molecules 375
References 380

Chapter 13. Recent Studies on Valence Isomerization between Norbornadiene and Quadricyclane 383
Ken-ichi Hirao, Asami Yamashita, and Osamu Yonemitsu

13.1. Introduction 383
13.2. Acylnorbornadiene and Acylquadricyclane System 384
13.3. The Photochemical Isomerization of Norbornadienes into Quadricyclanes 387
13.4. The Reversion of Quadricyclanes into Norbornadienes 390
13.5. Miscellaneous Studies 395
Acknowledgments 396
References 396

Index 401

Contributors

Wendell L. Dilling, 1810 Norwood Drive, Midland, Michigan 48640
Josette Fournier, Laboratoire de Chimie Bioorganique, Université d'Angers, IUT Belle-Beille, 49000 Angers, France
Camille Ganter, Swiss Federal Institute of Technology, Laboratory of Chemistry, ETH-Zentrum, CH-8092 Zürich, Switzerland
Hiroyuki Higuchi, Institute of Scientific and Industrial Research, Osaka University, 8-1 Mihogaoka, Ibaraki, Osaka 567, Japan
Ken-ichi Hirao, Department of Chemistry, Faculty of Science, Hokkaido University, Sapporo 060, Japan
Harry W. Kroto, School of Chemistry and Molecular Sciences, University of Sussex, Falmer, Brighton BN1 9QJ, United Kingdom
Alan P. Marchand, Department of Chemistry, University of North Texas, Denton, Texas 76203-5068
Fuyuhiko Matsuda, Department of Chemistry, Faculty of Science, Hokkaido University, Sapporo 060, Japan
Goverdhan Mehta, School of Chemistry, University of Hyderabad, Hyderabad 500134, and Jawaharlal Nehru Centre for Advanced Scientific Research, Indian Institute of Science Campus, Bangalore 560012, India
Tsutomu Miyashi, Department of Chemistry, Tohoku University, Sendai, Miyagi 980, Japan
Toshio Mukai, Department of Chemistry, Tohoku University, Sendai, Miyagi 980, Japan
Koichiro Naemura, Department of Synthetic Chemistry, Faculty of Basic Engineering, Osaka University, Toyonaka, Osaka 560, Japan
Jun Nishimura, Department of Chemistry, Faculty of Engineering, Gunma University, Tenjin-cho, Kiryu, Gunma 376, Japan
S. Padma, School of Chemistry, University of Hyderabad, Hyderabad 500314, India
Haruhisa Shirahama, Department of Chemistry, Faculty of Science, Hokkaido University, Sapporo 060, Japan
Yoshito Tobe, Department of Applied Fine Chemistry, Faculty of Engineering, Osaka University, 2-1 Yamada-Oka, Suita, Osaka 565, Japan
Ikuo Ueda, Institute of Scientific and Industrial Research, Osaka University, 8-1 Mihogaoka, Ibaraki, Osaka 567, Japan
Bernard Waegell, Laboratoire de Stéréochimie Associé au CNRS, LASCO, Faculté des Sciences St-Jérôme, Université d'Aix-Marseille III, 13397 Marseille Cedex 13, France

D. R. M. Walton, School of Chemistry and Molecular Sciences, University of Sussex, Falmer, Brighton BN1 9QJ, United Kingdom
Asami Yamashita, Hokkaido Institute of Pharmaceutical Sciences, Katsuraoka, Otaru 047-02, Japan
Yoshiro Yamashita, Department of Chemistry, Tohoku University, Sendai, Miyagi 980, Japan
Osamu Yonemitsu, Faculty of Pharmaceutical Sciences, Hokkaido University, Sapporo 060, Japan

1

Polycyclic Cage Molecules: Useful Intermediates in Organic Synthesis and an Emerging Class of Substrates for Mechanistic Studies

Alan P. Marchand

University of North Texas, Denton, Texas

Contents

1.1. Introduction
1.2. Synthetically Useful Cage Fragmentation Processes
    1.2A. Thermal Processes
    1.2B. Reductive Processes
    1.2C. Oxidative Processes
    1.2D. Lewis and Brønsted Acid Promoted (Cationic) Processes
    1.2E. Base-Promoted (Anionic) Processes
    1.2F. Transition-Metal-Promoted Reactions
1.3. Synthetic Applications of Cage Fragmentation Products
    1.3A. Synthesis of Triquinane Natural Products
    1.3B. Synthesis of Crown Ethers and Molecular Clefts
    1.3C. Use of Cage Molecules as Synthetic Templates
1.4. Caged Reactive Intermediates
    1.4A. Caged Pyramidalized Alkenes
    1.4B. Caged Cationic Intermediates
    1.4C. Caged Radical Intermediates
    1.4D. Caged Anionic Intermediates
    1.4E. Caged Divalent Carbon Species (Carbenes and Carbenoids)

1.5. Cage Compounds as Substrates for Mechanistic, Spectral, and Physical–Organic Studies
1.5A. Mechanism of Transmission of Electronic Substituent Effects
1.5B. Studies on Stereoelectronic Control of Reactions in Cage Systems; Evidence for the Importance of σ Hyperconjugation
1.6. Epilog
Acknowledgments
References

1.1. Introduction

Polycarbocyclic "cage" molecules comprise a unique class of organic compounds. Compounds of this type generally possess rigid, compact, and often highly strained structures whose mere existence is an apparent affront to the second law of thermodynamics. Many carbocyclic cage compounds possess aesthetically pleasing and often unusual symmetry characteristics that add to their attractiveness as synthetic targets. The "challenge and reward"[1] inherent in studies of the synthesis and chemistry of novel polycyclic cage molecules has been assessed in several recent reviews.[2–5] One by one, classic target molecules continue to yield to rational synthesis. Among the more prominent examples in this regard are cubane[6] and dodecahedrane,[7,8] both of which are carbocyclic analogs of Platonic solids.

In addition, the availability of polycyclic cage molecules has provided physical–organic chemists with a novel class of substrates for mechanistic studies. Several interesting and unusual rearrangements have been reported in cage systems. The driving force for these rearrangements often can be related to relief of steric strain and/or proximity effects associated with the compact structures so characteristic of cage molecules. Also, because most cage molecules possess rigid carbocyclic frameworks, bond distances and bond angles can be estimated reliably in these systems.[9] Thus, they constitute an attractive potential class of substrates for a variety of physical–organic studies such as, for example, the mechanism of transmission of electronic substituent effects in saturated systems[10] and the question of through-bond versus through-space orbital interactions.[11]

The successes that synthetic and physical–organic chemists have enjoyed in their pursuit of these often elusive target molecules have, on occasion, engendered considerable interest and excitement within the community of theoretical chemists.[12,13] Nevertheless, for the most part, cage molecules themselves and the methods employed in their synthesis only recently have begun to move beyond the realm of scientific curiosity. This situation is both surprising and regrettable. The routes employed to synthesize cage molecules, which rely generally upon kinetic control to circumvent the vagaries of thermodynamics, are often highly stereoselective and/or regioselective. Frequently, such reactions possess considerable generality; thus, they constitute promising additions to the growing arsenal of synthetic organic methods. Furthermore, the cage molecules themselves can serve as valuable intermediates and as "templates" for synthesizing structurally complex, noncage polycyclic molecules.

Although cage molecules continue to be an underutilized resource, the weight of current evidence confirms their value as substrates for mechanistic studies and as intermediates in the synthesis of complex organic molecules and/or natural products. The purpose of this review is to call attention to some particularly interesting and useful applications of polycyclic cage molecules and, in so doing, hopefully to stimulate new work that will further exploit their use as substrates for mechanistic studies and as intermediates in organic synthesis.

1.2. Synthetically Useful Cage Fragmentation Processes

1.2A. Thermal Processes

The pentacyclo[$5.4.0.0^{2,6}.0^{3,10}.0^{5,9}$]undecane (PCUD) skeleton has long been recognized as being a "rich repository of five-membered rings."[14] Access to the cyclopentanoid systems contained therein has been gained, for example, by employing functionalized PCUDs as intermediates in a two-step "photothermal metathesis" scheme[15] for the synthesis of cis,syn,cis linear triquinanes.[16] The first step of this reaction sequence involves intramolecular [2 + 2] photocyclization of the endo cycloadduct, **1**, formed via Diels–Alder addition of cyclopentadiene to appropriately substituted *p*-benzoquinones. The resulting cage compounds (i.e., substituted PCUD-8,11-diones, **2**) are subjected subsequently to flash vacuum pyrolysis. Under these conditions, fragmentation of the cyclobutane ring in **2** occurs via cleavage of the C(1)—C(7) and C(2)—C(6) σ bonds, thereby affording the corresponding linear triquinane, **3** (Scheme 1.1).[15] The overall two-step transformation of **1** to **3** is a net olefin metathesis.[15] A similar reaction sequence has been performed on other PCUD derivatives (e.g., **4** and **5**, Scheme 1.2).[16,17]

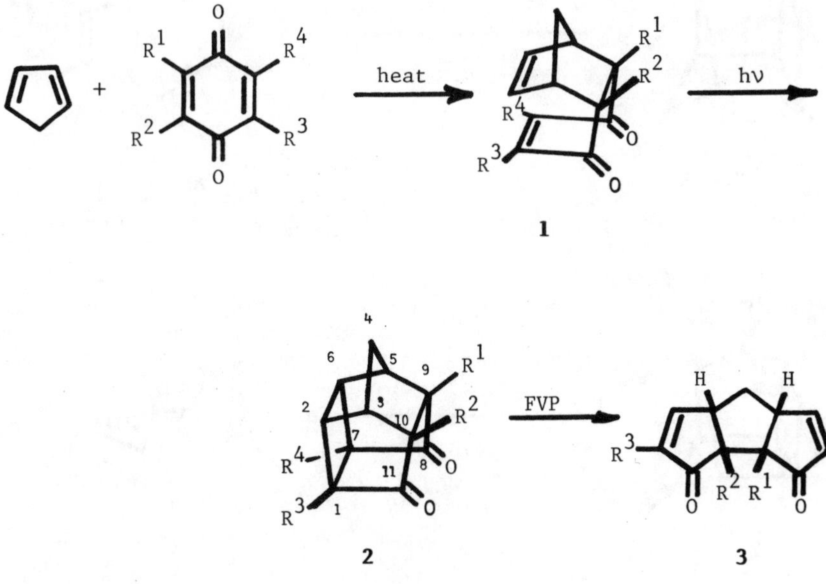

Scheme 1.1 ■

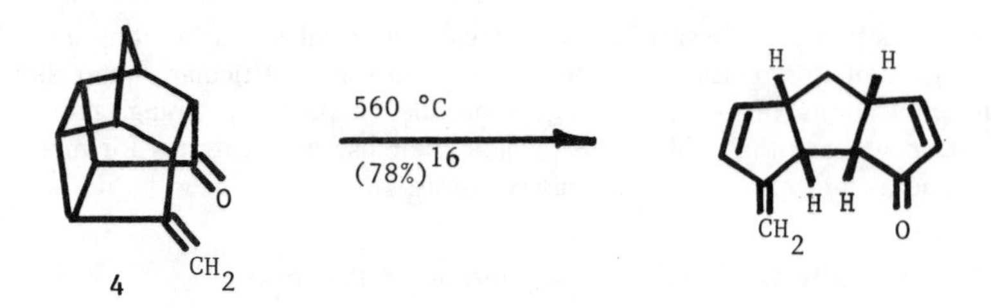

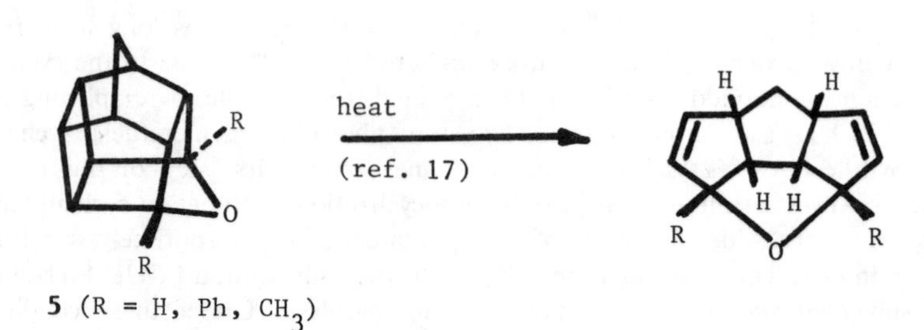

Scheme 1.2 ■

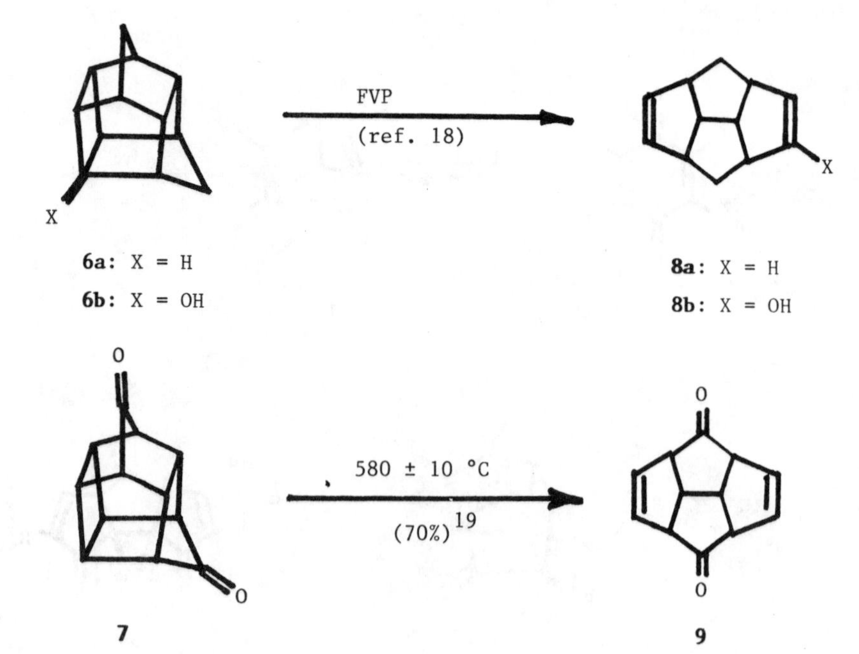

Scheme 1.3 ■

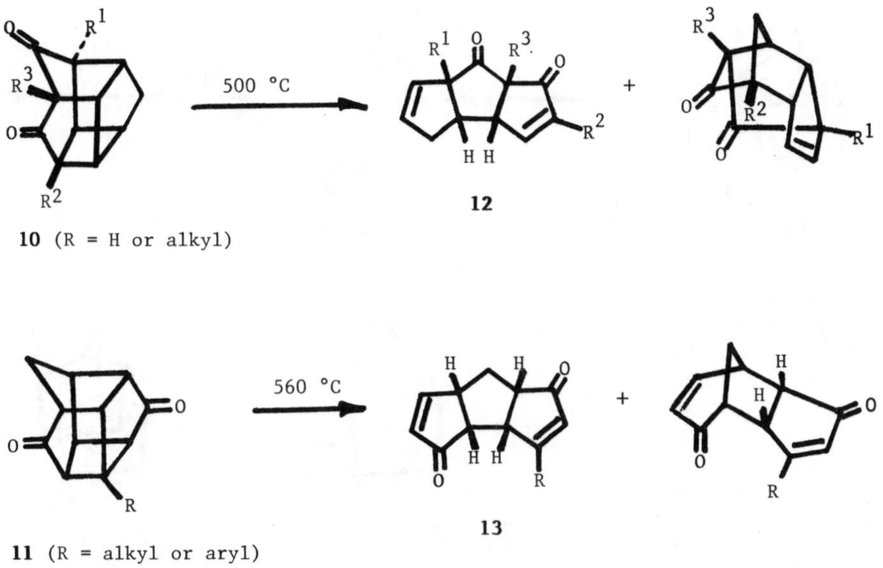

500 °C

10 (R = H or alkyl) **12**

560 °C

11 (R = alkyl or aryl) **13**

Scheme 1.4 ■

Gas-phase pyrolysis of 1,3-bishomopentaprismane (**6**)[18] and of 1,3-bishomopentaprismanedione (**7**)[19] has been employed to synthesize novel, substituted tetraquinanes (**8** and **9**, respectively, Scheme 1.3). Similarly, gas-phase pyrolysis of substituted 1,2,4-trishomocubanes has been reported to afford cis,cisoid,cis linear triquinanes (e.g., thermal conversion of **10** and **11** to **12** and **13**, respectively, Scheme 1.4).[20] Additional examples wherein flash vacuum pyrolysis or reductive cleavage of cage molecules has been utilized to synthesize polyquinane derivatives have been cited in recent reviews.[3,4,15,17,21]

1.2B. Reductive Processes

As an alternative to thermal fragmentation of the cyclobutane ring, compounds of type **2** can be converted into cis,syn,cis linear triquinanes by promoting reductive cleavage of carbon–carbon σ bonds in this system. Thus, for example, selective reduction of various σ bonds in PCUD-8,11-dione (**14**) can be achieved in the manner shown in Scheme 1.5.[16,17,22]

Hydrogenolyses of strained carbon–carbon σ bonds in cage systems have been the subject of numerous experimental and theoretical investigations.[4,5] In general, hydrogenolysis occurs preferentially at those C—C bonds whose cleavage results in the greatest relief of strain. Thus, stepwise hydrogenolysis of strained C—C bonds occurs when methanolic cubane **15** is reacted with hydrogen (1 atm) over palladized charcoal at 20°C, leading ultimately to the formation of bicyclo[2.2.2]octane (**16**).[23] Other processes compete with simple hydrogenolysis in this system. Thus, valence isomeriza-

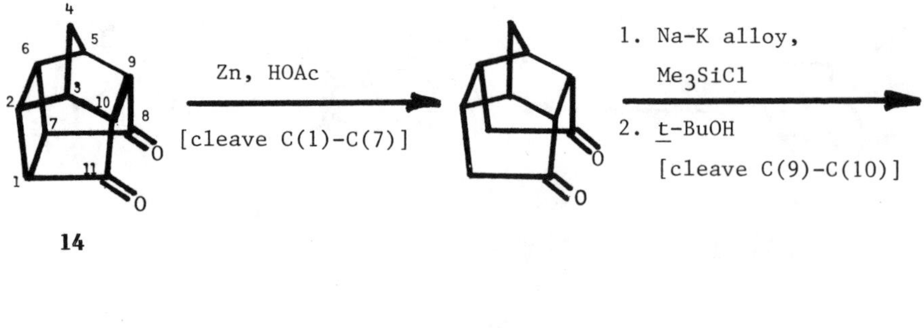

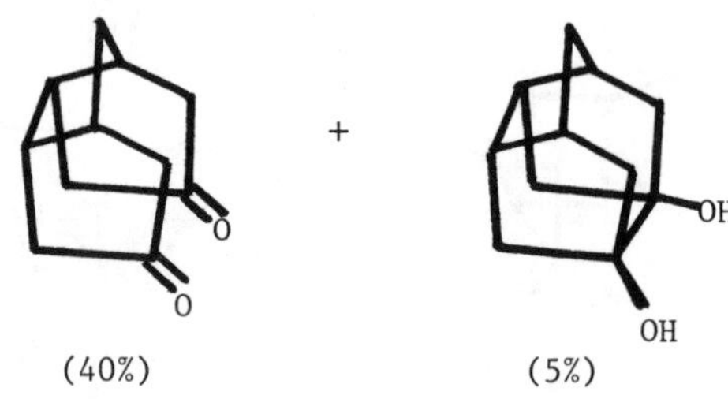

Scheme 1.5 ■

tion of cubane and of secocubane (**17**, formed from cubane via uptake of one equivalent of hydrogen) to cuneane (**18**) and "secocuneane" (**19**), respectively, also occurs. Subsequent hydrogenolysis of **18** and of **19** leads ultimately to the formation of bicyclo[3.2.1]octane (**20**, Scheme 1.6). Hydrogenolysis of dimethyl cubane-1,4-dicarboxylate (**21**) proceeds via a single dihydro- and a single tetrahydro-intermediate, ultimately affording pure dimethyl bicyclo[2.2.2]octane-1,4-dicarboxylate (**22**, Scheme 1.6).[23]

Hydrogenolysis of homocubane **23** in the presence of Pd, Pt, and Rh catalysts displays similar behavior. This reaction affords two dihydrohomocubanes, **24** and **25**, only one of which suffers further hydrogenolysis to afford a tetrahydrohomocubane, that is, twistbrendane **26**. Another process competes with simple hydrogenolysis, that is, valence isomerization of homocubane to norsnoutane (i.e., homocuneane **27**) also occurs, followed by subsequent hydrogenolysis of **27** to brendane **28** (Scheme 1.7).[23b, 24]

The results of catalytic hydrogenolyses of several other strained cage hydrocarbon systems have been reported. For example, hydrogenolysis of basketane (i.e., 1,8-bishomocubane, **29**) affords twistane **30** as the major reaction product.[25] Similarly, hydrogenolysis of **31** can be used to synthesize tetracyclo[6.2.2.0^{2,7}.0^{4,9}]dodecane-5,11-dione (i.e., ditwistane-5,11-dione, **32**).[26] In addition, hydrogenolysis of substituted 1,3-bishomocubanes (e.g., **33** and **34**) has been used as a method for synthesizing substituted tetracyclo[5.2.1.0^{2,6}.0^{4,8}]decanes [i.e., bisnorditwistanes **35** (ref. 26) and **36** (ref. 27)]. The foregoing reactions are summarized in Scheme 1.8.

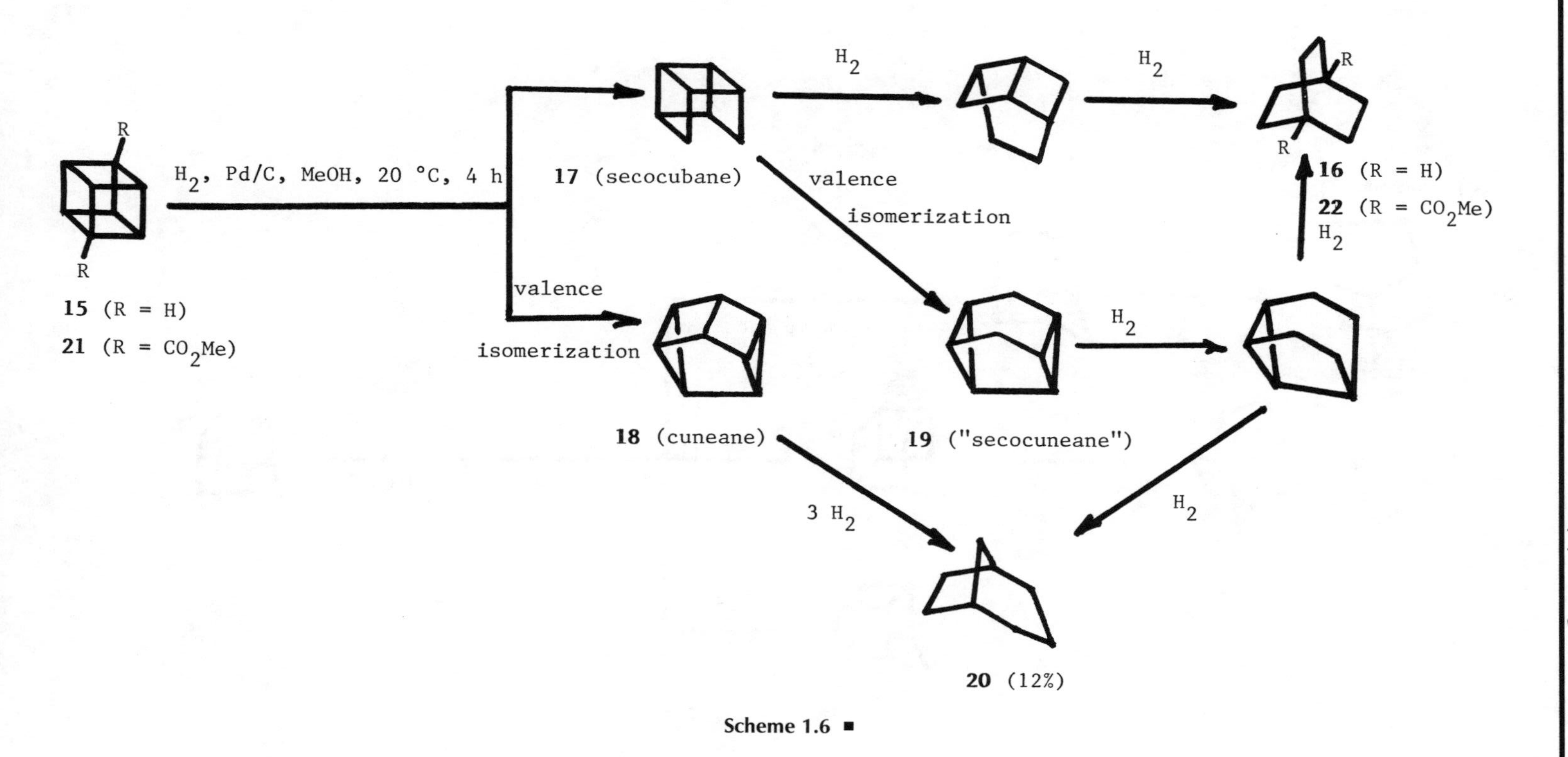
R
R
H_2, Pd/C, MeOH, 20 °C, 4 h
15 (R = H)
21 (R = CO_2Me)
17 (secocubane)
valence
isomerization
valence
isomerization
H_2
H_2
R
R
16 (R = H)
22 (R = CO_2Me)
H_2
H_2
18 (cuneane)
19 ("secocuneane")
3 H_2
H_2
20 (12%)

Scheme 1.6 ■

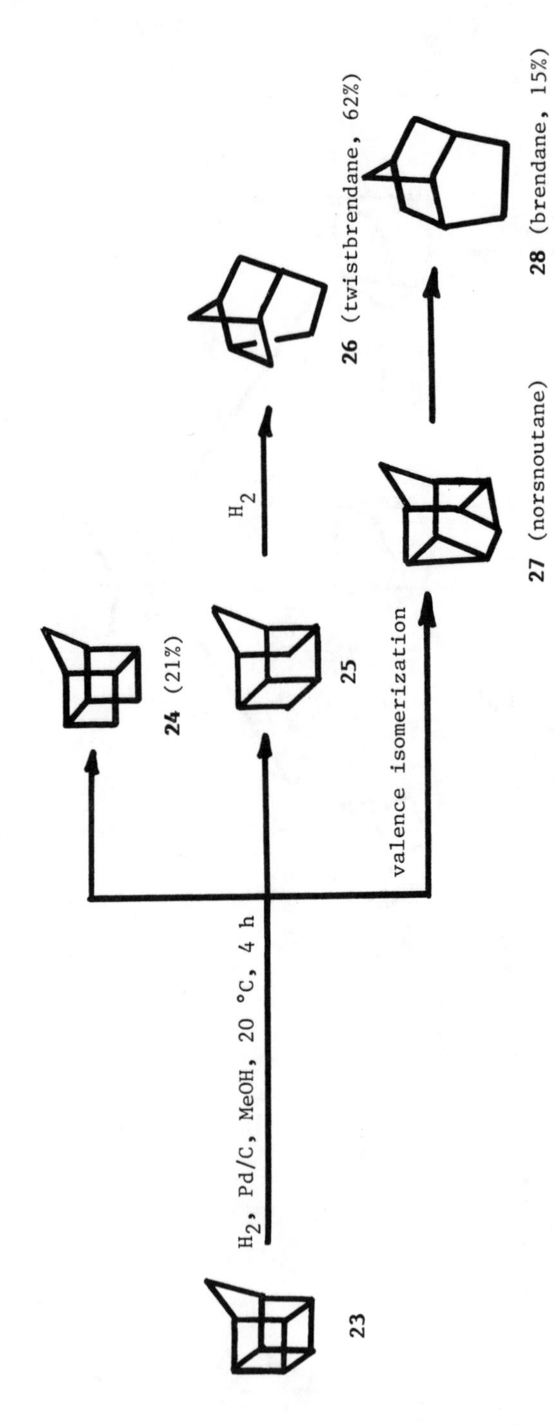
23
H2, Pd/C, MeOH, 20 °C, 4 h
24 (21%)
25
H2
26 (twistbrendane, 62%)
valence isomerization
27 (norsnoutane)
28 (brendane, 15%)

Scheme 1.7 ■

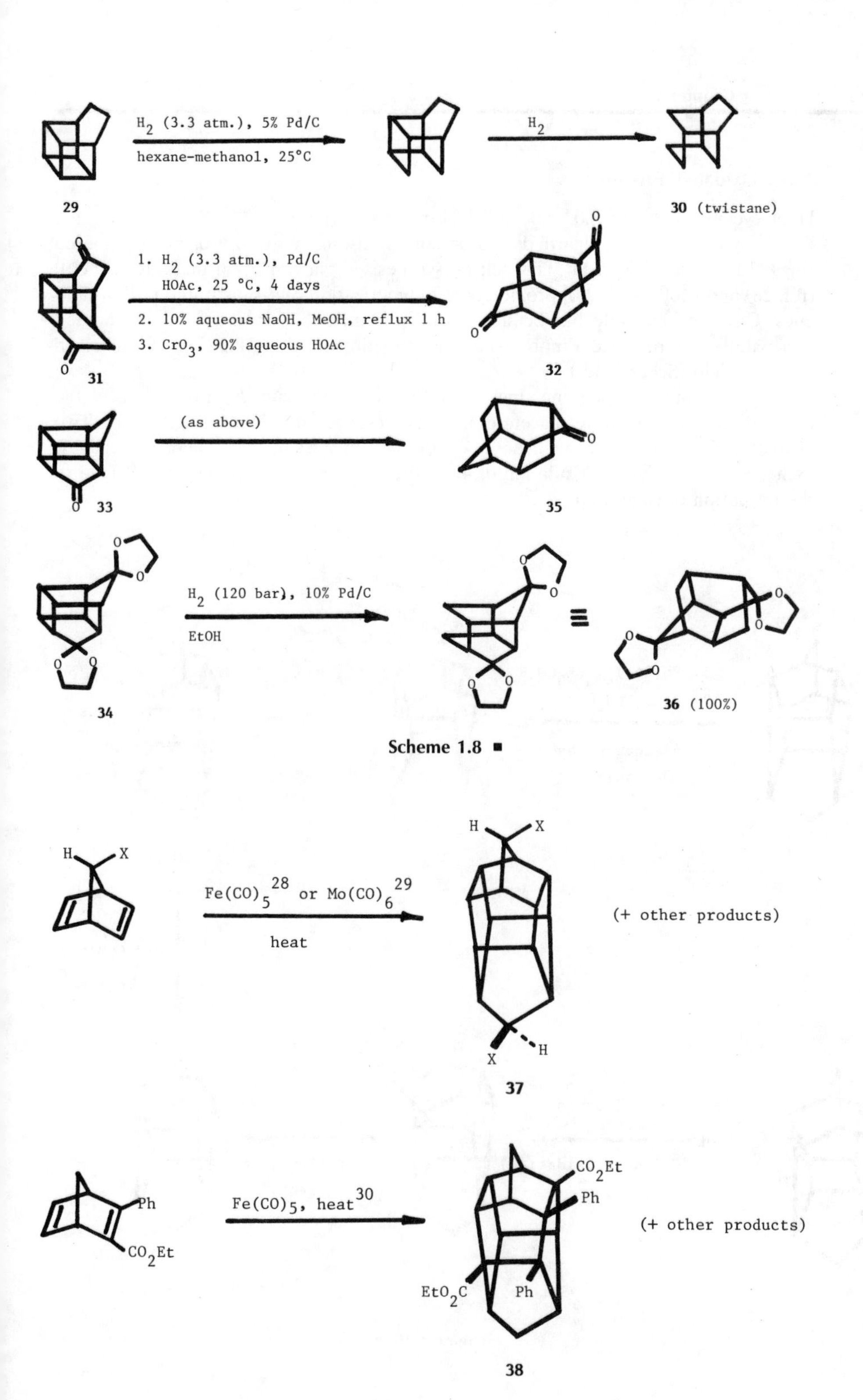

Scheme 1.8 ■

Scheme 1.9 ■

1.2C. Oxidative Processes

Heptacyclo[$6.6.0.0^{2,6}.0^{3,13}.0^{4,11}.0^{5,9}.0^{10,14}$]tetradecane [HCTD, **37** (X = H)], like the PCUD system, may be regarded as a potentially useful repository of five-membered rings. 7,12-Disubstituted HCTDs can be synthesized via transition metal [e.g., Fe(0) (ref. 28) and Mo(0) (ref. 29)] promoted cyclodimerization of 7-substituted norbornadienes. Only one example has been reported in which this approach has been used successfully to synthesize a more extensively functionalized HCTD (i.e., **38**), albeit in very low yield (Scheme 1.9).[30]

Access to the cyclopentane rings in HCTD has been gained via oxidation of the cage system with lead tetraacetate in the presence of trifluoroacetic acid. Two alcohols, **39** and **40** have been obtained thereby in 20 and 70% yield, respectively (Scheme 1.10).[31] Further oxidation of **40** in the manner shown in Scheme 1.10 led to the formation of cage dione **41**.[31]

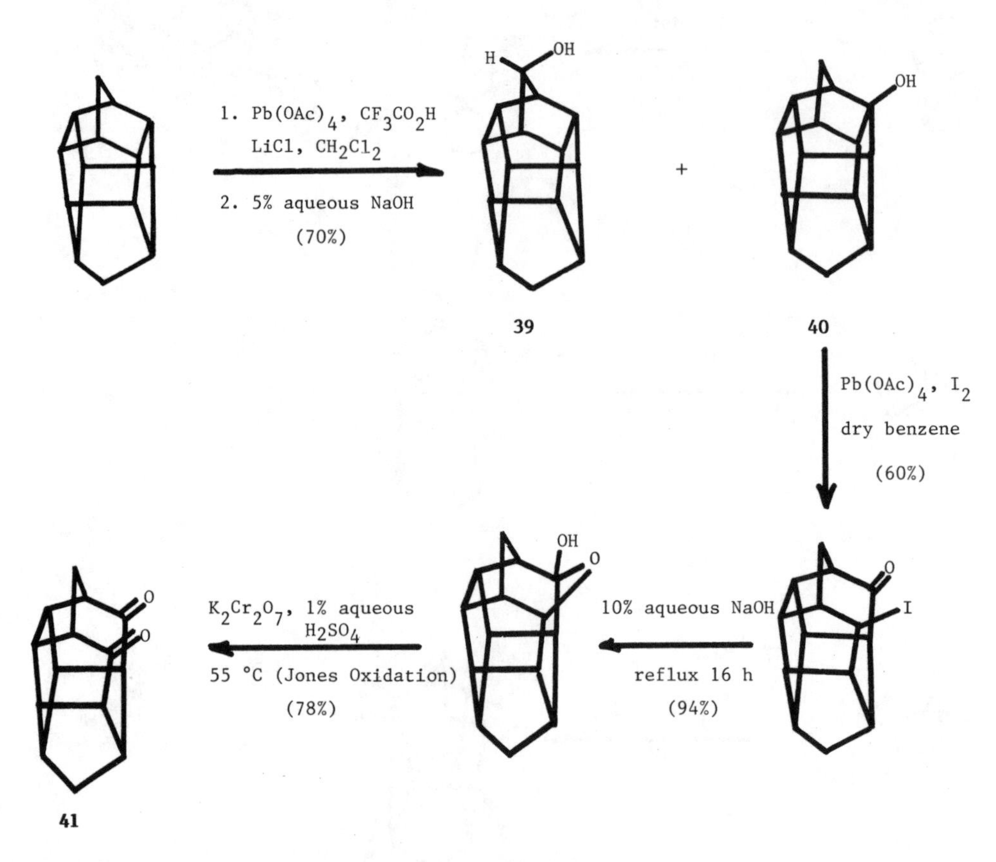

Scheme 1.10 ■

1.2D. Lewis and Brønsted Acid Promoted (Cationic) Processes

Appropriately substituted 1,3-bishomocubanones, in addition to undergoing thermal cyclobutane fragmentations, have also been found to afford cyclopentanoid products when treated with Lewis or proton acid catalysts. Thus, treatment of **42a** and **42b** with dry hydrogen chloride gas results in cleavage of one of the cyclobutane rings in the substrate, thereby affording **43a** and **43b**, respectively.[32,33] The presence of an electron-donating substituent at C-4 in the 1,3-bishomocubanone substrate is required for rearrangement to occur. Accordingly, a push–pull cationic mechanism was invoked to account for cleavage of the C(4)—C(5) bond in the acid-promoted ring opening of **42a** and of **42b** (Scheme 1.11).[32,33]

Interestingly, treatment of a suspension of **42b** in 20% aqueous potassium hydroxide solution with excess silver nitrate at 60°C for 4 h resulted in cationic rearrangement to **44**. A mechanism for this rearrangement is shown in Scheme 1.11.[32,33] The reaction is initiated via regiospecific electrophilic attack of Ag^+ upon the C(3)—C(4) bond of the substrate. The resulting cationic intermediate, **45**, is sufficiently long-lived to permit efficient trapping by hydroxide ion, thereby affording the corresponding hemiketal, **46**. Retro-Claisen scission of the C(4)—C(5) bond in **46** followed by reductive intramolecular cyclopropanation [which results in expulsion of Ag(0)] completes the mechanistic sequence leading to the observed rearrangement product, **44**.

A push–pull cationic mechanism also has been invoked to explain the ease with which 1-methoxy-PCUD-8,11-dione and related compounds undergo Lewis acid promoted [2 + 2] cycloreversion.[34,35] Thus, conversion of **47a** and **47b** to **48a** and **48b**, respectively, occurs quantitatively in the presence of boron trifluoride etherate at 25–30°C within 1 min (Scheme 1.12).[34]

Efficient [2 + 2] cycloreversions of a variety of substituted trishomocubanes by protic acids and/or Lewis acids have been reported.[36–38] Examples are shown in Scheme 1.13.

Aromatization can provide an important driving force for cationic rearrangement in suitably constructed cage systems. An example is provided by the boron trifluoride mediated reaction of 1,9-dihalo-PCUD-8,11-diones with ethyl diazoacetate. This reaction provides a synthetic entry into the cyclopent[*a*]indene ring system (see **57a** and **57b**, Scheme 1.14).[39]

Lewis and Brønsted acids have been found to promote thermodynamic rearrangements in complex cage hydrocarbon systems. Frequently, this reaction can be utilized to promote rearrangement to the most stable C_nH_m isomer, termed the stabilomer.[40] Thus, cationic rearrangements of this type have been used to great advantage to synthesize adamantane and other diamondoid hydrocarbons from appropriate polycyclic precursors.[41] In addition, D_3-trishomocubanes have been synthesized via cationic rearrangement of suitably substituted PCUDs.[42]

Mehta and collaborators have reported a number of cationic rearrangements in strained cage systems that lead to the formation of a variety of interesting products. Thus, 1,3-bishomocubanone (**58**), when treated with sodium azide in methanesulfonic acid (i.e., Schmidt reaction conditions), rearranges to a brendane derivative, **59** (Scheme 1.15).[43a] The corresponding reactions of substituted 1,3-bishomocubanones **60a** and **60b** with sodium azide in aqueous methanesulfonic acid afford a variety of products that arise via cationic intermediates (Scheme 1.16).[43b]

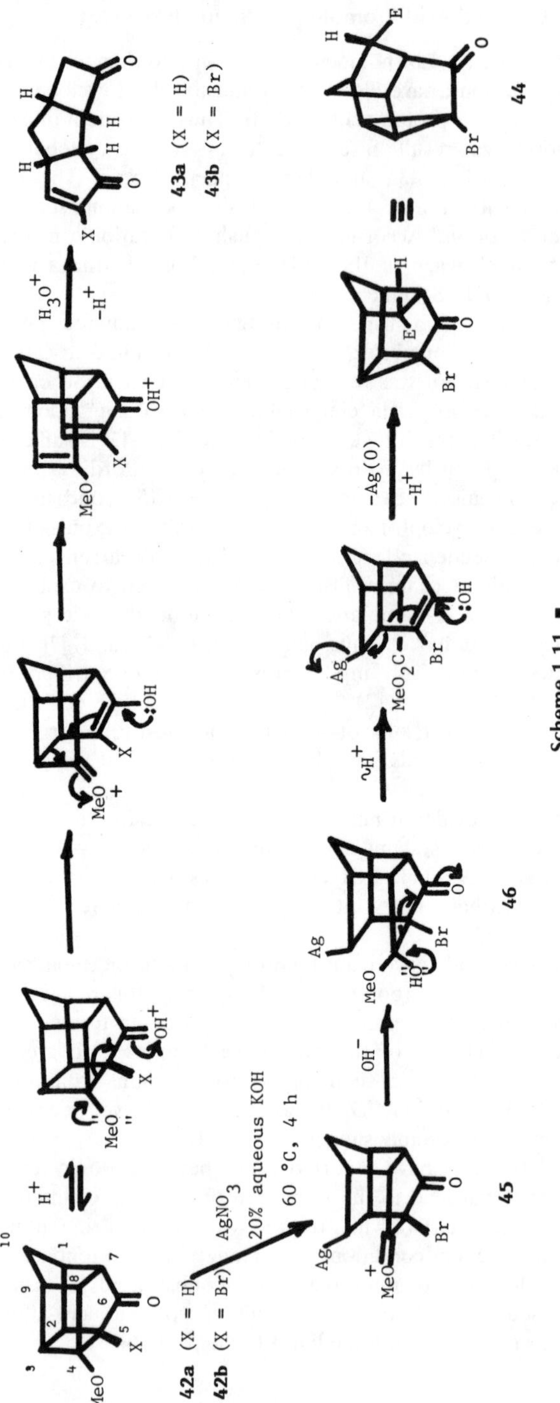
42a (X = H)
42b (X = Br)
H+
AgNO3
20% aqueous KOH
60 °C, 4 h
H3O+
-H+
43a (X = H)
43b (X = Br)
45
OH−
46
∿H+
-Ag(0)
-H+
44

Scheme 1.11 ■

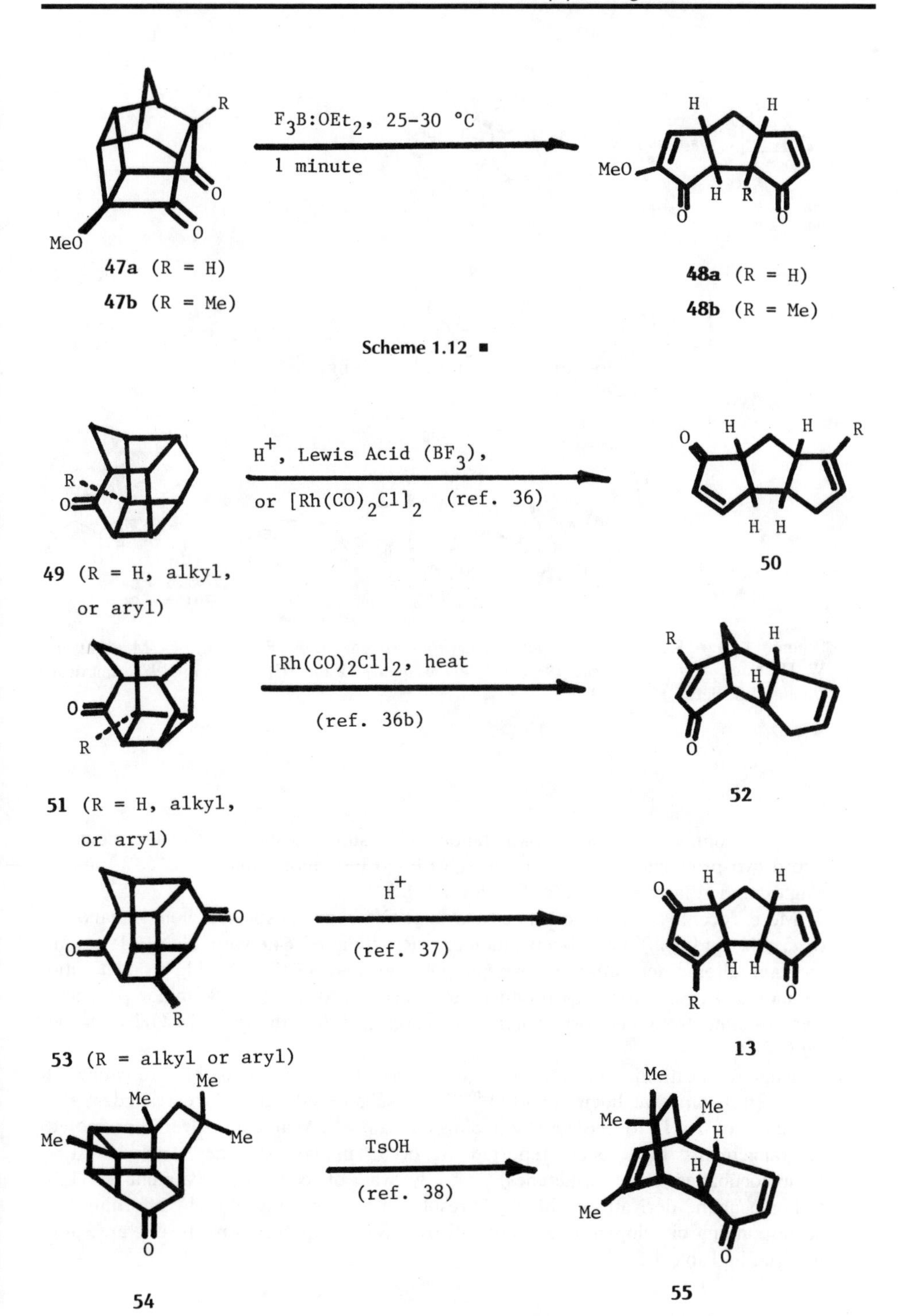
R
$F_3B:OEt_2$, 25-30 °C
1 minute
H H
MeO
H R
O
O
O O
MeO
47a (R = H)
47b (R = Me)
48a (R = H)
48b (R = Me)
Scheme 1.12
H^+, Lewis Acid (BF_3),
or $[Rh(CO)_2Cl]_2$ (ref. 36)
O H H R
R
O
H H
50
49 (R = H, alkyl,
or aryl)
$[Rh(CO)_2Cl]_2$, heat
(ref. 36b)
R H
H
O
R
O
52
51 (R = H, alkyl,
or aryl)
H^+
(ref. 37)
O H H
O
O
H H
R
O
R
13
53 (R = alkyl or aryl)
Me
Me
Me
Me
TsOH
(ref. 38)
Me
Me
Me
H
H
Me
O
O
54
55

Scheme 1.13 ■

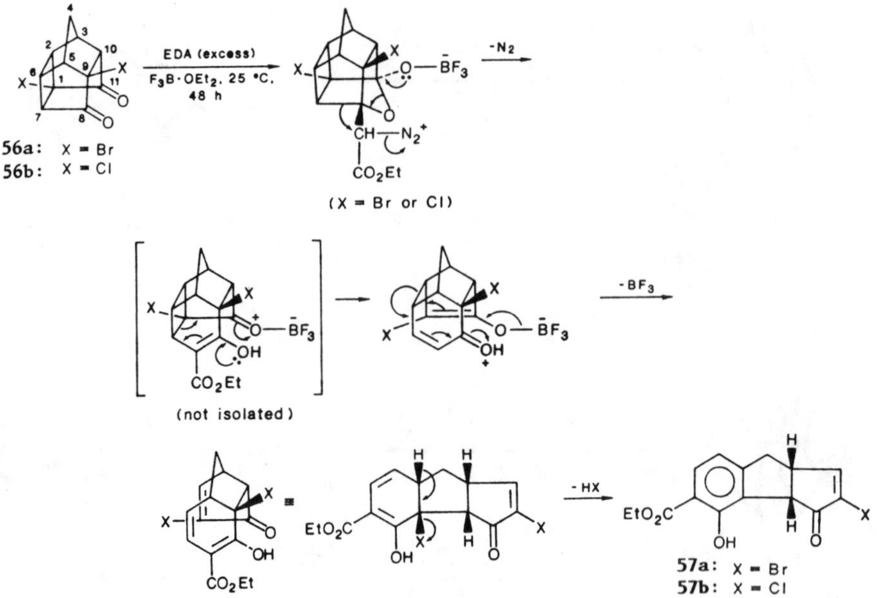

Scheme 1.14 ▪ Reprinted with permission from Marchand, A. P.; Reddy, G. M.; Watson, W. H.; Nagl, A. *J. Org. Chem.* **1988**, *53*, 5969 (Schemes II and III). Copyright 1988 American Chemical Society.

Under similar conditions, basketanone (**61**) suffers cationic rearrangement to afford two products, that is, a novel, substituted bishomoprismane (**62**, 60%) and a ring expanded lactam (**63**, 32%, Scheme 1.17).[44]

More recently, similar cationic rearrangements of substituted homocubanones (e.g., **64a** and **64b**) have been studied.[45] Reaction of **64a** with NaN_3 (1 eq) in methanesulfonic acid afforded two products, **65** and **66** (75% yield, product ratio **65** : **66** = 5 : 1). Similarly, two products, **67** (major product) and **68** (minor product), were isolated from the corresponding reaction of **64b** with NaN_3–MsOH (Scheme 1.18).[45]

Numerous cationic rearrangements of strained C—C σ bonds that are catalyzed by silver(I) ion have been reported.[46] Thus, substituted cubanes rearrange in the presence of Ag(I) to the corresponding cuneanes.[47] Analogous silver(I)-promoted rearrangements have been reported to occur in homocubanes[48] and in 1,8-bishomocubanes (i.e., basketanes).[48a, 49] By way of contrast, treatment of 1,3-bishomocubane derivatives with Ag(I) results in ring opening, thereby affording the corresponding dicyclopentadiene derivatives.[50] Some representative results are summarized in Table 1.1.

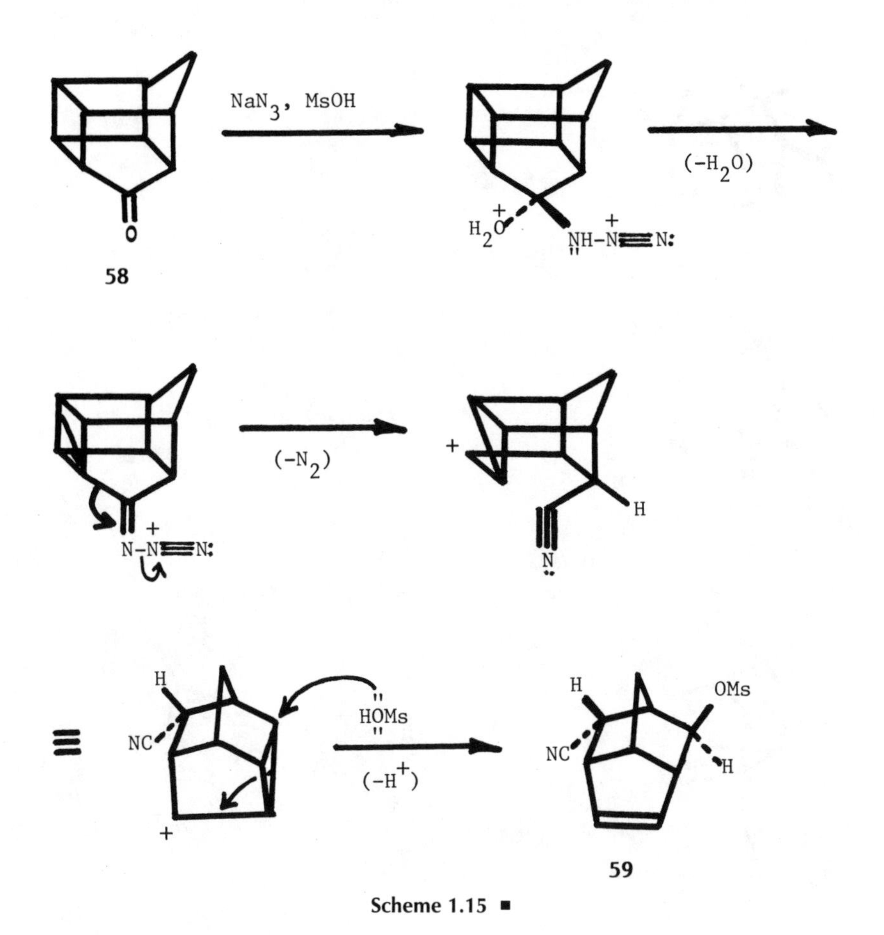

Scheme 1.15 ■

1.2E. Base-Promoted (Anionic) Processes

Reactions of appropriately substituted cage compounds with base that result in cleavage of a framework carbon–carbon σ bond in the cage system with concomitant ring opening have been the subject of several studies.[4,51] Examples of this are the base-induced homoketonizations of pentacyclic cage alcohols and acetates that have been studied extensively by Zwanenburg and collaborators (Table 1.2).[52–56]

The stereochemistry of the process by which the anion produced via base-promoted cleavage of strained cage σ bonds is captured by solvent has been shown to depend upon the structure of the cage substrate.[52] Thus, treatment of 4,5-dihydroxyhomocubane (**69**) with NaOMe–MeOD results in cleavage of the C(3)—C(4) σ bond with retention of configuration. However, the resulting secohomocubane, **70**, is not isolated. Instead, **70** undergoes further base-promoted ring fragmentation, thereby

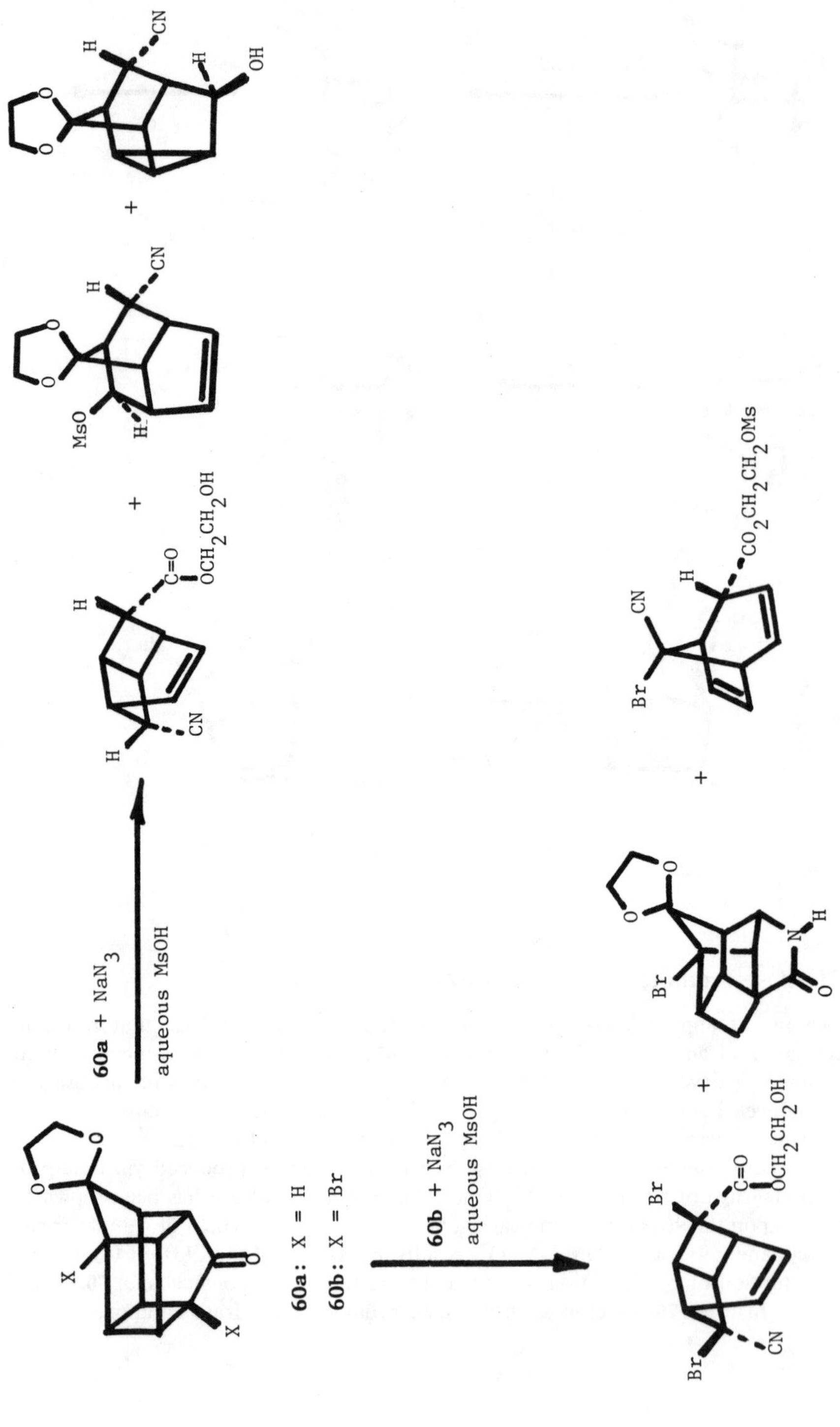
60a + NaN3
aqueous MsOH
60a: X = H
60b: X = Br
60b + NaN3
aqueous MsOH
MsO
OCH2CH2OH
CO2CH2CH2OMs

Scheme 1.16 ■

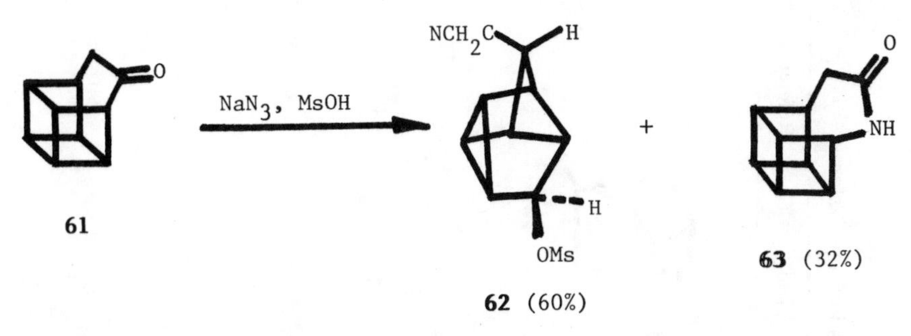

Scheme 1.17 ■

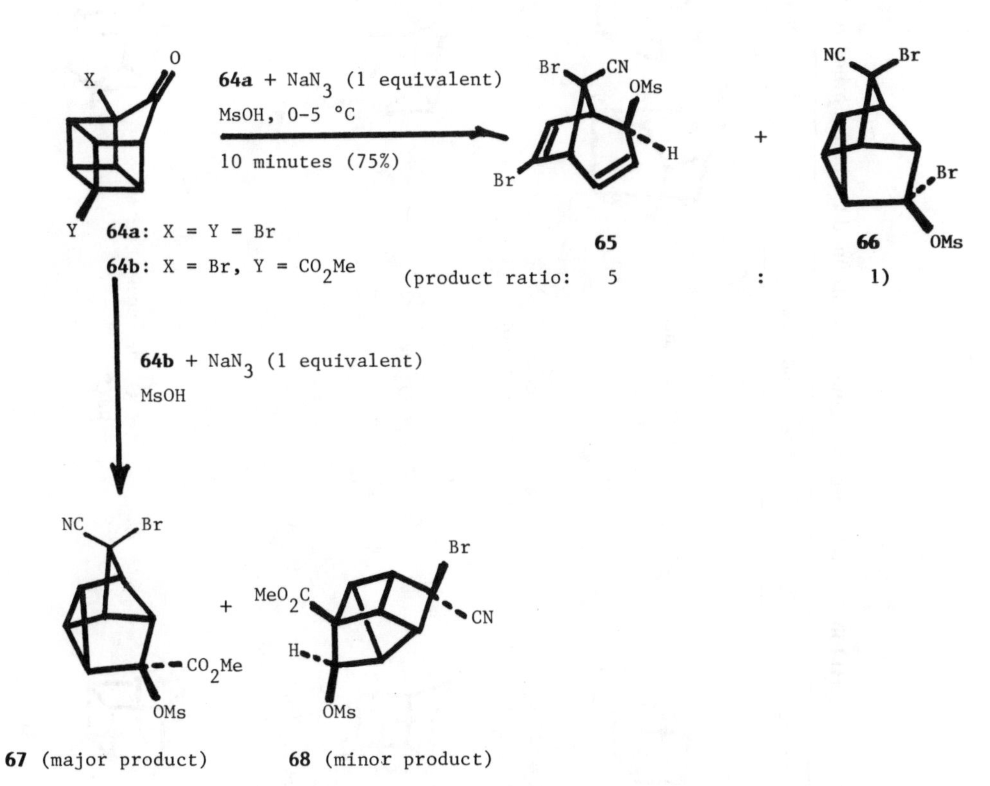

Scheme 1.18 ■

Table 1.1 ▪ Silver(I)-Ion-Promoted Rearrangement of Strained Carbon–Carbon σ Bonds in Cage Systems

Substrate	Conditions	Products (Percentage Yield)	References
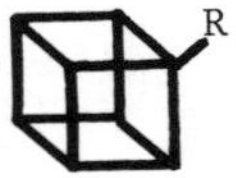 R	$AgClO_4$, benzene, 40°C	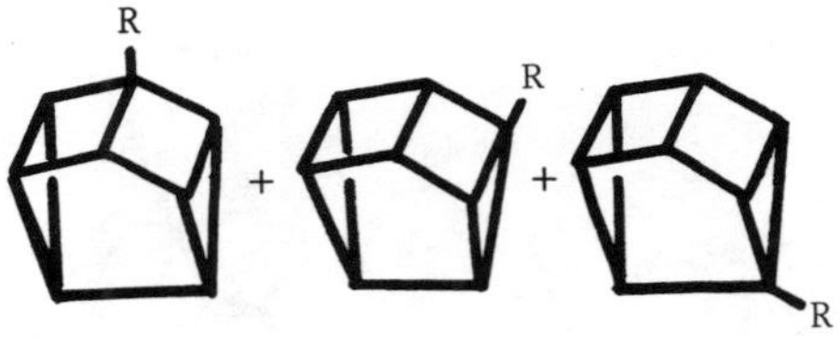 R + R + R	47a
R = CH_2OAc R = CO_2Me	Product ratio: Product ratio:	6 : 2 : 1 2.5 : 1 : 0	
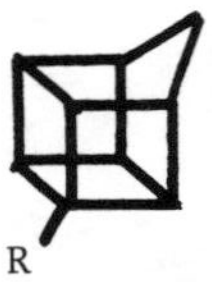 R	$AgClO_4$, benzene, heat	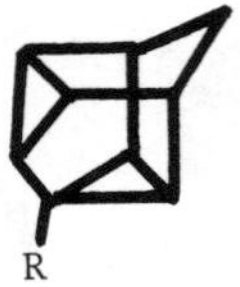 R (45–93%)	48c
R = H, D, CH_3, CD_3, CH_2OH, CH_2OAc, CH_2OCH_3, CO_2Me, vinyl, phenyl, cyclopropyl, CMe_3, OEt, $OSiMe_3$			
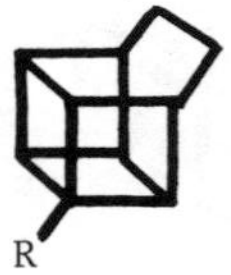 R	$AgClO_4$, benzene, 40°C	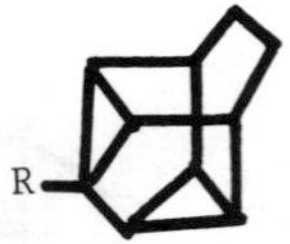 R (80–90%)	49b, 49c
R = H, CH_3			

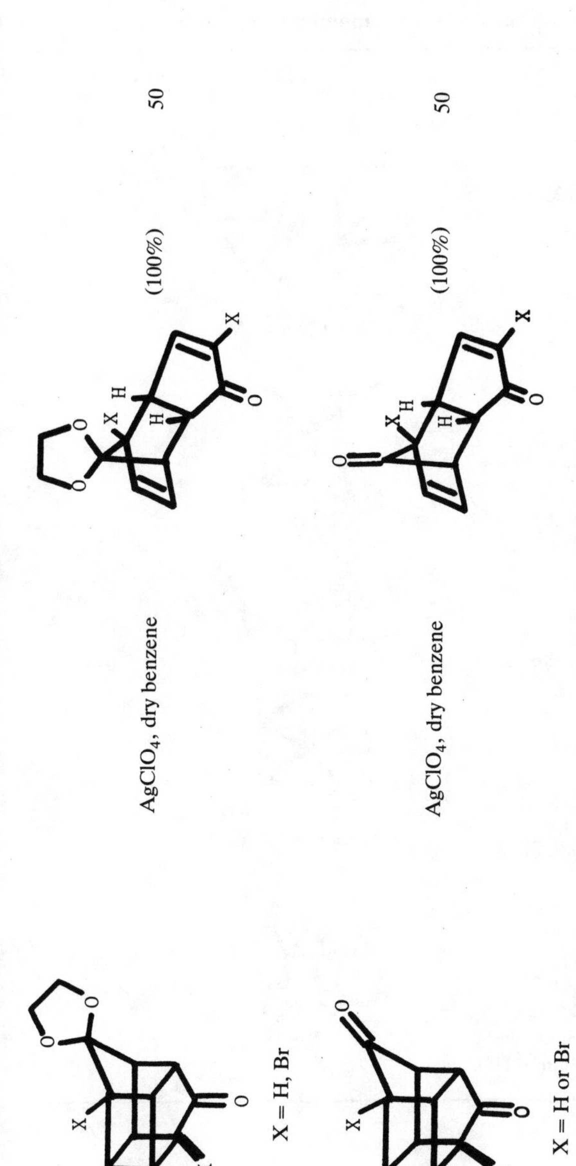
X = H, Br
AgClO$_4$, dry benzene
(100%)
50
X = H or Br
AgClO$_4$, dry benzene
(100%)
50

Table 1.2 ▪ Base-Promoted Cage-Opening Reactions

Substrate	Conditions	Products	References
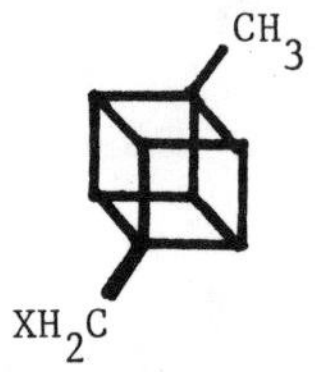 XH_2C (CH$_3$) cage; X = CN, CO_2Me	Excess LDA, dry THF, −30°C	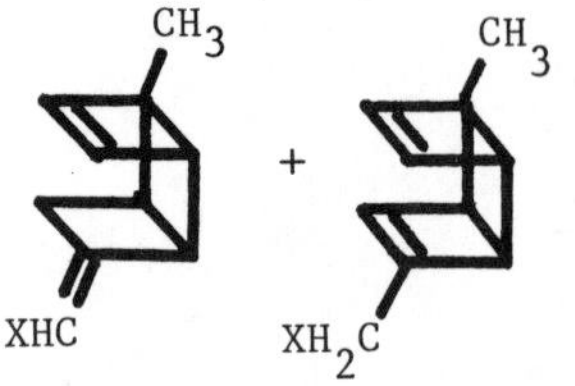 XHC (CH$_3$) + XH_2C (CH$_3$); X = CN, CO_2Me	53
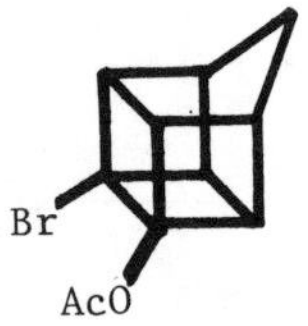 Br, AcO cage	Methanolic NaOMe, room temperature; stir 1 h, quench with NH_4Cl	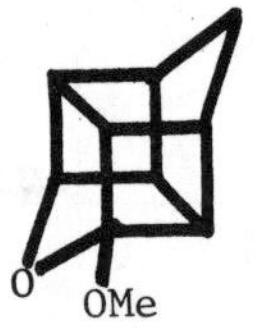 O, OMe	54
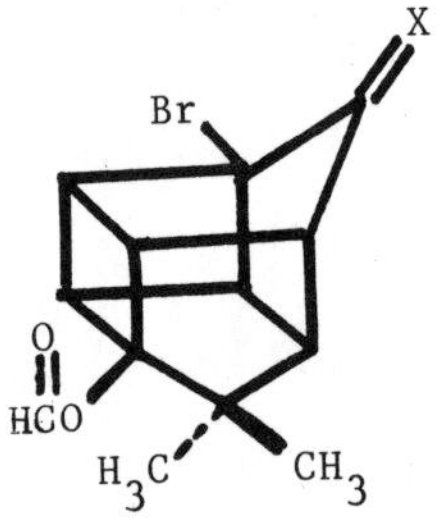 Br, HCO (O), H_3C, CH_3, X; X = H_2, OCH_2CH_2O	NaOMe, EtOD; reflux 16 h	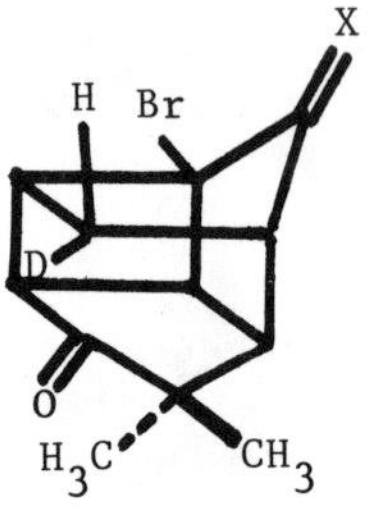 H, Br, D, O, H_3C, CH_3, X; X = H_2, OCH_2CH_2O	55
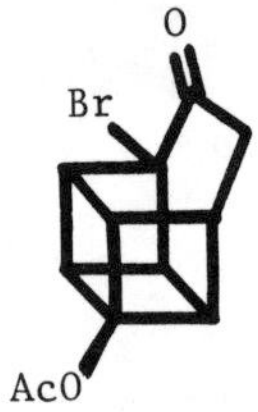 Br, O, AcO	NaOD, D_2O, CH_3OD; stir 15 min; then quench via addition of HOAc	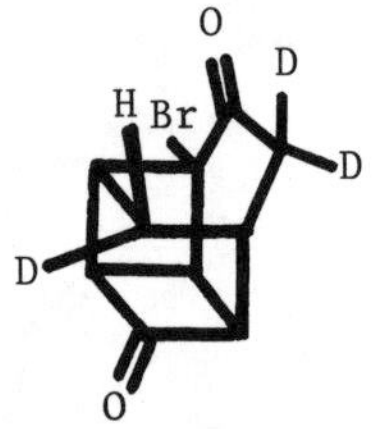 O, D, H, Br, D, D, O	56

affording **72**, which results via *inversion* of configuration of the reacting C(3)—C(4) σ bond in intermediate **71** (see Scheme 1.19).[57]

Generally, when more than one σ bond can be cleaved during a homoketonization process, the course of the reaction is determined by thermodynamic considerations, that is, the relative strengths of the various bonds that might be involved in the reaction.[58] However, it has been demonstrated recently that the mode of ring C—C bond cleavage that results in cage-opening can be influenced dramatically by the

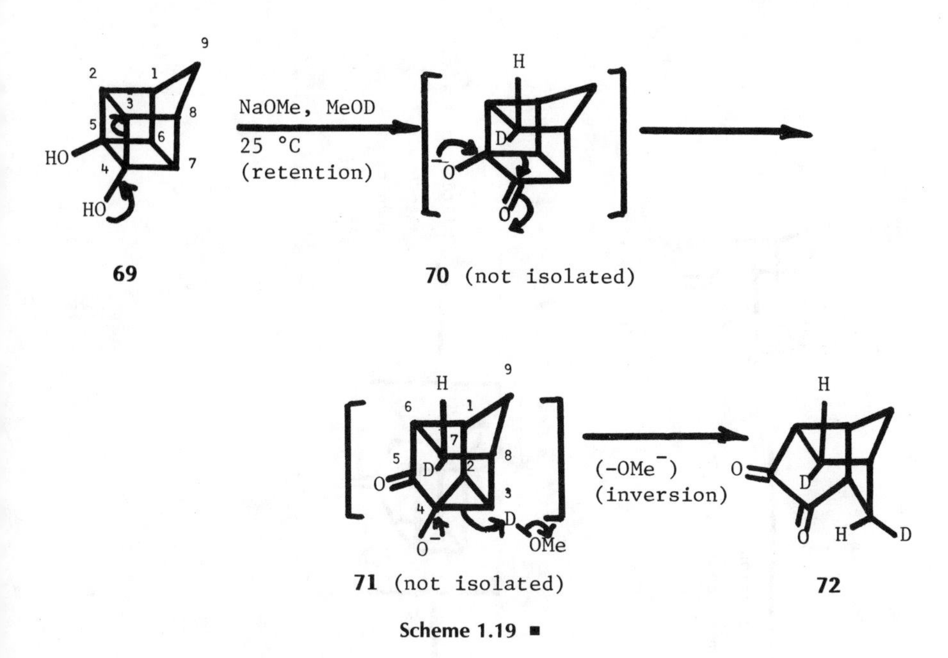

Scheme 1.19 ■

presence of remote substituents.[52] As an example, Zwanenburg and co-workers[59,60] have demonstrated that the presence of a carbanion-stabilizing substituent at C-6 in 4-acetoxy- and in 8-acetoxypentacyclo[5.3.0.0^{2,5}.0^{3,9}.0^{4,8}]decanes can change the regiochemistry of homoketonization. Thus, homoketonization of **73** (Scheme 1.20) proceeds with rupture of the C(4)—C(5) σ bond in the substrate to afford **75**. The corresponding process in **74** requires the application of higher reaction temperature and produces **76** via C(3)—C(4) bond cleavage.[59] Similarly, base-promoted homoketonization of **77** occurs under very mild conditions to afford a mixture of cage-opened products derived from the corresponding diketone, **79**, via C(7)—C(8) bond rupture. By way of contrast, **78** suffers C(4)—C(8) bond rupture to afford **80** only when treated with base at elevated temperature (see Scheme 1.20).[60]

Ring cleavage of strained cage compounds often results when intended base-promoted semibenzilic acid rearrangement of cage α-haloketones goes awry (i.e., Haller–Bauer type cleavage[61] intervenes).[4,6,62] Several such examples have been reported in chlorinated pentacyclo[5.4.0.0^{2,6}.0^{3,10}.0^{5,9}]undecane-8,11-diones.[63] An extreme example occurs when a toluene solution of pentacyclic cage triketone **81** is heated with solid sodium hydroxide. The ensuing reaction results in extensive rearrangement with concomitant aromatization, ultimately affording the corresponding substituted benzocyclobutene, **82**, in low yield (Scheme 1.21).[64]

Another unusual example of a base-promoted ring fragmentation reaction that results in extensive rearrangement is depicted in Scheme 1.22. Thus, reaction of a solution of perchlorinated homocubylamine **83** in aqueous *tert*-butyl alcohol with potassium hydroxide at room temperature results in rearrangement to afford **84**.[65]

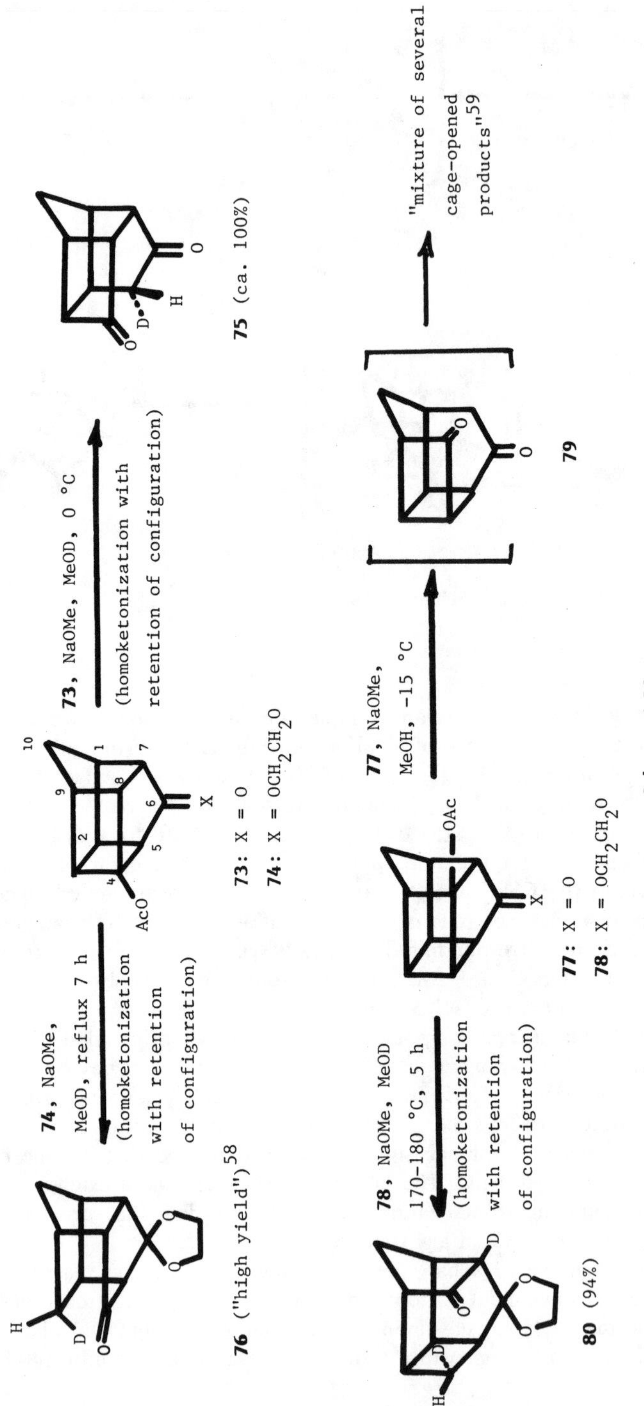
74, NaOMe,
MeOD, reflux 7 h
(homoketonization
with retention
of configuration)
73, NaOMe, MeOD, 0 °C
(homoketonization with
retention of configuration)
AcO
X
D
H
O
76 ("high yield")58
73: X = O
74: X = OCH2CH2O
75 (ca. 100%)
78, NaOMe, MeOD
170–180 °C, 5 h
(homoketonization
with retention
of configuration)
77, NaOMe,
MeOH, -15 °C
OAc
77: X = O
78: X = OCH2CH2O
79
"mixture of several
cage-opened
products"59
80 (94%)

Scheme 1.20 ■

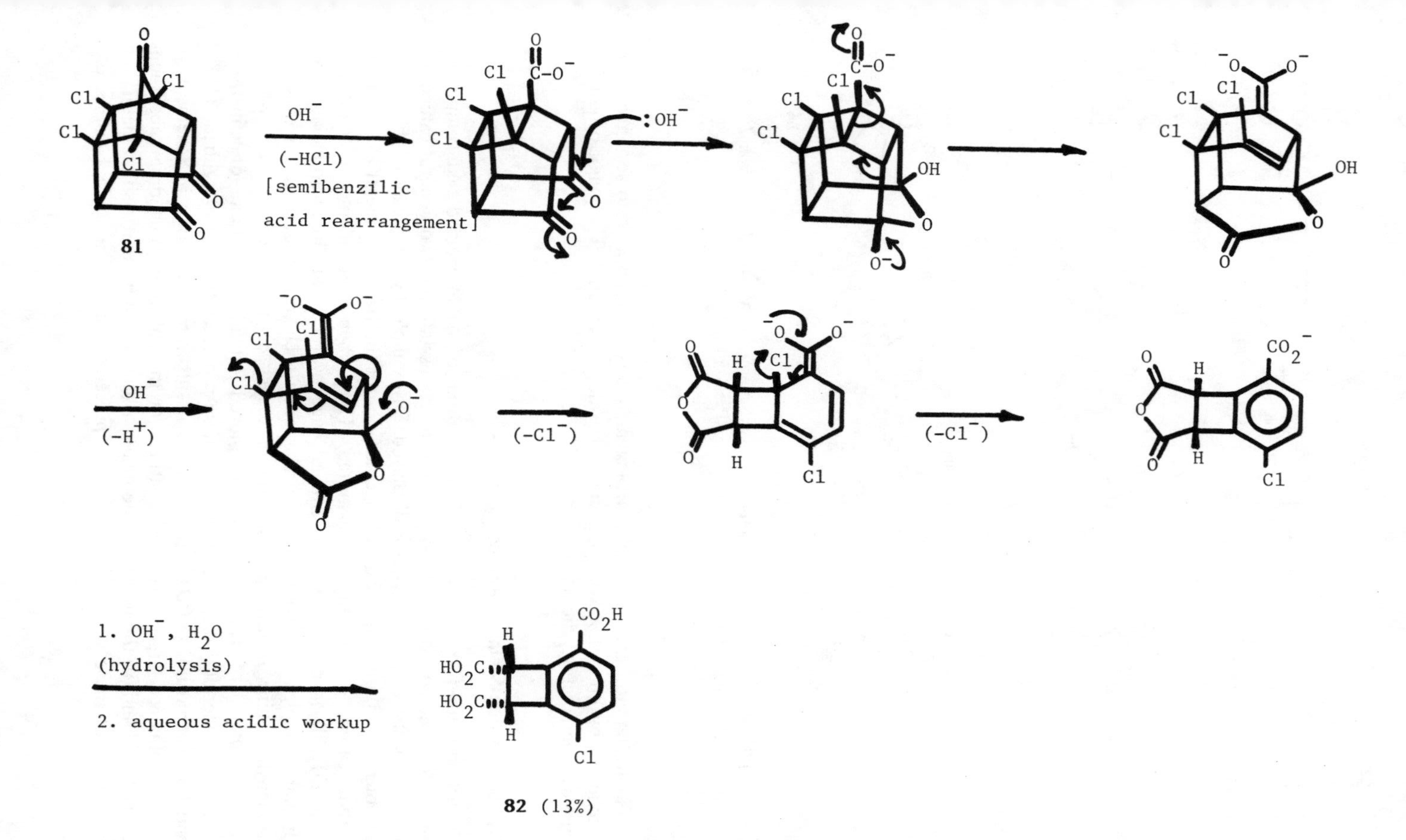

Scheme 1.21 ■ Reprinted with permission from Marchand, A. P.; Chou, T.-C. *J. Chem. Soc., Perkin Trans. 1* **1973**, 1948 (scheme, p. 1973). Copyright 1973 Royal Society of Chemistry.

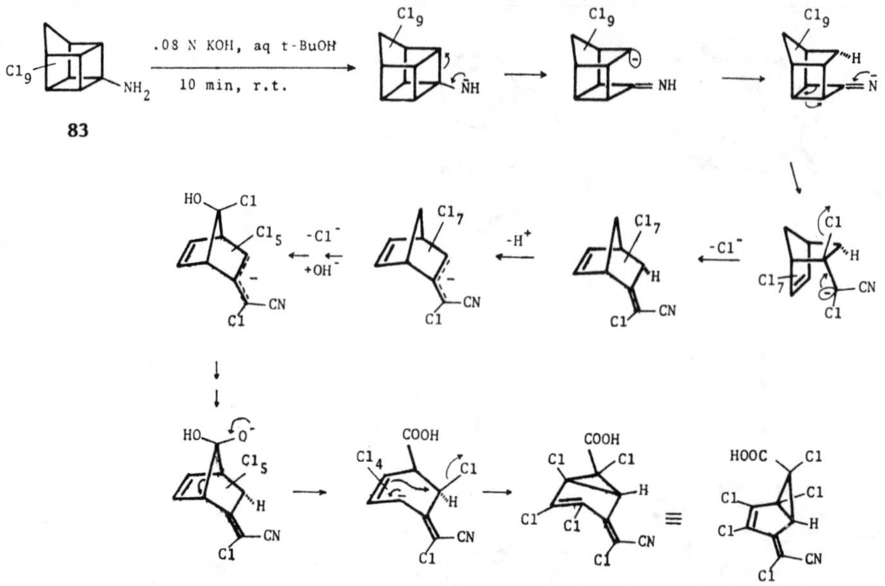

Scheme 1.22 ■ Reprinted with permission from Scherer, K. V., Jr., *Tetrahedron Lett.* **1972**, 2077 (Chart I, p. 2079). Copyright 1972 Pergamon Press, Ltd.

Finally, an interesting skeletal rearrangement has been reported to accompany the reaction of bis(enone) **85** with sodium methoxide in methanol. The major product of this reaction is tetracyclic diketone **86** (Scheme 1.23).[66]

1.2F. Transition-Metal-Promoted Reactions

Transition-metal-promoted valence isomerizations of highly strained cage compounds have been studied extensively.[4,5,46] These reactions fall into two general categories. The first of these reaction types is displayed, for example, by complexes of rhodium(I) and of nickel(0) and involves oxidative addition of the transition metal to a strained carbon–carbon σ bond in the substrate.[67] Thus, when a stoichiometric quantity of $[Rh(CO)_2Cl]_2$ is reacted with a strained cage hydrocarbon, an intermediate acyl-rhodium complex is formed. Ligand exchange of such complexes with, for example, triphenylphoshine results in formation of a new cage ketone. However, when a catalytic amount of $[Rh(diene)Cl]_2$ is employed instead, cyclobutane–diolefin transformation generally occurs. Examples of these processes that involve cubane (**15**)[47a] and 1,3-bishomocubane (**87**),[68] respectively, as substrates are shown in Scheme 1.24.

The influence of (i) structural features in the substrate (including substituent effects)[49a,69] and (ii) the nature of the ligand associated with the transition metal[70] on the course of these valence isomerization processes have been studied extensively.

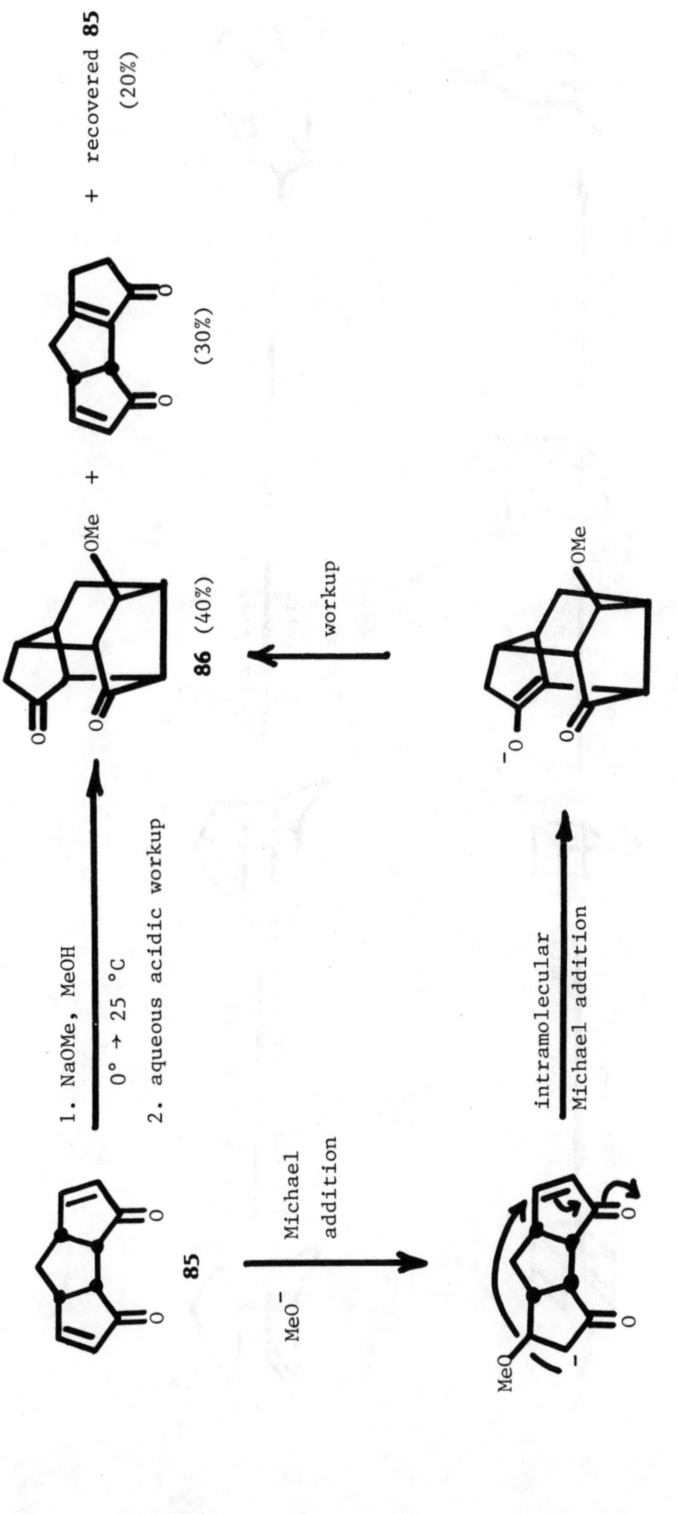

85

1. NaOMe, MeOH
0° → 25 °C
2. aqueous acidic workup

86 (40%) + (30%) + recovered 85 (20%)

MeO⁻ Michael addition

workup

intramolecular Michael addition

Scheme 1.23 ■

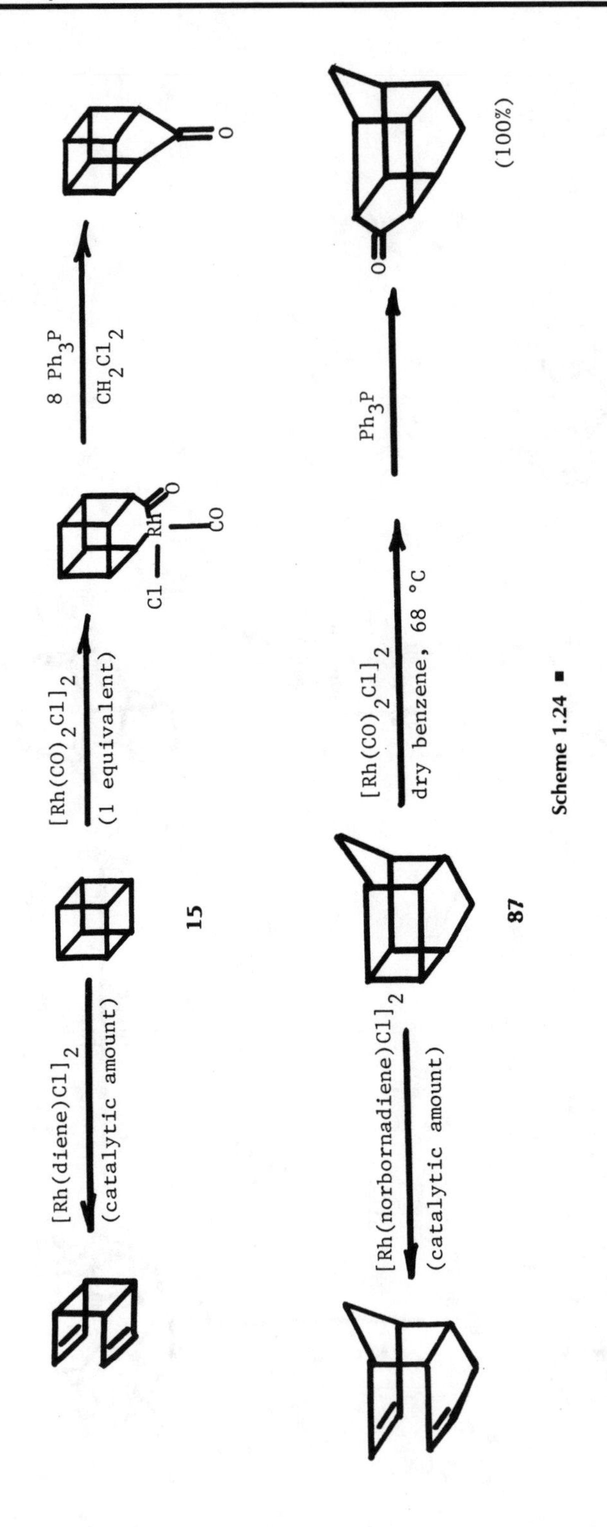
[Rh(diene)Cl]$_2$
(catalytic amount)
15
[Rh(CO)$_2$Cl]$_2$
(1 equivalent)
Cl
Rh
CO
O
8 Ph$_3$P
CH$_2$Cl$_2$
O
[Rh(norbornadiene)Cl]$_2$
(catalytic amount)
87
[Rh(CO)$_2$Cl]$_2$
dry benzene, 68 °C
Ph$_3$P
O
(100%)

Scheme 1.24 ■

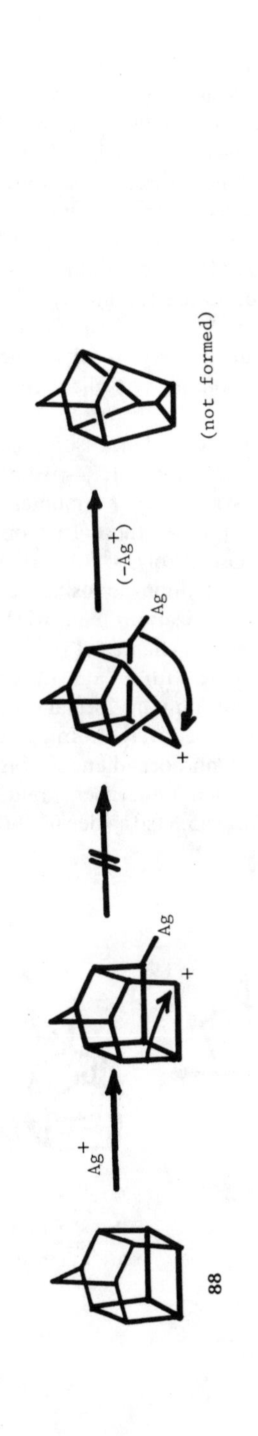
88
Ag+
Ag
+
(−Ag+)
(not formed)

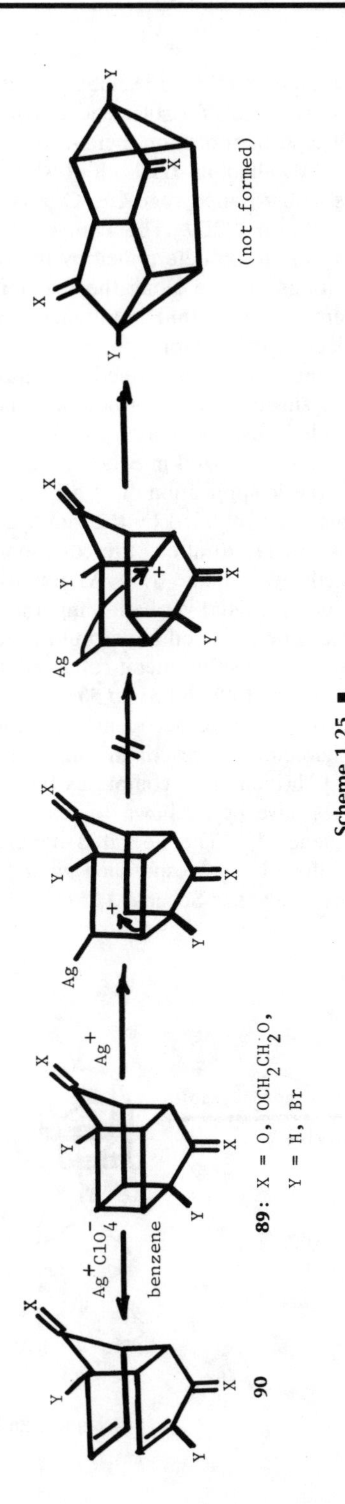
Ag+ClO4−
benzene
90
89: X = O, OCH2CH2O,
Y = H, Br
Ag+
Ag
X
Y
+
(not formed)

Scheme 1.25 ■

Such transition-metal-mediated cyclobutane–diolefin transformations are of intense current interest as models for solar energy conversion and storage.[71]

The second type of transition-metal-promoted valence isomerization is displayed by silver(I) and by palladium(II), both of which function essentially as Lewis acids in their reactions with strained cage C—C σ bonds[47b,48] (see Table 1.1 and pertinent discussion in Section 1.2D). The course of Ag(I)-promoted reactions with cage compounds appears to be determined by the relative energetics among intermediate argentocarbocations that lie along the reaction coordinate for rearrangement.[42d,42e] Thus, homopentaprismane (**88**) is unreactive toward Ag(I), a result that has been ascribed to a lack of driving force for rearrangement due to the cumulative effects of steric strain in the requisite intermediate cationic species.[42d,42e] By way of contrast, the reaction of substituted 1,3-bishomocubanes **89** with silver(I) results simply in cyclobutane–diolefin transformation to afford the corresponding tricyclic diene, **90**.[72] These results are summarized in Scheme 1.25.

A useful synthetic application of transition-metal-promoted valence isomerizations of cage systems is exemplified by the ring cleavage of substituted 1,2,4-trishomocubanes (e.g., **10** and **11**) to afford the corresponding cis,syn,cis linear triquinanes (**12** and **13**, respectively, Scheme 1.4). As noted in Section 1.2A, these ring opening processes can be promoted via high-temperature gas-phase pyrolysis.[20] Alternatively, similar results can be obtained under mild experimental conditions by using Lewis or proton acids or via transition-metal-catalyzed valence isomerizations [e.g., Rh(I)-promoted isomerization of **49** and **51** to **50** and **52**, respectively, Scheme 1.13].[36]

Treatment of substituted secopentaprismanones **91a–91c** with a catalytic amount of $[Rh(norbornadiene)Cl]_2$ results in ring opening with concomitant formation of the corresponding $[Rh(diene)Cl]_2$ complexes (i.e., **92a–92c**, respectively).[73] Interestingly, these complexes have been shown to react readily with norbornadiene to displace $[Rh(norbornadiene)Cl]_2$. The liberated diolefin ligand then undergoes rapid rearrangement to afford the corresponding *cis,cisoid,cis*-tricyclo[5.3.0.0^{2,6}]deca-3,9-dienes (**93a–93c**, respectively; see Scheme 1.26).[73]

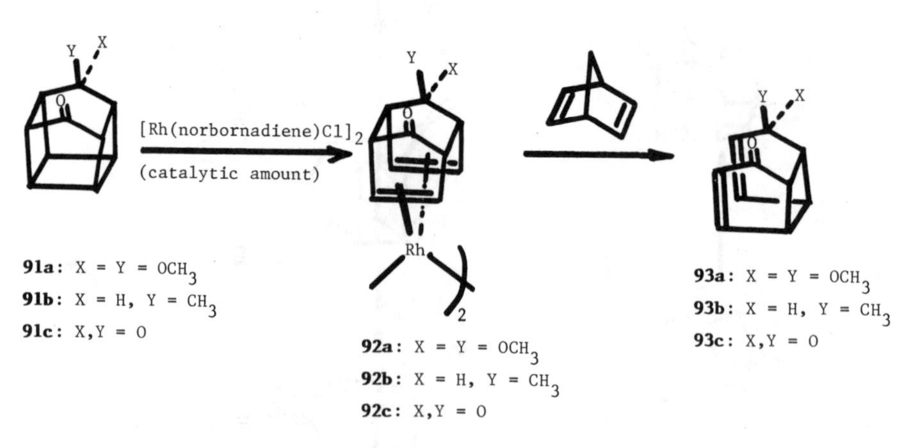

Scheme 1.26 ■

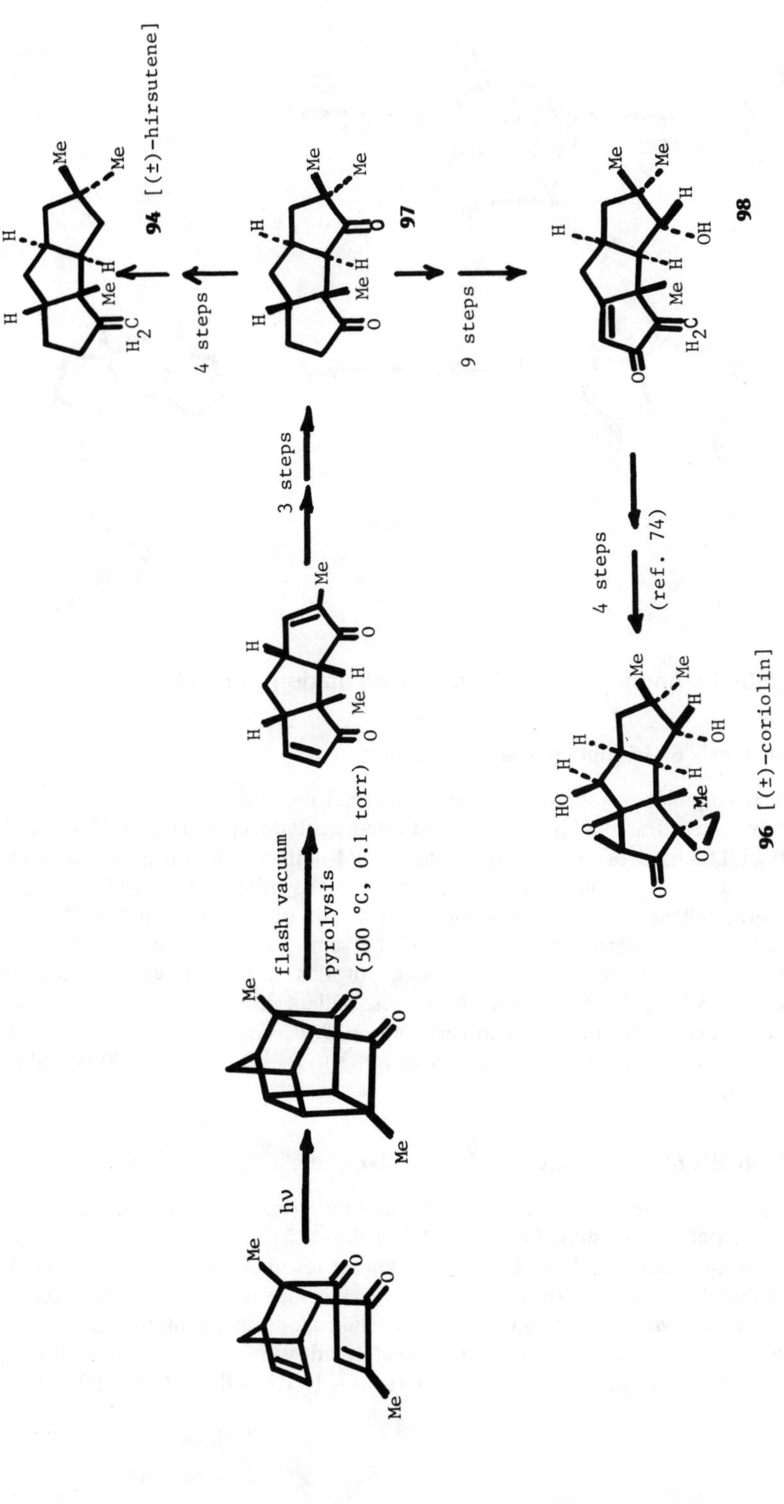

hν

flash vacuum pyrolysis (500 °C, 0.1 torr)

3 steps

97

4 steps

94 [(±)-hirsutene]

9 steps

98

4 steps (ref. 74)

96 [(±)-coriolin]

Scheme 1.27 ■

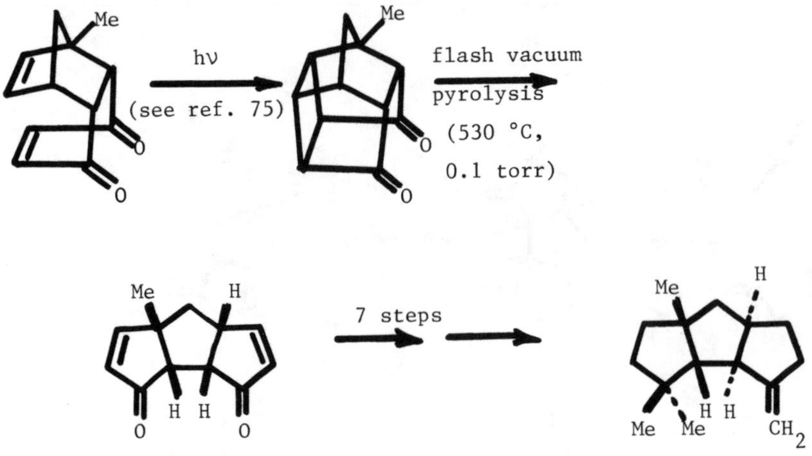

hν (see ref. 75)

flash vacuum pyrolysis (530 °C, 0.1 torr)

7 steps

95 [(±)-capnellene]

Scheme 1.28 ■

1.3. Synthetic Applications of Cage Fragmentation Products

1.3A. Synthesis of Triquinane Natural Products

The linear triquinanes generated via photothermal metathesis (see Scheme 1.1)[15,17] and via reductive fragmentation[16] of substituted pentacyclo[$5.4.0.0^{2,6}.0^{3,10}.0^{5,9}$]undecanes (PCUDs) have been used to synthesize polyquinane natural products. Mehta and co-workers have employed this approach in total syntheses of (±)-hirsutene **(94)** and (±)-capnellene **(95)** and in a formal synthesis of (±)-coriolin **(96)**.[14] Hirsutene and coriolin consist fundamentally of the same carbocyclic system. The photothermal metathesis route takes advantage of this fact by utilizing a common intermediate, **97**, in their respective syntheses. Salient features of the syntheses of hirsutene and of a near precursor to coriolin (i.e., **98**)[74] are shown in Scheme 1.27.[14] The corresponding photothermal metathesis route to (±)-capnellene **(95)** is outlined in Scheme 1.28.[14,75]

1.3B. Synthesis of Crown Ethers and Molecular Clefts

The cis,syn,cis linear triquinanedione **99** can be prepared readily in essentially quantitative yield in two steps from PCUD-8,11-dione.[16] Due to the close proximity of the two carbonyl functionalities in **99**, transannular reactions are observed frequently in this system that result in formation of an oxygen bridge between C-3 and C-11 (see Scheme 1.29).[17] Mehta and co-workers have taken advantage of this situation to synthesize a series of crown ethers via acid-catalyzed reaction of **99** with a variety of glycols of the type $HOCH_2CH_2—(OCH_2CH_2)_nOCH_2CH_2OH$ (Scheme 1.30).[76]

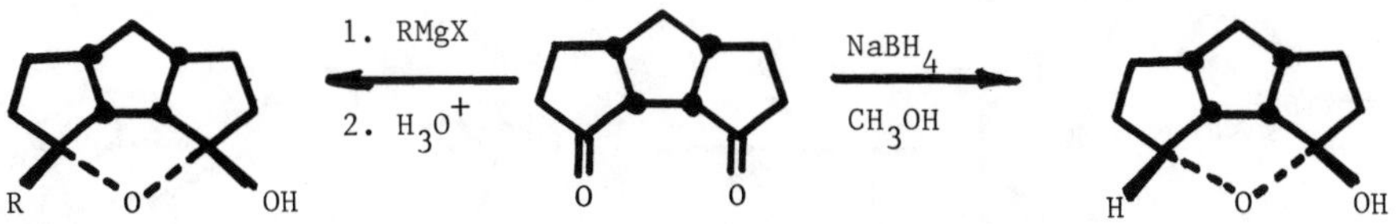

99

Scheme 1.29 ■

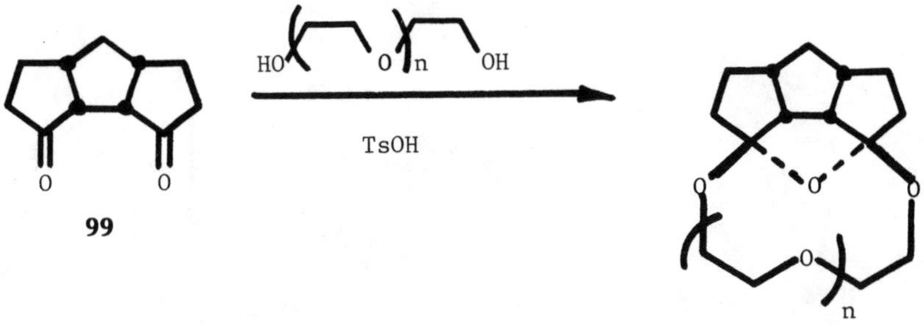

99

100a: n = 0
100b: n = 1
100c: n = 2
100d: n = 3

Scheme 1.30 ■

Cation binding studies have been performed with two of the resulting crown ethers, that is, **100c** and **100d**. Only **100d** showed potentially useful avidity toward, for example, Na^+ and K^+, as measured by its ability to promote phase transfer of the corresponding metal picrate salt from water into dichloromethane. In this regard, the complexing ability of **100d** was found to be comparable to 15-crown-5.

Additionally, the ability of **100c** and **100d** to function as ionophores by translocating metal ions across liposomes was studied.[76] Compound **100d** showed a slight preference for K^+ vis-à-vis Na^+ in these ion transport experiments. Interestingly, **100d** displayed the opposite selectivity in the corresponding metal picrate phase transfer experiments described previously. Compound **100c** proved to be capable of translocating Li^+ in the ion transport studies.

Crowned *para*-benzoquinones (15-crown-5 and 18-crown-6, **101a** and **101b**, respectively) have been synthesized by Japanese workers.[77] Interestingly, these novel crown ethers readily enter into Diels–Alder reactions with cyclopentadiene, thereby affording cycloadducts **102a** and **102b**, respectively. Intramolecular [2 + 2] photocyclization of **102a** and of **102b** with Pyrex-filtered light afforded the corresponding PCUD-8,11-dione derivatives, **103a** and **103b**, respectively, each in quantitative yield (Scheme 1.31).[77]

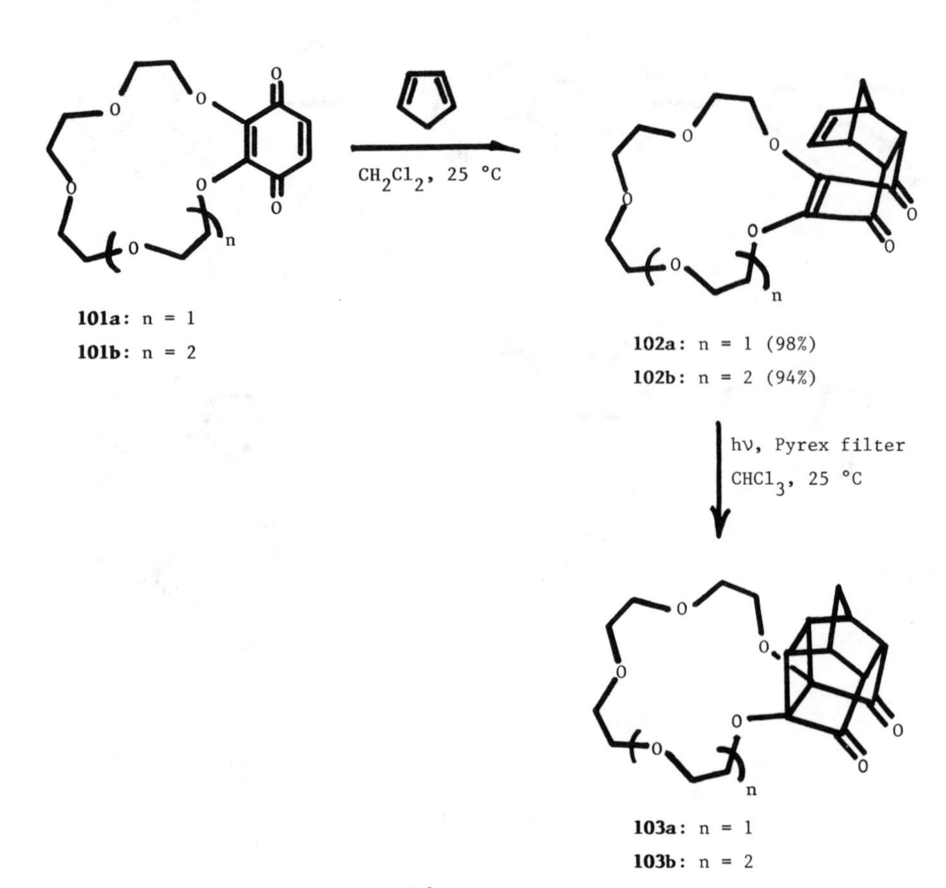

Scheme 1.31 ■

Thummel and co-workers have synthesized a series of polyaza cavity-shaped molecules by utilizing sequential Friedländer condensations between polycyclic 1,2-diketones and aromatic (or heteroaromatic) *ortho*-aminoaldehydes.[78] In an extension of this work, Thummel and Lim[79] obtained a rigid *syn*-orthocyclophane, **105**, via two sequential Friedländer condensations of tetracyclo[6.3.0.0^{4,11}.0^{5,9}]undecane-3,8-dione (**104**) with *ortho*-aminobenzaldehyde. More recently, Thummel and Hegde[80] utilized the bis(hydrazone), **106**, derived from dione **104** to generate another interesting *syn*-orthocyclophane system, that is, **107** (Scheme 1.32).

Additional examples have been reported that illustrate the use of the sequential Friedländer condensation methodology described previously to prepare novel molecular clefts. Thus, Marchand et al.[81] have prepared 2,3:6,7-bis(2′,3′-quinolino)pentacyclo-[6.5.0.0^{4,12}.0^{5,10}.0^{9,13}]tridecane (**108**, Scheme 1.33).

The synthesis and X-ray crystal structure of 3,4:10,11-bis(2′,3′-quinolino)-tricyclo[6.3.0.0^{2,6}]undecane (**109a**) have been reported by two laboratories.[82,83] In addition, the X-ray crystal structure of a cationic Rh(I) complex of **109** (i.e., **110**) has been reported[84] (Scheme 1.34).

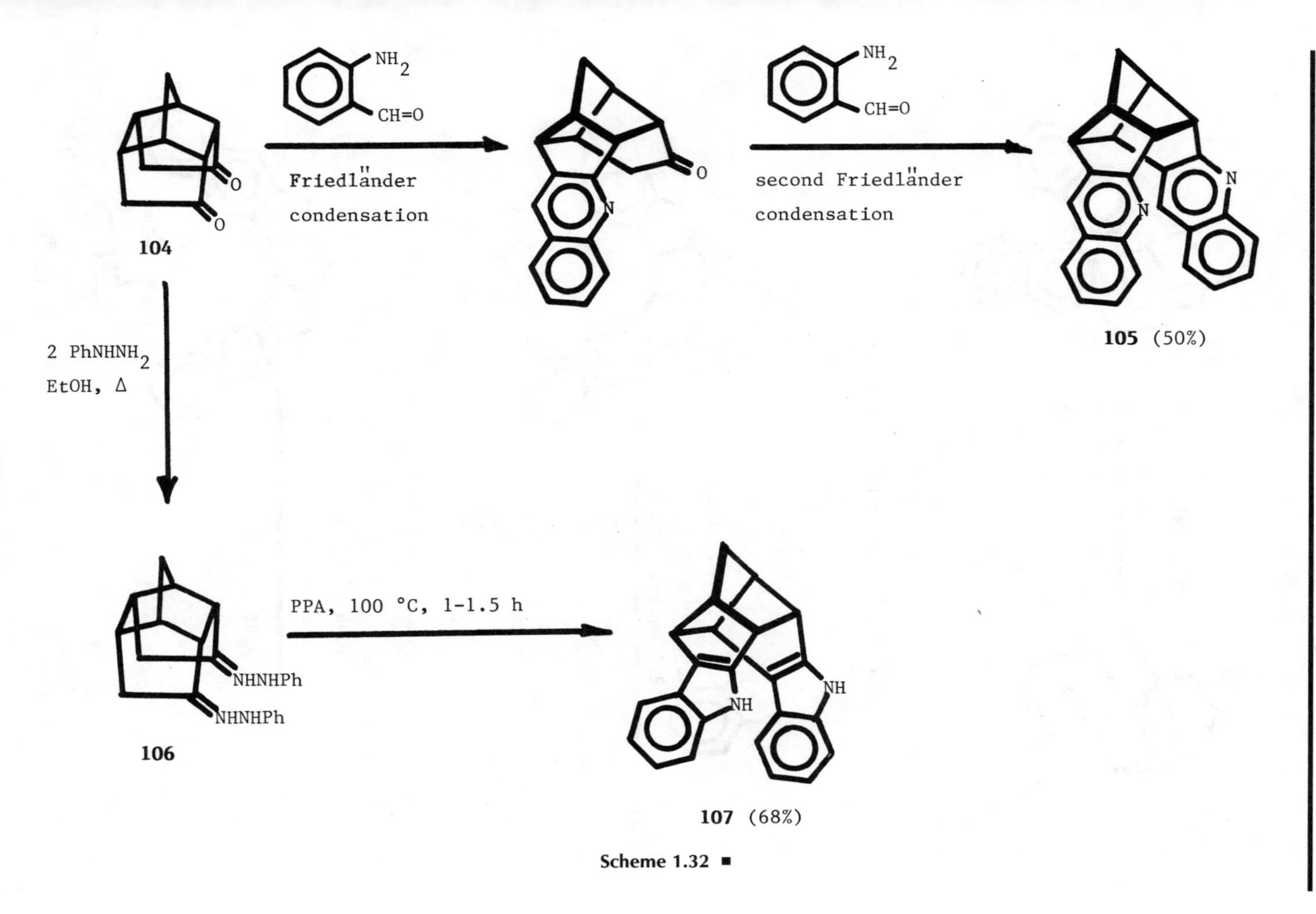
NH2
CH=O
Friedländer
condensation
O
O
104
N
O
NH2
CH=O
second Friedländer
condensation
N
N
105 (50%)
2 PhNHNH2
EtOH, Δ
NHNHPh
NHNHPh
106
PPA, 100 °C, 1-1.5 h
NH
NH
107 (68%)

Scheme 1.32 ■

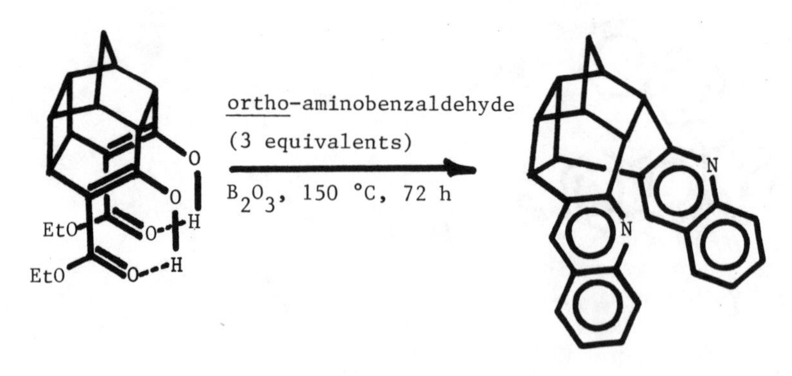

Scheme 1.33 ■

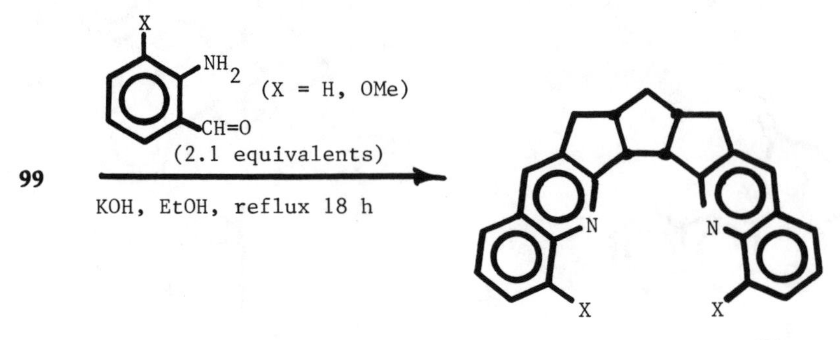

109a: X = H (73%)[84]
109b: X = OMe (62%)[84]

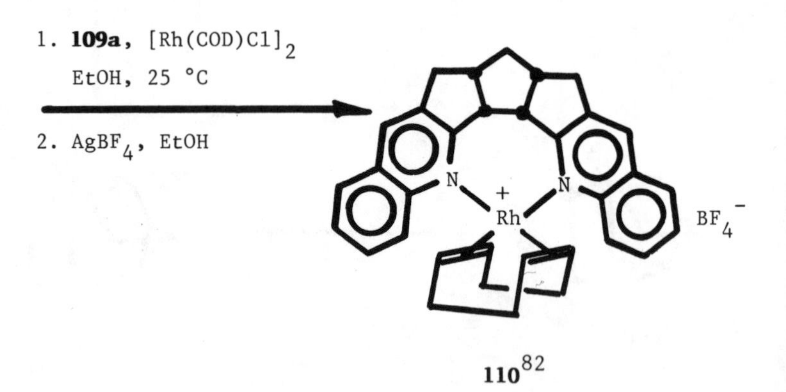

110[82]

Scheme 1.34 ■

Several features of the X-ray crystal structures of the novel molecular clefts described previously are noteworthy. The X-ray crystal structure of a monohydrate of **105** in which the water molecule is hydrogen-bonded simultaneously to both quinoline nitrogen atoms has been reported. The dihedral angle between the two quinoline rings in this system was found to be 50.5°, and the nonbonded N----N distance therein was found to be 3.65 Å.[79] In contrast, the dihedral angle between the corresponding quinoline rings in **108** is 76.4°, and the nonbonded N----N distance between quinoline nitrogen atoms in this system is 4.32 Å.[81] The corresponding dihedral angle in cationic Rh(I) complex **110** is 87°, whereas the N----N separation is only 2.73 Å.[82] Numerous aspects of the host–guest chemistry of molecular clefts **105**, **107**, **108**, and **109** are of considerable theoretical and potential practical interest and remain to be fully explored.

Unusual behavior has been noted upon attempted monoprotonation of **109a** with triflic acid. Examination of molecular models suggests that **109a** can "adopt a conformation in which the two quinoline nitrogen atoms come into close mutual proximity."[84] Although this observation suggests that it might be possible for **109a** to function as a "proton sponge"[85] to afford a symmetrical, proton-bridged species (i.e., **111**), it was demonstrated that only symmetrically diprotonated **109a** (i.e., **112**, Scheme 1.35) could be isolated upon reaction of **109a** with one equivalent of triflic

109a $\xrightarrow[\text{(1 equivalent)}]{CF_3SO_3H}$

N H + H N +

$2\ CF_3SO_3^-$

112 (two polymorphic forms of the monohydrate were isolated)[84]

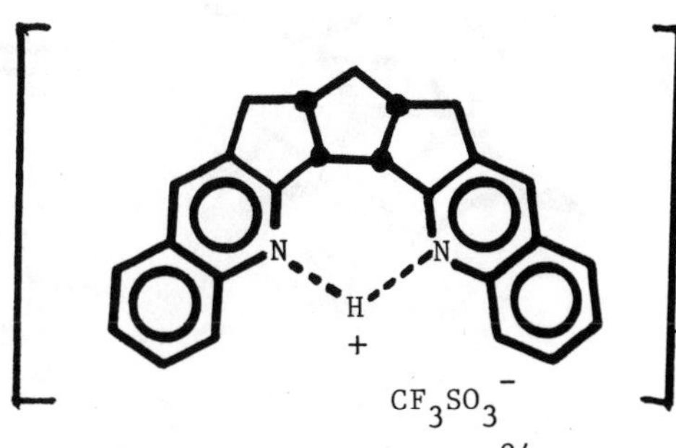

111 (not isolated)[84]

Scheme 1.35 ▪

acid. In fact, *two* polymorphic forms of hydrated **112** were isolated in this study, both of which were characterized via X-ray crystallographic methods.[84]

1.3C. Use of Cage Molecules as Synthetic Templates

Cage molecules themselves and various cage fragmentation products derived from these molecules often can provide the fundamental carbocyclic backbones from which more highly complex cage or noncage molecules ultimately can be constructed. One such example has already been seen in the application of the photothermal metathesis methodology for synthesizing triquinane natural products (Section 1.3A).

In a similar vein, relatively simple and readily accessible cage molecules have been utilized as starting materials to synthesize much more complex cage systems. Examples of this can be found in some of the highly imaginative potential routes that have been suggested for synthesizing dodecahedrane (in addition to the well-known Paquette synthesis of this molecule).[7,8] Thus, Prinzbach and co-workers have elaborated a relatively simple cage system, **113**, first into "pagodane" (i.e., **114**)[86] and then into pentagonal dodecahedrane (**115**, Scheme 1.36).[87]

Another example is provided by the work of Mehta and co-workers, who have employed derivatives of simple cage compounds as potential precursors to dodeca-

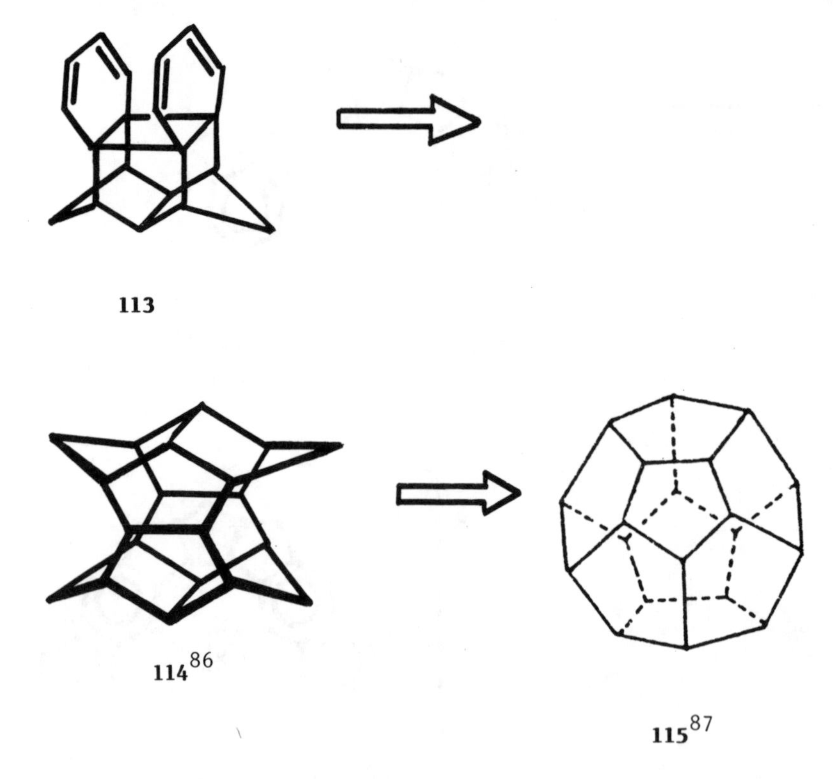

Scheme 1.36 ■

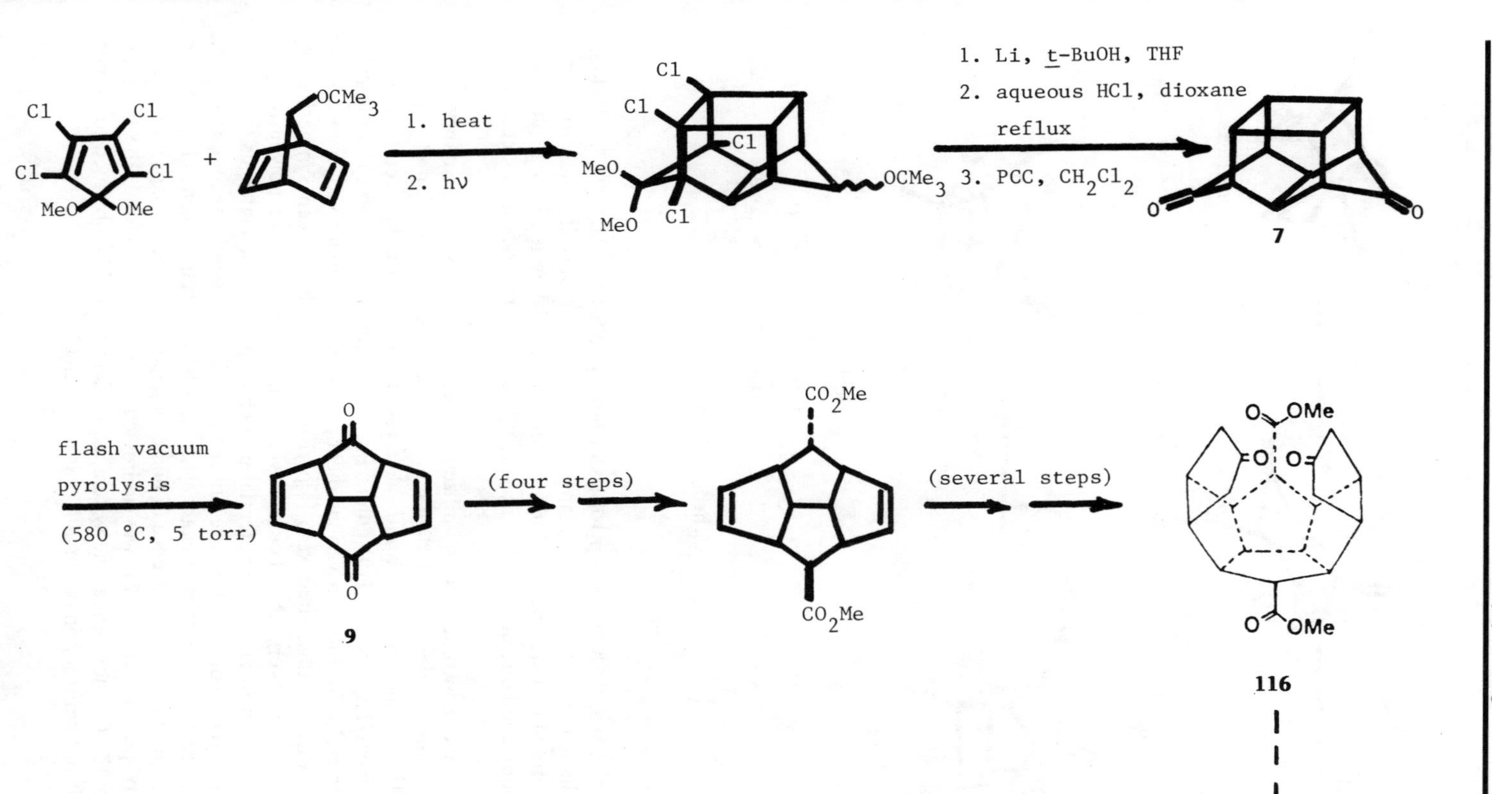

Cl
Cl
Cl
Cl
MeO
OMe
+
OCMe3
1. heat
2. hν
Cl
Cl
Cl
Cl
MeO
MeO
OCMe3
1. Li, t-BuOH, THF
2. aqueous HCl, dioxane
reflux
3. PCC, CH2Cl2
O
O
7
flash vacuum
pyrolysis
(580 °C, 5 torr)
O
O
9
(four steps)
CO2Me
CO2Me
(several steps)
OMe
O
O
O
O
OMe
116
115

Scheme 1.37 ■

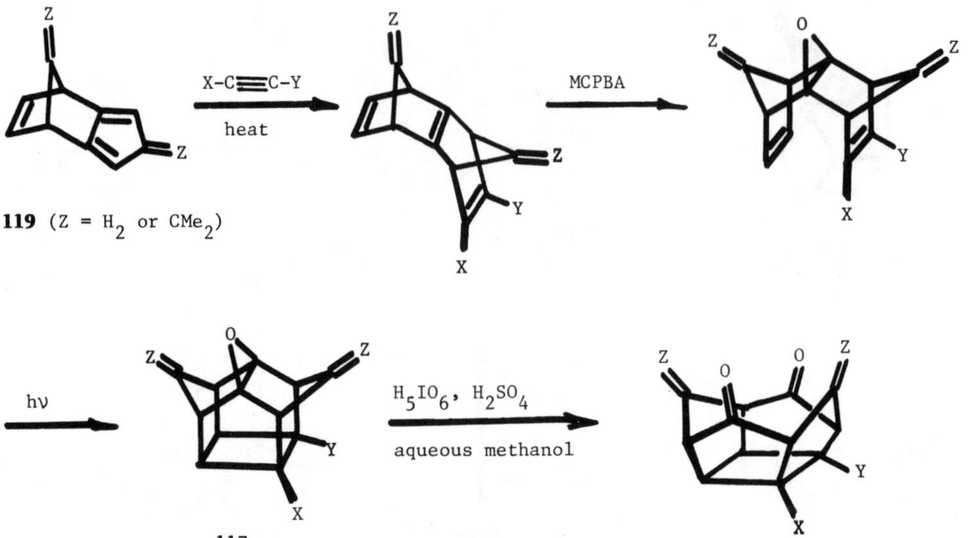

X-C≡C-Y

heat

MCPBA

119 ($Z = H_2$ or CMe_2)

hν

H_5IO_6, H_2SO_4

aqueous methanol

117

118a:[89a] $X = H$, $Y = Ts$, $Z = H_2$

118b:[89c] $X = SiMe_3$, $Y = NO_2$, $Z = H_2$

118c:[89d] $X = Y = SO_2Ph$, $Z = CMe_2$

Scheme 1.38 ■

hedrane.[19,88] Thus, flash vacuum pyrolysis of hexacyclo[5.4.1.$0^{2,6}$.$0^{3,10}$.$0^{5,9}$.$0^{8,11}$]dodecane-4,12-dione (i.e., 1,3-bishomopentaprismanedione, **7**)[88a] afforded the corresponding all-cis tetraquinanendione, **9**.[19] Subsequently, **9** was elaborated into a spheroidal, all-cis hexaquinanedione diester, that is, **116**,[8b,88c] a near precursor of dodecahedrane (Scheme 1.37).

Paquette and co-workers have utilized substituted 1,3-bishomopentaprismanes, **117**, as versatile reaction intermediates in their syntheses of substituted [4]peristylanes (**118**).[89] Ultimately, their approach owes its success to the fact that tricyclo[5.2.1.$0^{2,6}$]deca-2,5,8-triene (**119**) enters into facile regiospecific and stereospecific Diels–Adler cycloaddition with various dienophiles in the manner shown in Scheme 1.38.[90] The various substituted keto [4]peristylanes that have been synthesized by Paquette and co-workers have been converted into a number of polynitro[4]peristylanes that are of interest as a potential new class of energetic materials.[91]

Paquette and co-workers also have utilized a readily accessible cage molecule (i.e., **120**) as an intermediate in their elegant synthesis of C_{16}-hexaquinacene (**121**, Scheme 1.39).[92] The question of the existence of homoconjugative interactions, presumably via σ overlap of pπ orbitals in **121**, was addressed in this study. Despite the favorable geometry of the C_{16}-hexaquinacene system, no evidence could be obtained to support the concept of "neutral homoaromatic character" in **121**.[92]

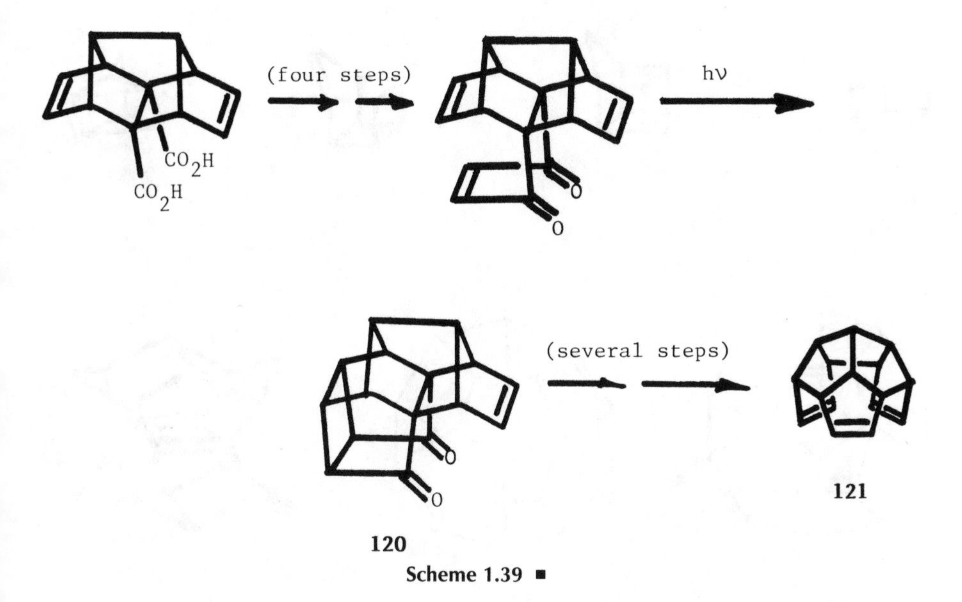

Scheme 1.39 ■

1.4. Caged Reactive Intermediates

1.4A. Caged Pyramidalized Alkenes

Pyramidalization of alkenes results when C=C double bonds are distorted in such a manner that "one or both of the doubly bonded carbons do not lie in the same plane as the three atoms attached to it."[93] The unusual skeletal bond angles plus the molecular rigidity that generally are associated with carbocyclic cage molecules render them of interest as a potential new class of substrates for alkene pyramidalization studies. Indeed, the utility of cage molecules in this regard has recently been recognized and exploited by several investigators. Thus, evidence for the transient existence of a number of very highly strained and pyramidalized cage alkenes has appeared. Representative examples in this regard, which include such exotic species as 1,2-dehydrocubane ("cubene," **122**),[94] homocub-4(5)-ene (**123a**),[95] homocub-1(9)-ene (**123b**),[96] quadricycl-1(7)-ene (**124**),[97] dodecahedrene (**125**),[8e, 98] dodecahedradienes (e.g., **126**),[99] and bissecododecahedradienes (e.g., **127**)[100] are depicted in Scheme 1.40.

Eaton and co-workers[96a, 96b] have investigated the detailed mechanism of the "olefin-to-carbene" rearrangement in 9-phenyl-1(9)-homocubene. Thus, an ethanol solution of carbonyl ^{13}C-labeled cubyl phenyl ketone tosylhydrazone (**128**), when heated in the presence of sodium ethoxide, afforded a mixture of two homocubyl ethers (i.e., 9-ethoxy-1-phenyl[1-^{13}C]- and 9-ethoxy-9-phenyl[9-^{13}C]pentacyclo-[4.3.0.0^{2,5}.0^{3,8}.0^{4,7}]nonane, **129** and **130**, respectively, product ratio ca. 1.7 : 1).[96b] The ^{13}C—Ph bond present in the starting tosylhydrazone remains intact in both products. Thus, formation of **129** and **130** both must occur with concomitant skeletal C—C

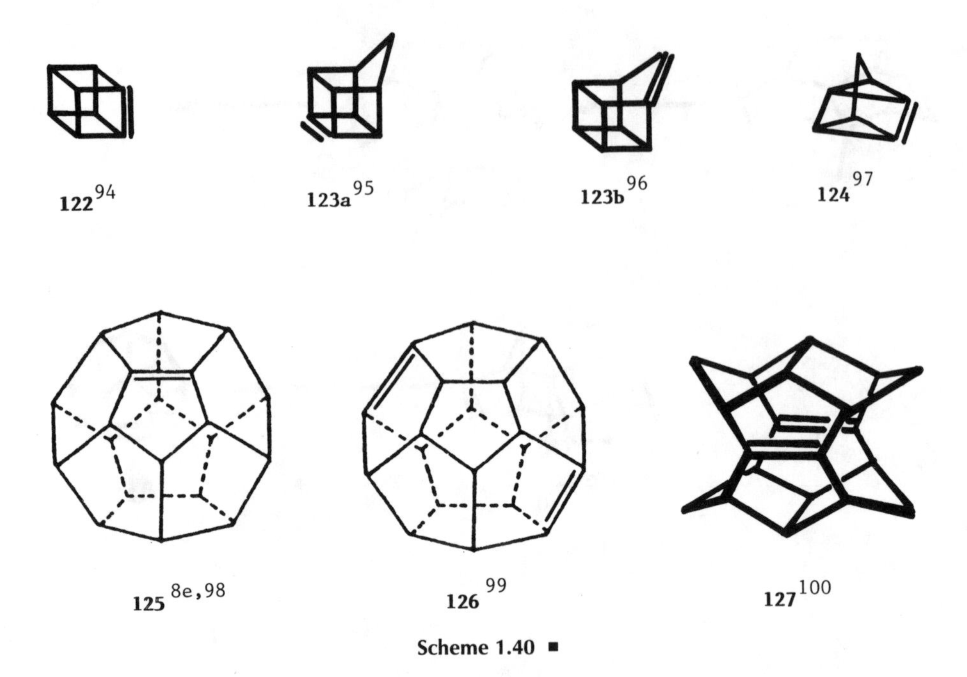

Scheme 1.40 ■

bond reorganization rather than via a mechanism that involves a 1,2-phenyl shift (which would afford **131**; see mechanism in Scheme 1.41).

1.4B. Caged Cationic Intermediates

The recent renaissance in the synthesis and chemistry of functionalized cubanes[5, 101, 102] has provided important new evidence regarding the potential intermediacy of cubyl cation. Recently, Eaton and co-workers[103] have utilized Kropp's photolytic procedure[104] to generate cubyl cations from the corresponding iodocubanes in solution. Thus, irradiation of 1,4-diiodocubane affords the putative intermediate 4-iodocubyl cation, **132**, which can be trapped by nucleophilic solvents, for example, wet acetonitrile or dry methanol, thereby affording **133** and **134**, respectively (Scheme 1.42).[103]

In addition, Eaton and Cunkle[105] have studied the oxidative deiodination of cubyl iodides. Their approach involves oxidation of the substituted iodocubane to a corresponding intermediate hypervalent iodine species that subsequently undergoes thermal decomposition to afford new cubanes with concomitant substitution at the C—I bond in the substrate. This approach also has been utilized by Moriarty and co-workers to synthesize new substituted cubanes.[106] Although other mechanistic pathways can be formulated, it nevertheless is possible to envision these reactions as proceeding via intermediate cubyl cations.[105, 106]

The geometry of cubyl cation is such that it certainly must be regarded as being a highly energetic carbocationic intermediate.[107] The results of molecular mechanics

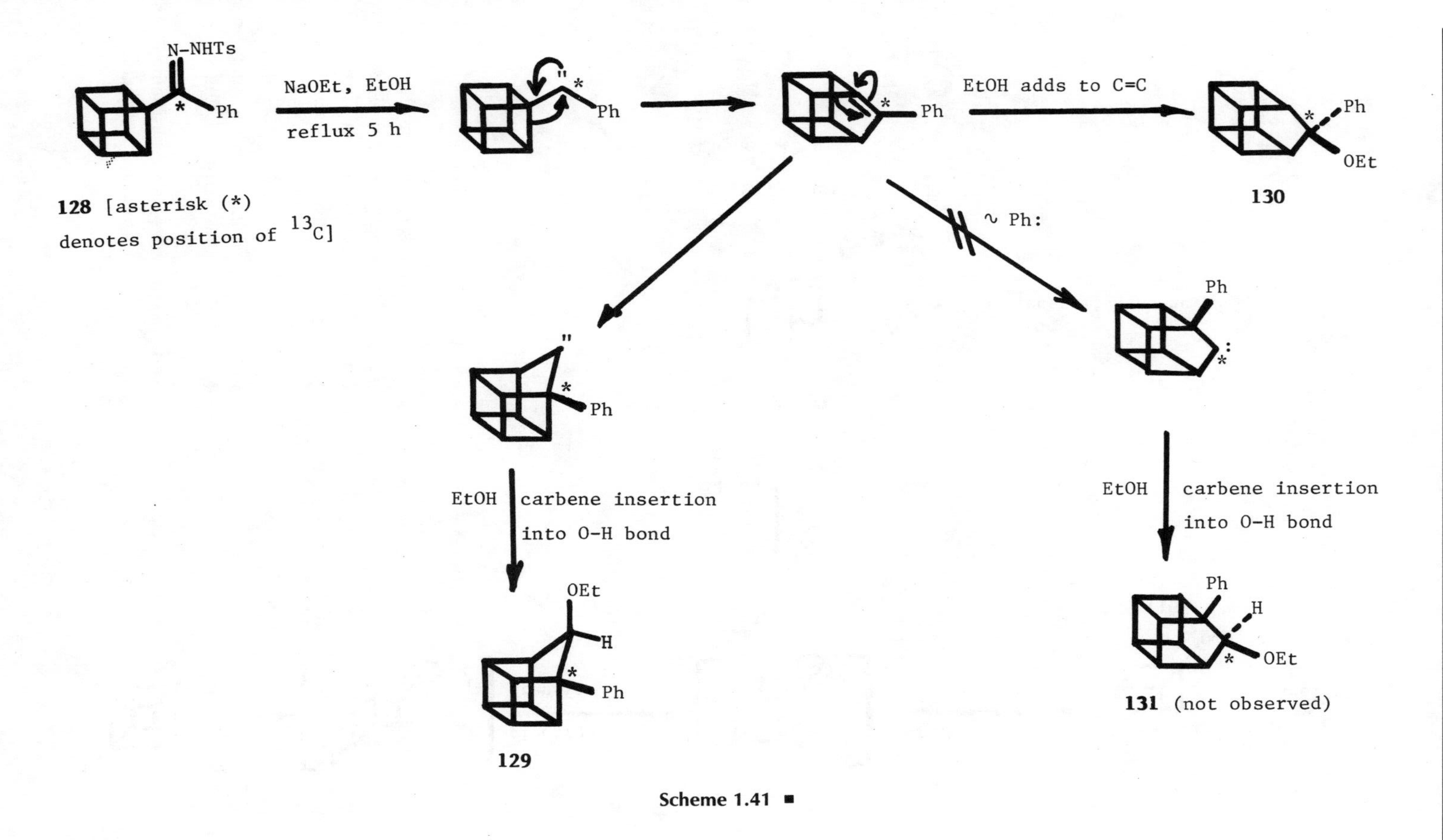
N-NHTs
Ph
NaOEt, EtOH
reflux 5 h
Ph
Ph
EtOH adds to C=C
Ph
OEt
130
128 [asterisk (*) denotes position of ^{13}C]
~ Ph:
Ph
Ph
EtOH
carbene insertion into O-H bond
EtOH
carbene insertion into O-H bond
OEt
H
Ph
129
Ph
H
OEt
131 (not observed)

Scheme 1.41 ■

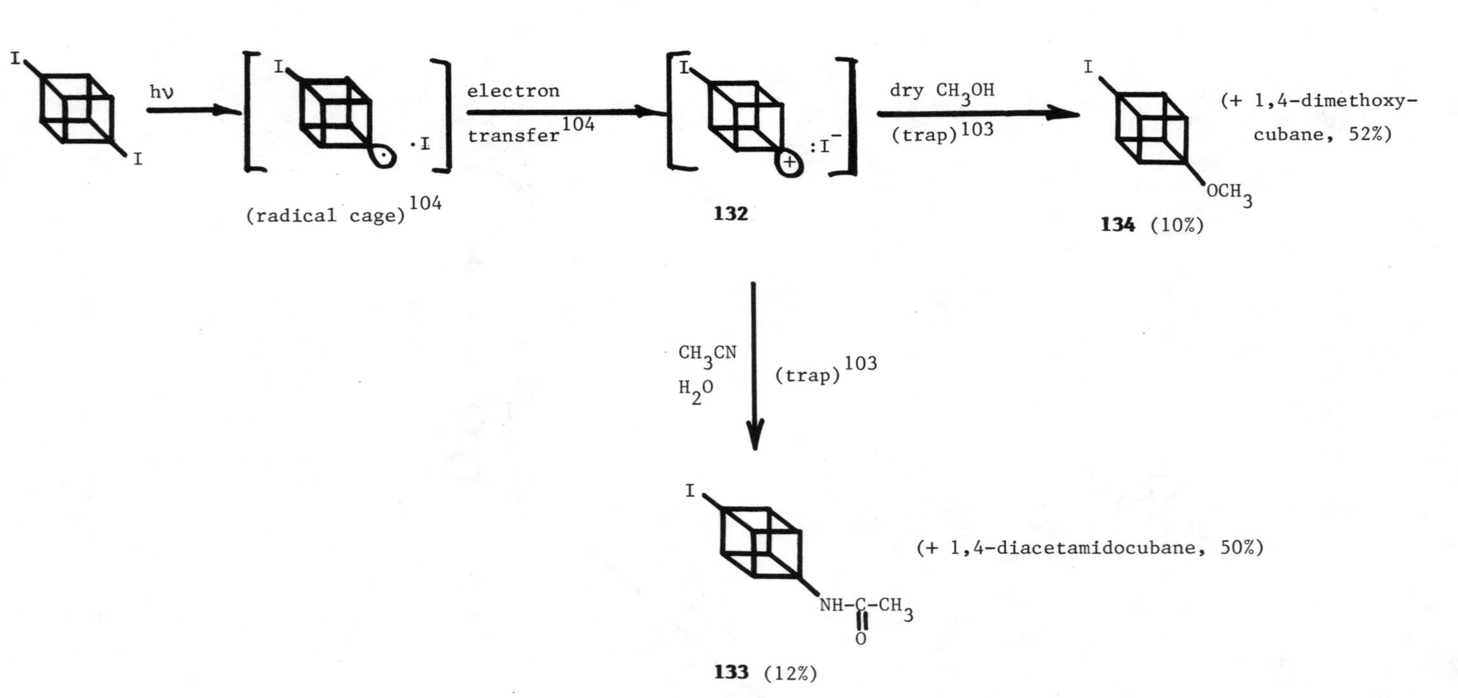

I
I
hν
I
·I
(radical cage)104
electron
transfer104
I
:I⁻
132
dry CH3OH
(trap)103
I
OCH3
134 (10%)
(+ 1,4-dimethoxy-
cubane, 52%)
CH3CN
H2O
(trap)103
I
NH-C-CH3
O
133 (12%)
(+ 1,4-diacetamidocubane, 50%)

Scheme 1.42 ■

calculations of bridgehead carbonium ion reactivities performed many years ago by Bingham and Schleyer[108] bear out this expectation. By their reckoning, cubyl triflate is predicted to be inert toward S_N1 acetolysis, even when this solvolysis is performed at temperatures as high as 250°C.[108]

In view of the foregoing, one might question whether it is indeed appropriate to consider cubyl cations as potential reaction intermediates when formulating mechanisms that conceivably might invoke them. Pertinent to this issue, compelling evidence has been obtained recently that suggests that the earlier calculations[108] seriously underestimate the accessibility of cubyl cations. Eaton et al.[109] have measured the rate of solvolysis of cubyl triflate in dry absolute methanol. This reaction displays pseudo-first-order kinetics (followed for at least three half-lives) and affords cubyl methyl ether as the only product; for solvolysis at 50.1°C, $k_1 = 7.72 \times 10^{-5}\ s^{-1}$ ($\Delta H = 24.6$ kcal mol^{-1}, $\Delta S = 1.5$ e.u. over the temperature range 50–70°C).[109]

Thus far, it has not been possible to find a set of environmental conditions (i.e., temperature, solvent system) that would make possible a direct comparison of the rates of solvolysis of cubyl triflate and 1-norbornyl triflate. However, it is possible to draw meaningful conclusions from the observation[109] that 1-norbornyl triflate is 90% solvolyzed in hexafluoroisopropanol (HFIP) at 60°C in 85 h, whereas cubyl triflate solvolyzes completely in HFIP after only five minutes at room temperature. This astounding result extrapolates to a factor of at least 10^5 in relative reactivities that energetically favors the cubyl cation vis-à-vis the corresponding 1-norbornyl cation![109,110]

In view of this important and unexpected result, additional experimental and theoretical scrutiny of the electronic and structural factors that might contribute to the surprising kinetic accessibility of cubyl cation is warranted. In this vein, it has been suggested[109] that cyclobutyl–cyclopropylcarbinyl type resonance interactions might contribute significantly toward stabilizing cubyl cation.[111]

Another unusual caged cationic system whose synthesis has been reported recently is the "pagodane dication." This species, which is obtained when pagodane (**114**) is dissolved in excess SbF_5–SO_2ClF at −78°C, can be represented by either of two D_{2h} structures (i.e., **135a** or **135b**, Scheme 1.43).[112] Thus prepared, solutions containing this dication were found to be unusually stable (i.e., capable of surviving for several hours at 0°C). The unusual stability of the pagodane dication has been ascribed to the operation of "pseudo-2π aromatic overlap"[112] that renders this species aromatic.[113]

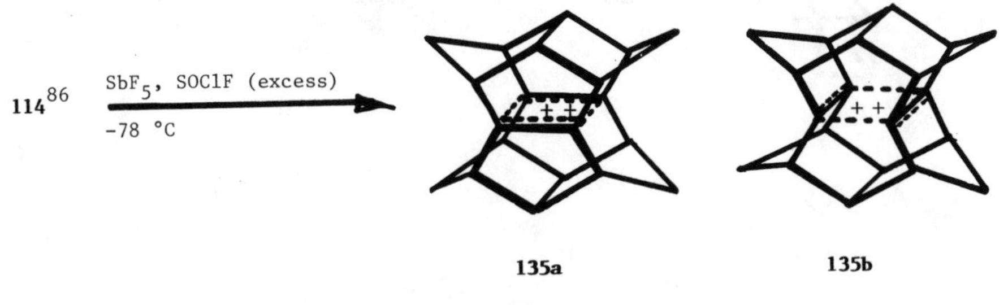

Scheme 1.43 ■

Table 1.3 ■ Methods for Free Radical Substitution on the Cubyl Ring System

Substrate	Conditions	Products (Percentage Yield)	References
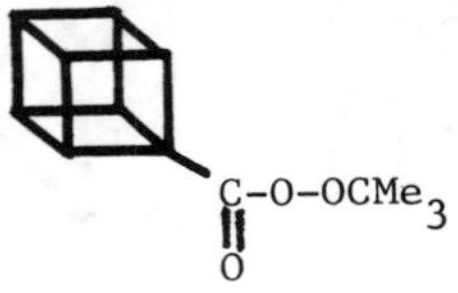 C–O–OCMe$_3$ / ‖ / O	Diisopropylbenzene, 150°C	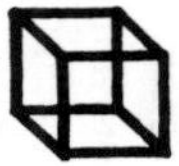 (ca. 30%)	6
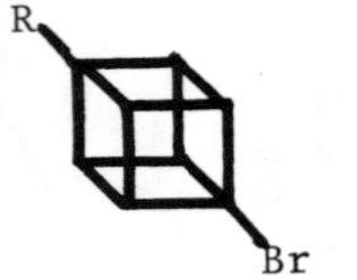 R, Br; R = H, Br	$(n\text{-Bu})_3SnH$, AIBN; hν	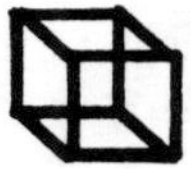 (80–92%)	114a
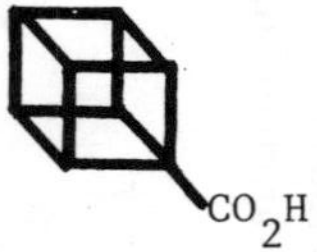 CO_2H	t-BuOOI (generated in situ), Freon 113®, 0°C, followed by hν, 40°C	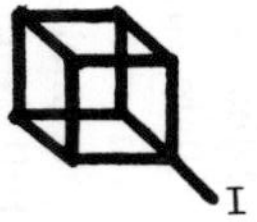 I (53%)	115b
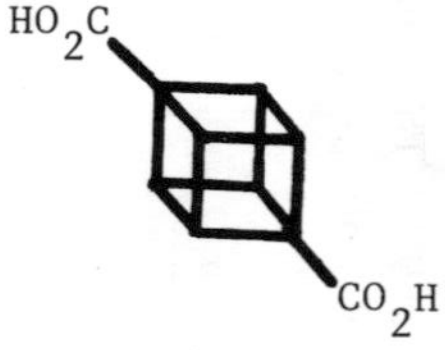 HO_2C, CO_2H	t-BuOOI (generated in situ), Freon 113®, hν, 24 h	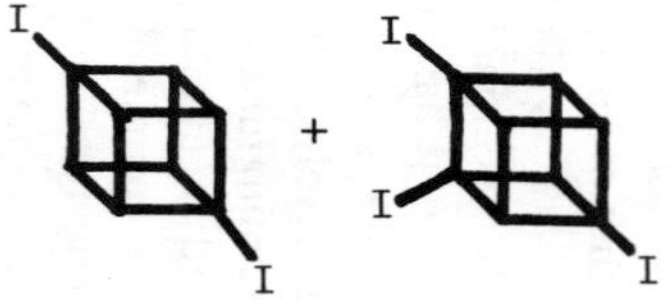 I, I (35%) + I, I, I (45%)	117

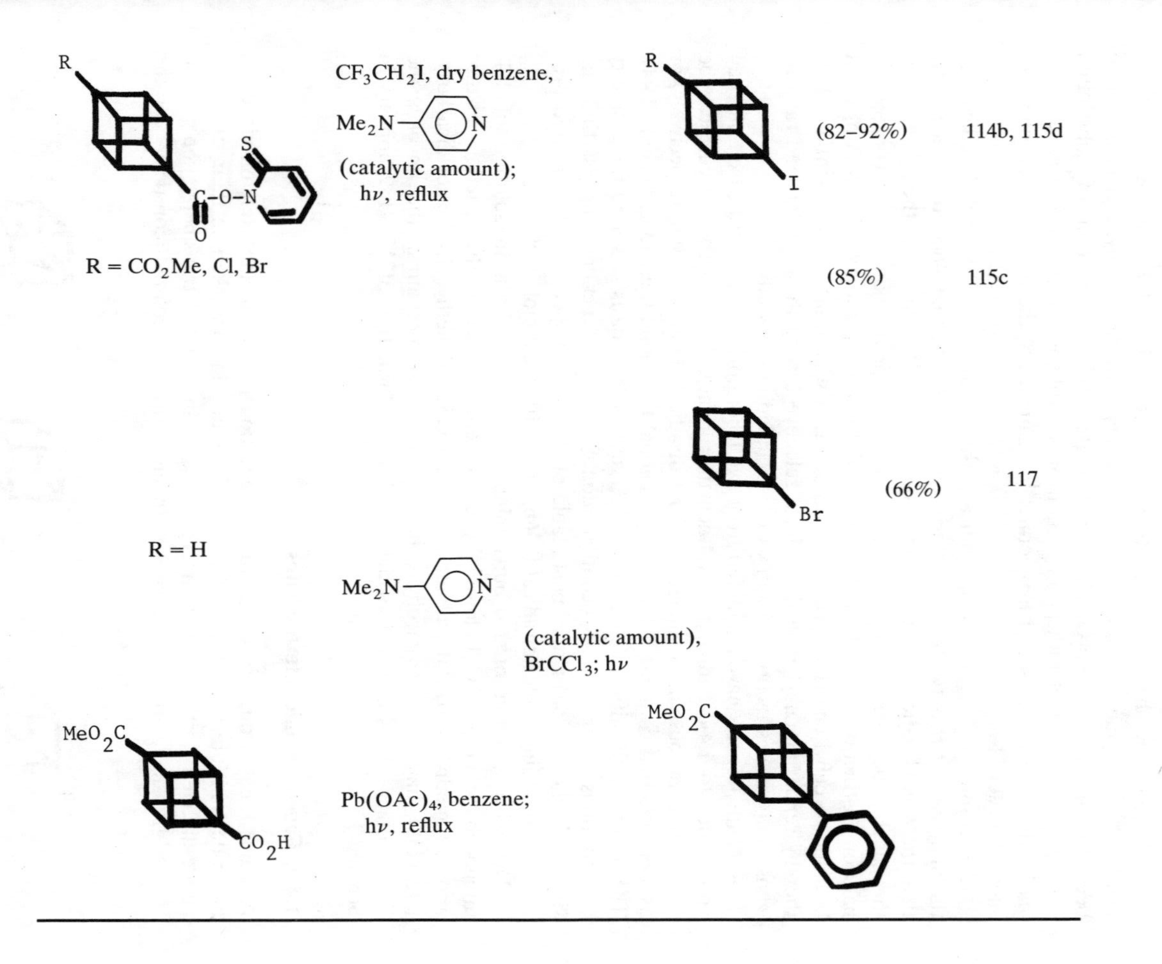

R
S
C–O–N
O
R = CO$_2$Me, Cl, Br
CF$_3$CH$_2$I, dry benzene,
Me$_2$N
N
(catalytic amount);
hν, reflux
R
I
(82–92%)
114b, 115d
(85%)
115c
Br
(66%)
117
R = H
Me$_2$N
N
(catalytic amount),
BrCCl$_3$; hν
MeO$_2$C
CO$_2$H
Pb(OAc)$_4$, benzene;
hν, reflux
MeO$_2$C

1.4C. Caged Radical Intermediates

Cubyl radicals have been generated by reductive decarboxylation (i.e., $RCO_2H \rightarrow RH$)[6,114] and by halodecarboxylation (i.e., $RCO_2H \rightarrow RX$, where X = halogen)[114a,94b,115] of substituted cubanecarboxylic acids. More recently, arylcubanes have been synthesized via $Pb(OAc)_4$-promoted free radical arylation of substituted cubanes.[116] Also, direct radical iodination of the cubane skeleton has been effected via irradiation with *tert*-butyl hypoiodite.[117] Examples of these various reactions are shown in Table 1.3.

The results of kinetic studies indicate that formation of the cubyl radical via thermolysis of *tert*-butyl cubanepercarboxylate occurs "about 4600-fold less rapidly than the *t*-butyl radical under the same conditions."[115a] In addition, the selectivity of the cubyl radical generated via cubyl perester decomposition toward competitive halogen abstraction with $XCCl_3$ (X = Br or Cl) has been studied. The ratio of cubyl bromide to cubyl chloride formed in this reaction was found to be essentially 1:1, thereby rendering cubyl radical the least selective among those bridgehead radicals whose halogen abstraction reactions were included in this study.[115a]

Recently, reports have emanated from two laboratories that describe experimental approaches to 1,4-cubadiyl (i.e., 1,4-dehydrocubane).[118a,118b] This molecule can be represented variously as a singlet or triplet diradical (**136a** and **136b**, respectively) or as a cubane that contains a 1,4 ("body-diagonal')[118a] bond (i.e., **136c**, Scheme 1.44). The results of ab initio calculations (3-21G and 6-31G* basis sets used for geometry optimizations along with two-configuration SCF wave functions) suggest that the singlet diradical (**136a**) is the most stable of these three species and is favored over the corresponding triplet diradical (**136b**) by ca. 10.5 kcal mol^{-1}.[118]

Dropwise addition of *tert*-butyllithium to a tetrahydrofuran solution of 1,4-diiodocubane at −78°C followed by carbomethoxylation of the reaction mixture afforded methyl 4-*tert*-butylcubanecarboxylate (**137**), dimethyl 4,4′-bicubyldicarboxylate (**138**), and methyl 4-bicubylcarboxylate (**139**). A mechanism that invokes the intermediacy of 1,4-cubadiyl to account for the formation of **137–139** appears in Scheme 1.45.[118c]

1.4D. Caged Anionic Intermediates

A major breakthrough in direct functionalization of the cubane skeleton via cubyl anion-derived intermediates was achieved by Eaton in co-workers in their studies of amide-directed "ortho-metalation" reactions[119] and related transmetalation[120] (and reverse transmetalation)[121] processes. In an effort to account theoretically for the

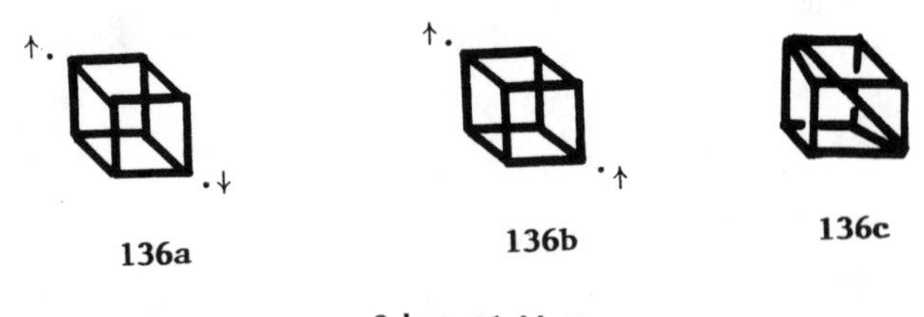

Scheme 1.44 ■

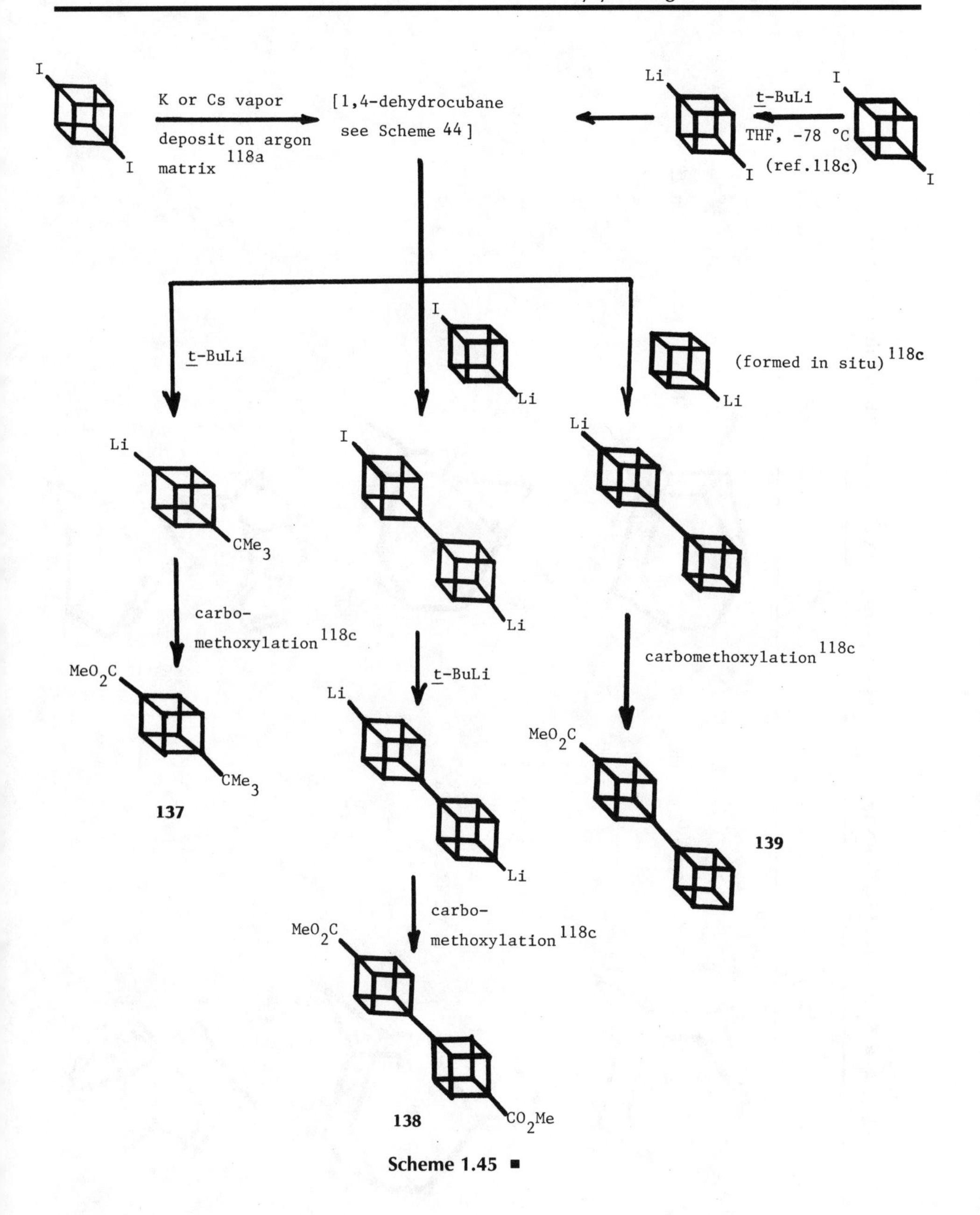
I
I
K or Cs vapor
deposit on argon matrix 118a
[1,4-dehydrocubane see Scheme 44]
Li
I
I
I
t-BuLi
THF, -78 °C
(ref.118c)
t-BuLi
I
Li
(formed in situ) 118c
Li
Li
CMe3
I
Li
Li
carbo-methoxylation 118c
MeO2C
CMe3
137
t-BuLi
Li
Li
carbomethoxylation 118c
MeO2C
139
carbo-methoxylation 118c
MeO2C
CO2Me
138

Scheme 1.45 ■

Table 1.4 ■ Synthesis of Highly Strained Cage Compounds via Divalent Carbon Intermediates

Divalent Carbon Precursor	Conditions	Products (Percentage Yield)	References
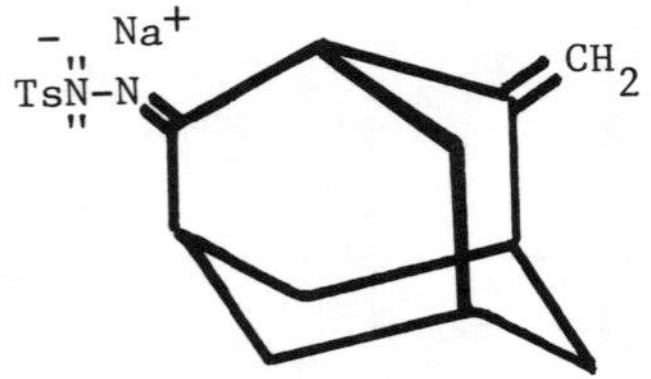 TsN^{-}–N=, Na^{+}, CH_2	180°C, 0.02 mm Hg	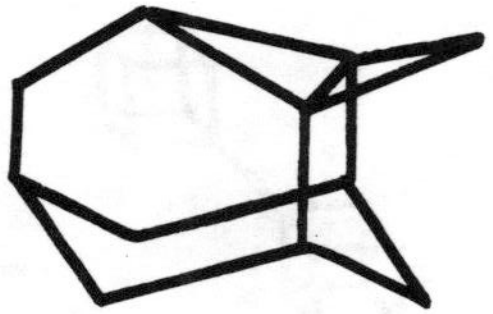 (ca. 70%)	125a, 125b
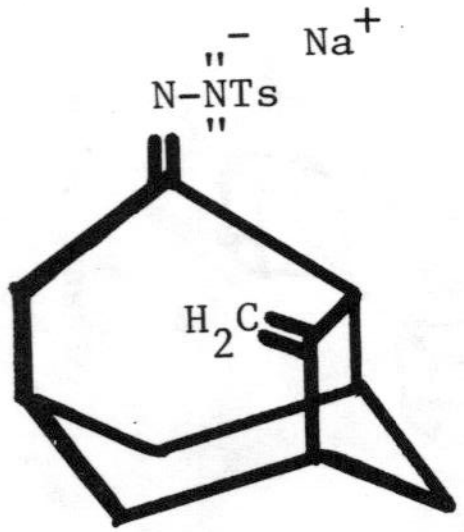 N–N^{-}Ts, Na^{+}, H_2C	200°C, ca. 1 Pa	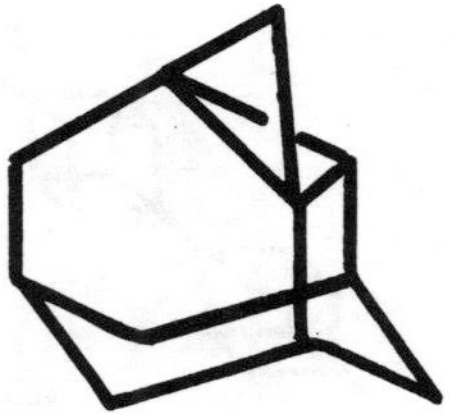 (75%)	125c
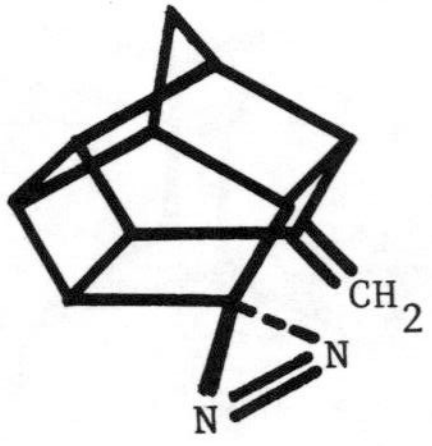 CH_2, N, N	310°C, 100 Pa	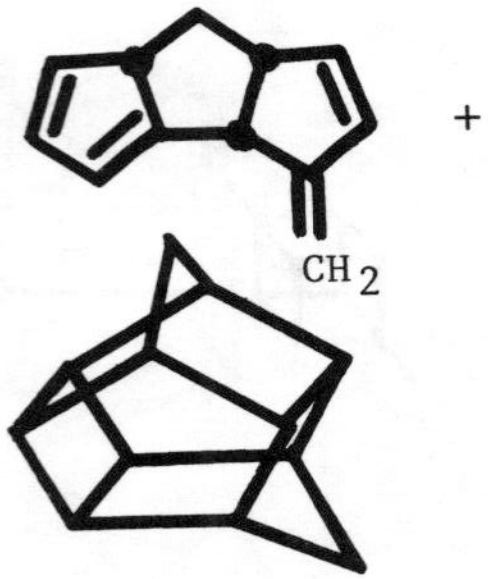 CH_2 + (15%)	125d

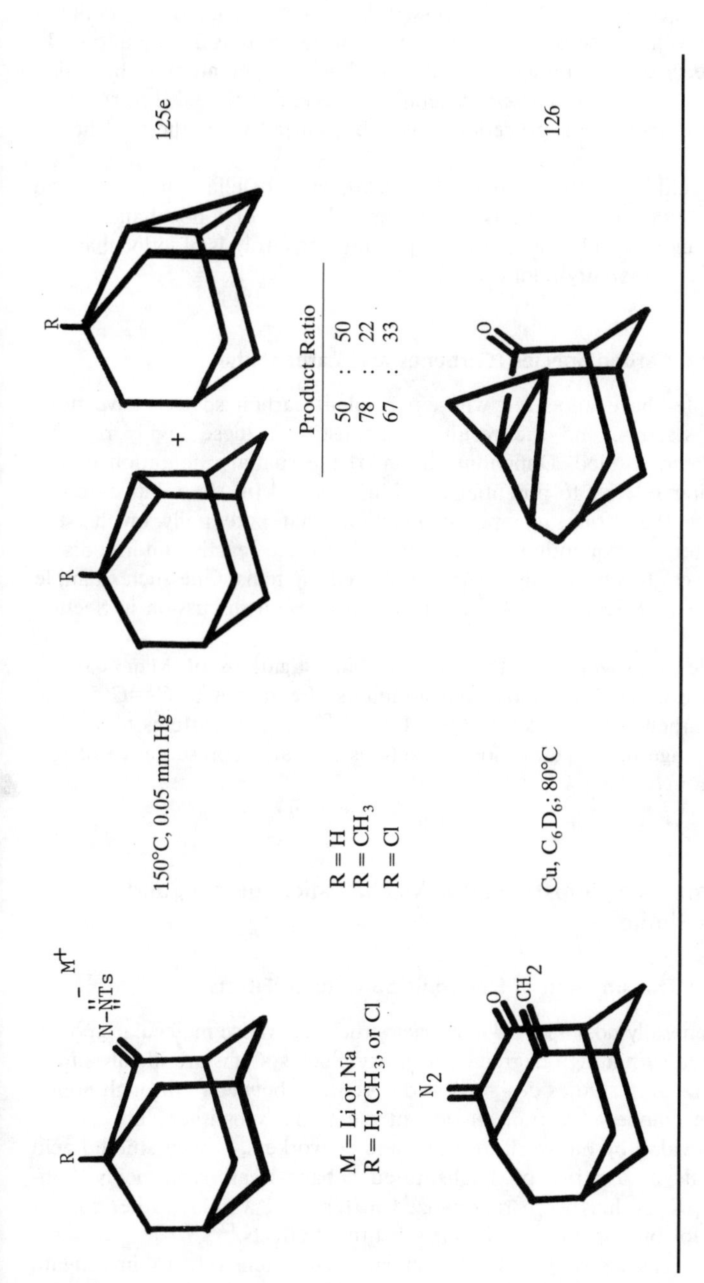

N–N̈Ts M^+
R
150°C, 0.05 mm Hg
R
+
R
125e
M = Li or Na
R = H, CH_3, or Cl
R = H
R = CH_3
R = Cl
Product Ratio
50 : 50
78 : 22
67 : 33
N_2
O
CH_2
Cu, C_6D_6; 80°C
O
126

"ortho-directing" ability of the amide group in these anionic cubane functionalization reactions, Politzer and co-workers[122] have stressed the importance of (i) the acidity of the ortho C—H bonds in the substrate and (ii) the stabilization that the metal-coordinated cubyl anion receives via interaction with the carbonyl oxygen atom in the amide group. These topics have been discussed in detail in a recent review.[5] More recently, a variant of the ortho-metalation procedure has been utilized to synthesize phenylcubanes.[123]

In addition, it should be mentioned that Abeywickrema and Della[124] have accessed the cubyl anion electrochemically via two-electron reduction of iodocubane. Thus, cubane was isolated in 65% yield via controlled-potential electrolysis of iodocubane in dimethylformamide at a mercury pool cathode.[124]

1.4E. Caged Divalent Carbon Species (Carbenes and Carbenoids)

Some recent examples have appeared wherein divalent carbon species have been generated in cage systems, and the resulting chemistry of these highly reactive intermediates has been studied. Depending upon (i) the spatial orientation of the divalent carbon center relative to potential reaction sites and (ii) the relative activation energies among the various competing reactions that potentially involve the divalent carbon center, it frequently is possible to utilize these reactive intermediates to gain synthetic access to very highly strained polycyclic systems. One such example was discussed earlier (see Schemes 1.40 and 1.41 and relevant discussion in Section 1.4A).

Another example is provided by the extensive investigations of Majerski and co-workers, who have utilized intramolecular additions of carbenes to C=C double bonds[125] and also carbene insertions into C—H bonds[126] in cage systems to synthesize highly strained cage propellanes. Some reactions that are representative of this approach are shown in Table 1.4.

1.5. Cage Compounds as Substrates for Mechanistic, Spectral, and Physical–Organic Studies

1.5A. Mechanism of Transmission of Electronic Substituent Effects

Cage compounds generally possess rigid, compact structures whose molecular geometries can be estimated with a high degree of accuracy. Such systems are ideally suited to function as substrates in studies designed to differentiate between "through-bond" and "through-space" models for transmission of electronic substituent effects. An early example is provided by the work of Stock and co-workers,[127] who studied acid dissociation constants of a series of 4-substituted cubane-1-carboxylic acids. Substituent effects on pK_a values were rationalized in terms of a field (rather than a σ-inductive) model for propagation of dipolar substituent effects.[128]

Nuclear magnetic resonance (NMR) has been used extensively to investigate mechanisms of transmission of substituent effects in rigid, saturated polycyclic systems.[129] Bishop[130] has utilized ^{13}C NMR chemical shifts to study interactions between

140a: X = Y = O
140b: X = Y = CH_2
140c: X = O, Y = CH_2
140d: X = O, Y = H_2
140e: X = CH_2, Y = H_2

Scheme 1.46 ▪

nonconjugated unsaturated groups in such alicyclic systems. Chow et al.[131] have extended this approach in a related study of transannular interactions in substituted heptacyclo[$6.6.0.0^{2,6}.0^{3,13}.0^{4,11}.0^{5,9}.0^{10,14}$]tetradecanes (**140a–140e**, Scheme 1.46). Although the existence of weak transannular interactions could be detected by correspondingly small changes in ^{13}C NMR chemical shifts along the series of compounds shown in Scheme 1.46, it was not possible to assess accurately the relative importance of through-bond versus through-space interactions in system **140** by this technique.

Photoelectron (PE) spectroscopy also has been utilized extensively to study transannular "proximity effects" of substituents in rigid polycyclic systems.[130a, 132] Such an approach has been applied to the study of a series of substituted pentacyclo[$5.4.0.0^{2,6}.0^{3,10}.0^{5,9}$]undecanes [i.e., **141a–141e** (Scheme 1.47)].[133] Analysis of vertical ionization potential data obtained via PE spectroscopic investigation of **141a–141e**[133] permitted assessment of the relative importance of through-bond versus through-space mechanisms for transmission of electronic effects between substituents at the 8 and 11 positions in this system.[133]

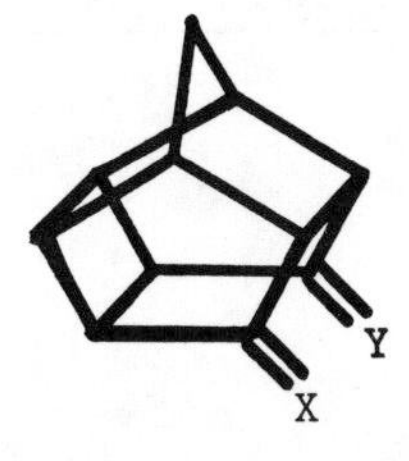

141a: X = Y = O
141b: X = Y = CH_2
141c: X = O, Y = CH_2
141d: X = O, Y = H_2
141e: X = CH_2, Y = H_2

Scheme 1.47 ▪

A. Nucleophilic attack on C=O:

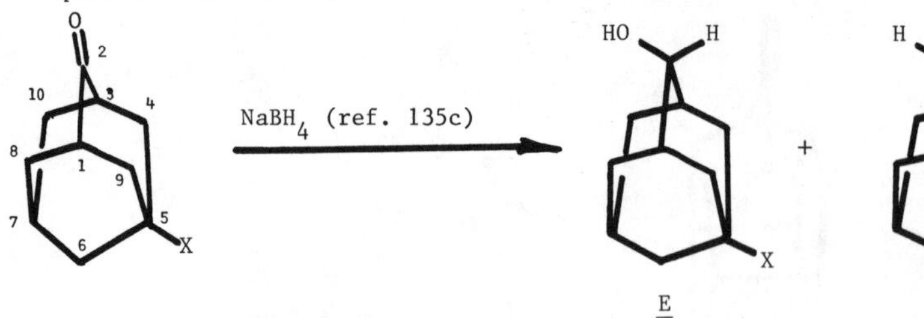

	Conditions	Product ratio E:Z	
142a: X = p-$C_6H_4NO_2$	Me_2CHOH, 25 °C	66	44
142b: X = F	Me_2CHOH, 25 °C	62	38
142c: X = OH	MeOH, 0 °C	43	57

B. Diels-Alder cycloaddition:

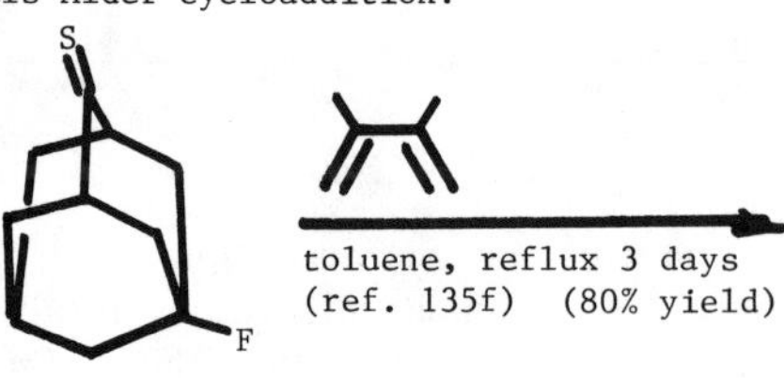

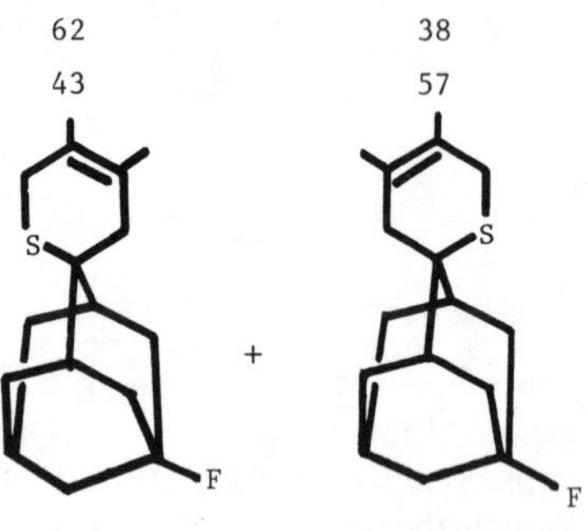

	E	Z
	Product ratio E:Z	
143	67	33

C. Electrophilic attack on C=C:

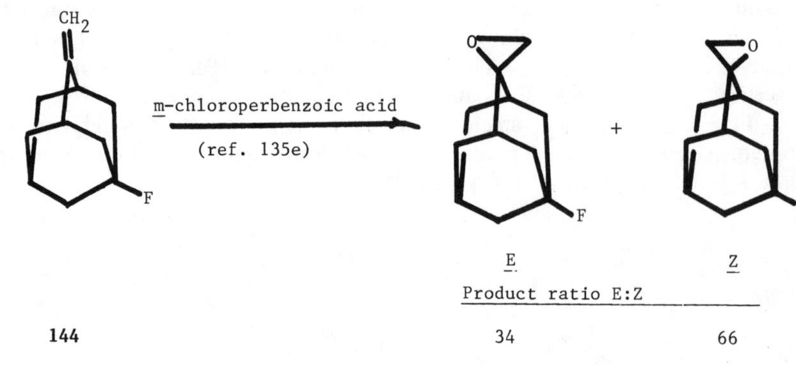

	E	Z
	Product ratio E:Z	
144	34	66

Scheme 1.48 ■

1.5B. Studies on Stereoelectronic Control of Reactions in Cage Systems; Evidence for the Importance of σ Hyperconjugation

The adamantane system, one of the few carbocyclic cage molecules that is virtually strain-free,[134] provides a rigid, saturated C—C framework with which to study stereoelectronic control by distant substituents. Thus, le Noble and co-workers[135] have investigated a number of reactions of 2,5-disubstituted adamantanes in an effort to probe the mechanism by which distant polar substituents affect the stereoselectivity of additions to trigonal carbon.[135] Indeed, in the stereochemical course of attack by nucleophiles [e.g., $LiAlH_4$, $NaBH_4$, Li(O-*t*-Bu)$_3$H, MeLi] upon the carbonyl group in 5-substituted adamantanones was found to be subject to "virtually exclusive electronic control by the substituent."[135c] Similarly, Diels–Alder cycloaddition of 2,3-dimethylbutadiene to 5-fluoroadamantane-2-thione was found to proceed with predominant syn facial selectivity of the diene upon the dienophile.[135f] In addition, the approach of electrophiles (e.g., RCO_3H, $:CX_2$, BH_3) to the C=C double bond in 5-fluoro-2-methyleneadamantane has been investigated.[135e] Representative reactions of this type appear in Scheme 1.48.

The foregoing results are consistent with predictions based upon Cieplak's theoretical model.[136] Thus, π-facial selectivity of approach by the reagent is directed by "hyperconjugation of antiperiplanar σ-bonds with the incipient σ* orbital,"[135f] and, hence, approach occurs from the direction anti to the most electron-rich σ bond in the substrate.[135e]

1.6. Epilog

To the extent that the stated raison d'être of this review has been fulfilled, such success is due unequivocally to the experimental and theoretical results of studies that reflect the ingenuity and creativity of the many investigators in laboratories worldwide whose work is cited herein. Because the purpose of the review as stated in the Introduction was to provide a prognosticative overview of vast areas of synthetic and mechanistic organic chemistry, the omission of many valuable and unique contributions was, sadly, unavoidable. I apologize for such omissions, which reflect solely the inevitable constraints of space and time. Specifically, readers are advised that, when they occur, such omissions do *not* imply negative judgment on the part of the reviewer.

Acknowledgments. I gratefully acknowledge receipt of financial support of our research on the synthesis and chemistry of novel cage molecules from the following agencies: the Robert A. Welch Foundation (Grant B-963), the Air Force Office of Scientific Research (Grants AFOSR-84-0085 and AFOSR-88-0132), the U.S. Army Armament Research, Development and Engineering Center, Picatinny Arsenal, NJ, and the University of North Texas Faculty Research Committee. I thank Dr. Gowaravaram Madhusudhan Reddy, Dr. Dayananda Rajapaksa, and Dr. Vuligonda Vidyasagar for having proofread and constructively criticized the manuscript of this review. Finally, I express thanks to Professor Philip E. Eaton for helpful discussions and for his permission to cite his research results prior to their appearance in primary literature publications.

References

1. Eaton, P. E., Ed.; *Synthesis of Non-Natural Products: Challenge and Reward*, Tetrahedron Symposia-in-Print Number 26, *Tetrahedron* **1986**, *42*, 1549–1915.
2. Greenberg, A.; Liebman, J. F. *Strained Organic Molecules*; Academic Press: New York, 1978.
3. Marchand, A. P. In *Advances in Theoretically Interesting Molecules*; Thummel, R. P., Ed.; JAI: Greenwich, CT, 1989; Vol. 1, pp. 357–399 and references cited therein.
4. Marchand, A. P. *Chem. Rev.* **1989**, *89*, 1011.
5. Griffin, G. W.; Marchand, A. P. *Chem. Rev.*, **1989**, *89*, 997.
6. Eaton, P. E.; Cole, T. W. *J. Am. Chem. Soc.* **1964**, *86*, 962, 3157.
7. (a) Paquette, L. A. In *Strategies and Tactics of Organic Synthesis*; Lindberg, T., Ed.; Academic Press: New York, 1984; pp. 175–200. (b) Paquette, L. A. *Chem. Aust.* **1983**, *50*, 138. (c) Paquette, L. A. *Proc. Natl. Acad. Sci. U.S.A.* **1982**, *79*, 4495.
8. (a) Paquette, L. A.; Weber, J. C.; Kobayashi, T.; Miyahara, Y. *J. Am. Chem. Soc.* **1988**, *110*, 8591. (b) Paquette, L. A.; Ternansky, R. J.; Balogh, D. W.; Kentgen, G. *J. Am. Chem. Soc.* **1983**, *105*, 5446. (c) Ternansky, R. J.; Balogh, D. W.; Paquette, L. A. *J. Am. Chem. Soc.* **1982**, *104*, 4503. (d) Paquette, L. A.; Balogh, D. W.; Usha, R.; Kountz, D.; Christoph, G. G. *Science* **1981**, *211*, 575. (e) Paquette, L. A. *Chem. Rev.* **1989**, *89*, 1051.
9. Marchand, A. P. *Stereochemical Applications of NMR Studies in Rigid Bicyclic Systems*; Verlag Chemie: Deerfield Beach, FL, 1982.
10. For example, the mechanism of transmission of electronic substituent effects has been studied extensively in substituted cubanes. (a) Edward, J. T.; Farrell, P. G.; Langford, G. E. *J. Am. Chem. Soc.* **1976**, *98*, 3075, 3085. (b) Edward, J. T.; Farrell, P. G.; Langford, G. E. *J. Org. Chem.* **1977**, *42*, 1957. (c) Cole, T. W., Jr.; Mayers, C. J.; Stock, L. M. *J. Am. Chem. Soc.* **1974**, *96*, 4555.
11. Marchand, A. P.; Huang, C.; Kaya, R.; Baker, A. D.; Jemmis, E. D.; Dixon, D. A. *J. Am. Chem. Soc.* **1987**, *109*, 7095.
12. Greenberg, A.; Stevenson, T. A. In *Molecular Structure and Energetics*; Liebman, J. F.; Greenberg, A., Eds.; VCH: Deerfield Beach, FL, 1986; Vol. 3, pp. 193–266.
13. Ōsawa, E.; Kanematsu, K. In *Molecular Structure and Energetics*; Liebman, J. F.; Greenberg, A., Eds.; VCH: Deerfield Beach, FL, 1986; Vol. 3, pp. 329–369.
14. Mehta, G.; Murthy, A. N.; Reddy, D. S.; Reddy, A. V. *J. Am. Chem. Soc.* **1986**, *108*, 3443.
15. Mehta, G. *J. Chem. Educ.* **1982**, *59*, 313.
16. (a) Mehta, G.; Rao, K. S. *Tetrahedron Lett.* **1983**, *24*, 809. (b) Mehta, G.; Rao, K. S.; Marchand, A. P.; Kaya, R. *J. Org. Chem.* **1984**, *49*, 3848. (c) Mehta, G.; Rao, K. S. *J. Org. Chem.* **1985**, *50*, 5537.
17. Mehta, G.; Srikrishna, A.; Reddy, A. V.; Nair, M. S. *Tetrahedron* **1981**, *37*, 4543.
18. Fukunaga, T.; Clement, R. A. *J. Org. Chem.* **1977**, *42*, 270.
19. Mehta, G.; Nair, M. S. *J. Am. Chem. Soc.* **1985**, *107*, 7519.
20. Mehta, G.; Reddy, A. V.; Srikrishna, A. *J. Chem. Soc., Perkin Trans. 1* **1986**, 291.
21. Paquette, L. A.; Doherty, A. M. *Polyquinane Chemistry*; Springer-Verlag: New York, 1987; pp. 37–40.
22. Wenkert, E.; Yoder, J. S. *J. Org. Chem.* **1980**, *35*, 2986.
23. (a) Stober, R.; Musso, H. *Angew. Chem., Int. Ed. Engl.* **1977**, *16*, 415. (b) Stober, R.; Musso, H.; Ōsawa, E. *Tetrahedron* **1986**, *42*, 1757.
24. (a) Toyne, K. J. *J. Chem. Soc., Perkin Trans. 1* **1976**, 1346. (b) Ōsawa, E.; Schneider, I.; Toyne, K. J.; Musso, H. *Chem. Ber.* **1986**, *119*, 2350.
25. (a) Sasaki, N. A.; Zunker, R.; Musso, H. *Chem. Ber.* **1973**, *106*, 2992. (b) Ōsawa, E.; Schleyer, P. von R.; Chang, L. W. K.; Kane, V. V. *Tetrahedron Lett.* **1974**, 4189. (c) Musso, H. *Chem. Ber.* **1975**, *108*, 337.

26. (a) Kirao, K.-i.; Iwakuma, T.; Taniguchi, M.; Abe, E.; Yonemitsu, O.; Date, T.; Kotera, K. *J. Chem. Soc., Chem. Commun.* **1976**, 691. (b) Hirao, K.-i.; Iwakuma, T.; Taniguchi, M.; Yonemitsu, O.; Date, T.; Kotera, K. *J. Chem. Soc., Perkin Trans. 1* **1980**, 163.
27. (a) Bosse, D.; de Meijere, A. *Tetrahedron Lett.* **1977**, 1155. (b) Bosse, D.; de Meijere, A. *Chem. Ber.* **1978**, *111*, 2233.
28. (a) Lemal, D. M.; Shim, K. S. *Tetrahedron Lett.* **1961**, 368. (b) Bird, C. W.; Colinese, D. L.; Cookson, R. C.; Hudec, J.; Williams, R. O. *Tetrahedron Lett.* **1961**, 373. (c) Schrauzer, G. N. *Adv. Catal.* **1968**, *18*, 373. (d) Marchand, A. P.; Hayes, B. R. *Tetrahedron Lett.* **1977**, 1027. (e) Ealick, S. E.; van der Helm, D.; Hayes, B. R.; Marchand, A. P. *Acta Crystallogr., Sect. B* **1978**, *34*, 3219. (f) Marchand, A. P.; Earlywine, A. D. *J. Org. Chem.* **1984**, *49*, 1660. (g) Marchand, A. P.; Wu, A.-H. *J. Org. Chem.* **1985**, *50*, 396. (h) Marchand, A. P.; Earlywine, A. D.; Heeg, M. J. *J. Org. Chem.* **1986**, *51*, 4096. (i) Barden, T. J.; Paddon-Row, M. N. *Aust. J. Chem.* **1988**, *41*, 817. (j) Marchand, A. P.; Dave, P. R. *J. Org. Chem.* **1989**, *54*, 2775.
29. (a) Chow, T. J.; Wu, M.-Y.; Liu, L.-K. *J. Organomet. Chem.* **1985**, *281*, C33. (b) Chow, T. J.; Chao, Y.-S. *J. Organomet. Chem.* **1985**, *296*, C23. (c) Chow, T. J.; Liu, L.-K.; Chao, Y.-S. *J. Chem. Soc., Chem. Commun.* **1985**, 700. (d) Chow, T. J.; Chao, Y.-S.; Liu, L.-K. *J. Am. Chem. Soc.* **1987**, *109*, 797. (e) Chow, T. J.; Wu, T.-K.; Shih, H.-J. *J. Chem. Soc., Chem. Commun.* **1989**, 490.
30. Flippen-Anderson, J. L.; Gilardi, R.; George, C.; Marchand, A. P.; Dave, P. R. *Acta Crystallogr., Sect. C: Cryst. Struct. Commun.* **1989**, *45*, 1171.
31. (a) Chow, T. J.; Wu, T.-K. *J. Org. Chem.* **1988**, *53*, 1102. (b) Chow, T. J.; Feng, J.-J.; Shih, H.-J.; Wu, T.-K.; Tseng, L.-H.; Wang, C.-Y.; Yu, C. *J. Chinese Chem. Soc.* **1988**, *35*, 291.
32. Klunder, A. J. H.; Ariaans, G. J. A.; Zwanenburg, B. *Tetrahedron Lett.* **1984**, *25*, 5457.
33. Klunder, A. J. H.; Ariaans, G. J. A.; van der Loop, E. A. R. M.; Zwanenburg, B. *Tetrahedron* **1986**, *42*, 1903.
34. Mehta, G.; Reddy, D. S.; Reddy, A. V. *Tetrahedron Lett.* **1984**, *25*, 2275.
35. Push–pull acceleration of thermal [2 + 2] cycloreversions in methoxy-substituted PCUD-8,11-diones also has been reported. See Okamoto, Y.; Kanematsu, K.; Fujiyoshi, T.; Ōsawa, E. *Tetrahedron Lett.* **1983**, *24*, 5645.
36. (a) Ogino, T.; Awano, K.; Ogihara, T.; Isogai, K. *Tetrahedron Lett.* **1983**, *24*, 2781. (b) Ogino, T.; Awano, K.; Ogihara, T.; Isogai, K. *Chem. Lett.* **1984**, 2023. (c) Ogino, T.; Awano, K. *Bull. Chem. Soc. Jpn.* **1986**, *59*, 2811.
37. (a) Mehta, G.; Srikrishna, A. *J. Chem. Soc., Chem. Commun.* **1982**, 218. (b) Mehta, G.; Reddy, A. V.; Srikrishna, A. *J. Chem. Soc., Perkin Trans. 1* **1986**, 291.
38. (a) Hamada, T.; Iijima, H.; Yamamoto, T.; Numao, N.; Hirao, K.-i.; Yonemitsu, O. *J. Chem. Soc., Chem. Commun.* **1980**, 696. (b) Hamada, T.; Iijima, H.; Yamamoto, T.; Numao, N.; Yonemitsu, O. *Kokagaku Toronkai Koen Yoshishu* **1979**, 2. See *Chem. Abstr.* **1980**, *93*, 70445t.
39. Marchand, A. P.; Reddy, G. M.; Watson, W. H.; Nagl, A. *J. Org. Chem.* **1988**, *53*, 5969.
40. Review: Godleski, S. A.; Schleyer, P. von R.; Ōsawa, E.; Wipke, W. T. *Progr. Phys. Org. Chem.* **1981**, *13*, 63–117.
41. See Fort, R. C., Jr. *Adamantane, The Chemistry of Diamond Molecules*; Dekker: New York, 1976, and references cited therein.
42. (a) Kent, G. J.; Godleski, S. A.; Ōsawa, E.; Schleyer, P. von R. *J. Org. Chem.* **1977**, *42*, 3852. (b) Underwood, G. R.; Ramamurthy, B. *Tetrahedron Lett.* **1970**, 4125. (c) Smith, E. C.; Barborak, J. C. *J. Org. Chem.* **1976**, *41*, 1433. (d) Marchand, A. P.; Chou, T.-C.; Ekstrand, J. D.; van der Helm, D. *J. Org. Chem.* **1976**, *41*, 1438. (e) Eaton, P. E.; Cassar, L.; Hudson, R. A.; Hwang, D. R. *J. Org. Chem.* **1976**, *41*, 1445. (f) Nakazaki, M.; Naemura, K.; Arashiba, N. *J. Org. Chem.* **1978**, *43*, 689. (g) Helmchen, G.; Staiger, G. *Angew. Chem., Int. Ed. Engl.* **1977**, *16*, 116. (h) Mehta, G.; Chaudhury, B. *Indian J. Chem., Sect. B* **1979**,

17, 421. (i) Tolstikov, G. A.; Lerman, B. M.; Galin, F. Z.; Struchkov, Y. T.; Andrianov, V. G. *Tetrahedron Lett.* **1978**, 1147. (j) Dekker, T. G.; Oliver, D. W. *S. Afr. J. Chem.* **1979**, *32*, 45. (k) Dekker, T. G.; Oliver, D. W.; Venter, A. *Tetrahedron Lett.* **1980**, 3101.
43. (a) Mehta, G.; Pandey, P. N.; Usha, R.; Venkatesan, K. *Tetrahedron Lett.* **1976**, 4209. (b) Mehta, G.; Suri, S. C. *Tetrahedron Let.* **1980**, *21*, 2093.
44. Mehta, G.; Srikrishna, A. *Indian J. Chem., Sect. B* **1980**, *19*, 997.
45. (a) Mehta, G.; Suri, S. C. *Tetrahedron Lett.* **1980**, *21*, 3821. (b) Mehta, G.; Suri, S. C. *Tetrahedron Lett.* **1980**, *21*, 3825. (c) Mehta, G.; Rao, K. S.; Suri, S. C.; Cameron, T. S.; Chan, C. *J. Chem. Soc., Chem. Commun.* **1980**, 650.
46. (a) Paquette, L. A. *Acc. Chem. Res.* **1971**, *4*, 280. (b) Bishop, K. C., III *Chem. Rev.* **1976**, *76*, 461. (c) Halpern, J. In *Organic Synthesis via Metal Carbonyls*; Wender, I.; Pino, P., Eds.; Wiley-Interscience: New York, 1977; Vol. 2, pp. 705–730.
47. (a) Cassar, L.; Eaton, P. E.; Halpern, J. *J. Am. Chem. Soc.* **1970**, *92*, 3515, 6366. (b) Koser, G. F. *J. Chem. Soc., Chem. Commun.* **1971**, 388.
48. (a) Paquette, L. A.; Stowell, J. C. *J. Am. Chem. Soc.* **1970**, *92*, 2584. (b) Paquette, L. A.; Stowell, J. C. *J. Am. Chem. Soc.* **1971**, *93*, 2459. (c) Paquette, L. A.; Ward, J. S.; Boggs, R. A.; Farnham, W. B. *J. Am. Chem. Soc.* **1985**, *97*, 1101.
49. (a) Dauben, W. G.; Schallhorn, C. H.; Whalen, D. L. *J. Am. Chem. Soc.* **1971**, *93*, 1446. (b) Paquette, L. A.; Beckley, R. S. *J. Am. Chem. Soc.* **1975**, *97*, 1084. (c) Paquette, L. A.; Beckley, R. S.; Farnham, W. B. *J. Am. Chem. Soc.* **1975**, *97*, 1089. (d) Yokoyama, K.; Saegusa, Y.; Miyashi, T.; Kabuto, C.; Mukai, T. *Chem. Lett.* **1984**, 89.
50. Luh, T.-Y. *Tetrahedron Lett.* **1977**, 2951.
51. Werstiuk, N. H. *Tetrahedron* **1983**, *39*, 205.
52. For a review, see Klunder, A. J. H.; Zwanenburg, B. *Chem. Rev.* **1989**, *89*, 1035, and references cited therein.
53. Klunder, A. J. H.; Zwanenburg, B. *Tetrahedron* **1975**, *31*, 1419.
54. Klunder, A. J. H.; Ariaans, G. J. A.; van der Loop, E. A. R. M.; Zwanenburg, B. *Tetrahedron* **1986**, *42*, 1903.
55. Klunder, A. J. H.; Zwanenburg, B. *Tetrahedron* **1973**, *29*, 1683.
56. Klunder, A. J. H.; van Seters, A. J. C.; Buza, M.; Zwanenburg, B. *Tetrahedron* **1981**, *37*, 1601.
57. Miller, R. D.; Dolce, D. L. *Tetrahedron Lett.* **1973**, 1151.
58. Ōsawa, W.; Aigami, K.; Inamoto, Y. *J. Chem. Soc., Perkin Trans. 2* **1979**, 181.
59. de Valk, W. C. G. M.; Klunder, A. J. H.; Zwanenburg, B. *Tetrahedron Lett.* **1980**, *21*, 971.
60. Klunder, A. J. H.; de Valk, W. C. G. M.; Verlaak, J. M. J.; Schellekens, J. W. M.; Noordik, J. H.; Parthasarathi, V.; Zwanenburg, B. *Tetrahedron* **1985**, *41*, 963.
61. Hamlin, K. E.; Weston, A. W. *Org. React.* **1957**, *9*, 1.
62. For a review, see Chenier, P. J. *J. Chem. Educ.* **1978**, *55*, 286.
63. Marchand, A. P.; Chou, T.-C. *Tetrahedron* **1975**, *31*, 2655.
64. Marchand, A. P.; Chou, T.-C. *J. Chem. Soc., Perkin Trans. 1* **1973**, 1948.
65. Scherer, K. V., Jr. *Tetrahedron Lett.* **1972**, 2077.
66. Mehta, G.; Srikrishna, A.; Suri, S. C.; Nair, M. S. *J. Org. Chem.* **1983**, *48*, 5107.
67. (a) Collman, J. P. *Acc. Chem. Res.* **1968**, *1*, 136. (b) Halpern, J. *Acc. Chem. Res.* **1970**, *3*, 376. (c) Halpern, J. *Adv. Chem. Ser.* **1968**, *No. 70*, 1.
68. Zlotogorski, C.; Blum, J.; Ōsawa, E.; Schwarz, H.; Höhne, G. *J. Org. Chem.* **1984**, *25*, 5457.
69. Paquette, L. A.; Boggs, R. A.; Farnham, W. B.; Beckley, R. S. *J. Am. Chem. Soc.* **1975**, *97*, 1112.
70. Dauben, W. G.; Kielbania, A. J., Jr. *J. Am. Chem. Soc.* **1971**, *93*, 7345.
71. For example, see the following: (a) Jones, G., II; Ramachandran, B. R. *J. Org. Chem.* **1976**, *41*, 798. (b) Scharf, H.-D.; Fleischhauer, J.; Leismann, H.; Ressler, I.; Schleker, W.; Weitz, R. *Angew. Chem., Int. Ed. Engl.* **1979**, *18*, 652. (c) Hirao, K.-i.; Ando, A.; Hamada, T.; Yonemitsu, O. *J. Chem. Soc., Chem. Commun.* **1984**, 300. (d) Yamashita, Y.; Mukai, T.

Chem. Lett. **1984**, 1741. (e) Hirao, K.-i.; Yamashita, A.; Ando, A.; Iijima, H.; Yamamoto, T.; Hamada, T.; Yonemitsu, O. *J. Chem. Res.* (*S*) **1987**, 162; *J. Chem. Res.* (*M*) **1987**, 1344.
72. Luh, T.-Y. *Tetrahedron Lett.* **1977**, 2951.
73. Eaton, P. E.; Patterson, D. R. *J. Am. Chem. Soc.* **1978**, *100*, 2573.
74. Shibasaki, M.; Iseki, K.; Ikegami, S. *Tetrahedron* **1981**, *37*, 4411.
75. (a) Marchand, A. P.; Suri, S. C.; Earlywine, A. D.; Powell, D. R.; van der Helm, D. *J. Org. Chem.* **1984**, *49*, 670. This reference was cited incorrectly in ref. 14 (see footnote 20 in that publication). (b) Recently, photothermal metathesis has been used to synthesize angular tetraquinanes; see Griesbeck, A. G. *Chem. Ber.* **1990**, *123*, 549.
76. Mehta, G.; Rao, K. S.; Krishnamurthy, N.; Srinivas, V.; Balasubramanian, D. *Tetrahedron* **1989**, *45*, 2743.
77. (a) Hayakawa, K.; Kido, K.; Kanematsu, K. *J. Chem. Soc., Chem. Commun.* **1986**, 268. (b) Hayakawa, K.; Kido, K.; Kanematsu, K. *J. Chem. Soc., Perkin Trans. 1* **1988**, 511. (c) Hayakawa, K.; Naito, R.; Kanematsu, K. *Heterocycles* **1988**, *27*, 2293.
78. (a) Thummel, R. P.; Lefoulon, F.; Cantu, D.; Mahadeven, R. *J. Org. Chem.* **1984**, *49*, 2208. (b) Thummel, R. P.; Lefoulon, F. *J. Org. Chem.* **1985**, *50*, 2407. (c) Thummel, R. P. *Tetrahedron* **1991**, *47*, 6851.
79. Thummel, R. P.; Lim, J.-L. *Tetrahedron Lett.* **1987**, *28*, 3319.
80. Thummel, R. P.; Hegde, V. *J. Org. Chem.* **1989**, *54*, 1720.
81. Marchand, A. P.; Annapurna, P.; Flippen-Anderson, J. L.; Gilardi, R.; George, C. *Tetrahedron Lett.* **1988**, *29*, 6681.
82. Mehta, G.; Prabhakar, C.; Padmaja, N.; Ramakumar, S.; Viswamitra, M. A. *Tetrahedron Lett.* **1989**, *30*, 6895.
83. Marchand, A. P.; Annapurna, P.; Taylor, R. W.; Simmons, D. L.; Watson, W. H.; Nagl, A.; Flippen-Anderson, J. L.; Gilardi, R.; George, C. *Tetrahedron* **1990**, *46*, 5077.
84. Watson, W. H.; Nagl, A.; Marchand, A. P.; Annapurna, P. *Acta Crystallogr., Sect. C: Cryst. Struct. Commun.* **1989**, *45*, 856.
85. "Proton Sponge®" is a registered trademark of Aldrich Chemical Company for 1,8-bis(dimethylamino)naphthalene. Here, the term "proton sponge" embraces the concept that **106**, like 1,8-bis(dimethylamino)naphthalene, might react with one equivalent of a protic acid to form a stable, symmetrical proton-bridged species of the type $R_2N \cdots H^+ \cdots NR_2$. See Staab, H. A.; Saupe, T. *Angew. Chem., Int. Ed. Engl.* **1988**, *27*, 865.
86. (a) Fessner, W.-D.; Prinzbach, H.; Rihs, G. *Tetrahedron Lett.* **1983**, *24*, 5857. (b) Fessner, W.-D.; Sedelmeier, G.; Spurr, P. R.; Rihs, G.; Prinzbach, H. *J. Am. Chem. Soc.* **1987**, *109*, 4626. (c) Melder, J.-P.; Prinzbach, H. *Chem. Ber.* **1991**, *124*, 1271.
87. (a) Fessner, W.-D.; Murty, B. A. R. C.; Worth, J.; Hunkler, D.; Fritz, H.; Prinzbach, H.; Roth, W. D.; Schleyer, P. von R.; McEwen, A.B.; Maier, W. F. *Angew. Chem., Int. Ed. Engl.* **1987**, *26*, 452. (b) Prinzbach, H.; Fessner, W.-D. In *Organic Synthesis: Modern Trends*; Chizhov, O., Ed.; Blackwell: Oxford, 1987; p. 23.
88. (a) Mehta, G.; Nair, M. S. *J. Chem. Soc., Chem. Commun.* **1985**, 629. (b) Mehta, G.; Reddy, K. R.; Nair, M. S. *Proc. Indian Acad. Sci.* (*Chem. Sci.*) **1988**, *100*, 223. (c) Mehta, G.; Reddy, K. R.*Tetrahedron Lett.* **1988**, 3607. (c) Mehta, G.; Nair, M. S.; Reddy, K. R. *J. Chem. Soc., Perkin Trans. 1* **1991**, 1297.
89. (a) Paquette, L. A.; Fischer, J. W.; Browne, A. R.; Doecke, C. W. *J. Am. Chem. Soc.* **1985**, *107*, 686. (b) Paquette, L. A.; Shen, C.-C. *Tetrahedron Lett.* **1988**, *29*, 4069. (c) Shen, C.-C.; Paquette, L. A. *J. Org. Chem.* **1989**, *54*, 3324. (d) Paquette, L. A.; Shen, C.-C.; Engel, P. *J. Org. Chem.* **1989**, *54*, 3329.
90. (a) Paquette, L. A.; Carr, R. V. C.; Böhm, M. C.; Gleiter, R. *J. Am. Chem. Soc.* **1980**, *102*, 1186. (b) Böhm, M.; Carr, R. V. C.; Gleiter, R.; Paquette, L. A. *J. Am. Chem. Soc.* **1980**, *102*, 7218. (c) Paquette, L. A.; Carr, R. V. C.; Charumilind, P.; Blount, J. F. *J. Org. Chem.* **1980**, *45*, 4922. (d) Paquette, L. A.; Kravetz, T. M.; Böhm, M. C.; Gleiter, R. *J. Org. Chem.* **1983**, *48*, 1250.

91. Marchand, A. P. *Tetrahedron* **1988**, *44*, 2377.
92. (a) Paquette, L. A.; Snow, R. A.; Muthard, J. L.; Cynkowski, T. *J. Am. Chem. Soc.* **1978**, *100*, 1600. (b) Paquette, L. A.; Snow, R. A.; Muthard, J. L.; Cynkowski, T. *J. Am. Chem. Soc.* **1979**, *101*, 6991.
93. For a review of pyramidalized alkenes, see Borden, W. T. *Chem. Rev.* **1989**, *89*, 1095.
94. (a) Hrovat, D. A.; Borden, W. T. *J. Am. Chem. Soc.* **1988**, *110*, 4710. (b) Eaton, P. E.; Maggini, M. *J. Am. Chem. Soc.* **1988**, *110*, 7230. (c) Gilardi, R.; Eaton, P. E.; Maggini, M. *J. Am. Chem. Soc.* **1988**, *110*, 7232.
95. (a) Hrovat, D. A.; Borden, W. T. *J. Am. Chem. Soc.* **1988**, *110*, 7229. (b) Schäfer, J.; Szeimies, G. *Tetrahedron Lett.* **1988**, *29*, 5253.
96. (a) Eaton, P. E.; Hoffmann, K. L. *J. Am. Chem. Soc.* **1987**, *109*, 5285. (b) Eaton, P. E.; White, A. J. *J. Org. Chem.* **1990**, *55*, 1321. (c) Chen, N.; Jones, M., Jr. *Tetrahedron Lett.* **1989**, *30*, 6969. (d) Chen, N.; Jones, M., Jr.; White, W. R.; Platz, M. S. *J. Am. Chem. Soc.* **1991**, *113*, 4981.
97. (a) Szeimies, G. In *Reactive Intermediates*; Abramovitch, R. A., Ed.; Plenum Press: New York, 1983; Vol. 3, pp. 299–366. (b) Szeimies, G. *Chimia* **1981**, *35*, 243. (c) Harnisch, J.; Baumgartel, O.; Szeimies, G.; Van Meerssche, M.; Germain, G.; Declerq, J.-P. *J. Am. Chem. Soc.* **1979**, *101*, 3370. (d) Baumgartel, O.; Szeimies, G. *Chem. Ber.* **1983**, 2180.
98. Kiplinger, J. P.; Tollens, F. R.; Marshall, A. G.; Kobayashi, T.; Lagerwall, D. R.; Paquette, L. A.; Bartmess, J. E. *J. Am. Chem. Soc.* **1989**, *111*, 6914.
99. Melder, J.-P.; Pinkos, R.; Fritz, H.; Prinzbach, H. *Angew. Chem., Int. Ed. Engl.* **1990**, *29*, 95.
100. (a) Spurr, R. P.; Murty, B. A. R. C.; Fessner, W.-D.; Fritz, H.; Prinzbach, H. *Angew. Chem., Int. Ed. Engl.* **1987**, *26*, 455. (b) Fessner, W.-D.; Murty, B. A. R. C.; Prinzbach, H. *Angew. Chem., Int. Ed. Engl.* **1987**, *26*, 451. (c) Prinzbach, H.; Murty, B. A. R. C.; Fessner, W.-D.; Mortensen, J.; Heinze, J.; Gescheidt, G.; Gerson, F. *Angew. Chem., Int. Ed. Engl.* **1987**, *26*, 457.
101. Worthy, W. Activation is key to preparing cubane derivatives, *Chem. Eng. News*, February 25, 1985, p. 31–32.
102. Chemistry of cubane enters flourishing new era, *Chem. Eng. News*, November 14, 1988, pp. 45ff.
103. Reddy, D. S.; Sollott, G. P.; Eaton, P. E. *J. Org. Chem.* **1989**, *54*, 722.
104. (a) Kropp, P. J.; Poindexter, G. S.; Pienta, N. J.; Hamilton, D. C. *J. Am. Chem. Soc.* **1976**, *98*, 8135. (b) Kropp, P. J. *Acc. Chem. Res.* **1984**, *17*, 131.
105. Eaton, P. E.; Cunkle, G. T. *Tetrahedron Lett.* **1986**, *27*, 6055.
106. (a) Moriarty, R. M.; Khosrowshahi, J. S.; Dalecki, T. J. *J. Chem. Soc., Chem. Commun.* **1987**, 675. (b) Moriarty, R. M.; Khosrowshahi, J. S.; Penmasta, R. *Tetrahedron Lett.* **1989**, *30*, 791. (c) Moriarty, R. M.; Khosrowshahi, J. S. *Syn. Commun.* **1989**, *19*, 1395.
107. (a) Fort, R. C.; Schleyer, P. von R. In *Advances in Alicyclic Chemistry*; Hart, H., Ed.; Wiley-Interscience: New York, 1966; Vol. 1, pp. 284ff. (b) Greenberg, A.; Liebman, J. F. *Strained Organic Molecules*; Academic Press: New York, 1978; pp. 77–81.
108. See footnote 73 in Bingham, R. C.; Schleyer, P. von R. *J. Am. Chem. Soc.* **1971**, *93*, 3189.
109. Eaton, P. E.; Yang, C.-X.; Xiong, Y. *J. Am. Chem. Soc.* **1990**, *112*, 3225.
110. Similarly, Rüchardt and co-workers have measured the rate of S_N1 solvolysis 4-bromohomocubane in HFIP at 120°C. This solvolysis affords 4-(hexafluoroisopropoxy)homocubane with pseudo-first-order rate constant $k_1 = 1.3 \times 10^{-7}\ s^{-1}$). 1-Bromonorbornane is inert to HFIP under these conditions. See Mergelsberg, I.; Langhals, H.; Rüchardt, C. *Chem. Ber.* **1983**, *16*, 360.
111. Wiberg, K. B.; Hess, B. B., Jr.; Ashe, A. A., III. In *Carbonium Ions*; Olah, G. A.; Schleyer, P. von R., Eds.; Wiley-Interscience: New York, 1972; Vol. 3, pp. 1295ff.
112. Surya Prakash, G. K.; Krishnamurthy, V. V.; Herges, R.; Bau, R.; Yuan, H.; Olah, G. A.; Fessner, W.-D.; Prinzbach, H. *J. Am. Chem. Soc.* **1986**, *108*, 836.

113. Olah, G. A.; Liang, G.; Schleyer, P. von R.; Engler, E. M.; Dewar, M. J. S.; Bingham, R. C. *J. Am. Chem. Soc.* **1973**, *95*, 6829.
114. (a) Della, E. W.; Patney, H. K. *Synthesis* **1976**, 251. (b) Della, E. W.; Tsanaktsidis, J. *Aust. J. Chem.* **1986**, *39*, 2061.
115. (a) Luh, T.-Y.; Stock, L. M. *J. Org. Chem.* **1978**, *43*, 3271. (b) Abeywickrema, R. S.; Della, E. W. *J. Org. Chem.* **1980**, *45*, 4226. (c) Della, E. W.; Tsanaktsidis, J. *Aust. J. Chem.* **1989**, *42*, 61. (d) Tsanaktsidis, J.; Eaton, P. E. *Tetrahedron Lett.* **1989**, *30*, 6967.
116. Moriarty, R. M.; Khosrowshahi, J. S.; Miller, R. S.; Flippen-Anderson, J.; Gilardi, R. *J. Am. Chem. Soc.* **1989**, *111*, 8943.
117. Reddy, D. S.; Maggini, M.; Tsanaktsidis, J.; Eaton, P. E. *Tetrahedron Lett.* **1990**, *31*, 805.
118. (a) Hassenrück, K.; Radziszewski, J. G.; Balaji, V.; Murthy, G. S.; McKinley, A. J.; David, D. E.; Lynch, V. M.; Martin, H.-D.; Michl, J. *J. Am. Chem. Soc.* **1990**, *112*, 873. (b) Hrovat, D. A.; Borden, W. T. *J. Am. Chem. Soc.* **1990**, *112*, 875. (c) Eaton, P. E.; Tsanaktsidis, J. *J. Am. Chem. Soc.* **1990**, *112*, 876.
119. (a) Eaton, P. E.; Castaldi, G. *J. Am. Chem. Soc.* **1985**, *107*, 724.
120. (a) Eaton, P. E.; Higuchi, H.; Millikan, R. *Tetrahedron Lett.* **1987**, 1055.
121. Eaton, P. E.; Cunkle, G. T.; Marchioro, G.; Martin, R. M. *J. Am. Chem. Soc.* **1987**, *109*, 948.
122. Jayasuriya, K.; Alster, J.; Politzer, P. *J. Org. Chem.* **1987**, *52*, 2306.
123. (a) Bashir-Hashemi, A. *J. Am. Chem. Soc.* **1988**, *110*, 7234. (b) Bashir-Hashemi, A.; Ammon, H. L.; Choi, C. S. *J. Org. Chem.* **1990**, *55*, 416.
124. Abeywickrema, R. S.; Della, E. W. *J. Org. Chem.* **1981**, *46*, 2352.
125. See, for example, the following: (a) Mlinarić-Majerski, K.; Majerski, Z. *J. Am. Chem. Soc.* **1980**, *102*, 1418. (b) Mlinarić-Majerski, K.; Majerski, Z. *J. Am. Chem. Soc.* **1983**, *105*, 7389. (c) Majerski, Z.; Zuanic, M. *J. Am. Chem. Soc.* **1987**, *109*, 3496. (d) Majerski, Z.; Veljković, J.; Kaselj, M. *J. Org. Chem.* **1988**, *53*, 2662. (e) Hiršl-Starčević, S.; Majerski, Z. *J. Org. Chem.* **1982**, *47*, 2520.
126. See, for example, Majerski, Z.; Kostove, V.; Hibšer, M.; Mlinarić-Majerski, K. *Tetrahedron Lett.* **1990**, *31*, 915.
127. Cole, T. W., Jr.; Mayers, C. J.; Stock, L. M. *J. Am. Chem. Soc.* **1974**, *96*, 4555.
128. Westheimer, F. H.; Kirkwood, J. G. *J. Chem. Phys.* **1938**, *6*, 513.
129. See ref. 9, pp. 29–42, and references cited therein.
130. (a) Bishop, R. *Aust. J. Chem.* **1984**, *37*, 319. (b) Bishop, R.; Lee, G.-H. *Aust. J. Chem.* **1987**, *40*, 249.
131. Chow, T. J.; Wu, T.-K.; Shih, H.-J. *J. Chem. Soc., Chem. Commun.* **1989**, 490.
132. (a) For a review, see Martin, H.-D.; Mayer, B. *Angew Chem., Int. Ed. Engl.* **1983**, *22*, 283. (b) Albert, B.; Elsässer, D.; Martin, H.-D.; Mayer, B.; Chow, T. J.; Marchand, A. P.; Ren, C.-T.; Paddon-Row, M.N. *Chem. Ber.* **1991**, *124*, 2871.
133. Marchand, A. P.; Huang, C.; Kaya, R.; Baker, A. D.; Jemmis, E. D.; Dixon, D. *J. Am. Chem. Soc.* **1987**, *109*, 7095.
134. Fort, R. C. *Adamantane: The Chemistry of Diamond Molecules*; Dekker: New York, 1976.
135. (a) le Noble, W. J.; Chiou, D.-M.; Okaya, Y. *Tetrahedron Lett.* **1978**, 1961. (b) le Noble, W. J.; Chiou, D. M.; Okaya, Y. *J. Am. Chem. Soc.* **1979**, *101*, 3244. (c) Cheung, C. K.; Tseng, L. T.; Lin, M. H.; Srivastava, S.; le Noble, W. J. *J. Am. Chem. Soc.* **1986**, *108*, 1598. (d) Cheung, C. K.; Tseng, L. T.; Lin, M. H.; Srivastava, S.; le Noble, W. J. *J. Am. Chem. Soc.* **1987**, *109*, 7239. (e) Srivastava, S.; le Noble, W. J. *J. Am. Chem. Soc.* **1987**, *109*, 5874. (f) Chung, W.-S.; Turro, N. J.; Srivastava, S.; Li, H.; le Noble, W. J. *J. Am. Chem. Soc.* **1988**, *110*, 7882. (g) Lin, M.; le Noble, W. J. *J. Org. Chem.* **1989**, *54*, 997. (h) Xie, M.; le Noble, W. J. *J. Org. Chem.* **1989**, *54*, 3836. (i) Lin, M.-H.; Boyd, M. K.; le Noble, W. J. *J. Am. Chem. Soc.* **1989**, *111*, 8746. (j) Chung, W.-S.; Turro, N. J.; Silver, J.; le Noble, W. J. *J. Am. Chem. Soc.* **1990**, *112*, 1202.
136. (a) Cieplak, A. S. *J. Am. Chem. Soc.* **1981**, *103*, 4540. (b) However, see also Meyers, A. I.; Wallace, R. H. *J. Org. Chem.* **1989**, *54*, 2509.

2

High-Symmetry Chiral Cage-Shaped Molecules

Koichiro Naemura

Osaka University, Toyonaka, Osaka, Japan

Contents

2.1. Introduction
2.2. Chiral Tricyclic Cage-Shaped Molecules
 2.2A. C_2-Bissecocubane, Tricyclo[4.2.0.0^{3,6}]octane
 2.2B. Twist-brendane, Tricyclo[4.3.0.0^{3,8}]nonane
 2.2C. Twistane, Tricyclo[4.4.0.0^{3,8}]decane
 2.2D. Brexane, Tricyclo[4.3.0.0^{3,7}]nonane
2.3. Chiral Tetracyclic Cage-Shaped Molecules
 2.3A. Dehydrotwist-brendane, Tetracyclo[4.3.0.0^{2,5}.0^{3,8}]nonane
 2.3B. Ditwist-brendane, Tetracyclo[5.2.1.0^{2,6}.0^{4,8}]decane
 2.3C. Ditwistane, Tetracyclo[6.2.2.0^{2,7}.0^{4,9}]dodecane
2.4. Chiral Pentacyclic Cage-Shaped Molecules
 2.4A. C_2-Bishomocubane, Pentacyclo[5.3.0.0^{2,5}.0^{3,9}.0^{4,8}]decane; C_1-Homobasketane, Pentacyclo[5.4.0.0^{2,6}.0^{3,9}.0^{5,8}]undecane
 2.4B. C_2-Dehydroditwistane, Pentacyclo[6.4.0.0^{2,7}.0^{3,10}.0^{6,9}]dodecane
 2.4C. D_3-Trishomocubane; Pentacyclo[6.3.0.0^{2,6}.0^{3,10}.0^{5,9}]undecane
 2.4D. Ethanoditwist-brendane, Pentacyclo[7.3.0.0^{2,7}.0^{3,11}.0^{6,10}]dodecane; Methanoditwistane, Pentacyclo[7.4.0.0^{2,6}.0^{3,11}.0^{5,10}tridecane
 2.4E. D_3-Tritwistane, Pentacyclo[8.4.0.0^{2,7}.0^{3,12}.0^{6,11}]tetradecane
2.5. Molecules with Chiral Polyhedral Symmetry
2.6. Chiral Cage-Shaped Molecules of Adamantane Homolog
 2.6A. D_{2d}-Bisnoradamantane, Tricyclo[3.3.0.0^{3,7}]octane
 2.6B. Protoadamantane, Tricyclo[4.3.1.0^{3,8}]decane
2.7. Chiroptical Properties of Cage-Shaped Molecules
2.8. Biocatalysts in Syntheses of Optically Active Cage-Shaped Molecules
References

2.1. Introduction

Every molecule is either chiral or achiral. Most chiral molecules lack all elements of symmetry except for C_1 axes and are defined to be asymmetric. Asymmetry is a necessary but insufficient criterion for a chiral molecule to have a nonsuperposable enantiomer. Four-bladed windmills that are constantly rotated in the definite direction of the wind are chiral, but they are not asymmetric because of their fourfold axes (C_4 axes) of symmetry. The term "high-symmetry chiral"[1,2] has been proposed to differentiate a chiral object with C_n ($n \geq 2$) axes of symmetry from an asymmetric one.

Classification of molecular symmetry and of chiral point groups can be shown as follows:

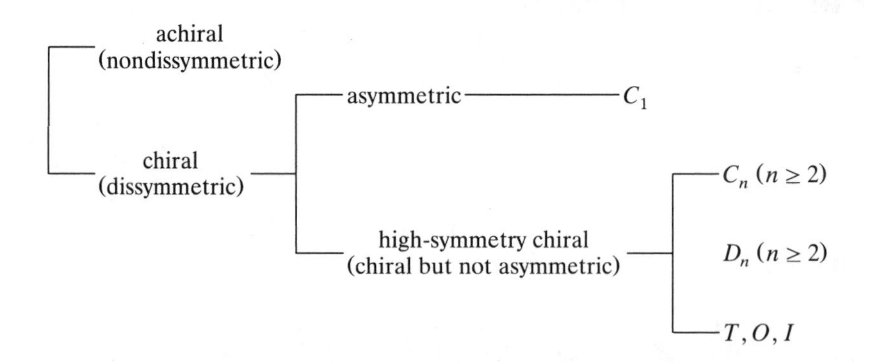

Many kinds of molecules with high-symmetry chirality were known very early in the history of stereochemistry, and we can find a variety of molecules of this type in the family of three-dimensional polycyclic molecules—cage-shaped molecules.

A force field calculation[3] showed that the D_3-twisted conformation of bicyclo[2.2.2]octane (**1**) is 0.1 kcal mol^{-1} more stable than the eclipsed conformation **2** of D_{3h} symmetry. In the case of this molecule, the interconversion of two D_3-twisted conformers, **1-M** and **1-P** via an achiral D_{3h}-conformer **2** is very facile. However, C(2)–C(8) diagonal bridging of this framework with a single bond or a short polymethylene group (CH_2 or CH_2CH_2) freezes the twisted conformation to furnish a tricyclic cage-shaped molecule **3** of high-symmetry chirality.

Further C(5)–C(7) diagonal bridging of **3** with a short bridge provides a tetracyclic cage-shaped molecule **4**, and final C(3)–C(6) diagonal bridging with a short bridge furnishes a pentacyclic molecule **5**.

A topological characteristic common to these cage-shaped molecules **3**, **4**, and **5** is the D_3-twisted bicyclo[2.2.2]octane moiety with rigid conformation. (See Chart 2.1.)

Since the preparation of hexamethylenetetramine was reported by Butlerow[4] in 1860, a large number of papers related to the preparation and properties of cage-shaped molecules have been published. I have selected only certain species that possess the D_3-twisted bicyclo[2.2.2]octane moiety as a common structural feature and have described the synthesis, chiroptical properties, and biological transformations of these chiral molecules in this chapter.

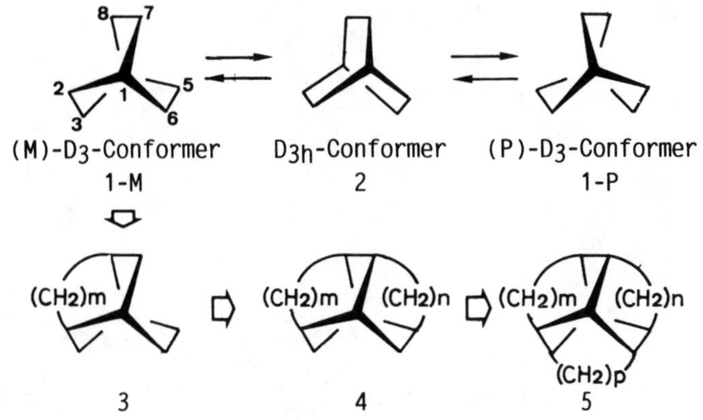

(M)-D_3-Conformer 1-M D_{3h}-Conformer 2   (P)-D_3-Conformer 1-P

$(CH_2)m$ 3 $(CH_2)m$ $(CH_2)n$ 4 $(CH_2)m$ $(CH_2)n$ $(CH_2)p$ 5

Chart 2.1 ■

2.2. Chiral Tricyclic Cage-Shaped Molecules

2.2A. C_2-Bissecocubane, Tricyclo[4.2.0.0^{3,6}]octane

Among the four possible tricyclic hydrocarbons attainable by two-bond fission of cubane (**17**), which can be regarded as being composed of two enantiomeric D_3-bicyclo[2.2.2]octane moieties (the hatched and the dotted ones illustrated in **17**), tricyclo[4.2.0.0^{3,6}]octane (**9**) [C(1)—C(6) and C(2)—C(3) bond fission] is the only chiral compound and the other three possible bissecocubanes, tricyclo[3.1.1.1^{2,4}]octane, tricyclo[4.2.0.0^{2,5}]octane, and 2-methyltricyclo[3.1.1.0^{3,6}]heptane, are all achiral. In order to differentiate **9** with C_2 symmetry from the other achiral bissecocubanes, **9** is called C_2-bissecocubane and is the smallest tricyclic molecule possessing the D_3-twisted bicyclo[2.2.2]octane moiety as a central core (the hatched part).

Hydrogenolysis of the diazabasketane derivative **6** with Pd on carbon in acetic acid gave **7**, which was transformed to diazatwistene (**8**) by hydrolysis followed by treatment with $CuCl_2$ and NH_4OH. Pyrolytic denitrogenation of **8** at 240–250 °C yielded **9**, in a racemic form, mp 91°C (in a sealed tube).[5] (See Scheme 2.1.)

Paquette and co-workers[6] were interested in studying the chiroptical property of **8**, whose molecular rigidity fixed the —N=N— unit into an essentially planar cisoid geometry, and synthesized **8** and **9** in optically active forms. By means of regiocontrolled hydrogenolysis, **10** was converted into a mixture of **11** and its diastereomer, which was separated by recrystallization. Hydrolysis and oxidation of (+)-**11** by modification of the conditions of Askani and co-workers[5] gave (+)-**8**, $[\alpha]_D = +201°$, which was converted into (+)-**9**, $[\alpha]_D = +21.5°$. Paquette estimated the absolute rotations ($[\alpha]_{D\,abs}$) of (+)-**9** and (+)-**8** to be +125° and +1165°, respectively. The circular dichroism (CD) spectrum of (−)-**8** is characterized by a rather intense band at 370 nm showing a negative Cotton effect and a weaker absorption of opposite sign around 220 nm.

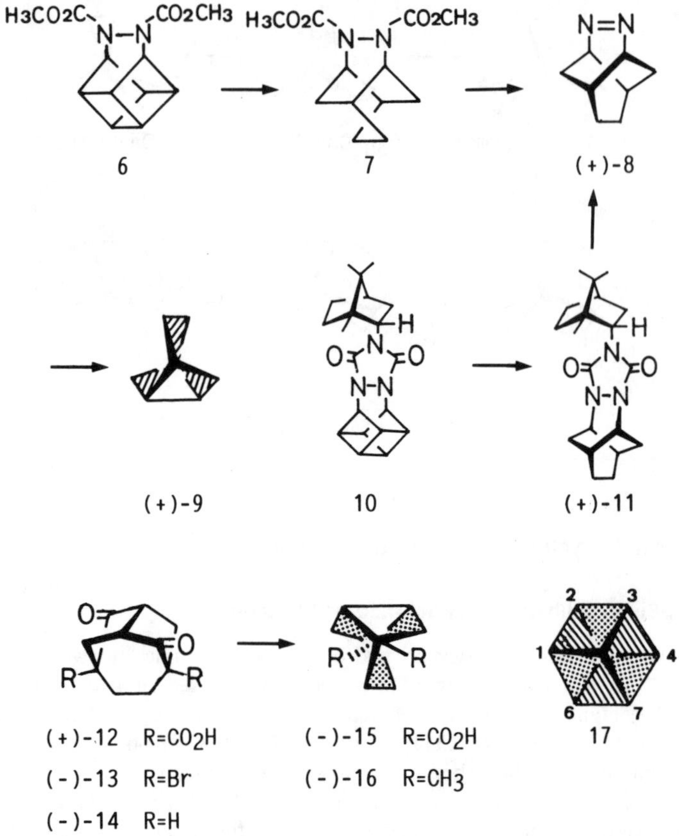
H3CO2C
CO2CH3
N–N
6
7
N=N
(+)-8
H
O
N
(+)-9
10
(+)-11
O
R
R'''
2
3
1
4
6
7
(+)-12 R=CO2H
(-)-13 R=Br
(-)-14 R=H
(-)-15 R=CO2H
(-)-16 R=CH3
17

Scheme 2.1 ■

Our optical resolution of **12** via its cinchonidine salt furnished (+)-**12**, and double Favorskii ring construction of (−)-**13**, prepared from (+)-**12**, yielded (−)-**15**, $[\alpha]_D = -97.9°$, which was transformed to (−)-**16**, $[\alpha]_D = -85.0°$. Their absolute configurations were determined on the basis of the CD spectrum of (−)-**14**, prepared by Hunsdicker reaction of (+)-**12** followed by reduction with Zn–Hg, exhibiting negative Cotton effects around 300 nm.[7]

2.2B. Twist-brendane, Tricyclo[4.3.0.0^{3,8}]nonane

Nickon et al.[8] proposed the trivial name brexane for tricyclo[4.3.0.0^{3,7}]nonane (**18**), because of the existence of a bridge involving an *exo*-norbornyl bond in this molecule. Analogously, tricyclo[4.2.1.0^{3,7}]nonane (**19**), in which an ethano bridge involving an *endo*-norbornyl bond exists, is named brendane. Replacement of the ethano bridge in brendane to a diagonal position (C-2 in **19**) leads tricyclo[4.2.1.0^{3,8}]nonane (**20**)

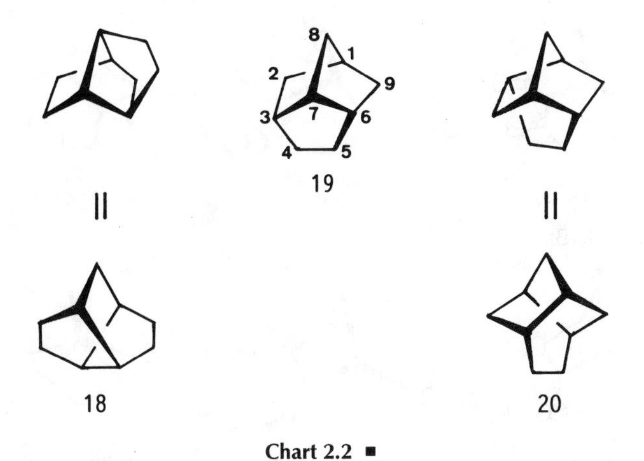

Chart 2.2 ■

possessing the D_3-twisted bicyclo[2.2.2]octane moiety, which belongs to the C_2 point group and is called twist-brendane. (See Chart 2.2.)

The first synthesis of **20** began with the homoadamantane derivative **21**, the disilver salt of which was converted to **22** by Hunsdicker reaction. Double Favorskii rearrangement of **22** provided **23**, which in turn was transformed to **20**, mp 165–166°C (in a sealed tube), via **24**.[9] Sauers and co-workers developed a method based on a reductive ring fission of the tetracyclic oxetane **31**, and they synthesized twist-brendane derivatives bearing a functional group at C-9 in a racemic form.[10]

The first optically active **20** was prepared from (−)-*endo*-bicyclo[2.2.1]hept-5-ene-2-carboxylic acid (**25**) with known absolute configuration and optical purity.[11] The carboxylic acid (−)-**25** in turn was transformed to **27** via (+)-**26** and intramolecular alkylation with NaH in DMF converted **27** into (−)-twist-brendan-2-one (**28**), $[\alpha]_{D\,abs} = -290°$, Wolff–Kishner reduction of which gave (−)-(1R,3S,6S,8R)-**20**, $[\alpha]_{D\,abs} = -284°$.[12] Another synthesis of optically active **20** began with (+)-bicyclo[2.2.2]oct-5-ene-2-carboxylic acid (**29**). According to Sauers' procedure, (+)-**30**, prepared from (+)-**29**, was converted into (+)-**31** by intramolecular Paterno–Bűchi reaction[13] and reductive cleavage of (+)-**31** with $LiAlH_4$ in *N*-methylmorpholine gave (+)-**32**, which was converted to (+)-**20** via (+)-twist-brendan-9-one (**33**), $[\alpha]_{D\,abs} = +282°$.[14] (See Scheme 2.2.)

As **20** is topologically formed by C(2)–C(5) diagonal bridging of bicyclo[2.2.1]heptane with an ethano bridge and also by bridging of cicyclo[2.2.2]octane with a methano bridge, the syntheses of (−)-**20** and (+)-**20** just described revealed unambiguously the configurational correlation between (−)-**25** and (+)-**29**.

2.2C. Twistane, Tricyclo[4.4.0.0^{3,8}]decane

Twistane (**39**) of D_2 symmetry is composed of four twist-boat cyclohexane moieties with the same chirality and is a twist-boat isomer of adamantane (**160**) composed of chair-form cyclohexane moieties. The peculiarity of this structure prompted Whitlock[15] to prepare twistane. Intramolecular alkylation with NaH in DMF converted **37**,

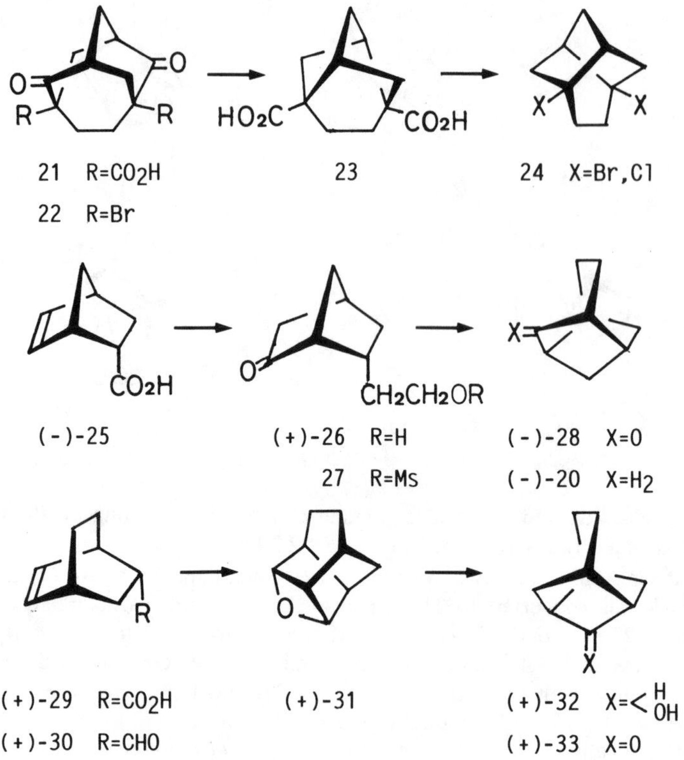

21 R=CO_2H
22 R=Br

23

24 X=Br,Cl

(−)-25

(+)-26 R=H
27 R=Ms

(−)-28 X=O
(−)-20 X=H_2

(+)-29 R=CO_2H
(+)-30 R=CHO

(+)-31

(+)-32 X=<$^{H}_{OH}$
(+)-33 X=O

Scheme 2.2 ■

prepared from **29** via **36**, to twistan-2-one (**38**), which was reduced to give **39**, mp 163–164.8°C (in a sealed tube).[15] (See Scheme 2.3.)

As force field calculation[16] shows that **39** is not completely free of angle strain, **39** and twistan-2-ol (**34**) were easily isomerized to **160** and adamantan-1-ol (**35**) with Lewis acid, respectively.[17]

A short synthesis of **39**, developed by Deslongchamps and co-workers,[18] began with *cis*-decaline-2,7-dione (**44**), which in turn was converted to **45**. Treatment with NaH in 1,4-dioxane converted **45** to twistan-4-one (**46**), reduction of which gave **39**. Treatment of **44** with BF_3-etherate, acetic anhydride, and acetic acid gave **40**, from which 1-substituted twistane derivatives such as **41**, **42**, and **43** were prepared.

Three groups synthesized independently the interesting hydrocarbon in an optically active form. Our first synthesis of optically active **39** began with (−)-**36**, which was obtained by resolution of (±)-**36** via its cinchonidine salt. According to Whitlock's procedure, (−)-**36** was transformed to (+)-**39**, $[\alpha]_D = +414°$, via (+)-**38**, $[\alpha]_D = +412°$.[19]

Another synthesis, reported by Tichý and Sicher,[20] began with **47**, which was resolved via its brucine salt. Acyloin condensation of **48**, derived from (−)-**47**, gave

34 35

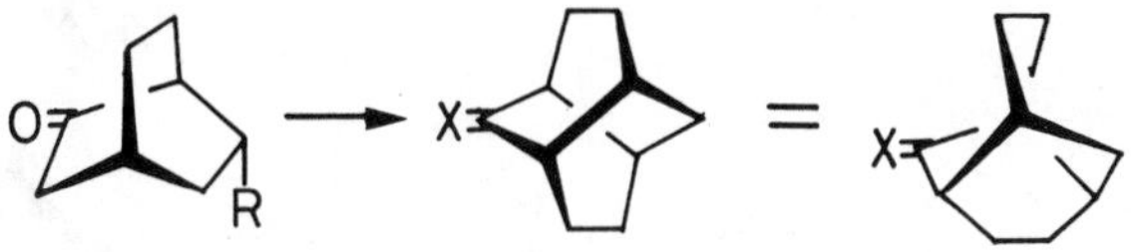

(-)-36 R=CH_2CO_2H (+)-38 X=O

37 R=CH_2CH_2OMs (+)-39 X=H_2

Scheme 2.3 ■

(+)-**49**, which was transformed to (+)-twistene (**50**), $[\alpha]_D = +417°$, by treatment with bis(thiocarbonyl)imidazole followed by heating with trimethylphospine. Hydroboration of (+)-**50** followed by oxidation gave (+)-**46**, $[\alpha]_D = +295°$, and catalytic hydrogenation converted (+)-**50** to (+)-**39**, $[\alpha]_D = +434°$. (See Scheme 2.4.)

Our application of the empirical rule of Djerassi and Klyne[21] to (+)-**38** led us to an erroneous conclusion as to the absolute configuration; Tichý and Sicher were also led to the same erroneous configurational assignment.

The correct absolute configuration, (+)-(1*R*,3*R*,6*R*,8*R*)-**39**,[22] was confirmed on the basis of the absolute configuration of (+)-**29**, which was determined by a chemical correlation with (−)-(2*R*)-bicyclo[2.2.2]octan-2-ol.[23]

The unique synthesis of 4,9-twistadiene (**55**), developed by Ganter and co-workers,[24] began with **51**. Pyrolysis of 7-allyloxycycloheptatriene gave **51**, which was in turn transformed to **52**. Resolution of **52** was accomplished via **53**. Treatment with $KOBu^t$ in DMSO converted (−)-**54**, derived from (−)-**52**, into (−)-**55**, $[\alpha]_D = -458°$, catalytic hydrogenation of which provided (−)-**39**, $[\alpha]_D = -437°$. (See Scheme 2.5.)

A characteristic feature of chiroptical properties of twistane being drawn on the basis of these results is that the sign of the specific rotation of twistane derivatives is specified by the chirality of its twisted framework, the D_3-bicyclo[2.2.2]octane moiety and independent of a variety of functional groups. This feature is generally found in all cage-shaped molecules having the D_3-twisted bicyclo[2.2.2]octane moiety except for a few cases.

Rigid conformations together with the high-symmetry characteristics of chiral cage-shaped molecules make this family of compounds a testing ground for theories of optical activity. Brewster [25] was the first to apply his "helical model"[26] to correct the erroneous absolute configuration previously assigned to **39** and his calculated $[M]_D = 484°$ for **39** was found to be close to the absolute molecular rotation of 598°, which was experimentally determined by ^1H NMR analysis of the acetate of (−)-**34**.[27]

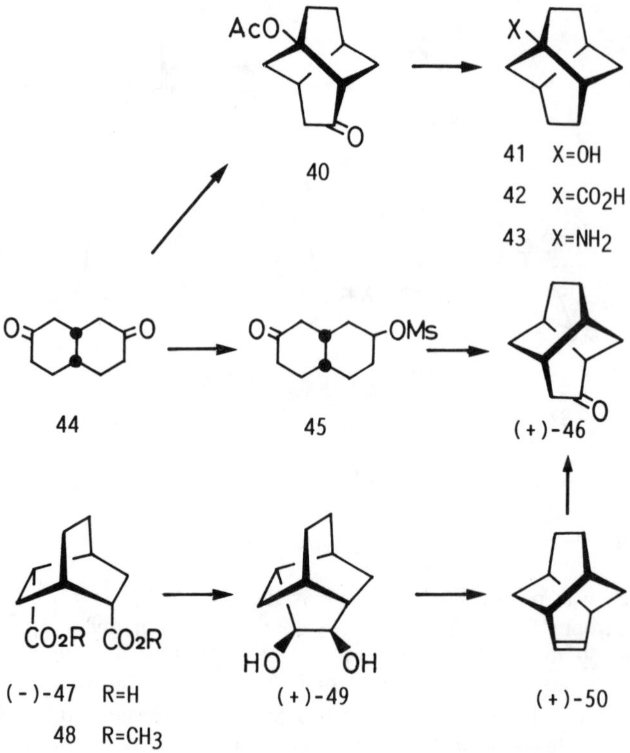

AcO
X
40
41 X=OH
42 X=CO_2H
43 X=NH_2
O O
O OMs
44
45
(+)-46
CO_2R CO_2R
HO OH
(-)-47 R=H
48 R=CH_3
(+)-49
(+)-50

Scheme 2.4 ■

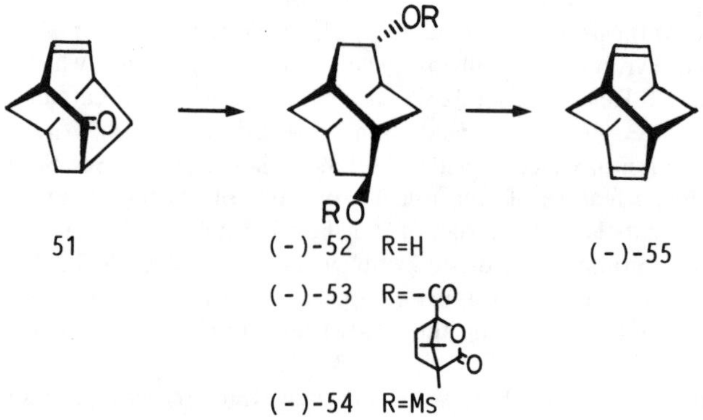

OR
O
RO
51
(-)-52 R=H
(-)-53 R=-CO
(-)-54 R=Ms
(-)-55

Scheme 2.5 ■

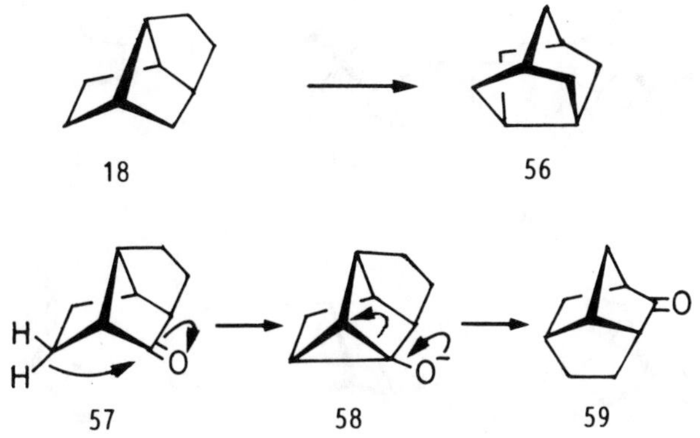

18 56

57 58 59

Scheme 2.6 ■

2.2D. Brexane, Tricyclo[4.3.0.0^{3,7}]nonane

Brexane (**18**) is more strained than brendane (**19**) and noradamantane (**56**). This instability of the brexane molecular framework seems to be nicely demonstrated in a facile rearrangement of **18** into **56**.[28] Brexan-2-one (**57**) was also transformed to brendan-2-one (**59**) through the species **58** on heating with KOBu^t in tert-butyl alcohol at 185°C.[29] (See Scheme 2.6.)

The diol **61**, prepared by a reductive cyclization of **60** with Na–Hg in water, was the first compound with this strained framework.[30] The parent hydrocarbon **18** was first synthesized by Nickon and co-workers.[8,29] Intramolecular alkylation of **64** with NaH in DMF gave **57**, Wolff–Kishner reduction of which provided **18** in a racemic form as a liquid.[8,29]

Our synthesis[31] of optically active **18** began with (+)-**25**, which in turn was converted into (+)-**63**. According to Nickon's procedure, (+)-**63** was converted to (−)-**57**, $[\alpha]_{D\,abs} = -286°$, from which (−)-**18**, $[\alpha]_{D\,abs} = -134°$ and (−)-**65** were prepared. (See Scheme 2.7.)

Brieger and Anderson[32] developed a facile synthesis of the brexane ring system. Heating a benzene solution of 1-(3-butenyl)cyclopentane-1,3-diene at 180°C in a sealed tube gave **62**.

2.3. Chiral Tetracyclic Cage-Shaped Molecules

A topological characteristic common to the tetracyclic cage-shaped molecules discussed here is also the D_3-twisted bicyclo[2.2.2]octane moiety. Among them, the smallest molecule is secocubane (**4**, $m = n = 0$), which is conceptually formed by fission of any single bond of cubane (**17**). Masamune and co-workers[33] prepared the secocubane derivative **69** from basketene (**68**) by oxidative cleavage of the unsaturated bond. This molecule possesses the D_3-twisted bicyclo[2.2.2]octane moiety, but is achiral. (See Scheme 2.8.)

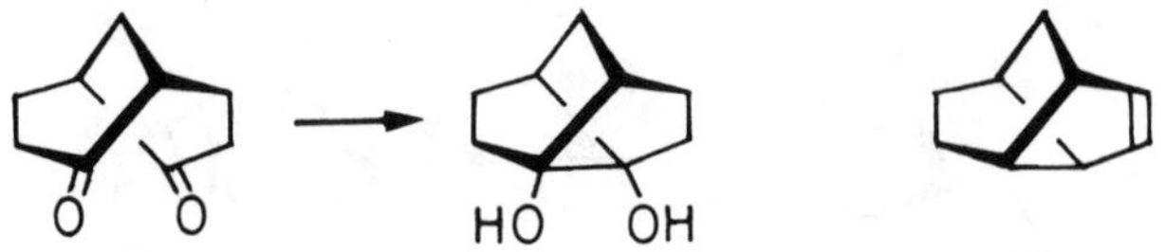

O O
60

HO OH
61

62

(+)-63 R=CO_2CH_3
64 R=CH_2CH_2OMs

(−)-18 X=H_2
(−)-57 X=O
(−)-65 X=<H OH

Scheme 2.7 ■

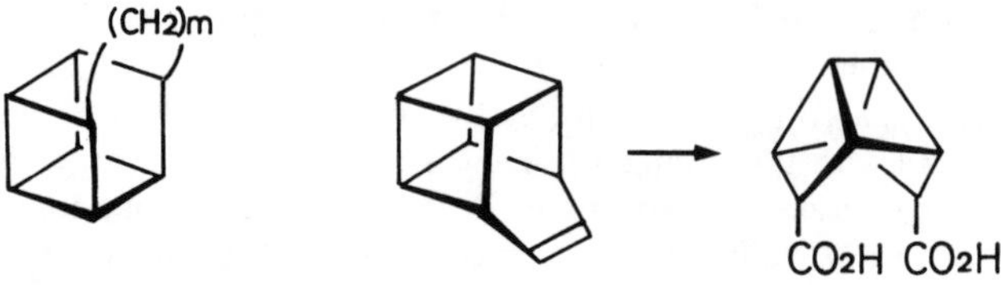

$(CH_2)_m$

66 n=1
67 n=2

68

CO_2H CO_2H
69

Scheme 2.8 ■

2.3A. Dehydrotwist-brendane, Tetracyclo[4.3.0.0^{2,5}.0^{3,8}]nonane

Both homocubane (**66**) and basketane (**67**) possess bicyclo[2.2.0]hexane moieties, a central single bond of which is easily cleaved on hydrogenolysis. Muso[34] reported that it is the C(2)–C(5) bond that is cleaved initially by hydrogenolysis of **67**, not the C(4)–C(5) bond.

Hydrogenolysis with 10% Pd on carbon in methanol cleaved the C(2)–C(5) bond of **72** furnishing **70**, which was transformed to **74** with Li in *tert*-butyl alcohol and tetrahydrofuran.[35a] The C(4)–C(7) bond of **74** was further cleaved to give the twist-brendane derivative on hydrogenolysis with 5% Rh on alumina in cyclohexane. The C(4)–C(5) bond fission was first demonstrated in Toyne.[35a] Treatment of **71** with 5% Rh on alumina in cyclohexane provided **73** being achiral. In addition, the various reactions of homocubanes and basketanes have been reported.[35b]

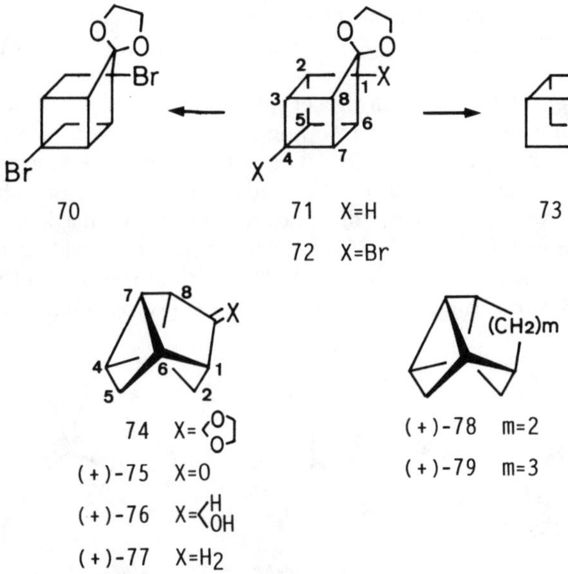

70

71 X=H

72 X=Br

73

74

(+)-75 X=O

(+)-76

(+)-77 $X=H_2$

(+)-78 m=2

(+)-79 m=3

Scheme 2.9 ■

Utilization of enzymes in organic synthesis to prepare optically active compounds is well documented.[36] Horse liver alcohol dehydrogenase (HLADH) mediated enantioselective reduction of (±)-**75** gave (+)-**75** and (−)-**76**. Wolff–Kishner reduction of (+)-**75** gave (+)-**77**, $[\alpha]_{D\,abs} = +105°$, which is conveniently called dehydrotwist-brendane, and the optical purity and the absolute configuration of (+)-**75** were confirmed by a chemical correlation with (+)-**32**.[37]

A ring expansion of ketones provides facile routes to higher homologous cage-shaped compounds. Treatment of (+)-**75** with diazomethane and BF_3-etherate in ether gave a mixture of the ketones, Wolff–Kishner reduction of which furnished (+)-**78**, $[\alpha]_{D\,abs} = +92.3°$, and (+)-**79**, $[\alpha]_{D\,abs} = +3.3°$. (See Scheme 2.9.)

2.3B. Ditwist-brendane, Tetracyclo[5.2.1.0^{2,6}.0^{4,8}]decane

Schleyer[38] proposed to designate **145** as [6]diadamantane, because this molecule has two adamantane units with six carbon atoms in common. According to Schleyer's nomenclature, tetracyclo[5.2.1.0^{2,6}.0^{4,8}] decane (**85**) containing two twist-brendane units with eight carbon atoms in common may be called [8]-ditwist-brendane or, more conveniently, ditwist-brendane. The synthesis of **85** in a racemic form, developed by Yonemitsu and co-workers,[39] was based on hydrogenolysis of the bishomocubane skeleton with Pd on carbon in acetic acid. Hydrogenolysis in acetic acid followed by hydrolysis and oxidation converted **92** into ditwist-brendan-5-one (**95**), which was then transformed to **85**, mp 160–161°C (in a sealed tube). Hydrogenolysis of **93** also gave

85. Rothberg and co-workers[40] developed a short route to ditwist-brendanes bearing a functional group at C-3. Acetolysis of **81**, prepared from the Diels–Alder product **80**, gave **82**.

Optical resolution of **83** via the hydrogen phthalate was accomplished by using (+)-2-(1-aminoethyl)naphthalene as a resolving agent. The optical purity of (−)-**83** was determined by ^{1}H NMR analysis. Oxidation of (−)-**83** gave (−)-ditwist-brendan-3-one (**84**), $[\alpha]_{D\,abs} = -277°$, Wolff–Kishner reduction of which provided (−)-**85**, $[\alpha]_{D\,abs} = -304°$.[41] Another synthesis of optically active **85** of the known absolute configuration began with (−)-**86**.[27,42] Heating with sodium methoxide converted the 2,5-dibromocyclopentanone ethylene ketal into **87**,[43] partial hydrolysis of which gave **88**. Reduction of **88** with $LiAlH_4$ gave exclusively **86**, the hydrogen phthalate of which

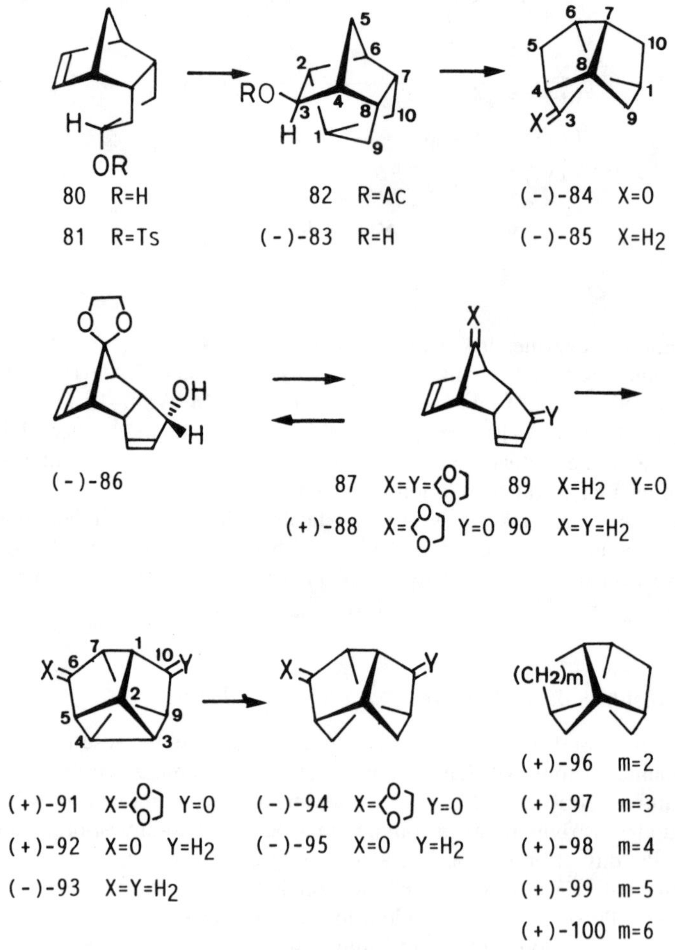

80 R=H
81 R=Ts

82 R=Ac
(−)-83 R=H

(−)-84 X=O
(−)-85 X=H_2

(−)-86

87 X=Y=(ethylenedioxy)
(+)-88 X=(ethylenedioxy) Y=O
89 X=H_2 Y=O
90 X=Y=H_2

(+)-91 X=(ethylenedioxy) Y=O
(+)-92 X=O Y=H_2
(−)-93 X=Y=H_2

(−)-94 X=(ethylenedioxy) Y=O
(−)-95 X=O Y=H_2

(+)-96 m=2
(+)-97 m=3
(+)-98 m=4
(+)-99 m=5
(+)-100 m=6

Scheme 2.10 ■

was resolved via the (+)-2-(1-aminoethyl)naphthalene salt. Oxidation of (−)-**86** gave (+)-**88**. Irradiation of (+)-**88** gave (+)-**91**, which was in turn converted into (−)-**93**, $[\alpha]_D = -33.8°$, via (+)-**92**, $[\alpha]_{D\,abs} = +11.0°$. The absolute configuration of (+)-**88** was determined by a chemical correlation with (−)-**25** via (+)-methyl 3-(*endo*-2-norbornyl)propionate, and the correlation eventually led to the assignment of the 1*S*,2*S*,3*S*,4*S*,5*R*,7*S*,8*S*,9*R* configuration to (−)-**93**. Catalytic hydrogenolysis of (+)-**91** afforded (−)-**94**, which was converted into (−)-(1*R*,2*R*,4*R*,6*R*,7*R*,8*R*)-**85** via (−)-ditwist-brendan-5-one (**95**), $[\alpha]_{D\,abs} = -237°$. Direct conversion of (−)-**93**, $[\alpha]_D = -33.8°$ into (−)-**85**, $[\alpha]_D = -232°$ (76% ee)[41] by hydrogenolysis determined the maximum rotation of **93** to be 44°.

Diazomethane ring expansion followed by reduction, when applied to (+)-**95**, led to the formation of a mixture of hydrocarbons, (+)-**96**–(+)-**100** (see Scheme 2.10),[37] $[M]_{D\,abs}$ values of which are shown in Section 2.7 (Figure 2.2).

2.3C. Ditwistane, Tetracyclo[6.2.2.0^{2,7}.0^{4,9}]dodecane

Tetracyclo[6.2.2.0^{2,7}.0^{4,9}]dodecane (**107**) is called ditwistane, which was first synthesized in a racemic form by Yonemitsu and co-workers.[39] Catalytic hydrogenolysis in acetic acid followed by hydrolysis and oxidation converted **104** into **106**, which was transformed to **107**, mp 117–118.5°C (in a sealed tube).

The synthesis of optically active (−)-**107** began with (+)-**91**.[44] Diazomethane ring expansion of (+)-**91** with known absolute configuration and optical purity gave **101**, which was transformed to (−)-**102** by Wolff–Kishner reduction followed by acid hydrolysis. A similar sequence of conversions transformed (−)-**102** to (−)-C_2-dehydroditwistane (**105**), $[\alpha]_{D\,abs} = -225°$, catalytic hydrogenolysis of which gave (−)-(1*S*,2*S*,4*S*,7*R*,8*R*,9*S*)-**107**, $[\alpha]_{D\,abs} = -695°$. The conversion dramatically demonstrated that significant increase in a specific rotation was observed accompanying the single bond rupture in going from a pentacyclic cage-shaped hydrocarbon to a tetracyclic one.

2.4. Chiral Pentacyclic Cage-Shaped Molecules

The smallest pentacyclic cage-shaped molecule possessing twisted bicyclo[2.2.2]octane moiety is cubane (**17**), which was first prepared by Eaton and Cole.[45]

Theoretically, there are four ways to desymmetrize this polyhedral molecule furnishing four types of bishomocubanes[46] by insertion of each of two methylene groups between its eight methine groups situated on the eight corners. Among four bishomocubanes, only pentacyclo[5.3.0.0^{2,5}.0^{3,9}.0^{4,8}]decane (**93**) is chiral (C_2 symmetry) and called C_2-bishomocubane.

2.4A. C_2-Bishomocubane, Pentacyclo[5.3.0.0^{2,5}.0^{3,9}.0^{4,8}]decane; C_1-Homobasketane, Pentacyclo[5.4.0.0^{2,6}.0^{3,9}.0^{5,8}]undecane

Photocyclization of dicyclopentadiene (**90**) in the presence of acetone as a photosensitizer provided **93**,[47] mp 134–136°C, and, similarly, **89** was converted to **92**.[48] The preparation of optically active C_2-bishomocubane derivatives[27,42] was described in

101 X=(ethylenedioxy) Y=O
(−)-102 X=O Y=H2
(−)-103 X=Y=H2
104 X=O
(−)-105 X=H2
106 X=O
(−)-107 X=H2
108 X=NNHTs
109

Scheme 2.11 ■

Section 2.3B (Scheme 2.10) and (−)-C_1-homobasketane (**103**), $[\alpha]_{D\,abs} = -120°$, was prepared by reduction of (−)-**102**.[44] (See Scheme 2.11.)

Marchand and co-workers[49] prepared 6,6,8-trinitro- and 6,6,10,10-tetranitro-C_2-bishomocubanes that are of interest as a potential new class of explosives and propellants.

2.4B. C_2-Dehydroditwistane, Pentacyclo[6.4.0.0^{2,7}.0^{3,10}.0^{6,9}] dodecane

The synthesis of C_2-dehydroditwistane (**105**) in an optically active form is previously described in Section 2.3C.[44] Another interesting molecule related to **105** is C_2-ansaradiene (**109**), a bisvinylog of cubane and a short of benzene dimer that was synthesized from **108** and was quantitatively split into two benzene molecules on refluxing in hexachlorobutadiene.[50]

2.4C. D_3-Trishomocubane, Pentacyclo[6.3.0.0^{2,6}.0^{3,10}.0^{5,9}]undecane

The noteworthy feature of D_3-trishomocubane (**117**) with D_3 symmetry is that it is composed of six twisted cyclopentane moieties with the same chirality and is the only framework possessing neither three- or four-membered rings nor other strained features among the large number of possible $C_{11}H_{14}$ pentacyclic hydrocarbons. D_3-Trishomocubane (**117**) is thus likely to be the pentacycloundecane "stabilomer."[51] The term "stabilomer" is defined as "that isomer possessing the lowest free energy of formation at 25°C in the gas phase." Indeed, treatment of the pentacyclic hydrocarbon **112** with $AlBr_3$ in CS_2 at room temperature gave **117** together with a small amount of 2,4- and 2,8-ethanoadamantanes.[52]

Among a variety of methods used for the preparation of this exquisite molecule, the first synthesis of **117** in a racemic form, reported by Underwood and Ramamoorthy,[53] began with **110**, which was prepared by Diels–Alder reaction of cyclopentadiene with 1,4-benzoquinone. Photocyclization of **110** gave **111**, which was in turn converted into **114** via **113**. Desymmetrization of **114** to the D_3-trishomocubane derivative **115** was accomplished on treatment with Br_2 in CCl_4 and debromination of **115** with Li in *tert*-butyl alcohol yielded **117**, mp 147–149°C (in a sealed tube).

D_3-Trishomocubane derivatives with functional groups that provide a ready handle for optical resolution were independently prepared by Eaton's group[54] and by Barborak's group.[55] Treatment of **111** with zinc in acetic acid followed by reduction with $NaBH_4$ gave **118**. Heating with 32% HBr in acetic acid converted **118** into **119**, which reacted with $KOBu^t$ in ether to give D_3-trishomocuban-4-one (**121**).[54] Desymmetrization of **113** with 98% H_2SO_4 in acetic acid at 120°C followed by hydrolysis and oxidation gave **122** via **116**.[55]

Three independent syntheses of **117** in an optically active form were reported in succession. Helmchen and Staiger[56] resolved **123** via its camphanate and reduction of

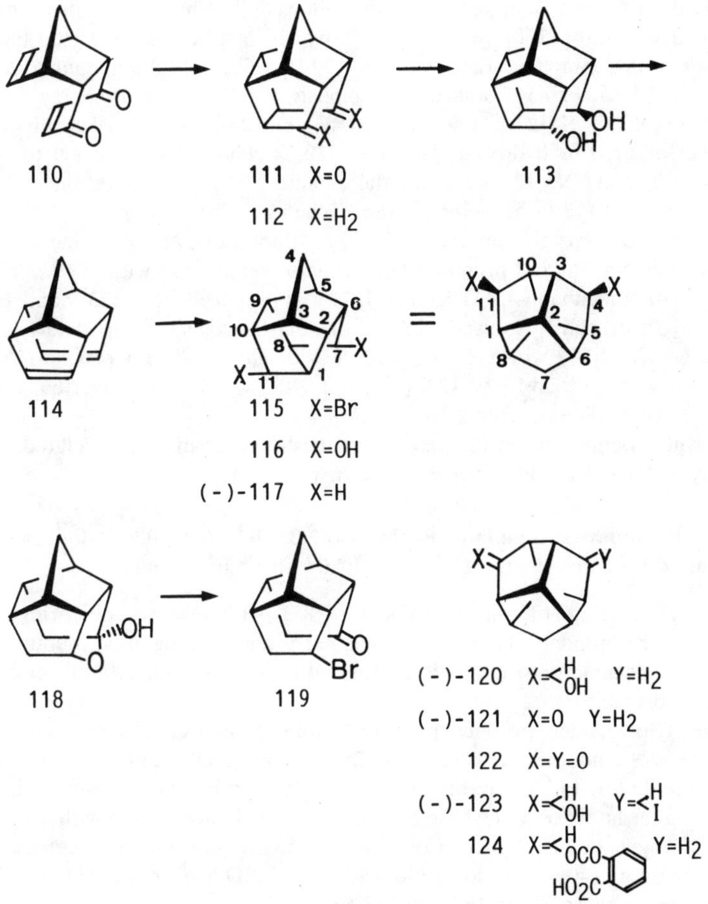
O
O
110
X
X
111 X=O
112 X=H2
OH
OH
113
114
4
5
9
6
3
2
10
8
7
X
X
11
1
115 X=Br
116 X=OH
(-)-117 X=H
10
3
X
X
11
2
4
1
5
8
6
7
OH
O
118
O
Br
119
X
Y
(-)-120 X=<H OH Y=H2
(-)-121 X=O Y=H2
122 X=Y=O
(-)-123 X=<H OH Y=<H I
124 X=<H OCO- Y=H2
HO2C

Scheme 2.12 ■

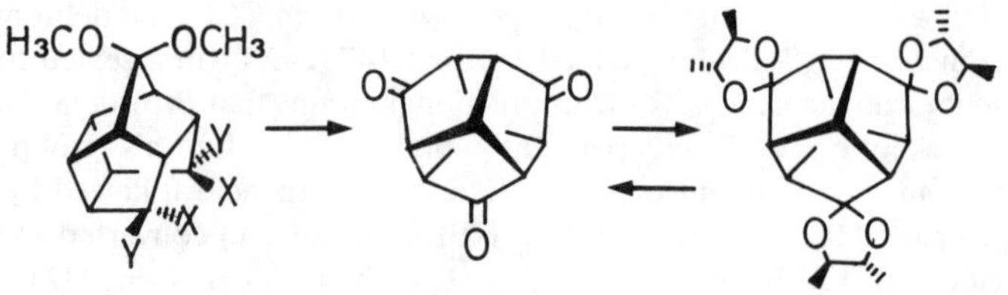

125 X=OH Y=H (+)-127 (-)-128

126 X=H Y=OH

Scheme 2.13 ■

(−)-**123** with zinc and acetic acid followed by hydrolysis provided (−)-D_3-trishomocuban-4-ol (**120**), $[\alpha]_D = -147°$, which in turn was converted into (−)-**117**, $[\alpha]_D = -164°$, via (−)-**121**, $[\alpha]_D = -99.1°$. The absolute configuration, (−)-(1*S*,3*S*,5*S*,6*S*,8*S*,10*S*)-**117** was unambiguously confirmed by an X-ray study on the (−)-camphanate of **123**.[56] Our resolution[57] of **124** with (+)-2-(1-aminoethyl)naphthalene followed by hydrolysis gave (+)-**120**, which was transformed to (+)-**117** via (+)-**121**. The ^{1}H NMR analysis of the acetate of (+)-**120** determined the absolute rotation of (+)-**117** to be +165°.[56] (See Scheme 2.12.)

Eaton and Leipzig[58] reported a facile resolution of **121** employing (−)-ephedrine as a resolving agent and prepared both enantiomers of **117** with > 98% ee.

D_3-Trishomocubane-4,7,11-trione (**127**) was prepared from **125**[59] and **126**.[60] Fessmer and Prinzbach[59] resolved **127** via the diastereoisomeric trisketals prepared by reaction of (±)-**127** with (*R*,*R*)-2,3-butanediol and treatment of (−)-**128** with 33% HBr in acetic acid gave (+)-**127**, $[\alpha]_D = +949°$, which was converted into (−)-**117**, $[\alpha]_D = -165°$. (See Scheme 2.13.)

Further details of syntheses of D_3-trishomocubane and related pentacyclic molecules have been described in a recent review.[61]

2.4D. Ethanoditwist-brendane, Pentacyclo[7.3.0.0^{2,7}.0^{3,11}.0^{6,10}] dodecane; Methanoditwistane, Pentacyclo[7.4.0.0^{2,6}.0^{3,1}.0^{5,10}]tridecane

Diagonal bridging of ditwist-brendane (**85**) with an ethano bridge furnishes ethanoditwist-brendane (**134**) of C_2 symmetry and, analogously, methanoditwistane (**138**) of C_2 symmetry is topologically formed from ditwistane (**107**) by bridging with a methano bridge.

The Diels–Alder product prepared from cyclohexa-1,3-diene and 1,4-benzoquinone was smoothly converted to **129** by photocyclization[62] and **129** was transformed to **130** by $LiAlH_4$ reduction. The facile rearrangement observed in **113** failed in the higher homolog **130** whose heating at 120°C in acetic acid with a small amount of H_2SO_4 gave exclusively **131**. Conversion to **132** required rather vigorous conditions, and heating at 140–150°C in acetic acid and H_2SO_4 converted **131** into **132**, which was in turn converted into **134**, mp 135°C, via **133**.[63]

Optically active **134**, $[\alpha]_{D\,abs} = -293°$, was prepared by Wolff–Kishner reduction of (−)-ethanoditwist-brendan-4-one (**143**), which was prepared by diazomethane ring

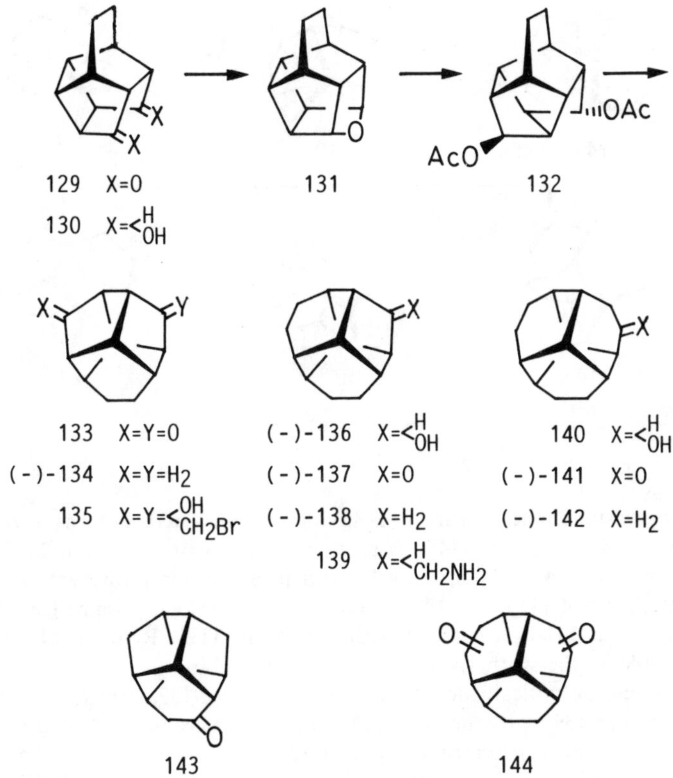

129 X=O
130 X=<H OH
131
132
133 X=Y=O
(−)-134 X=Y=H2
135 X=Y=<OH CH2Br
(−)-136 X=<H OH
(−)-137 X=O
(−)-138 X=H2
139 X=<H CH2NH2
140 X=<H OH
(−)-141 X=O
(−)-142 X=H2
143
144

Scheme 2.14 ■

expansion of (−)-**121**.[63]

Rhodotorula rubra mediated reduction of (±)-**137**, prepared from **132** via **136**, gave (−)-methanoditwistan-4-one (**137**) (22% ee) and (+)-**136** (33% ee). Wolff–Kishner reduction of (−)-**137** provided (−)-**138**, $[\alpha]_{D\,abs} = -404°$. Optical purity of (−)-**138** was determined by the ^{1}H NMR analysis of (−)-**136** and the CD spectral analysis of (−)-**137** led to the assignment of the 1*S*,2*S*,3*S*,5*S*,6*S*,9*S*,10*S*,11*S* configuration to (−)-**138**.[64] (See Scheme 2.14.)

2.4E. D_3-Tritwistane, Pentacyclo[8.4.0.0^{2,7}.0^{3,12}.0^{6,11}]tetradecane

From ditwistane (**107**), the route to the higher member diverges, giving three tritwistanes, **142**, **146**, and **147** (Scheme 2.15); **142** of D_3 symmetry is called D_3-tritwistane. The whole family of compounds constructed by diagonal ethano bridging starting from D_2-twist-boat cyclohexane can be represented with the molecular formula C_nH_{n+6} ($n = 2p + 6$, where $p = 0, 1, 2, \ldots$) and has ($n - 5$) D_2-twist-boat cyclohexane units. D_3-Tritwistane (**142**) can be seen to be composed of nine twist-boat cyclohexane units of the same chirality.

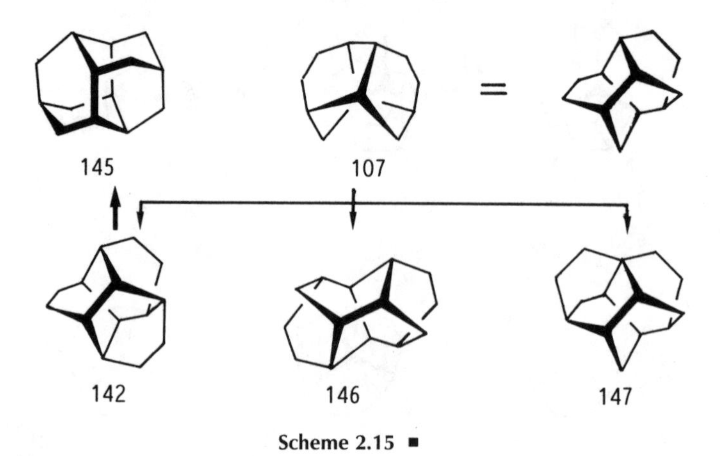

Scheme 2.15 ■

Because twistane readily rearranges to adamantane (**163**), **142** was also converted quantitatively to diamantane (**145**) by treatment with $AlBr_3$ in CS_2 at 25°C.[64]

D_3-Tritwistane (**142**) was first synthesized in a racemic form by Yonemitsu and Hirao.[65] Reaction of **133** with CH_2Br_2 and lithium dicyclohexylamine gave **135**, which was converted to **144** by treatment with *n*-BuLi in THF. Removal of the carbonyl groups completed the synthesis of **142**, mp 100–102°C.

Our synthesis of optically active **142** began with (−)-**137**, $[\alpha]_{D\,abs} = -247°$, which was converted to **139** by treatment with tosylmethyl isocyanide followed by reduction with $LiAlH_4$. Demjanow rearrangement of **139** gave **140**, which in turn was transformed to (−)-**142**, $[\alpha]_{D\,abs} = -567°$, via (−)-$D_3$-tritwistan-4-one (**141**), $[\alpha]_{D\,abs} = -463°$.[64]

2.5. Molecules with Chiral Polyhedral Symmetry

Tetrahedrane (T_d symmetry), cubane (O_h symmetry), and dodecahedrane (I_h symmetry) are geometrically equivalent to the Platonic solids tetrahedron, hexahedron, and dodecahedron; the remaining two Platonic solids are not likely to have hydrocarbon equivalents. Among three possible hydrocarbons, cubane[45] and dodecahedrane[66] have been synthesized. The derivatives of tetrahedrane were also reported in 1978,[67,68] but the parent hydrocarbon **148** is not yet known.

The desymmetrization that transfers a regular polyhedron into a chiral polyhedron (T, O, and I symmetry) can be achieved by a twist deformation around a C_n axis. A chiral polyhedron of T symmetry is generated by twisting a tetrahedron around C_3 axes passing through each vertex. The first experimental proof of the existence of a molecule of conformational T symmetry was discussed by Bartell et al.[69] The empirical force field calculation[70] showed that, in the case of tetra-*tert*-butyl-tetrahedrane (**149**),[68] the ground state is predicted to have T symmetry, with all 4 *tert*-butyl groups twisted by ca. 140° and all 12 methyl groups by ca. 2–6°, all in the same direction from a staggered T_d conformation. For this molecule, however, facile

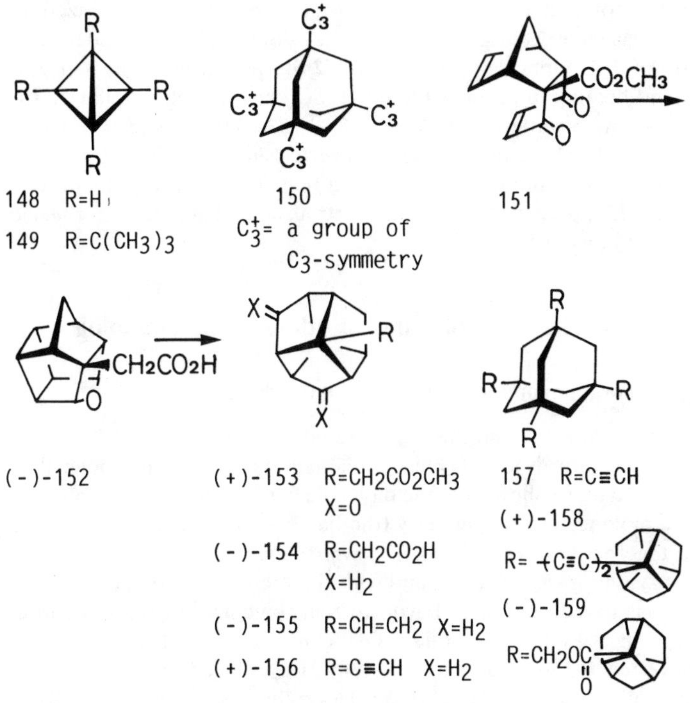

148 R=H

149 R=C(CH3)3

150

C_3^+= a group of C_3-symmetry

151

(−)-152

(+)-153 R=CH2CO2CH3 X=O

(−)-154 R=CH2CO2H X=H2

(−)-155 R=CH=CH2 X=H2

(+)-156 R=C≡CH X=H2

157 R=C≡CH

(+)-158

(−)-159

Scheme 2.16 ■

racemization is expected because T_d conformers, a possible achiral transition state, were shown to lie only 2–5 kcal mol^{-1} above the ground state of T symmetry.

Farina and Morandi[1] described that a molecule **150** with T symmetry may be obtained by replacing four tertiary hydrogen atoms of adamantane by four optically active subunits with threefold internal symmetry with the same chirality.

The only chiral polyhedral organic molecule prepared in an optically active form is the cage-shaped molecule (+)-**158** which is a molecule of the Farina-model type **150**. The highest attainable static and time-averaged dynamic symmetry of (+)-**158** are T and $(C_3)^4 \wedge T$, respectively, and the tetraester (−)-**159** belongs to the D_2 point group.[71]

Two kinds of building blocks, the C_3 subunit and the T_d core, are necessary for constructing (+)-**158** (Scheme 2.16). One of them is the optically active **156** of C_3 symmetry and the other is **157** of T_d symmetry. Strain value, as defined by (ΔH_f° cation) − (ΔH_f° hydrocarbon), is 55 kcal mol^{-1} for D_3-trishomocubanyl-2-cation, demonstrating that the C-2 position appears to be unreactive toward free-radical formation.[72a] Treatment of **117** with Br_2 and $AlBr_3$ gave 2-bromo-D_3-trishomocubane, but free radical chlorocarbonylation of **117** yielded exclusively the C(4)-substituted derivative.[72b] Our preparation of the D_3-trishomocubane derivative bearing a functional group at the C-2 position began with **151**, which was obtained by Diels–

Alder reaction of cyclopentadiene with 2-methoxycarbonyl-1,4-benzoquinone. The adduct **151** was converted into **152** by the similar procedure described for the preparation of **131**. Optical resolution of **152** via its cinchonidine salt gave (−)-**152** (97% ee), and heating in acetic acid and H_2SO_4 at 150–160°C followed by oxidation and esterification converted (−)-**152** into (+)-**153**. Wolff–Kishner reduction of (+)-**153** gave (−)-**154**, which in turn was converted into (+)-**156** of C_3 symmetry via (−)-**155**. Chodkiewicz and Cadiot's coupling reaction of the C_3-subunit (+)-**156** with the T_d-subunit **157**, prepared from 1,3,5,7-tetrabromoadamantane, furnished (+)-**158**, mp > 350°C; $[\alpha]_D = +65.3°$.[73]

2.6. Chiral Cage-Shaped Molecules of Adamantane Homolog

2.6A. D_{2d}-Bisnoradamantane, Tricyclo[3.3.0.0^{3,7}]octane

Removal of two methylene groups (e.g., C-2 and C-6) situated on the same C_2 axis in adamantane (**160**) furnishes D_{2d}-bisnoradamantane (**163**). The molecular model of D_{2d}-bisnoradamantane shows that the molecule consists of two enantiomeric twist-boat cyclohexane moieties of D_2 symmetry (the hatched and the dotted ones illustrated in **163**) fused together and has achiral D_{2d} symmetry.

The first synthesis of the compound with this framework, reported by Webster and Sommer,[74] was based on double Favorskii rearrangement of the adamantane derivative. Treatment with hot alcoholic KOH converted smoothly **161** into **162**. Vogt et al.[75] converted **162** into **163**, mp 103–104°C (in a sealed tube). Freeman and Rao[76] developed another route. Although **163** is less stable than its isomer **165**, hydrogenolysis of **166**, prepared from **164**, with Pt in methanol gave exclusively **163**. (See Scheme 2.17.)

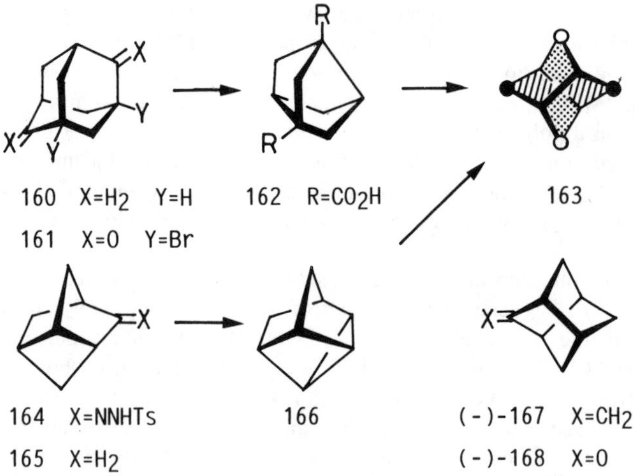

160 X=H2 Y=H
161 X=O Y=Br
162 R=CO2H
163
164 X=NNHTs
165 X=H2
166
(−)-167 X=CH2
(−)-168 X=O

Scheme 2.17

Interesting methods developed by Sauers and co-workers[77] are useful for the preparation of D_{2d}-bisnoradamantanes bearing functionalized substituents. Treatment of **169**, prepared by irradiation of 2-formylbicyclo[2.2.1]hept-5-ene, with $LiAlH_4$ in *N*-methylmorpholine followed by oxidation gave D_{2d}-bisnoradamantan-2-one (**168**), and ring fission with $KOBu^t$ in ether converted **177** [prepared from the acid chloride of 2-(carboxymethyl)bicyclo[2.2.1]hept-5-ene by heating with triethylamine in benzene] into **180**.

Coaxially disubstituted derivatives were prepared by modifying the conditions of Sauers. Treatment of **178** with $KOBu^t$ yielded **181**, which in turn was converted into **184** via **182**. Cope elimination of **184** followed by oxidation gave **175**, which was then converted into D_{2d}-bisnoradamantane-2,6-dione (**176**) by oxidation with OsO_4 followed by treatment with $NaIO_4$.[78] Analogously, **174** was prepared from **185** by Gleiter and co-workers.[79] They reported also the preparation of 2,4-dimethylene- and 2,4,6-trimethylene-D_{2d}-bisnoradamantanes.[79] The derivatives such as **174**, **175**, and **176** are an interesting model for studying the intramolecular interaction of two π fragments via a six-membered ring.

The molecular geometry of D_{2d}-bisnoradamantane results in two sets of homotopic methylene group (illustrated with the closed and the open circles in **163**), and the sets are in turn enantiotopic. The molecule can be desymmetrized by differentiation of one set of enantiotopic methylene groups from the other set. A variety of D_{2d}-bisnoradamantane derivatives were prepared in optically active forms. The most simple chiral D_{2d}-bisnoradamantane derivative is a monosubstituted one such as **167**,

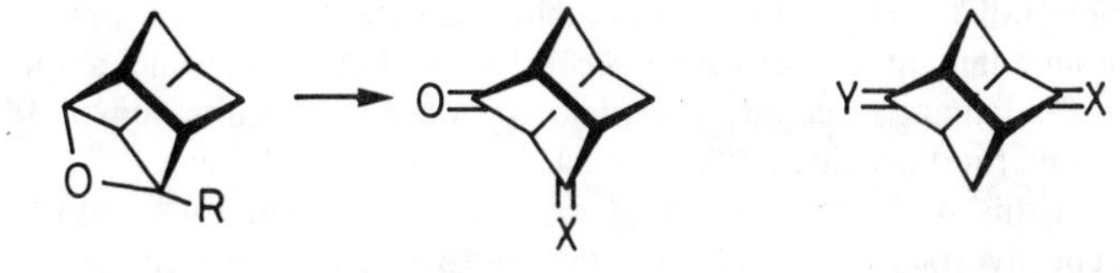

169 R=H
(−)-170 R=CH_3
(−)-171 R=$CH(CH_3)_2$

(−)-172 X=CH_2
(+)-173 X=$C(CH_3)_2$

174 X=Y=CH_2
175 X=O Y=CH_2
176 X=Y=O

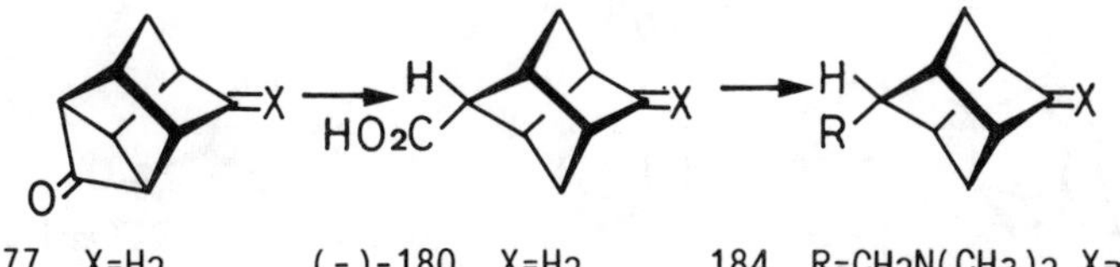

177 X=H_2
178 X=CH_2
179 X=$C(CH_3)_2$

(−)-180 X=H_2
181 X=CH_2
(−)-182 X=O
(−)-183 X=$C(CH_3)_2$

184 R=$CH_2N(CH_3)_2$(→O) X=C(H)(OH)
185 R=$CH_2N(CH_3)_2$(→O) X=CH_2

Scheme 2.18 ■

168, and **180**. Resolution of **180** via its (+)-2-(1-aminoethyl)naphthalene salt gave (−)-**180**, whose optical purity was determined by the ^{1}H NMR analysis of its methyl ester. The monosubstituted derivatives, (−)-**167**, $[\alpha]_{D\,abs} = -45.2°$, and (−)-**168**, $[\alpha]_{Dabs} = -78.7°$, were derived from (−)-**180**.[80] The molecules **172** and **173** are the chiral D_{2d}-bisnoradamantane derivatives of the other type. The oxetane derivatives (−)-**170** and (−)-**171** were prepared from (−)-**25**, and treatment with $LiN(C_2H_5)_2$ followed by oxidation converted (−)-**170** and (−)-**171** into (−)-**172**, $[\alpha]_{D\,abs} = -46.9°$, and (+)-**173**, $[\alpha]_{D\,abs} = +48.2°$, respectively.[80]

Coaxially disubstituted derivatives were also prepared in optically active forms. Optical resolution of **183**, prepared from **179**, was accomplished via the (+)-2-(1-aminoethyl)naphthalene salt to furnish (−)-**183**, which was converted into (−)-**182**, $[\alpha]_{D\,abs} = -70.2°$.[81] (See Scheme 2.18.)

The UV spectrum of **172** exhibiting a λ_{max} (isooctane) at 300.5 nm (ε 445) suggests an unusually strong homoconjugation. The UV spectra of **175** and **176** exhibit λ_{max} at 286 nm (24.3) and at 281 nm (28.7), respectively, and comparison with the UV spectrum of **168** with a λ_{max} at 282 nm (25.3) seems to show that much less intramolecular interactions between the unsaturated centers are noticeable in **175** and **176** than in **172**.

The S_4- and C_{2v}-tetraesters of D_{2d}-bisnoradamantane-2,4,6,8-tetracarboxylic acid were prepared by Park and Paquette.[82]

2.6B. Protoadamantane, Tricyclo[4.3.1.0^{3,8}]decane

Adamantane has been described as "a bottomless pit into which rearranging molecules may irreversibly fall."[17] However, a protoadamantane skeleton was formed by rearrangement of an adamantane derivative. Acetolysis of **186** gave a small amount of **188** together with 2-adamantyl acetate as a major product.[83] Deamination of **187** by the phenyltriazene method provided **188** in much larger yield.[83]

As shown in this study, protoadamantane (**190**) is a secondary stabilomer among $C_{10}H_{16}$ tricyclic hydrocarbons and has been prepared by a variety of procedures. According to Vogt's procedure,[84] optically active **190**, $[\alpha]_D = -118°$, was prepared from (+)-**191** via (−)-**189**.[85] (See Scheme 2.19.)

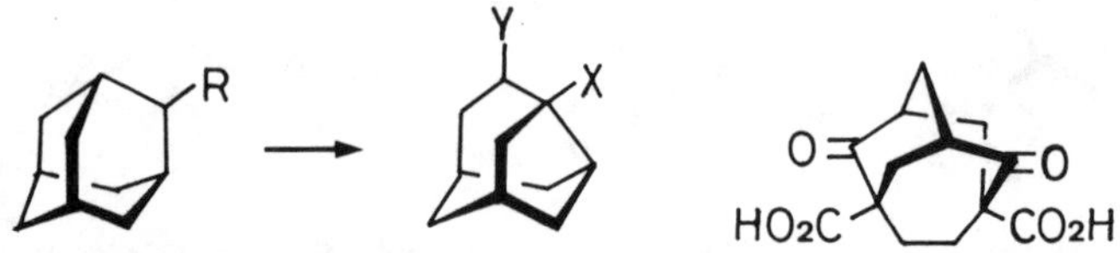

R
Y
X
O
O
HO_2C
CO_2H

186 R=OTs
187 R=NH_2

188 X=H Y=OAc
(−)-189 X=CO_2H Y=H
(−)-190 X=Y=H

(+)-191

Scheme 2.19 ■

2.7. Chiroptical Properties of Cage-Shaped Molecules

The octant rule for saturated ketones is the significant and successful attempt to correlate the three-dimensional structure of a chiral cage-shaped molecule with its experimental properties.

Early examples of application of the octant rule to the cage-shaped ketone were the determination of the absolute configurations of (+)-**38**[19] and (+)-**46**,[20] both of which exhibited a positive Cotton effect in their CD spectra, and their octant projections are illustrated in **192** and **193**, respectively (Chart 2.3). Snatzke and Werner-Zamojska[86] described that, in both cases of **192** and **193**, the greater part of the "outer rings" lies in the positive region (upper left and lower right octants) in the back octants, and the positive contribution coming from the "outer rings" overrides the negative contribution of the twisted middle ring (the hatched part in **192** and **193**). This interpretation is in agreement with the experimental properties exhibited by all the cage-shaped ketones as well as twistane derivatives.

Projection formulas of the ketones (+)-**28**, (−)-**33**, (−)-**95**, and (−)-**121**, all of which exhibit a positive Cotton effect around 300 nm in their CD spectra, can be illustrated as shown in **194**, **195**, **196**, and **197**, respectively, and the absolute configurations of these ketones were unambiguously confirmed by a chemical correlation or an X-ray analysis.

An inspection of Figure 2.1 will show a regularity in $[M]_D$ values for the hydrocarbons possessing the twisted bicyclo[2.2.2]octane moiety as a central core. First of all, the most conspicuous feature is the levorotation exhibited by every molecule having the bicyclo[2.2.2]octane moiety of M helicity; the regularity is held on the condition that the chirality of the twisted central core is frozen with short diagonal bridges.

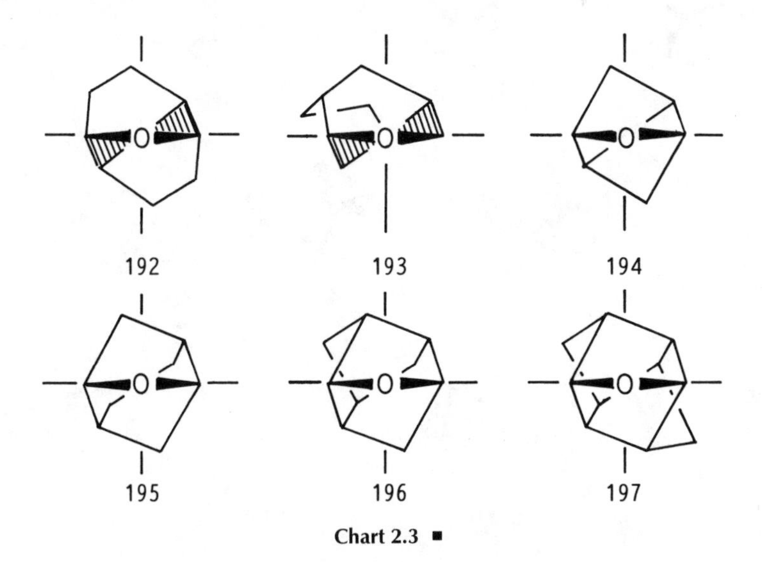

Chart 2.3 ■

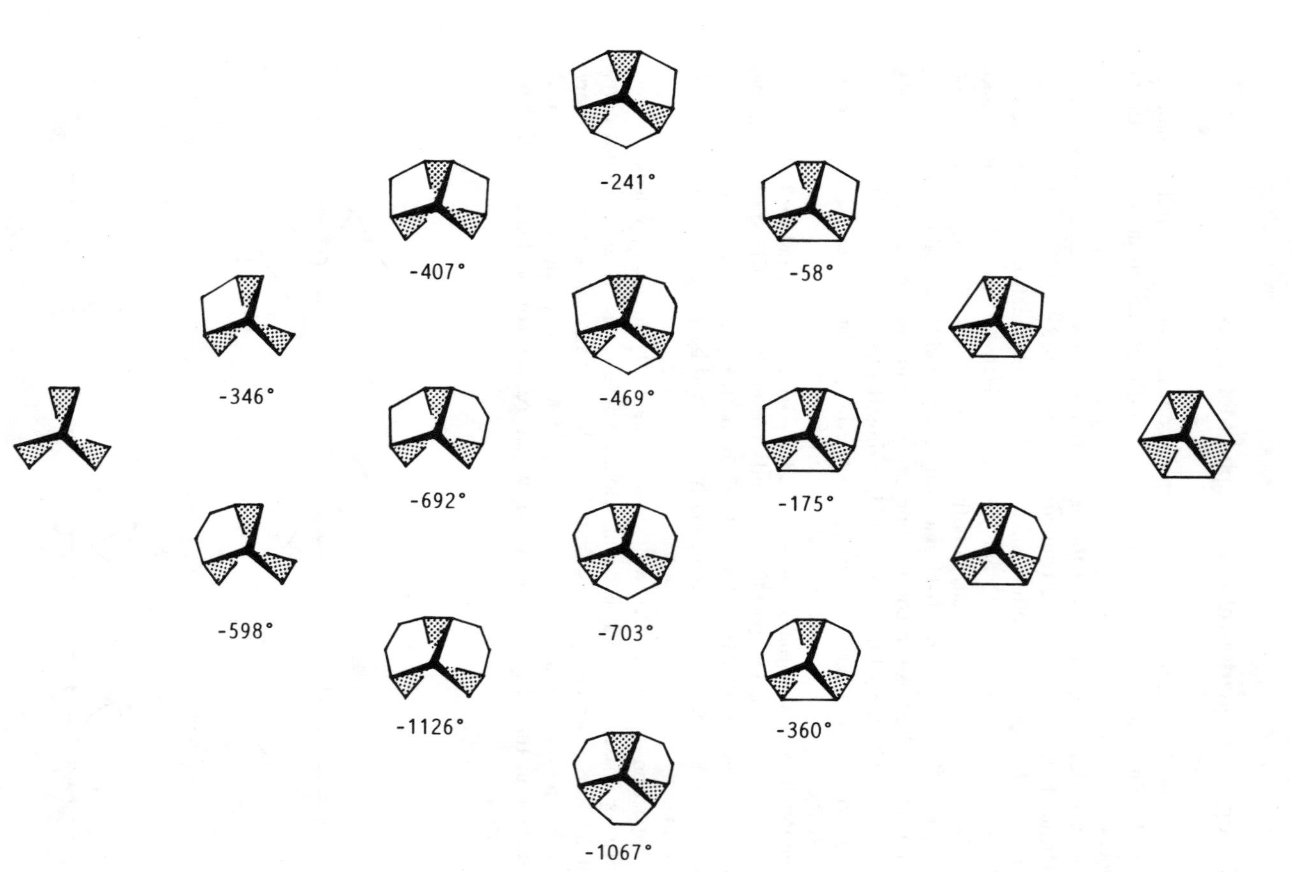

Figure 2.1 ■ Absolute configurations and absolute molecular rotations of the levorotatory tricyclic-, tetracyclic, and pentacyclic hydrocarbons having D_3-twisted bicyclo[2.2.2]octane molecular framework with M helicity.

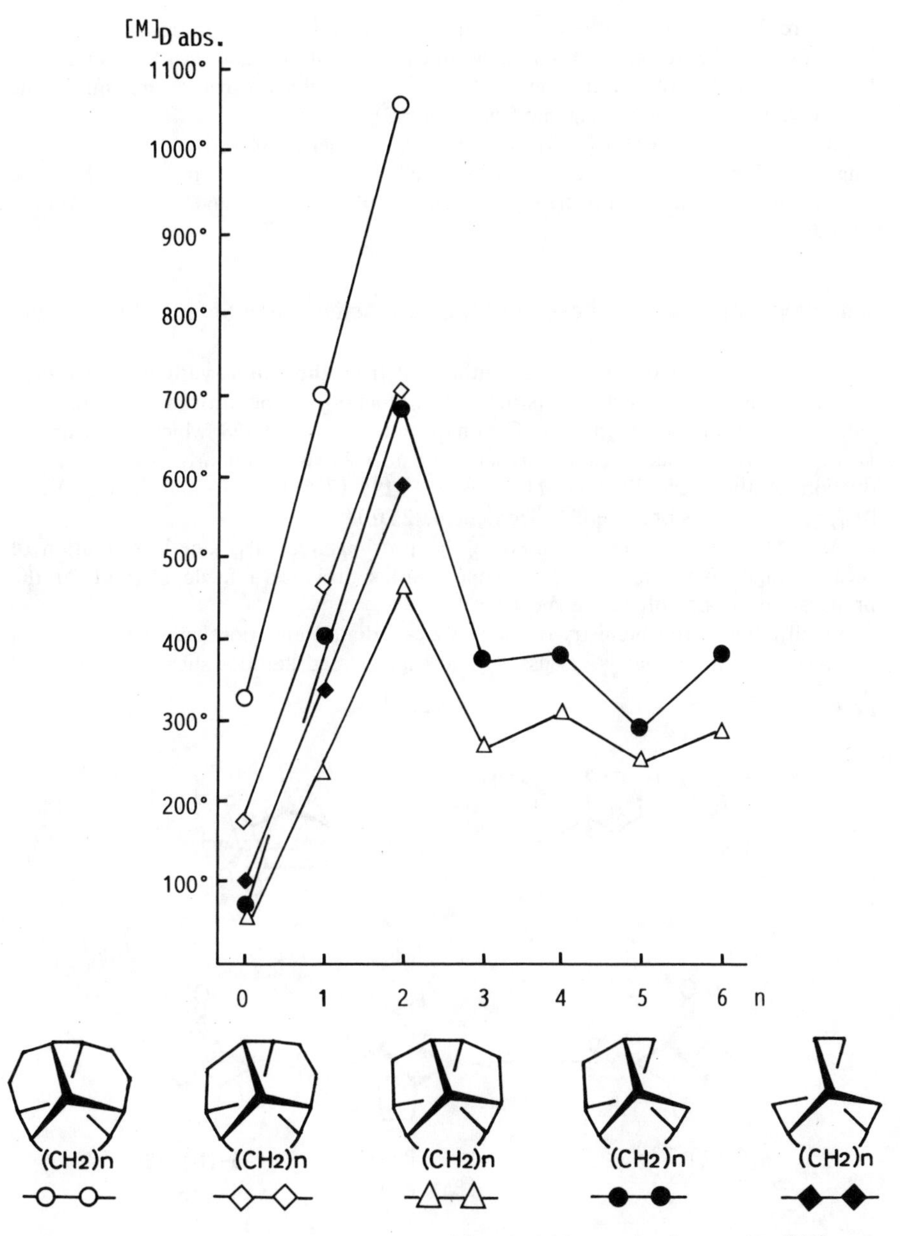

Figure 2.2 ■ Correlation between bridge span (n) and absolute molecular rotation ($[M]_{\mathrm{D\,abs}}$).

Figure 2.1 also shows a systematic increase in $[M]_D$ values on going toward the bottom of the figure, and a remarkably linear correlation between $[M]_D$ value and bridge span (n) is observed (Figure 2.2).[36,61,87] But, the linearity is missing in the hydrocarbons with a long diagonal bridge ($n \geq 3$).

Significant increases in $[M]_D$ value have also been observed accompanying the single bond rupture and the most dramatic example is shown by an $[M]_D$ value increase of 766° observed in hydrogenolysis of **105**, $[M]_D = -360°$, to **107**, $[M]_D = -1126°$.

2.8. Biocatalysts in Syntheses of Optically Active Cage-Shaped Molecules

Some cage-shaped molecules were synthesized from the chiral synthon prepared by biocatalyst-mediated reaction. Asymmetric reduction of the prochiral diketone **44** with HLADH and co-enzyme (NADH) gave optically active **198**, which, according to Deslongchamp's procedure, was transformed to (+)-**46**.[88] Similarly, asymmetric reduction of the meso-diketone **111** gave (−)-**199** (73% ee), from which (+)-**120**, $[\alpha]_D = +108°$, was prepared.[89] (See Scheme 2.20.)

As a few results were discussed in the previous sections, kinetic resolution of racemic cage-shaped molecules with biocatalysts provides a facile method for the preparation of optically active species.[36,61,90]

Finally, the stereochemistry of biocatalyst-mediated enantioselective reduction of the cage-shaped ketones is discussed. The cage-shaped ketones such as **33**, **57**, and

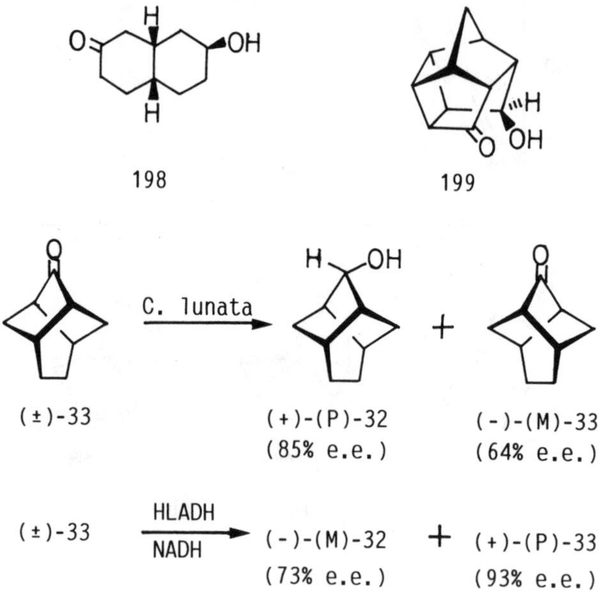

198 199

(±)-33 —C. lunata→ (+)-(P)-32 (85% e.e.) + (−)-(M)-33 (64% e.e.)

(±)-33 —HLADH / NADH→ (−)-(M)-32 (73% e.e.) + (+)-(P)-33 (93% e.e.)

Scheme 2.20 ■

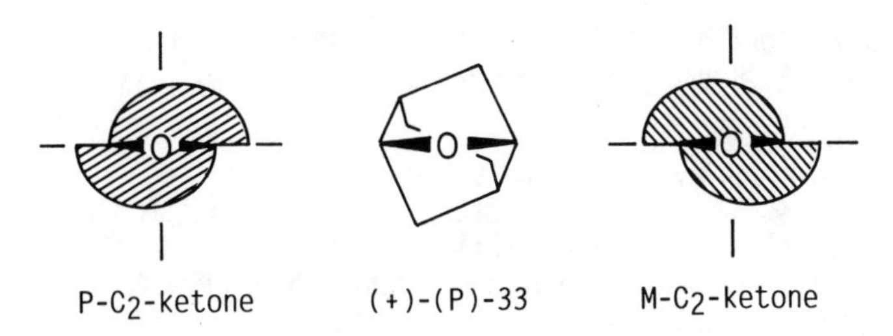

Figure 2.3 ■ The quadrant orientations for the enantiomers of C_2-ketones with *P* and *M* helicity.

121 belong to the C_2 point group and have a C_2 axis coincident with the carbonyl axis. They are conveniently called C_2-ketones and a common stereochemical feature of *P*-C_2-ketones such as (+)-**33** is that they have larger parts of a molecule in upper right and lower left quadrants in their projections (Figure 2.3). In the microbial enantioselective reduction (*Curvularia runata* and *Rhodotorula rubra*), the enantiomers of C_2-ketones with *P*-helicity were preferentially reduced to give *P*-alcohols such as (+)-**32** and C_2-ketones with *M*-helicity were recovered (Scheme 2.20). Nakazaki and co-workers[91] called this stereochemistry the microbial "*P*-C_2 ketone rule."

Incubation of C_2-ketones with HLADH and NADH gave a mixture of *M*-alcohols and *P*-ketones (Scheme 2.20). The demonstrated enantiomer selectivity of HLADH was completely opposite to that found in the microbial process, and the selectivity was called the "*M*-C_2 ketone rule."[92] Nakazaki and co-workers[93] further proposed the "quadrant rule" for biocatalyst-mediated enantioselective reductions of C_1-ketones such as **28**, **46**, and **59**, which have no symmetry element passing through the carbonyl group. This rule of thumb is useful for the determination of the absolute configuration of cage-shaped ketones.

References

1. Farina, M.; Morandi, C. *Tetrahedron* **1974**, *30*, 1819.
2. Nakazaki, M. *Topics Stereochem.* **1983**, *15*, 199.
3. Engler, E. M.; Chang, L.; Schleyer, P. v. R. *Tetrahedron Lett.* **1972**, 2525. Ermer, O.; Dunitz, J. D. *Helv. Chim. Acta* **1969**, *52*, 1861.
4. Butlerow, A. *Ann. Chem.* **1860**, *115*, 322.
5. Askani, R.; Gurang, J.; Schwertfeger, W. *Tetrahedron Lett.* **1975**, 1315. Askani, R.; Schwertfeger, W. *Chem. Ber.* **1977**, *110*, 3046.
6. Jenkins, J. A.; Doehner, R. E., Jr.; Paquette, L. A. *J. Am. Chem. Soc.* **1980**, *102*, 2131.
7. Nakazaki, M.; Naemura, K.; Sugano, Y.; Kataoka, Y. *J. Org. Chem.* **1980**, *45*, 3232.
8. Nickon, A.; Kwasnik, H.; Swartz, T.; Williams, R. O.; DiGiorgio, J. B. *J. Am. Chem. Soc.* **1965**, *87*, 1613.
9. Vogt, B. R. *Tetrahedron Lett.* **1968**, 1579.
10. Sauers, R. R.; Schinski, W.; Mason, M. M. *Tetrahedron Lett.* **1969**, 79. Sauers, R. R.; Whittle, J. A. *J. Org. Chem.* **1969**, *34*, 3579.

11. Berson, J. A.; Ben-Efraim, D. A. *J. Am. Chem. Soc.* **1959**, *81*, 4083. Berson, J. A.; Walia, J. S.; Remanick, A.; Suzuki, S.; Reynolds-Warnoff, P.; Willner, D. *J. Am. Chem. Soc.* **1961**, *83*, 3986.
12. Naemura, K.; Nakazaki, M. *Bull. Chem. Soc. Jpn.* **1973**, *46*, 888.
13. Srinivasan, R. *J. Am. Chem. Soc.* **1960**, *82*, 775.
14. Nakazaki, M.; Naemura, K.; Harita, S. *Bull. Chem. Soc. Jpn.* **1975**, *48*, 1907.
15. Whitlock, H. W., Jr. *J. Am. Chem. Soc.* **1962**, *84*, 3412.
16. Engler, E. M.; Farcasiu, M.; Sevin, A.; Cense, J. M.; Schleyer, P. v. R. *J. Am. Chem. Soc.* **1973**, *95*, 5769.
17. Whitlock, H. W., Jr.; Siefken, M. W. *J. Am. Chem. Soc.* **1968**, *90*, 4929.
18. Gauthier, J.; Deslongchamps, P. *Can. J. Chem.* **1967**, *45*, 297. Belanger, A.; Poupart, J.; Deslongchamps, P. *Tetrahedron Lett.* **1968**, 2127. Belanger, A.; Lambert, Y.; Deslongchamps, P. *Can. J. Chem.* **1969**, *47*, 495.
19. Adachi, K.; Naemura, K.; Nakazaki, M. *Tetrahedron Lett.* **1968**, 5467.
20. Tichý, M.; Sicher, J. *Tetrahedron Lett.* **1969**, 4609. Tichý, M.; Sicher, J. *Collect. Czech. Chem. Commun.* **1972**, *37*, 3106.
21. Moffitt, W.; Woodward, R. B.; Moscowitz, A.; Klyne, W.; Djerassi, C. *J. Am. Chem. Soc.* **1961**, *83*, 4013. Djerassi, C.; Klyne, W. *Proc. Natl. Acad. Sci. U.S.A.* **1962**, *48*, 1093.
22. Tichý, M. *Collect. Czech. Chem. Commun.* **1974**, *39*, 2673.
23. Walbarski, H. M.; Baum, M. E.; Yousseff, A. A. *J. Am. Chem. Soc.* **1961**, *83*, 988. Berson, J. A.; Willner, D. *J. Am. Chem. Soc.* **1962**, *84*, 675. Berson, J. A.; Luibrand, R. T.; Kundu, N. G.; Morris, D. G. *J. Am. Chem. Soc.* **1971**, *93*, 3075.
24. Capraro, H.-G.; Ganter, C. *Helv. Chim. Acta* **1976**, *59*, 97. Capraro, H.-G.; Ganter, C. *Helv. Chim. Acta* **1980**, *63*, 1347.
25. Brewster, J. H. *Tetrahedron Lett.* **1972**, 4355.
26. Brewster, J. H. *Topics Stereochem.* **1967**, *2*, 1.
27. Nakazaki, M.; Naemura, K.; Nakahara, S. *J. Org. Chem.* **1978**, *43*, 4745.
28. Schleyer, P. v. R.; Wiskott, E. *Tetrahedron Lett.* **1967**, 2845.
29. Nickon, A.; Kwasnik, H. R.; Mathew, C. T.; Swartz, T. D.; Williams, R. O.; DiGiorgio, J. B. *J. Org. Chem.* **1978**, *43*, 3904.
30. Meerwein, H.; Kiel, F.; Klosgen, G.; Schoch, E. *J. Prakt. Chem.* **1922**, *104*, 161.
31. Nakazaki, M.; Naemura, K.; Kadowaki, H. *J. Org. Chem.* **1976**, *41*, 3725. Nakazaki, M.; Naemura, K.; Kadowaki, H. *J. Org. Chem.* **1978**, *43*, 4947.
32. Brieger, G.; Anderson, D. R. *J. Org. Chem.* **1971**, *36*, 242.
33. Chin, C. G.; Cuts, H. W.; Masamune, S. *J. Chem. Soc., Chem. Commun.* **1966**, 880.
34. Sasaki, N. A.; Zunker, P.; Muso, H. *Chem. Ber.* **1973**, *106*, 2992. Muso, H. *Chem. Ber.* **1975**, *108*, 337.
35. (a) Toyne, K. J. *J. Chem. Soc., Perkin Trans. 1* **1976**, 1346. (b) Sasaki, N. A.; Zunker, R.; Musso, H. *Chem. Ber.* **1973**, *106*, 2992. Engler, E. M.; Andose, J. D.; Schleyer, P. v. R. *J. Am. Chem. Soc.* **1973**, *95*, 8005. Ōsawa, E.; Schleyer, P. v. R.; Chang, L. W. K.; Kane, V. V. *Tetrahedron Lett.* **1974**, 4174. Ōsawa, E.; Schleyer, P. v. R.; Chang, L. W. K.; Kane, V. V. *Tetrahedron Lett.* **1974**, 4189. Paquette, L. A.; Ward, J. S.; Boggs, R. A.; Farnhan, W. B. *J. Am. Chem. Soc.* **1975**, *97*, 1101. Paquette, L. A.; Boggs, R. A.; Ward, J. S. *J. Am. Chem. Soc.* **1975**, *97*, 1118. Musso, H. *Chem. Ber.* **1975**, *108*, 337. Ōsawa, E.; Schneider, I.; Toyne, K. J.; Musso, H. *Chem. Ber.* **1986**, *119*, 2350. Stober, R.; Musso, H.; Ōsawa, E. *Tetrahedron* **1986**, *41*, 1757.
36. Porter, R.; Clark, S. *Enzymes in Organic Synthesis*; Pitman: London, 1985. Tramper, J.; van der Plas, H. C.; Linko, P. *Biocatalysts in Organic Synthesis*; Elsevier: Amsterdam, 1985. Schneider, M. P. *Enzymes as Catalysts in Organic Synthesis*; Reidel: Dordrecht, 1986.
37. Naemura, K.; Katoh, T.; Fukunaga, R.; Chikamatsu, H.; Nakazaki, M. *Bull. Chem. Soc. Jpn.* **1985**, *58*, 1407.

38. Graham, W. D.; Schleyer, P. v. R.; Hagaman, E. W.; Wenkert, E. *J. Am. Chem. Soc.* **1973**, *95*, 5785.
39. Hirao, K.; Iwakuma, T.; Taniguchi, M.; Abe, E.; Yonemitsu, O. *J. Chem. Soc., Chem. Commun.* **1974**, 691. Hirao, K.; Iwakuma, T.; Taniguchi, M.; Yonemitsu, O.; Date, T.; Kotera, K. *J. Chem. Soc., Perkin Trans. 1* **1980**, 163.
40. Rothberg, I.; King, J. C.; Kirsch, S.; Skidanow, H. *J. Am. Chem. Soc.* **1970**, *92*, 2570. Rothberg, I.; Fraser, J.; Garnick, R.; King, J. C.; Kirsch, S.; Skidanow, H. *J. Org. Chem.* **1974**, *39*, 870.
41. Nakazaki, M.; Naemura, K.; Arashiba, N. *J. Org. Chem.* **1978**, *43*, 689.
42. Nakazaki, M.; Naemura, K. *J. Org. Chem.* **1977**, *42*, 2985.
43. Chapman, N. B.; Key, J. M.; Toyne, K. J. *J. Org. Chem.* **1970**, *35*, 3860.
44. Nakazaki, M.; Naemura, K.; Kondo, Y.; Nakahara, S.; Hashimoto, M. *J. Org. Chem.* **1980**. *45*, 4440.
45. Eaton, P. E.; Cole, T. W., Jr. *J. Am. Chem. Soc.* **1964**, *86*, 962. Eaton, P. E.; Cole, T. W., Jr. *J. Am. Chem. Soc.* **1964**, *86*, 3157.
46. Dilling, W. L.; Braendlin, H. P.; McBee, E. T. *Tetrahedron* **1967**, *23*, 121, and references cited therein.
47. Schenck, G. O.; Steinmetz, R. *Chem. Ber.* **1963**, *96*, 520.
48. Cookson, R. C.; Hudec, J.; Williams, R. O. *Tetrahedron Lett.* **1960**, 29.
49. Marchand, A. P.; Suri, S. C. *J. Org. Chem.* **1984**, *49*, 2041. Marchand, A. P.; Reddy, D. S. *J. Org. Chem.* **1984**, *49*, 4078.
50. Martin, H.-D.; Pfohler, P. *Angew, Chem., Int. Ed. Engl.* **1978**, *17*, 847.
51. Godleski, S. A.; Schleyer, P. v. R.; Ōsawa, E.; Inamoto, Y.; Fujikura, Y. *J. Org. Chem.* **1976**, *41*, 2596. Kent, G. J.; Godleski, S. A.; Ōsawa, E.; Schleyer, P. v. R. *J. Org. Chem.* **1977**, *42*, 3852.
52. Godleski, S. A.; Schleyer, P. v. R.; Ōsawa, E.; Kent, G. J. *J. Chem. Soc., Chem. Commun.* **1974**, 976.
53. Underwood, G. R.; Ramamorthy, B. *Tetrahedron Lett.* **1970**, 4125.
54. Eaton, P. E.; Hudson, R. A.; Giordano, C. *J. Chem. Soc., Chem. Commun.* **1974**, 978.
55. Smith, E. C.; Barborak, J. C. *J. Org. Chem.* **1976**, *41*, 1433.
56. Helmchen, G.; Staiger, G. *Angew., Chem., Int. Ed. Engl.* **1977**, *16*, 116.
57. Nakazaki, M.; Naemura, K.; Arashiba, N. *J. Org. Chem.* **1978**, *43*, 689.
58. Eaton, P. E.; Leipzig, B. *J. Org. Chem.* **1978**, *43*, 2483.
59. Fessner, W.-D.; Prinzbach, H. *Tetrahedron* **1986**, *42*, 1797.
60. Marchand, A. P.; Sharma, G. V. M.; Annapurna, G. S.; Pedneker, P. R. *J. Org. Chem.* **1987**, *52*, 4784.
61. Marchand, A. P. *Chem. Rev.* **1989**, *89*, 1011.
62. Cookson, R. C.; Crundwell, E.; Hill, R. R.; Hudec, J. *J. Chem. Soc.* **1964**, 3062. Valentine, D.; Turro, N. J.; Hammond, G. S. *J. Am. Chem. Soc.* **1964**, *86*, 5202.
63. Nakazaki, M.; Naemura, K.; Arashiba, N.; Iwasaki, M. *J. Org. Chem.* **1979**, *44*, 2433.
64. Nakazaki, M.; Naemura, K.; Chikamatsu, H.; Iwasaki, M.; Hashimoto, M. *Chem. Lett.* **1980**, 1571. Nakazaki, M.; Naemura, K.; Chikamatsu, H.; Iwasaki, M.; Hashimoto, M. *J. Org. Chem.* **1981**, *46*, 2300.
65. Hirao, K.; Yonemitsu, O. *J. Chem. Soc., Chem. Commun.* **1980**, 423.
66. Ternasky, R. J.; Balogh, D. W.; Paquette, L. A. *J. Am. Chem. Soc.* **1982**, *104*, 4503. Paquette, L. A.; Ternansky, R. J.; Balogh, D. W.; Kentgen, G. *J. Am. Chem. Soc.* **1983**, *105*, 5446.
67. Rauscher, G.; Clark, T.; Poppinger, D.; Schleyer, P. v. R. *Angew. Chem., Int. Ed. Engl.* **1978**, *17*, 276.
68. Maier, G.; Pfreim, S. *Angew. Chem., Int. Ed. Engl.* **1978**, *17*, 519; Maier, G.; Pfreim, S.; Schafer, U.; Matusch, R. *Angew. Chem., Int. Ed. Engl.* **1978**, *17*, 520.

69. Bartell, R. S.; Clippard, F. B., Jr.; Boates, T. L. *Inorg. Chem.* **1970**, *9*, 2436.
70. Hounshell, W. D.; Mislow, K. *Tetrahedron Lett.* **1979**, 1205.
71. Nakazaki, M.; Naemura, K. *J. Chem. Soc., Chem. Commun.* **1980**, 911. Nakazaki, M.; Naemura, K. *J. Org. Chem.* **1981**, *46*, 106.
72. (a) Kent, G. J.; Godleski, S. A.; Ōsawa, E.; Schleyer, P. v. R. *J. Org. Chem.*. **1977**, *42*, 3852. (b) Sorochinskii, A. E.; Alaksandrov, A. M.; Petrenko, A. E.; Kukhar, V. P. *J. Org. Chem. USSR* (*Engl. Transl.*) **1988**, *23*, 1987. Petrenko, A. E.; Aleksandrov, A. M.; Sorochinskii, A. E.; Kukhar, V. P. *J. Org. Chem. USSR* (*Engl. Transl.*) **1988**, *23*, 1988.
73. Nakazaki, M.; Naemura, K.; Hokura, Y. *J. Chem. Soc., Chem. Commun.* **1982**, 1245. Naemura, K.; Hokura, Y.; Nakazaki, M. *Tetrahedron* **1986**, *42*, 1763.
74. Webster, O. W.; Sommer, L. H. *J. Org. Chem.* **1964**, *29*, 3103.
75. Vogt, B. R.; Suter, S. R.; Hoover, R. E. *Tetrahedron Lett.* **1968**, 1609.
76. Freeman, P. K.; Rao, V. N. M. *J. Chem. Soc., Chem. Commun.* **1965**, 511.
77. Sauers R. R.; Kelly, K. W. *J. Org. Chem.* **1970**, *35*, 3286. Kelly, K. W.; Sickles, B. R. *J. Org. Chem.* **1972**, *37*, 537. Sauers, R. R.; Schinski, W.; Mason, M. M.; O'Hara, E.; Byrne, B. *J. Org. Chem.* **1973**, *38*, 642.
78. Nakazaki, M.; Naemura, K.; Harada, H.; Narutaki, H. *J. Org. Chem.* **1982**, *47*, 3470.
79. Kissler, B.; Gleiter, R. *Tetrahedron Lett.* **1985**, *26*, 185. Gleiter, R.; Kissler, B. *Tetrahedron Lett.* **1987**, *28*, 6151. Gleiter, R.; Kissler, B.; Glanter, C. *Angew., Chem. Int. Ed. Engl.* **1987**, *26*, 1252. Gleiter, R.; Sigwart, C.; Kissler, B. *Angew. Chem., Intl. Ed. Engl.* **1989**, *28*, 1526.
80. Nakazaki, M.; Naemura, K.; Arashiba, N. *J. Chem. Soc., Chem. Commun.* **1976**, 678. Nakazaki, M.; Naemura, K.; Arashiba, N. *J. Org. Chem.* **1978**, *43*, 888.
81. Naemura, K.; Komatsu, M.; Chikamatsu, H. *Bull. Chem. Soc. Jpn.* **1986**, *59*, 1265.
82. Park, P.; Paquette, L. A. *J. Org. Chem.* **1980**, *45*, 5378.
83. Sinnott, M. L.; Toresund, H. J. S.; Whiting, M. C. *J. Chem. Soc., Chem. Commun.* **1969**, 1000.
84. Vogt, B. R. *Tetrahedron Lett.* **1968**, 1575.
85. Nakazaki, M.; Naemura, K. *J. Org. Chem.* **1977**, *42*, 4108.
86. Snatzke, M.; Werner-Zamojska, F. *Tetrahedron Lett.* **1972**, 4275.
87. Nakazaki, M.; Naemura, K.; Hashimoto, M. *Bull. Chem. Soc. Jpn.* **1983**, *56*, 2543.
88. Dodds, D. R.; Jones, J. B. *J. Chem. Soc., Chem. Commun.* **1982**, 1080. Nakazaki, M.; Chikamatsu, H.; Taniguchi, M. *Chem. Lett.* **1982**, 1761.
89. Naemura, K.; Fujii, T.; Chikamatsu, H. *Chem. Lett.* **1986**, 923.
90. Naemura, K.; Katoh, T.; Chikamatsu, H.; Nakazaki, M. *Chem. Lett.* **1984**, 1371.
91. Nakazaki, M.; Chikamatsu, H.; Naemura, K.; Nishino, M.; Murakami, H.; Asao, M. *J. Chem. Soc., Chem. Commun.* **1978**, 667. Nakazaki, M.; Chikamatsu, H.; Naemura, K.; Nishino, M.; Murakami, H.; Asao, M. *J. Org. Chem.* **1979**, *44*, 4588.
92. Nakazaki, M.; Chikamatsu, H.; Naemura, K.; Sasaki, T.; Fujii, T. *J. Chem. Soc., Chem. Commun.* **1980**, 626. Nakazaki, M.; Chikamatsu, H.; Naemura, K.; Suzuki, T.; Iwasaki, M.; Sasaki, T.; Fujii, T. *J. Org. Chem.* **1981**, *46*, 2726.
93. Nakazaki, M.; Chikamatsu, H.; Naemura, K.; Hirose, Y.; Shimizu, T.; Asao, M. *J. Chem. Soc., Chem. Commun.* **1978**, 668. Nakazaki, M.; Chikamatsu, H.; Naemura, K.; Asao, M. *J. Org. Chem.* **1980**, *45*, 4432.

3

Postfullerene Organic Chemistry

H. W. Kroto and D. R. M. Walton

University of Sussex, Brighton, U.K.

Contents

3.1. Introduction
3.2. Spontaneous Fullerene Formation
3.3. C_{60} Formation in Flames
3.4. Pentagons among the Hexagons
3.5. The Significance of Barth and Lawton's Corannulene Synthesis
3.6. Routes to Nonplanar sp^2 Networks
3.7. Discussion
Acknowledgments
References

3.1. Introduction

In 1985, during laboratory experiments devised to simulate the conditions prevailing in circumstellar shells of red giant carbon stars, the stable C_{60} molecule was detected.[1] It was proposed that this observation could be rationalized if the molecule were to possess a truncated icosahedral cage structure.[1–3] Fifteen years earlier in 1970, however, the fascinating suggestion that such a species might be stable had been advanced by Osawa and Yoshida,[4,5] who discussed the likelihood that it might have superaromatic properties. The possibility that closed graphitelike hollow cages might be formed was in fact first suggested by Jones in 1966, writing under the pseudonym of Daedalus in the *New Scientist*.[6,7] Jones conjectured that large spheroidal networks might be created by modifying the high-temperature process used in graphite manufacture. The existence of a C_{60} molecule had also been contemplated by Bochvar and

Gal'pern[8] and by Davidson[9] prior to and by Haymet[10] coincidentally with the discovery. Various approaches to synthesis were also probed[11] and very recently in an important development Krätschmer et al.[12] obtained a crystalline material from discharge-processed graphite. Taylor et al.[13] have isolated C_{60} and C_{70} from a similarly processed graphite sample, whereas Diederich and co-workers have obtained results that imply that a rational synthesis may be feasible[14] (*vide infra*).

By analogy with the geodesic properties of the domes devised by Buckminster Fuller,[15] which guided the likely structure assignment, the stable C_{60} species was named Buckminsterfullerene.[1,16] Various general aspects of the discovery have been reviewed[2,3] and, in particular, circumstantial evidence favors the existence of a whole family of closed cages C_n ($n > 20$), with 12 pentagonal and an unlimited number of hexagonal rings[17,18]; the family has been named fullerenes and the very large ones giant fullerenes.[16–18]

That the highly symmetric and novel C_{60} molecule might have formed spontaneously was totally unexpected at the outset. With time, it has become clear that the fullerene structures, if correct (the work of Krätschmer et al.[12] and Taylor et al.[13] has unequivocally confirmed this fact), presents intriguing implications for organic chemistry, as there must be hitherto unrecognized routes leading to it. The discovery thus opens up the possibility that aspects of this chemical phenomenon might be exploited in synthesis. This article considers these possibilities and probes likely avenues leading to new types of nonplanar aromatic compounds.

3.2. Spontaneous Fullerene Formation

When it was realized that closed cages might have been formed during the graphite vaporization experiments, the way in which such an unexpected species could result from carbon clustering in a chaotic plasma was clearly the subject of speculation. The possibility that the flat sheets, traditionally associated with graphite, had somehow re-jigged themselves into the Buckminsterfullerene configuration was considered. However, because carbon particle formation was the main process taking place it seemed that nucleation must be involved in some way, and a novel scheme was developed[19–21] to account for the production of C_{60}; an analogous scheme might account for the generation of soot. Only physical aspects have been discussed in detail hitherto; this paper deals with the chemistry.

The key feature of the nucleation argument is that the morphology of large polycyclic aromatic hydrocarbons (PAHs) can, under certain circumstances, be governed by the properties of radical intermediates. For network creation at high temperature, the structural characteristics of these intermediates, which are different from those of stable PAHs, are locked into the framework of the resulting molecule. These ideas are intrinsic to the nucleation mechanism developed to explain the spontaneous formation of Buckminsterfullerene.[19–21]

Although the nucleation scheme was postulated initially to account for the formation of C_{60} and pure carbon graphitelike microparticles, many features associated with soot particles can also be explained by a closely related mechanism. The scheme was refined to allow for chemical perturbations caused by the presence of H and O atoms during the high-temperature network formation phase.[19–21] This suggestion has

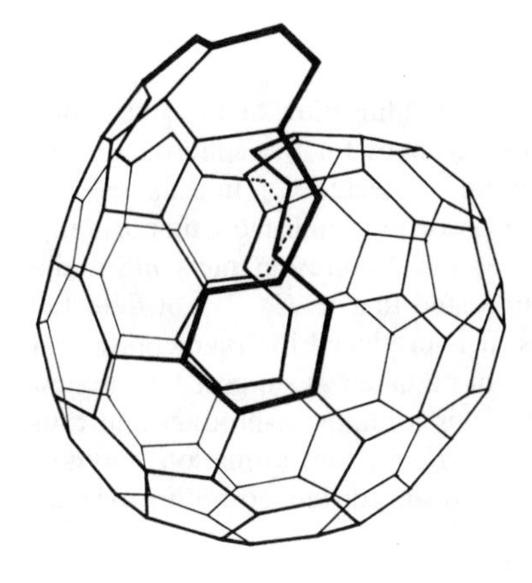

Figure 3.1 ■ Schematic diagram of a typical open-caged carbon cluster.

been criticized by Frenklach and Ebert,[22,23] who maintain that soot particles are formed primarily by condensation of flat PAHs into coagulating liquid drops. The new scheme, however, involves the formation of curved icospiral* PAH embryos that offer a "catalytic" surface for further growth via epitaxial accretion of mainly C_2- and C_4-type species.[24] It can be shown that Frenklach and Ebert's criticism is unsound and that our concept is consistent with many properties of soot: kinetics of soot formation, external shape and internal structure as well as the known chemistry.[21] In fact, the new scheme can explain many more characteristics of soot than can the PAH coagulation theory. During the early phases of soot formation it is known that PAHs form and that the intermediates must be either completely or partially dehydrogenated PAHs, that is, PA(H) radicals of all kinds. As a result of the C_{60} discovery we contend that these species are likely to have curved and/or closing shapes, Figure 3.1.

The fundamental concept is highlighted by a simple example. It is not at all clear that we understand the structural constraints that apply to radical intermediates. During synthetic network growth, one must consider all physicochemical characteristics and, in particular, the geometric configurations of species such as C_nH_m ($n = 2, \ldots, 100, \ldots,\ m = 0, \ldots$). Consequently, it is unwise to extrapolate the thermodynamic and structural properties of all intermediates, C_nH_m ($m = 0, \ldots$) from those of stable PAHs with a given n value at room temperature as has been done previously.[22] For example, the mere fact that $C_{24}H_{12}$, coronene, is flat should not be taken as evidence that all feasible $C_{24}H_n$ ($n = 0, \ldots, 50, \ldots$) intermediates are also flat—many structures are conceivable. Indeed, excited (transition) states often play important roles in reactions and, in this context, it is noteworthy that the excited state of pyridine is nonplanar.[25]

Furthermore, Thomas made a remarkably prescient observation,[26] namely, that the compounds isolated from sooting systems are *not* key intermediates because they are isolable and are therefore too stable to participate further in the growth process.

* Icosahedral spiral shells as discussed elsewhere.[20]

3.3. C_{60} Formation in Flames

Shortly after the publication of our proposal that a modification of the nucleation scheme responsible for C_{60} production might also account for the spheroidal morphology of soot,[19–21] Gerhardt, Löffler, and Homann[27] detected C_{60} in a flame. This observation might be seen as confirmation of the prediction and, more importantly, support for the idea that the icospiral-embryo–epitaxial-accretion mechanism did indeed apply to soot formation. Homann has suggested that as C_{60} is not detected prior to the formation of incipient soot particles, it is produced by fragmentation of existing particles. Thus the occurrence of C_{60} cannot be taken as support for the new proposal.[28] However, C_{60} is *not* an intermediate; it is a minor by-product and thus might not have built up into detectable concentrations until soot formation is almost complete.[21] This interesting question remains to be resolved; indeed both processes may occur.

3.4. Pentagons among the Hexagons

In his original article, Jones discussed the construction of graphite balloons and pointed out Euler's requirement, namely, that 12 pentagons are necessary in order to close a three-connected hexagonal network.[6,7] He also made the interesting suggestion that it might be possible to modify the high-temperature synthesis of graphite by introducing heteroatoms (such as B or P) that could favor pentagonal configurations. The spontaneous production of C_{60}, however, indicates that such doping is unnecessary and this is a particularly interesting aspect of the way in which conventional wisdom misleads us. Our traditional intuition, based on out-of-plane strain concepts, etc., leads us to overlook the fact that a *pure* carbon sheet with 60 atoms is subject to structural and stability constraints that are quite different from those prevailing in the analogous 60 carbon atom "stable" PAH, in which all the edges are capped by C—H bonds. In fact, a pure carbon "flat" C_{60} species constructed entirely from hexagonally linked carbon should have benzynelike short peripheral bonds. The effect of such bonds around the perimeter of an extended polycyclic network should cause the structure to adopt a saucer shape.[29] These factors are identical to those governing embryo production in the basic nucleation scheme (above), which is controlled by the energetics of the cluster–PAH radical intermediates. Indeed it seems fairly clear from recent theoretical studies that the most stable form of a 60-atom carbon cluster is Buckminsterfullerene.[30,31]

Thus we see that the edge defect of a flat sheet will become increasingly "defective" as the sheet area increases. However, if the edge is subsumed into the body of a closed cage, it is integrated in such a way that only 12 defective cusp sites arise, whatever the cluster size!

3.5. The Significance of Barth and Lawton's Corannulene Synthesis

In their synthesis of corannulene, Barth and Lawton[32] found that the final stage (Figure 3.2), which they described as "a rather optimistic experiment" went unexpectedly ("!") with relative ease.

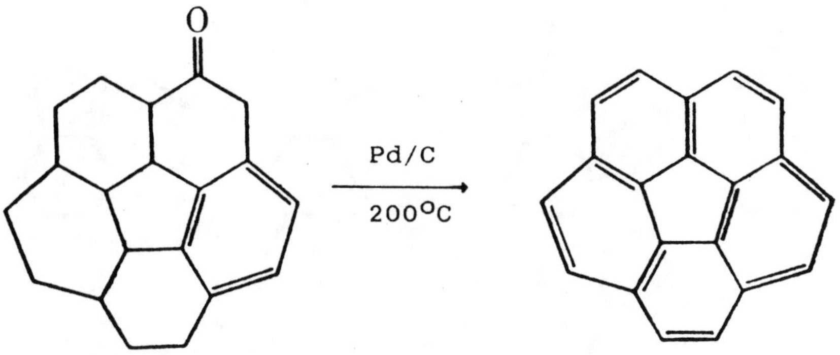

Figure 3.2 ■ Corannulene synthesis: final step.

In retrospect this observation can be recognized as the discovery of a "new round world" of post-Buckminsterfullerene chemistry to complement the pre-Buckminsterfullerene flat one.[27]

3.6. Routes to Nonplanar sp^2 Networks

Although it is now clear that the separation of C_{60} from copious quantities of sootlike material, produced during a carbon discharge, is a viable synthetic option,[12,13] goal-oriented synthesis is still a stimulating challenge for chemists. Indeed, even prior to its discovery, the synthesis of C_{60} was considered. The Barth and Lawton experiments, however, show that, given the right precursors in appropriate circumstances, curved products are preferred relative to planar alternatives. The spontaneous creation of C_{60} indicates that complete closure is also favored!

In our study of the propensity for particle formation during aromatic hydrocarbon thermolysis,[24] we showed that precursors, such as ethylbenzene, which fragment to give (acyclic?) C_4-based units, exhibit the greatest tendency to form spheroidal particles. Thomas has also noted that butadiene is the most efficient sooting agent, a significant result, probably consistent with the known behavior of butadiyne, which decomposes explosively in any appreciable concentration above 4° C to yield carbon black. Because soot consists essentially of PAH aggregates, a rapid sequence of cycloadditions has probably occurred. Thus we arrive at a simple pathway in which C_4 fragments cycloadd to form $(C_4)_n$ polymers containing five- and/or six-membered rings. The first likely steps in such a scheme are shown in Figure 3.3. All possible configurations for cis–trans C_4 unit cycloadditions generating five- and six-membered rings are shown in the scheme, together with analogous configurations for the next step $(C_8 + C_4)$ in the sequence. These rings grow by grafting on additional C_4 units leading eventually to the closed cage fullerenes, C_{60} in particular.

The number of conformational possibilities increases rapidly in Figure 3.3, where one can see that it is relatively simple to construct C_{60} itself and corannulene-like carbon intermediates as postulated in the original nucleation scheme, Figure 3.4.

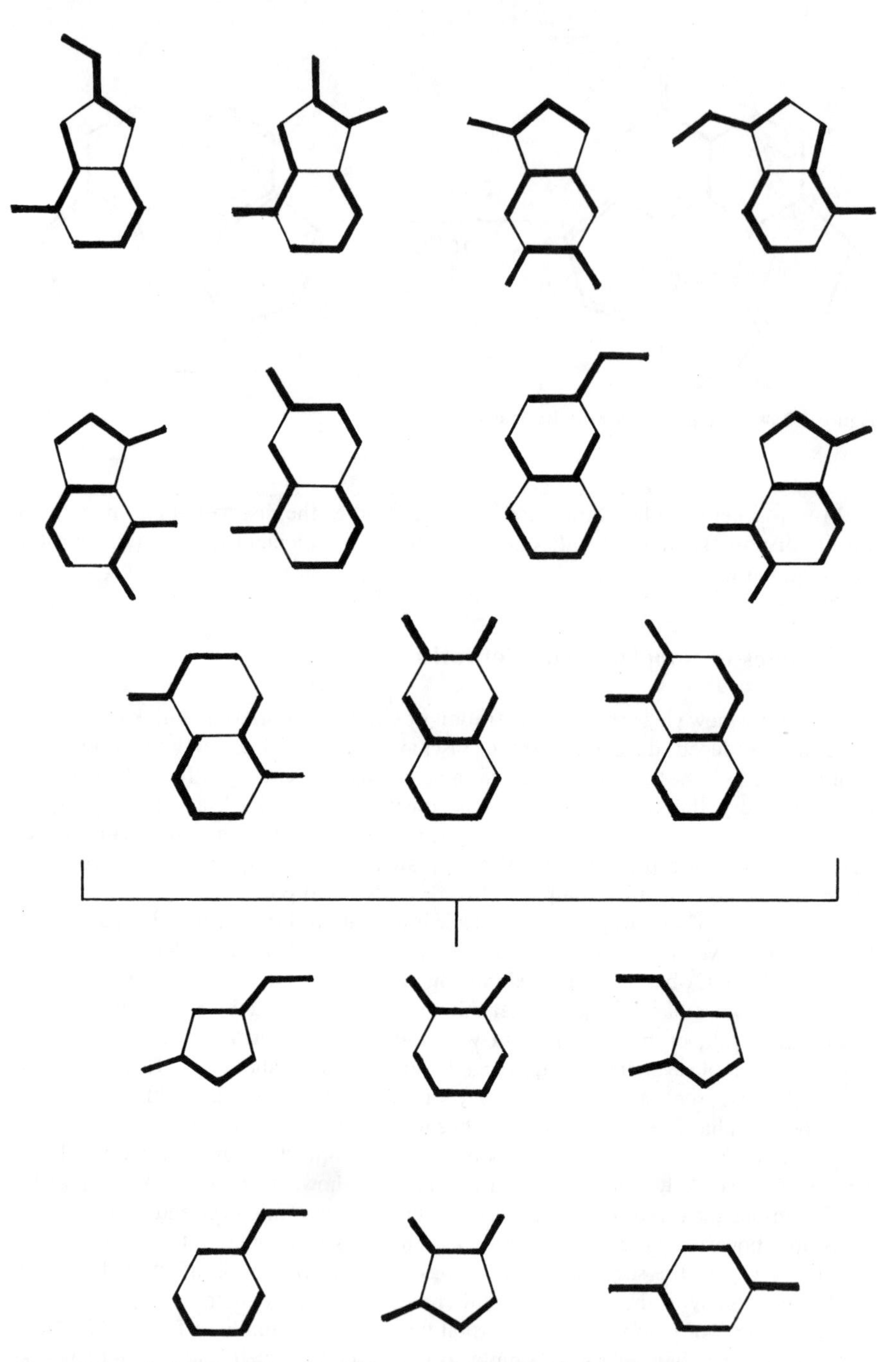

Figure 3.3 ▪ Initial C_4 unit cycloadditions generating nonplanar networks.

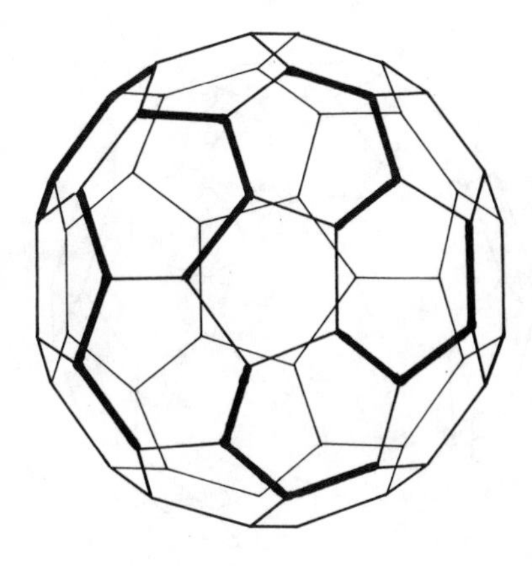

Figure 3.4 ▪ Typical C_4 cycloaddition sequence leading to C_{60}.

This process is almost certainly the key sequence responsible for surface growth and the major contribution to soot mass increase. The copious quantities, in sooting flames, of long-chain polyyne intermediates[28] implies that they also cyclize efficiently. The detection of C_{60} by Gerhardt et al. in such flames suggests that even this special configuration may be accessible. In Figure 3.5 we see that it is relatively straightforward to cross-link a long carbon chain (polyyne–cumulene) helix into such a configuration.

Diederich[14] and co-workers recently observed that laser desorption of a compound, which might be expected to decarbonylate yielding the planar monocyclic species C_{30}, led instead to C_{60} in almost quantitative yield. This result introduces a new (third!) dimension into the argument. Since the time of Hintenberger et al.[34] it has been recognized that species containing up to 30 carbon atoms can form in carbon vapor. Recent studies on these clusters have shown that they form C_nH_2 species ($n = 6, \ldots, 26$),[35,36] a result consistent with chain structures or, perhaps, monocyclic analogs. Indeed, acyclic polyynes with up to 32 carbons (16 conjugated triple bonds) were prepared two decades ago.[37] Diederich's results[14] suggest that there is an

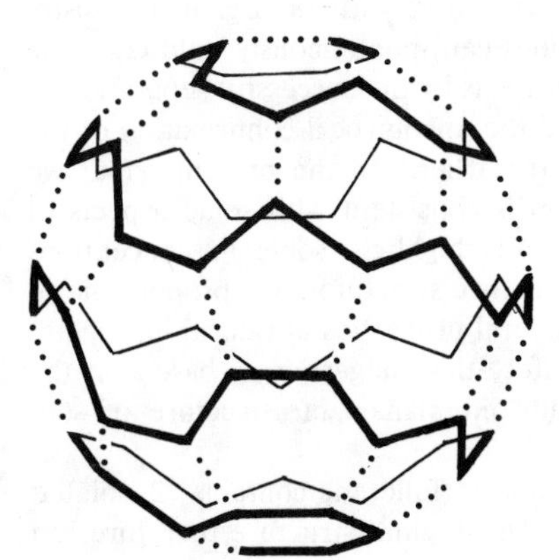

Figure 3.5 ▪ Spirocyclization of a C_{60} acyclic polyyne.

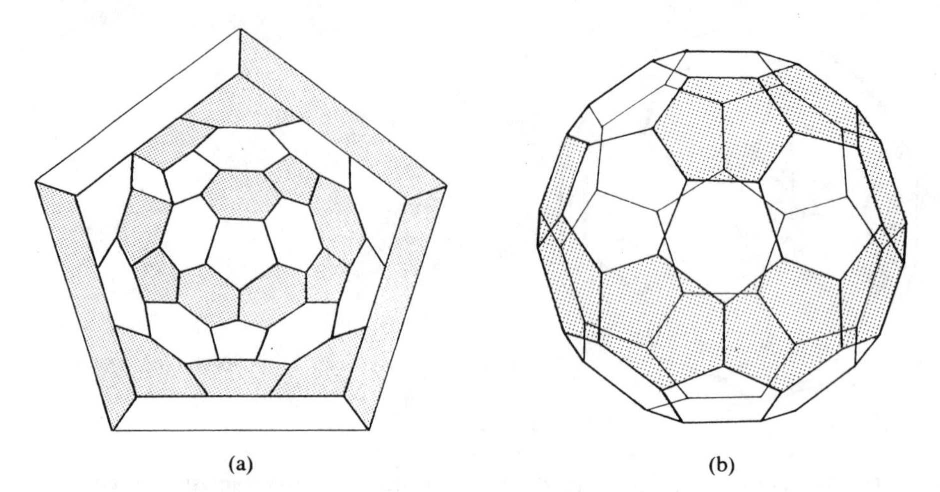

Figure 3.6 ■ Spirodimerization of C_{30} polyyne rings: (a) Schlegel diagram showing both units; (b) perspective diagram highlighting one C_{30} unit.

important route to long sp^2 networks via such chains or their monocyclic equivalents. The results show that a puckering condensation must proceed with extreme efficiency. Such a phenomenon would provide a missing link in the chain, indicating how the embryo might be formed. This may be a second or the primary way in which embryo formation can take place. Whatever the situation, as far as soot is concerned this result is the first major breakthrough in fullerene synthesis. We see in Figure 3.6 that two C_{30} rings can each be puckered to form complementary aromatic helices that can be zipped up to form the C_{60} sphere.

3.7. Discussion

We have drawn attention to the possible implications for synthetic organic chemistry of the unexpected result that carbon condensations can spontaneously yield C_{60}. The phenomenon has been thrown into even sharper focus by the successful separation of Buckminsterfullerene by Krätschmer et al.[12] and the unequivocal confirmation of the hollow closed-cage structure proposed five years earlier.[1] In the present paper we have pointed out that this spontaneous creation is consistent with some aspects of conventional organic chemistry. Indeed the results highlight some misconceptions about PAH morphology based on inadequate reference structures. We propose that it should be possible to take advantage of the structural properties of radical intermediates in order to create nonplanar PAHs. In effect this suggestion is based on the theory advanced to account for the intrinsically nonplanar infrastructure of soot particles.[20]

One significant conceptual point is that Buckminsterfullerene contains 12 isolated pentagons that account for all its carbon atoms. This regular structure therefore can be regarded as being constructed entirely by the fusion of 12 C_5 rings and therefore the presence of the 20 C_6 rings is, in a way, superfluous. It is interesting to speculate that if cyclopentadiene could be completely dehydrogenated, it might spontaneously polymerize to yield Buckminsterfullerene. This idea is most readily appreciated by reference to a soccer ball, in which the black pentagonal elements are all isolated.

Acknowledgments. We are most grateful to Eiji Osawa for encouraging us to write this chapter. We also thank John Cornforth, François Diederich, and Wolfgang Krätschmer for stimulating discussions and private communications.

References

1. Kroto, H. W.; Heath, J.; O'Brien, S. C.; Curl, R. F.; Smalley, R. E. *Nature* **1985**, *318*, 162.
2. Kroto, H. W. *Science* **1988**, *242*, 1139.
3. Curl, R. F.; Smalley, R. E. *Science* **1988**, *242*, 1017.
4. Osawa, E. *Kagaku* **1970**, *25*, 854.
5. Yoshida, Z.; Osawa, E. *Aromaticity*; Kegaku-Dojin: Kyoto, 1971; p. 175.
6. Jones, D. E. H. (Daedalus). *New Scientist* 3 Nov. **1966**, 245.
7. Jones, D. E. H. *The Inventions of Daedalus*; Freeman: Oxford, 1982; pp. 118–119.
8. Bochvar, D. A.; Gal'pern, E. G. *Proc. Acad. Sci. USSR* **1973**, *209*, 239.
9. Davidson, R. A. *Theor. Chim. Acta* **1986**, *58*, 193.
10. Haymet, A. D. J. *J. Am. Chem. Soc.* **1986**, *108*, 319.
11. Chapman, O. L. Private communication.
12. Krätschmer, W.; Lamb, L. D.; Fostiropoulos, K.; Huffman, D. R. *Nature* **1990**, *347*, 354.
13. Taylor, R.; Hare, J. P.; Abdul-Sada, A. K.; Kroto, H. W. *J. Chem. Soc., Chem. Commun.* **1990**, 1423.
14. Rubin, Y.; Kahr, M.; Knobler, C. B.; Diederich, F.; Wilkins, C. L. *J. Am. Chem. Soc.* **1991**, *113*, 495.
15. Buckminster Fuller, R. *Inventions—The Patented Works of Buckminster Fuller*; St. Martin's Press: New York, 1983.
16. Nickon, A.; Silversmith, E. F. *Organic Chemistry—The Name Game: Modern Coined Terms and their Origins*; Pergamon: New York, 1987.
17. Kroto, H. W. *Nature* **1987**, *329*, 529.
18. Kroto, H. W. *Chem. Britain* **1990**, *26*, 40, 45.
19. Zhang, Q. L.; O'Brien, S. C.; Heath, J. R.; Liu, J. R.; Kroto, H. W.; Curl, R. F.; Smalley, R. E. *J. Phys. Chem.* **1986**, *90*, 525.
20. Kroto, H. W.; McKay, K. G. *Nature* **1988**, *331*, 328.
21. Kroto, H. W. *Science* **1988**, *242*, 1139.
22. Frenklach, M.; Ebert, L. B. *J. Phys. Chem.* **1988**, *92*, 561.
23. Ebert, L. B. *Science* **1990**, *247*, 1468.
24. Kroto, H. W.; McKay, K. G.; Walton, D. R. M.; Wood, S. G. Unpublished.
25. Jesson, J. P.; Kroto, H. W.; Ramsay, D. A. *J. Chem. Phys.* **1972**, *56*, 6257.
26. Thomas, A. *Combustion and Flame* **1962**, *6*, 46.
27. Gerhardt, Ph.; Löffler, S.; Homann, K. H. *Chem. Phys. Lett.* **1987**, *137*, 306.
28. Gerhardt, Ph.; Homann, K. H.; Löffler, S.; Wolf, H. *AGARD Conf. Proc. No. 422*, PEP Symposium Chania, Crete, 1987; pp. 22-1–22-10.
29. Almlof, J. In *Carbon in the Galaxy*; Chang, S., Ed.; NASA Conference, 1987.
30. Luthi, H. P.; Almlof, J. *J. Chem. Phys. Lett.* **1987**, *137*, 357.
31. Slanina, Z.; Rudzinski, J. M.; Togashi, M.; Osawa, E. *Thermochim. Acta* **1989**, *140*, 87.

32. Barth, W. E.; Lawton, R. G. *J. Am. Chem. Soc.* **1971**, *93*, 1730.
33. Kroto, H. W. *Pure Appl. Chem.* **1990**, *62*, 407.
34. Hintenberger, V. H.; Franzen, J.; Schuy, K. D. *Z. Naturforsch., Teil A* **1963**, *18*, 1236.
35. Heath, J. R.; Zhang, Q. L.; O'Brien, S. C.; Curl, R. F.; Kroto, H. W.; Smalley, R. E. *J. Am. Chem. Soc.* **1987**, *109*, 359.
36. Kroto, H. W.; Heath, J. R.; O'Brien, S. C.; Curl, R. F.; Smalley, R. E. *Astrophys. J.* **1987**, *214*, 352.
37. Eastmond, R.; Johnson, T. R.; Walton, D. R. M. *Tetrahedron* **1972**, *28*, 4601.

4

A New Look at Natural Products Chemistry in Three Dimensions

Fuyuhiko Matsuda and Haruhisa Shirahama

Hokkaido University, Sapporo, Japan

Contents

4.1. Introduction
4.2. Stereocontrolled Transannular Cyclization of Macrocyclic Compounds
  4.2A. Biogenetic and Biomimetic Transannular Cyclization of Macrocyclic Terpenes
  4.2B. Transannular Diels–Alder Reaction
4.3. Stereoselective Intermolecular Reactions of Macrocycles
  4.3A. Total Synthesis of Periplanone B
  4.3B. Remote Asymmetric Induction Using Macrocyclic Molecules
References

4.1. Introduction

Natural macrocyclic compounds, such as macrocyclic terpenoids or macrolids, have long been of interest in natural product chemistry because of their unique structures as well as their remarkable biological activity. In terpene chemistry, some macrocyclic terpenes are important intermediates of biosynthesis of various polycyclic terpene skeletons, which are formed enzymatically via stereocontrolled transannular cyclizations.[1] In this connection three-dimensional structures of these molecules, especially macrocyclic terpenoids, have been investigated extensively. NMR studies[2] and X-ray crystallographic analyses[3] of macrocyclic compounds have revealed characteristic aspects of macrocyclic molecules mentioned in this chapter. Although macrocyclic compounds are usually capable of existence in a number of stable conformations, only a few of these conformations are low enough in energy to be appreciably populated at normal temperature. This conformational biasing arises from the pronounced ten-

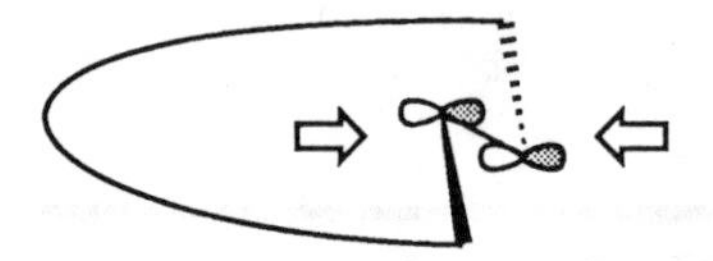

Intramolecular **Intermolecular** Figure 4.1 ▪

dency of these molecules to take conformations in which important transannular nonbonding interactions are minimized. Functionalized macrocycles have stereostructures (which are significantly different from normal rings) in that sp^2 centers tend to stand perpendicular to the plane of the ring to minimize transannular nonbonded repulsion.

Because it is clear from these views that two faces of sp^2 π systems in macrocyclic rings are sterically very different, intermolecular and intramolecular reactions should proceed exclusively from one side (Figure 4.1). In fact, it has often been reported that intramolecular and intermolecular reactions of macrocyclic compounds, especially germacrene- and humulene-type sesquiterpenes, occur in a highly diastereoselective manner.[4] Stereochemical courses of these reactions have been rationalized qualitatively on the basis of the characteristic features of macrocycles.[2c, i; 3c, g; 4e, k, p, s, t; 5] In these explanations, it has been assumed that the reactions take place smoothly without significant change in three-dimensional structures, because conformational similarity between a reactant and the product has been discussed in several papers describing the chemistry of germacrene, humulene, and other medium-sized cycloolefins.[2c, i; 3b, c, g, h; 4e, k, p, s, t; 5]

Quantitative conformational analysis of macrocyclic ring compounds should help to predict the course of the macrocyclic stereocontrolled reactions. Molecular mechanics calculations[6] are pragmatic and are clearly the method of choice for these compounds, to obtain the desired properties of the molecule (e.g., molecular geometry, heat of formation, or steric energy). In fact, recently, molecular mechanics calculations proved to be quite suitable for this purpose, especially in terpene chemistry. In many cases of stereoselective reactions of macrocyclic compounds, molecular mechanics calculations were employed effectively in quantitative prediction or analysis of stereochemistry of the products and stereoselectivity. Furthermore, these macrocyclic stereocontrolled reactions have turned out to be useful for total synthesis of natural products. This chapter reviews the recent trends of these stereoselective reactions of macrocycles in the field of natural product chemistry.

4.2. Stereocontrolled Transannular Cyclization of Macrocyclic Compounds

4.2A. Biogenetic and Biomimetic Transannular Cyclization of Macrocyclic Terpenes

Intramolecular reactions of macrocyclic compounds, such as transannular cyclizations, should prefer to occur to one π lobe of each sp^2 center, being intraannular and unexposed to attack by an external reagent, due to the conformational arrangements for macrocyclic rings shown previously. Stereoselective transannular cyclizations of macrocyclic terpenes such as humulene (**1**) or germacrene (**2**) have attracted much

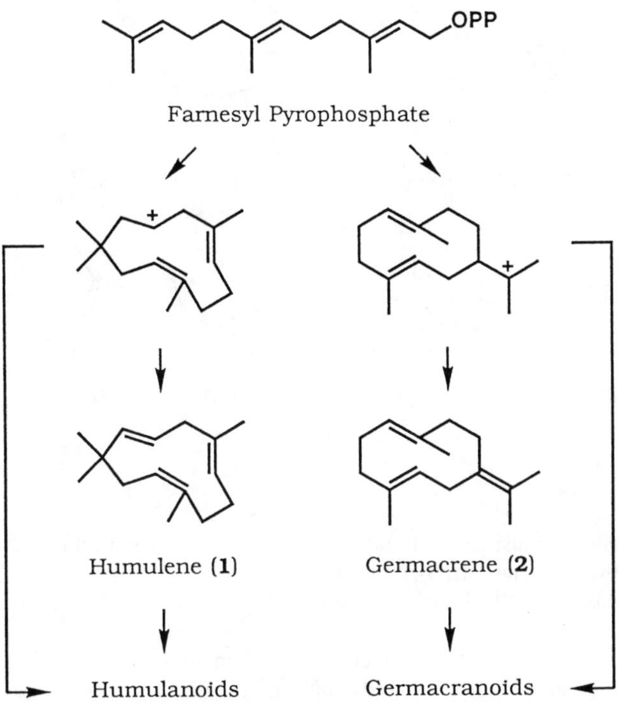

OPP

Farnesyl Pyrophosphate

Humulene (1)

Germacrene (2)

Humulanoids

Germacranoids

Figure 4.2 ■

attention from a viewpoint of studies on biosynthesis of polycyclic terpenoids,[2a–i,3a–j,4a–q] because various terpene frameworks are constructed biogenetically through these reactions.[1] Humulene (**1**) and germacrene (**2**) are fundamental compounds in sesquiterpene biosynthesis. Farnesyl pyrophosphate undergoes head-to-tail intramolecular cyclization to produce humulenic and germacrenic cations, from which versatile polycyclic sesquiterpene skeletons are derived (Figure 4.2).[1] In 1978, application of molecular mechanics calculations to quantitative analysis of transannular cyclizations of macrocyclic terpenes was first performed by Shirahama and co-workers from the biogenetic point of view.[7]

They investigated conformational behavior of humulene (**1**) by means of molecular mechanics calculations in the course of studies on biomimetic synthesis of illudoids.[8] As described previously, in the 11-membered ring of **1** containing three endocyclic trans double bonds, the planes of the latter should be almost perpendicular to the plane of the ring. Thus, a great number of stable conformations is limited to the number of combinations of the directions of the three double bonds. The four stable conformations **1-A**, **1-B**, **1-C**, and **1-D** then can readily be envisaged (Figure 4.3). The conformer **1-A** appears in crystalline silver nitrate complex.[3a] Energy minimizations of the four principal conformations were successfully achieved using the MMI program[9] to give their precise geometries and strain energies. No other stable conformers

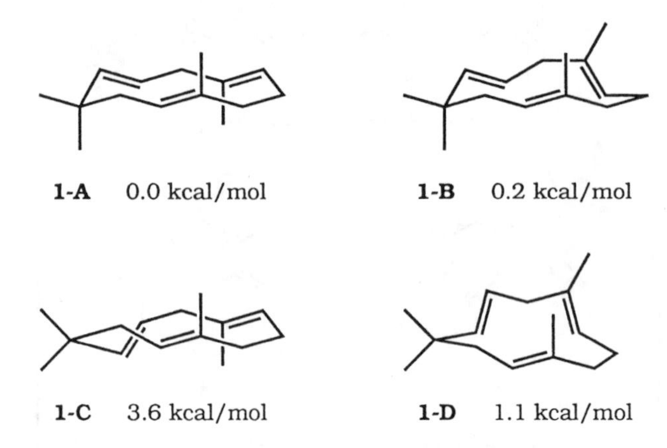

Figure 4.3 ■

appeared during exhaustive examinations of double bond rotation processes. According to the MMI calculations, **1-A** and **1-B** are significantly more stable than others. Thus, in addition to the known conformer **1-A**, a new conformer **1-B** should be equally stable.

This result coincides with the facts observed in biosynthesis of illudoids, hirsutanoids, and other cyclohumulanoids and in humulene chemistry, assuming that conformations of the macrocyclic ring are kept almost unchanged in the course of transannular cyclizations in a similar manner as previously described.[3c; 4e, q, r, s; 5; 10] In illudoids biosynthesis[10, 11] and also transannular cyclization of humulene with Hg(II),[8d, e, h–j] a C(2)—C(9) bond is formed exclusively so as to give a *cis*-bicyclo[6.3.0]undecane system that would originate from **1-A** and **1-B** (Figure 4.4). The conformers **1-C** and **1-D** would have led to a trans bicyclic compound that so far has been found neither in natural products nor in products of transannular reactions of humulene. Furthermore, **1-A** and **1-B** are well suited to the precursors of illudoids and hirsutanoids, respectively. The existence of two separated biosynthetic paths [e.g., **1-A** → protoilludane (a) and **1-B** → hirsutane (b)] can be suggested instead of assuming a single route [e.g., **1-A** → protoilludane → hirsutane (c)]. This supposition is supported by the fact that co-occurrence of illudoids and hirsutanoids has not been recognized and, moreover, by investigations on biosynthesis of illudoids and hirsutanoids performed by Cane and Nachbar[11] using labeled farnesyl pyrophosphate.

Two different types of cyclohumulanoids, africanol (**3**) and bicyclohumulenone (**4**), have been independently isolated from a marine animal, *Lemnalia africana*,[12] and a moss, *Plagiochila acanthophylla* subsp. *japonica*,[13] respectively (Figure 4.5). As described previously, humulene (**1**) is in equilibrium with its stable conformers **1-A** and **1-B**. Stereochemistry of these natural products implies that **3** and **4** can be derived from the conformers **1-A** and **1-B**, respectively. Thus, one explanation is that the two conformers of humulene **1-A** and **1-B** are separately trapped by two different enzymes so as to yield africanol (**3**) on the one hand and bicyclohumulenone (**4**) on the other hand. Most cyclohumulanoids are considered to stem from 9,10-dihydrohumulene-9-yl

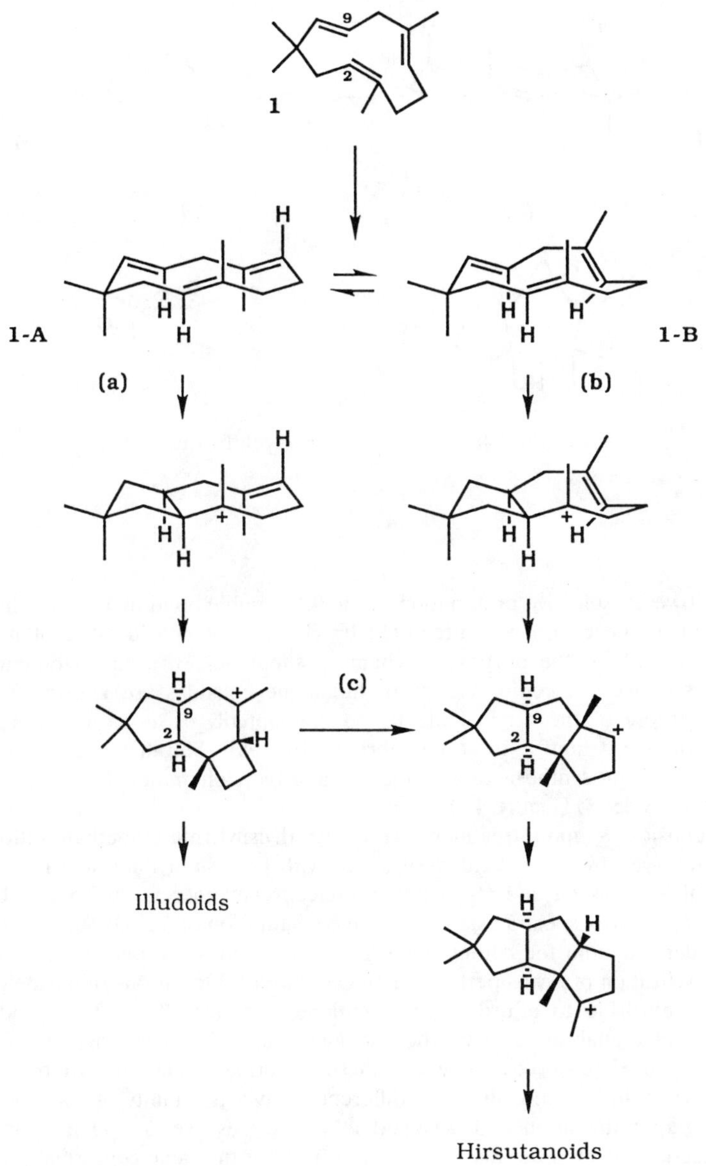

1

1-A

1-B

(a)

(b)

(c)

Illudoids

Hirsutanoids

Figure 4.4 ■

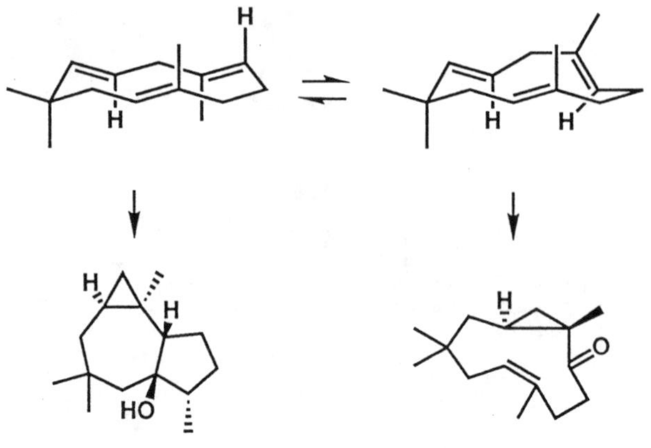

1-A

1-B

Africanol (**3**)

Bicyclohumulenone (**4**)

Figure 4.5 ■

cation. However, selective protonation of the 9,10 double bond of **1** is so difficult that generation of the cation was attempted by cleavage of epoxide ring of humulene 9,10-epoxide (**5**) for the purpose of chemical simulation of such cyclohumulanoids biosynthesis. Because conformations of humulene epoxides are known to be very similar to those of the original olefin,[3b] **5** undoubtedly exists as a mixture of the conformations **5-A** and **5-B** at equilibrium. In 1981, Shirahama and co-workers achieved divergent synthesis of africanol (**3**) and bicyclohumulenone (**4**) from humulene 9,10-epoxide (**5**) (Figure 4.6).[14]

The epoxide **5** upon treatment with trimethylsilyltrifluoromethanesulfonate in toluene followed by subsequent desilylation with potassium fluoride gave rise to a mixture of alcohols **6a** and **6b** in a stereoselective manner in 80% yield. Detailed analysis of ^{1}H NMR spectra established the configuration of **6a** and **6b**, both of which must be derived from the conformer **5-A**. Another conformationally selective transannular cyclization of **5** was performed by exposure of **5** to boron trifluoride etherate in acetic anhydride to afford bicyclohumulene diacetate (**7**) in 70% yield. X-ray crystallographic analysis revealed the configuration of **7**, which originates from the conformer **5-B**. High stereoselectivity of these conformationally selective transannular cyclizations could be rationalized by difference between stability of transition states depending upon the absence (**5** → **6a** and **6b**) or the presence (**5** → **7**) of a nucleophile in the reaction medium because **5** was shown to have the same conformation in both the solvents, namely, acetic anhydride and toluene, by ^{13}C NMR spectral studies. Moreover, MMI calculations on the model compounds corresponding to the possible intermediates **6, 8** and **7, 9** expected at the first stages of the two transannular reactions, respectively, revealed that ring closure reactions take place through pathways leading to the lower strain energy products in each case.[15] Total syntheses of africanol (**3**) and bicyclohumulenone (**4**) have been successfully performed by using each of the transannular cyclization products **6a** and **7**, respectively. Thus, two

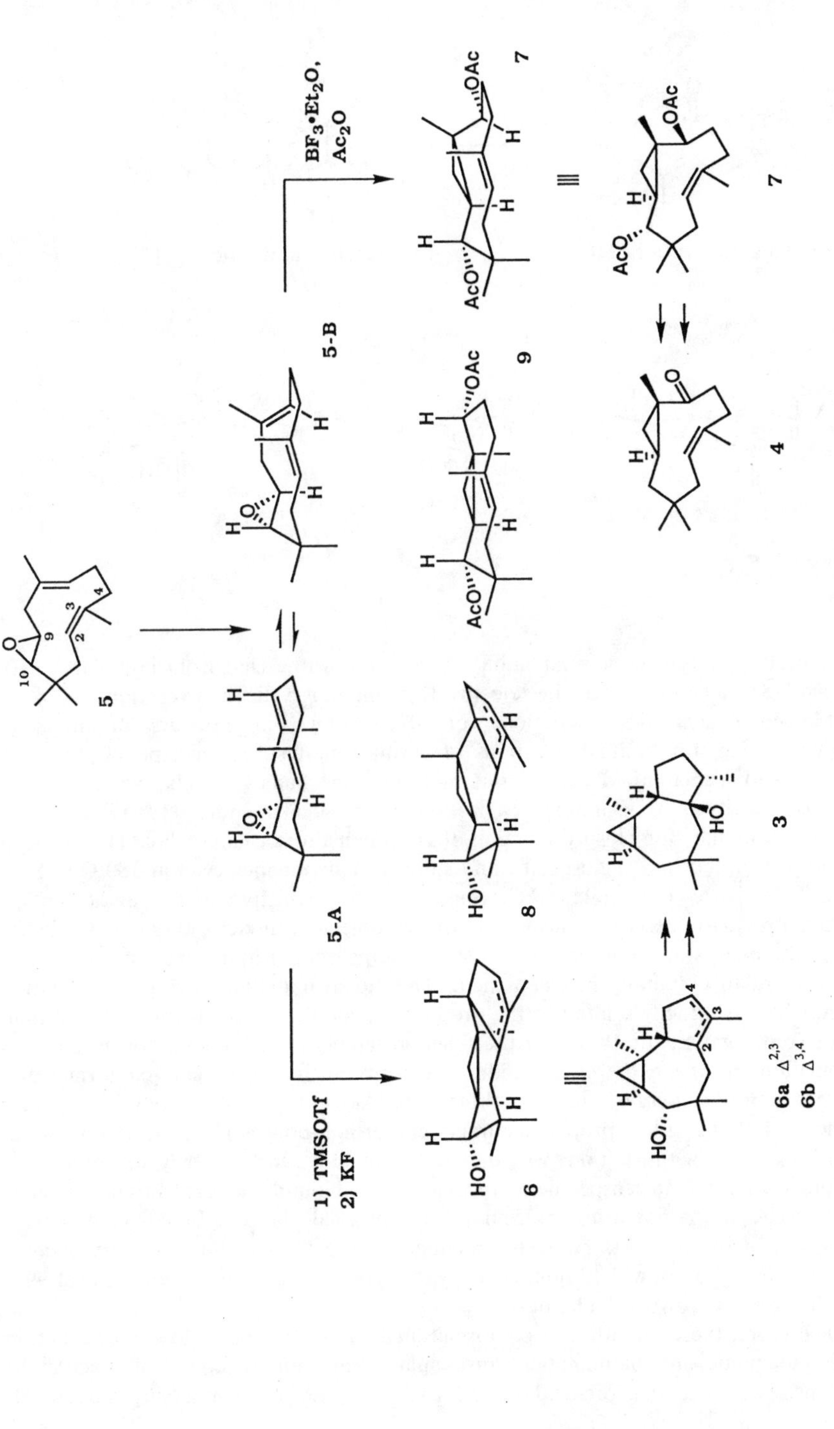

Figure 4.6 ■

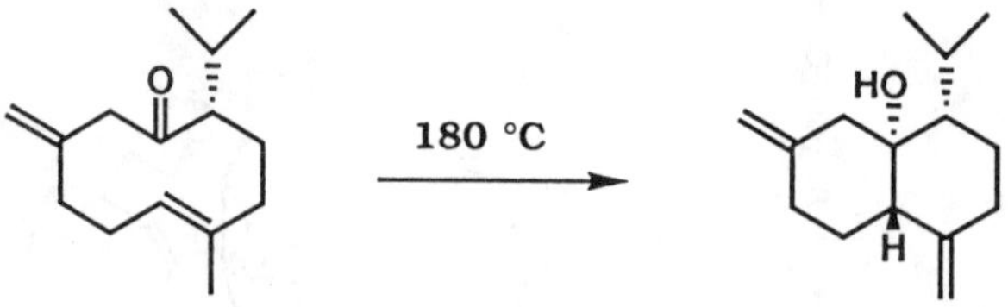

Preisocalamendiol (**10**)

180 °C

Dehydroisocalamendiol (**11**)

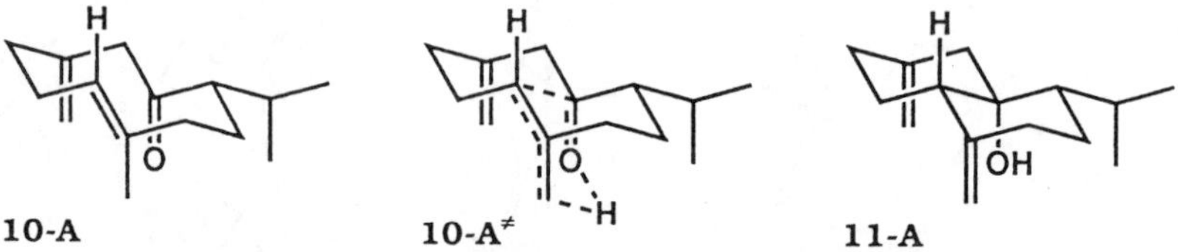

10-A

10-A$^{\neq}$

11-A

Figure 4.7 ■

structurally different cyclohumulanoids have been synthesized from humulene 9,10-epoxide (**5**) via conformationally selective transannular cyclization reactions.

Molecular mechanics calculations were also employed in the course of studies on biogenetic-like transannular reactions of germacrene-type sesquiterpenes. In 1979, Terada and Yamamura[16] carried out molecular mechanics calculations to explain stereoselectivity in transannular cyclization of preisocalamendiol (**10**) (Figure 4.7). The transannular ring closure reaction of **10** to dehydroisocalamendiol (**11**) has been found to proceed in a regiospecific and stereospecific manner even at 180°C in 86% yield.[4k] Similarly to humulene (**1**), considering the principle of perpendicular sp^2 centers previously described, inspection of the molecular model shows that **10** should adopt the eight stable conformations. MMI energy minimizations were performed to estimate relative stability between them. The global minimum-energy conformation **10-A**, which is directly related to the product **11**, is only marginally more stable than other conformations of **10** and at the reaction temperature accounts for 46% of the population. However, the energy difference between the most stable conformer and other conformers increases in a transition state like **10-A**‡, which was optimized under some restrictions with appropriate parameters for half-bonds. The product conformation **11-A** agrees with what one would expect from **10-A**‡. In 1984, they also reported[17] rationalization of stereospecificity recognized in biomimetic cyclizations of some macrocyclic diterpenes using molecular mechanics calculations. In this case, stereochemistry of the products is nicely explained on the basis of the geometry of most stable conformations, which should occupy the ground state to the extent of over 99% according to the MMI calculations.

Therefore, these quantitative conformational studies are likely to confirm that, in both biogenetic and biomimetic transannular cyclization reactions of macrocyclic terpenes, the reactions proceed via least-motion pathways connecting a relatively

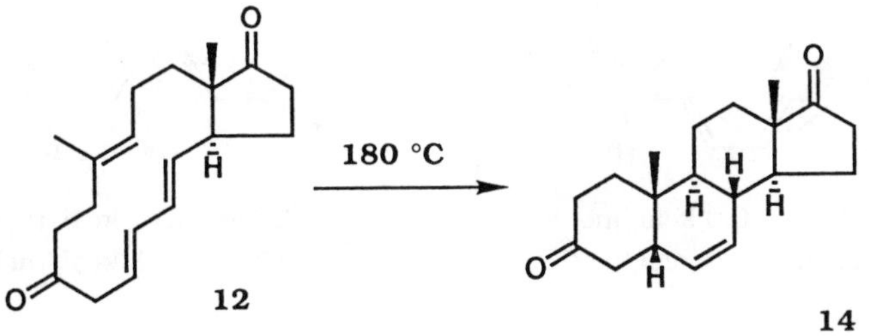

Figure 4.8 ▪

low-energy conformation of starting material with a directly related product conformation without significant alteration of stereostructure throughout the ring closure.

4.2B. Transannular Diels–Alder Reaction

The macrocyclically stereocontrolled transannular cyclization has much potential not only in biogenetic-like cyclization but also in synthetic organic chemistry. In 1988, Takahashi and co-workers[18] reported stereocontrolled construction of the steroid ABC-ring system **14** via transannular Diels–Alder reaction of the 14-membered (*E*,*E*,*E*)-macrocyclic triene **12** (Figure 4.8). They predicted high selectivity of macrocyclic Diels–Alder reaction of **12** and stereochemistry of the adduct **14** by molecular modeling using molecular mechanics calculations.

At first, stable conformations of the (*E*,*E*,*E*)-triene **12** and (*E*,*Z*,*E*)-triene **13** at the ground state were investigated to determine a suitable substrate for macrocyclic Diels–Alder reaction. Because Diels–Alder reaction should take place via conformations in which the diene moiety adopts s-cis form, fixing the geometry of the diene part for s-cis conformation, a macroring was constructed by using the MMRS program.[19] MM2 optimizations[20] on resulting conformations afforded the four minimum-energy conformations **12-A**, **12-B**, **12-C**, and **12-D** (Figure 4.9). Evaluating distance and orbital overlapping between the diene and dienophile parts, **12-A**, **12-C**, and **12-D** turned out to be favorable to Diels–Alder reaction. On the other hand, similar calculations on **13** revealed that **13** takes a large number of minimum-energy conformations and in each conformer the diene and dienophile moieties are so far apart that Diels–Alder reaction could not proceed.

Stereoselectivity of Diels–Alder reaction of **12** were predicted quantitatively by employing an MM2 transition structure model based on ab initio calculations explored by Houk and co-workers.[21] Ab initio quantum mechanical calculations are used to obtain quantitative information about the geometry and energy of the transition state. However, the time requirement is so large, especially when full geometry optimization is included, that molecules containing more than 10 heavy (nonhydrogen) atoms are practically impossible to handle at present. As described later, the MM2 transition structure model is the method of choice in this case. To calculate MM2 transition structures in the Diels–Alder reaction of **12**, synchronous STO-3G transition structure of butadiene–ethylene reaction reported by Brown and Houk[22]

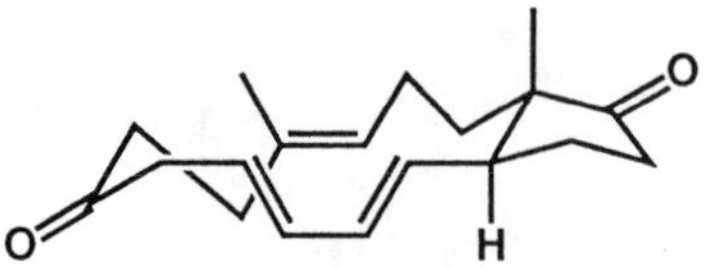

12-A 0.0 kcal/mol
12-A$^{\neq}$ 0.0 kcal/mol

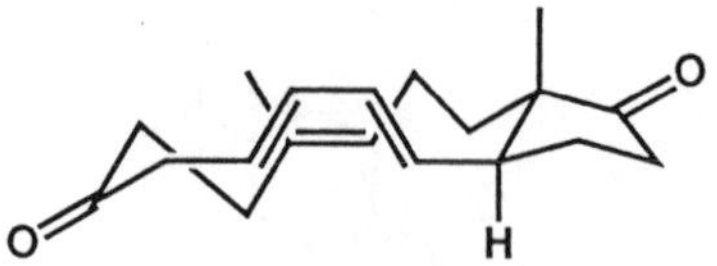

12-B 5.1 kcal/mol
12-B$^{\neq}$ 3.3 kcal/mol

12-C 0.8 kcal/mol
12-C$^{\neq}$ 6.4 kcal/mol

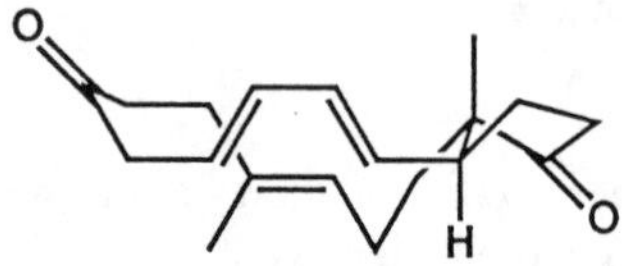

12-D 1.6 kcal/mol
12-D$^{\neq}$ 7.3 kcal/mol

Figure 4.9 ■

was employed as partial geometry of the reactive diene and dienophile in the macrocycle. Substituting appropriate hydrogens in the ab initio transition structure for two carbon chains corresponding to the 14-membered ring part, calculation of initial coordinates by means of the MMRS program followed by MM2 energy minimizations gave rise to precise three-dimensional structures and strain energies of the four transition structures **12-A**‡, **12-B**‡, **12-C**‡, and **12-D**‡, which corresponded to the ground-state conformers, **12-A**, **12-B**, **12-C**, and **12-D**, respectively (Figure 4.9). Among them, **12-A**‡ is extremely more stable than other transition structures and Boltzmann distribution of **12-A**‡ is estimated to be 98% at 180°C. Therefore, 6,(5β)-androstene-3,17-dione (**14**), which is derived directly from **12-A**‡, would be expected to be obtained through the transannular Diels–Alder reaction of **12** in high selectivity.

As forecast, the Diels–Alder reaction of the (E,E,E)-triene **12** was completed within 1 h in xylene at 180°C to give 6,(5β)-androstene-3,17-dione (**14**) in 84% yield in a completely diastereoselective manner (Figure 4.8). ^{1}H NMR spectral data of synthetic **14** were identical with those previously reported.[23] In contrast, the (E,Z,E)-triene **13** gave rise to a complex reaction mixture under the same conditions. Thus, the predictions of product stereochemistry and stereoselectivity based on calculations are in quantitative agreement with the experimental trends. It is remarkable that the three-dimensional structure of the transition state **12-A**‡ is very similar to that of the global energy minimum conformation **12-A**. Hence, in this case too, the transannular reaction also proceeds from the relatively stable conformation of the starting material to the directly related product conformations via a least-motion pathway, in which conformational arrangement is kept almost unchanged all through the reaction.

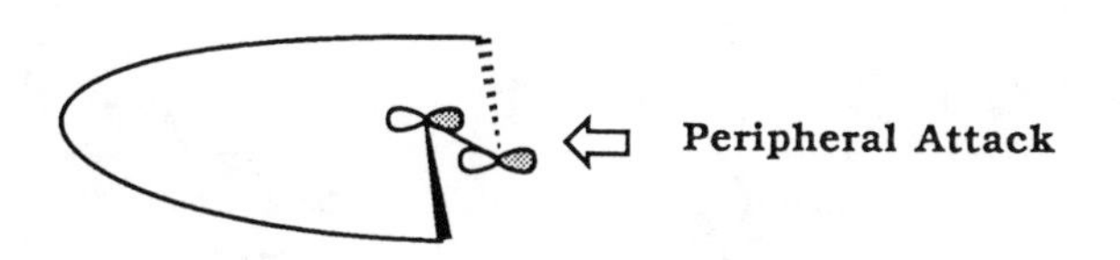

Figure 4.10 ▪

4.3. Stereoselective Intermolecular Reactions of Macrocycles

4.3A. Total Synthesis of Periplanone B

As mentioned previously, macrocyclic compounds exist in a limited number of minimum-energy conformations, in which π orbitals of sp^2 centers are oriented horizontally to the plane of the ring. This situation should strongly favor intermolecular approach of reagents from the less-hindered peripheral face of the π system (Figure 4.10). This principle of peripheral attack of external reagents was first pointed out by Still in the course of his elegant total synthesis of periplanone B (**15**), a sex excitant pheromone of an American cockroach, *Periplaneta americana*.[24]

Periplanone B (**15**), the highly oxygenated germacranoid, has been an attractive synthetic target because of its challenging structural feature and high level of biological activity in addition to the lack of complete assignment of stereochemistry (Figure 4.11).[25] In 1979, Still reported highly stereoselective synthesis of three of the four possible diastereomers of periplanone B and unambiguous establishment of its absolute stereochemistry represented as in **15**. Stereoselective epoxidation of germacrene ring systems were successfully performed according to a general strategy for stereochemical control based on the principle of peripheral attack. Because conformational similarity between a macrocyclic olefin and the corresponding epoxide has been recognized in germacrene[3h] as well as humulene chemistry shown previously, epoxidation reactions would proceed through least-motion pathways without serious change in conformations of the 10-membered rings. Thus, the reaction course of the epoxidation could be predicted qualitatively, considering stable conformations of starting macrocycles (which could be estimated as mentioned later) and the principle of peripheral stereocontrol. The required knowledge of stable conformations of the 10-membered rings, the substrates for epoxidation reactions, could be accessible from various information accumulated in the course of qualitative conformational studies on 1,4- and 1,5-cyclodecadienes, especially germacranoids.[2a–i, 3f–j, 4d–q, 5] This is to say, in the cases of 1,4- and 1,5-cyclodecadiene derivatives, typical lack of serious transannular nonbonded repulsions makes it possible to consider only those conformers composed totally of minimum-energy torsional fragments (A—C—C—B, staggered; A—C—C=B, eclipsed).[26] Consideration of these torsional constraints along with the usual nonbonded interactions for proximate atoms often points to a single conformation with a fair degree of certainty.

Syntheses of three diastereomers, **15**, **16**, and **17** (Figure 4.11), were begun from the 10-membered α,β-unsaturated ketone **18**. Because **18** should adopt the conformation **18-A**, anticipated according to the principle of conformation biasing mentioned

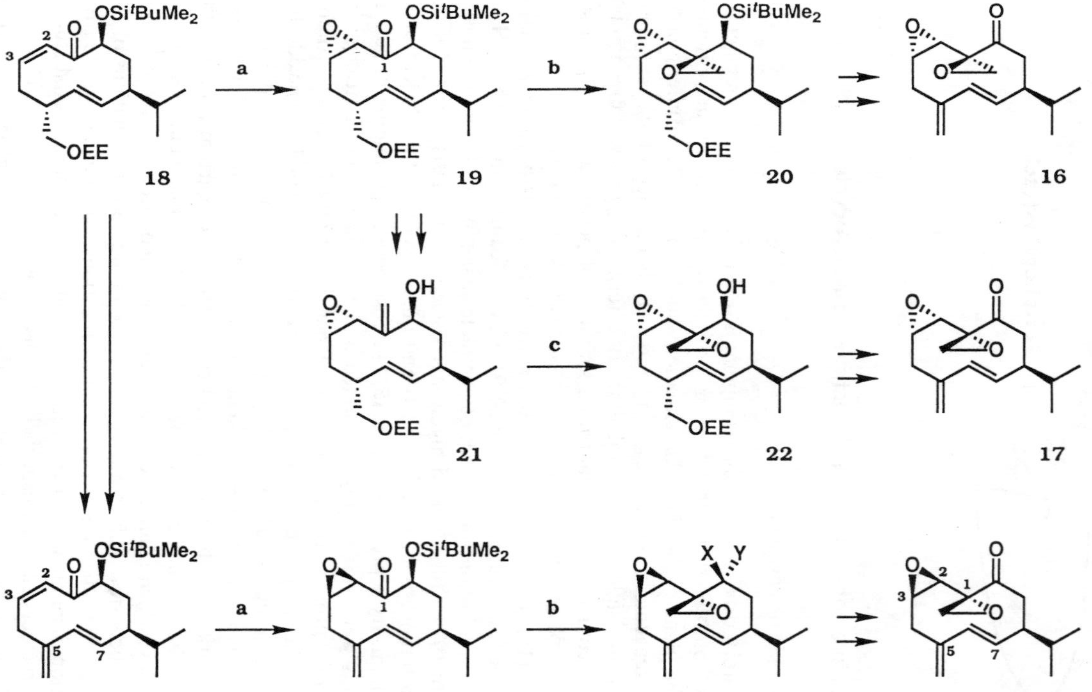

18 19 20 16

21 22 17

23 24 Periplanone B (**15**)

25 X=OSi^{t}BuMe$_2$, Y=H
26 X=OH, Y=H
27 X=H, Y=OBz

a) ^{t}BuO$_2$H, Triton B b) CH_2=SMe$_2$
c) ^{t}BuO$_2$H, VO(acac)

Figure 4.11 ■

EEO OX O 18-A

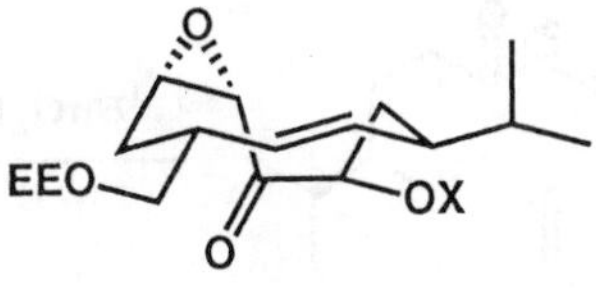

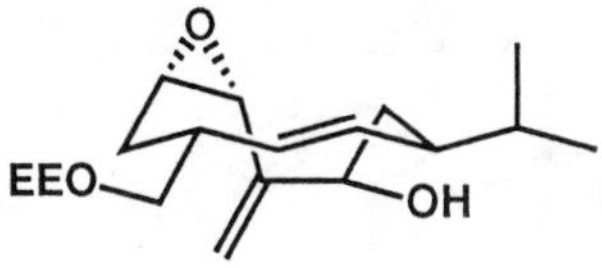

EEO OX O 19-A

EEO OH 21-A

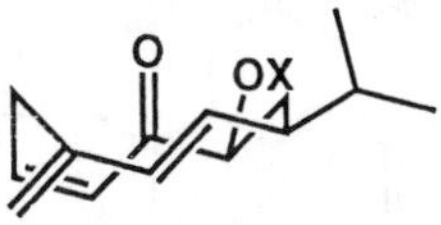

O OX 23-B

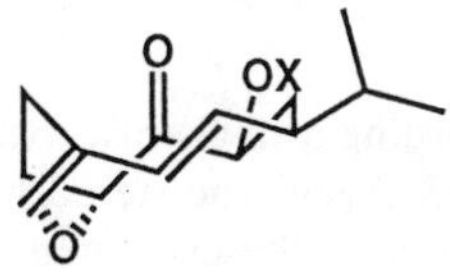

O OX O 24-B

X=Si^{t}BuMe$_2$

Figure 4.12 ■

previously, the α-epoxide **19** would be expected to be produced via peripheral epoxidation (Figure 4.12). In fact, epoxidation of **18** under basic conditions afforded exclusively **19** in 66% yield. The C-1 epimers **20** and **22** were prepared in a stereoselective manner through ketone epoxidation (75%) of **19** with dimethylsulfonium methylide and hydroxyl-directed epoxidation (95%) of the allyl alcohol **21** derived from **19**, respectively. Similarly to the epoxidation of **18**, in each case, stereochemistry of the products as predictable from the anticipated stable conformations **19-A** and **21-A**. The two diastereomers **16** and **17** were synthesized from **20** and **22**, respectively. Spectral comparison of **16** and **17** with authentic periplanone B shows the compounds to be not identical.

Syntheses of other diastereomers required a method for formation of C(2)–C(3) β-epoxide. For this purpose, C(5)–C(7) conjugated diene was constructed before epoxidation to afford the α,β-unsaturated ketone **23**, because inherent preference of 1,3-dienes for s-trans conformation is assumed to be enough to bring the medium ring into a new conformation **23-B**. Because the opposite face of the C(2)–C(3) olefin would be expected for epoxidation reaction in **23-B**, peripheral attack would give the desired β-epoxide **24**. As expected, basic epoxidation of **23** gave rise to a 4 : 1 mixture of epoxy ketones in 74% yield in which the major isomer was **24**. The β-epoxide **24**

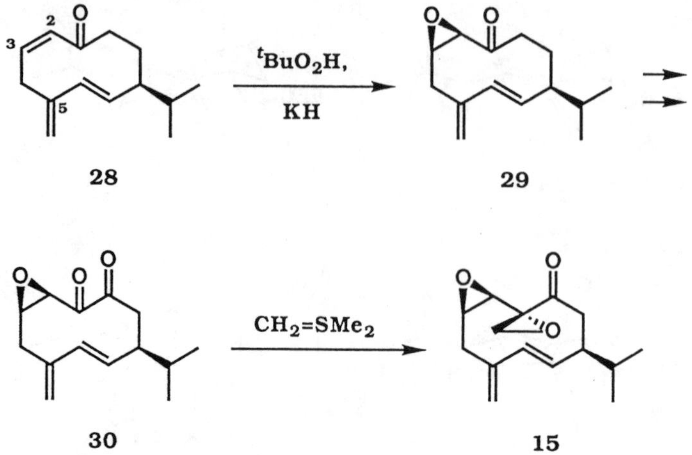

Figure 4.13 ■

was treated with dimethylsulfonium methylide affording a single bisepoxide **25** (69%), which was converted into the third diastereomer **15**. Again, the stereochemistry of **25** follows peripheral attack on the C-1 carbonyl group of the estimated conformation **24-B**. Comparison of **15** by various spectra with natural periplanone B revealed that the two substances were identical. Relative stereochemistry of **15**, **16**, and **17** was largely confirmed by ^{1}H NMR spectra of the synthetic intermediates and X-ray crystallographic study on periplanol B (**26**), prepared from **25**. After resolution by liquid chromatography of the MTPA esters of **26**, absolute configuration of 10-*epi*-periplanol B benzoate (**27**), derived from resolved MTPA ester of **26**, was determined by the exciton chirality method, concluding that stereochemistry including absolute configuration of periplanone B is pictured in **15**.

Schreiber and Santini[27] described total synthesis of periplanone B (**15**) in 1984. In this total synthesis, crucial epoxidation reactions were also subject to similar peripheral stereocontrol (Figure 4.13). Epoxidation of the α,β-unsaturated ketone **28** under basic conditions was found to take place in a stereocontrolled manner to give the desired β-epoxide **29** in a ratio of 4 : 1 in 83% yield. Periplanone B (**15**) was synthesized exclusively via macrocyclic stereocontrolled epoxidation of the α-diketone **30**, prepared from **29**, with dimethylsulfonium methylide (62%).

As shown previously, these qualitative predictions of the stereochemical courses of the macrocyclic epoxidations coincide with the experimental results. In 1986, Takahashi and co-workers[28, 18b] discussed diastereoselectivity of C(2)–C(3) epoxidations quantitatively based on molecular mechanics calculations in the course of their total synthesis of periplanone B (**15**). As described later, MM2 calculations on the model compound **31** suggested that changing C-5 stereochemistry to the β-siloxymethyl (the α,β-unsaturated ketone **32**) from the exocyclic methylene (Still's intermediate **23**, Schreiber's intermediate **28**) could provide a high stereoselectivity (Figure 4.14). Employing the MMRS program followed by MM2 optimization of geometry,

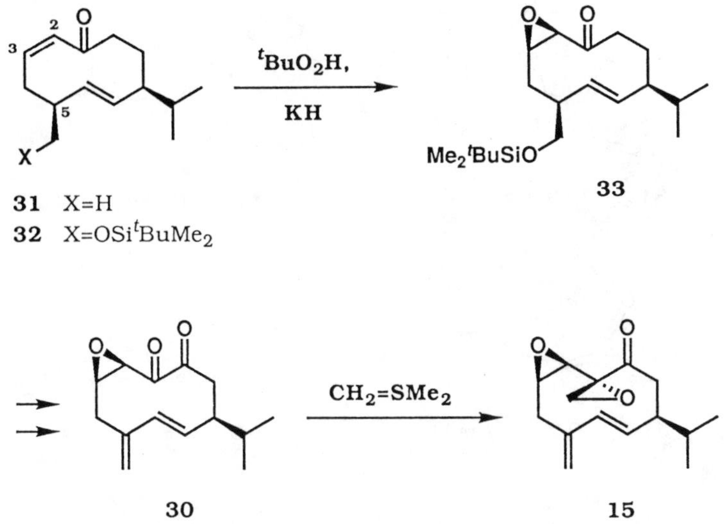

Figure 4.14 ■

the five minimum-energy conformations **31-A**, **31-B**, **31-C**, **31-D**, and **31-E** were constructed (Figure 4.15). A reactant-like transition state is anticipated in peripheral epoxidation, considering three-dimensional resemblance of a macrocyclic epoxide to the parent olefin described previously. Therefore, the conformations **31-A, 31-B, 31-C** and **31-D, 31-E** should lead to the α- and β-epoxides, respectively. According to the calculations, the two low-energy conformations **31-A** and **31-B** accounts for 88% of the population at normal temperature. This conformational distribution of **31** implies that the intermediate **32** should lead to the β-epoxide **33** in a highly selective manner. As expected, epoxidation of **32** occurred with complete selectivity to provide the β-epoxide **33**, which was transformed to periplanone B (**15**) through **30**, Schreiber's intermediate (Figure 4.14). Similar MM2 calculation on **28**, Schreiber's intermediate for C(2)–C(3) epoxidation, showed that population of the minimum-energy conformations derives the predicted diastereomer ratio of 5 : 1 ($\beta : \alpha$). Therefore, the calculated stereoselectivity shows good correspondence with observation.

Considering these results, it may be concluded that macrocyclic stereocontrolled epoxidations take place via least-motion pathways and that product ratios reflect conformational distributions of starting materials in the ground state.

4.3B. Remote Asymmetric Induction Using Macrocyclic Molecules

Based on preference of the peripheral attack of external reagents coupled with expectation of the efficient conformational biasing by ring substituents as observed in the course of total synthesis of periplanone B (**14**), Still and Galynker[29] developed a new type of remote asymmetric induction. In 1981, they reported high diastereoselectivity of several addition reactions on monomethylated 8-, 9-, and 10-membered cyclic

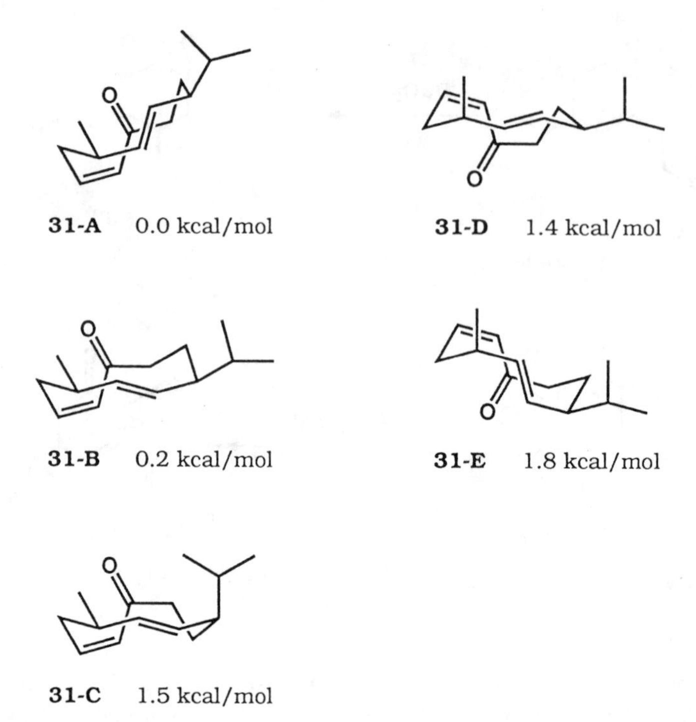

Figure 4.15 ■

ketones and 9- through 12-membered lactones. Kinetic enolate alkylation of the saturated ketones and lactones, and conjugated addition of dimethylcuprate and catalytic hydrogenation to the α,β-unsaturated ketones and lactones were found to occur in a highly stereocontrolled manner, leading to the conclusion that the macrocyclic stereocontrolled reaction is a common feature of these systems.

Computational molecular modeling by using molecular mechanics calculations could predict clearly product stereochemistry and stereoselectivity as shown later. Because most of the reactions are kinetically controlled, quantitative information about three-dimensional structure and energy of the transition state should be required. Because molecular mechanics calculations are not suitable for this purpose, however, a very simple model mentioned later was employed for comparing stereoisomerically related transition states. This is to say, on assumption that, in macrocyclic ring systems, least-motion reactions connecting a starting material conformation with a closely related product conformation without significant change of conformational features are preferred, those pathways leading smoothly from relatively low-energy starting geometries to relatively low-energy product conformations chosen and those reaction pathways either starting from or leading to highly strained conformations of reactants or products are ignored. Considering the principle of peripheral attack and strain energies of conformationally related starting materials and products, calculated

34	1) $^iPr_2NLi$ 2) MeI	36	100 : 0
35	1) $^iPr_2NLi$ 2) MeI	37	98 : 2
38	Me_2CuLi	40	100 : 0
39	H_2, $Rh\text{-}Al_2O_3$	40	91 : 9

Figure 4.16 ■

by using the MM2 program, the product ratio can be forecast on the basis of starting material energies (early transition state) or product energies (late transition state).

For example, kinetic alkylation of the enolates of 2-methylcyclooctanone (**34**) and 3-methylcyclooctanone (**35**) proceeds with high stereoselectivity in over 90%, giving *trans*-2,8-dimethylcyclooctanone (**36**) and *cis*-2,7-dimethylcyclooctanone (**37**), respectively (Figure 4.16). Base equilibration of **36** and **37** to 53 : 47 and 63 : 37 cis : trans diastereomeric mixtures, respectively, verified the kinetic nature of the product distributions. Stereochemistry of **36** and **37** was determined by reduction of **36** to a *dl*-alcohol showing two different doublet methyl groups in ^1H NMR spectrum, and by Baeyer–Villager oxidation of **37** to a nine-membered lactone whose configuration was proved by alternative synthesis from optically active fragments. The molecular modeling rationalized stereoselectivity of these alkylation reactions as described later. MM2 calculations showed stereostructures and strain energies of the four minimum-energy conformations of the enolate of **34** (**34-A**, **34-B**, **34-C**, and **34-D**) along with those of the closely related conformations of the product (**36 36-A**, **36-B**, **36-C**, and **36-D**, respectively) (Figure 4.17). Thus, the relatively low energies of **34-A** and **34-C**, which lead to the relatively low-energy product conformations **36-A** and **36-C**, respectively, could be used to explain the observed stereoselectivity for the trans product. Similarly, geometries and strain energies of the four stable conformations of the enolate of

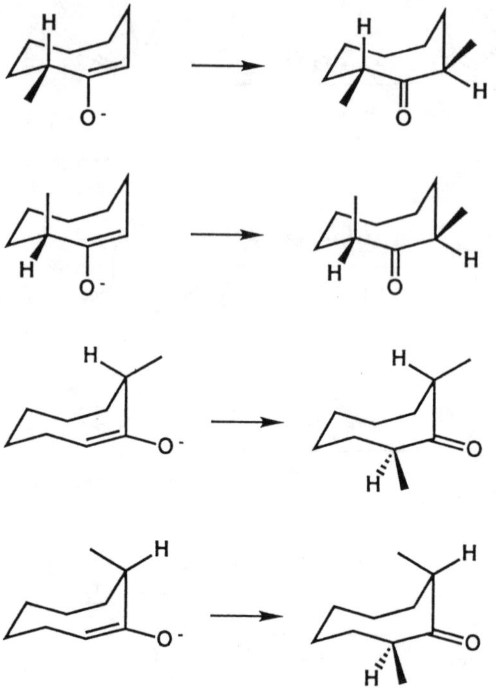

34-A
0.5 kcal/mol

36-A
0.0 kcal/mol

34-B
3.9 kcal/mol

36-B
4.3 kcal/mol

34-C
0.0 kcal/mol

36-C
1.3 kcal/mol

34-D
5.4 kcal/mol

36-D
7.0 kcal/mol

Figure 4.17 ■

35 (**35-A**, **35-B**, **35-C**, and **35-D**) and the corresponding minimum-energy conformation of the product **37** (**37-A**, **37-B**, **37-C**, and **37-D**, respectively) were also obtained by employing the MM2 program (Figure 4.18). Energy differences between the four conformations of the enolate are so small that 98% selectivity for the cis product is not explained based on the starting enolate energies. However, conformations **35-C** and **35-D** are essentially excluded, because the extremely strained product conformations **37-C** and **37-D** are derived from **35-C** and **35-D**, respectively, due to developing transannular repulsion associated with the alkylation. Thus, Boltzmann distribution between **35-A** and **35-B** at −60°C affords the observed ratio exactly. This rationalization may be quite reasonable, because an early reactant-like transition state was anticipated during studies on kinetic enolate methylations.[30] Furthermore, in kinetic alkylation reactions of other macrocyclic compounds, diastereomer ratios calculated on the basis of the enolate strain energies are in good agreement with observations.

High diastereoselectivity was also observed in conjugated addition reaction of 9-methylcyclooct-2-enone (**38**) with dimethylcuprate and hydrogenation of 7-methyl-2-methylenecyclooctanone (**39**), giving *trans*-2,7-dimethylcyclooctanone (**40**) in each case (Figure 4.16). In more large ring systems, these macrocyclic reactions also took place in a highly stereoselective manner. For typical examples, stereoselective reactions of the 9- and 10-membered lactones (**41–46**) are illustrated in Figure 4.19. The

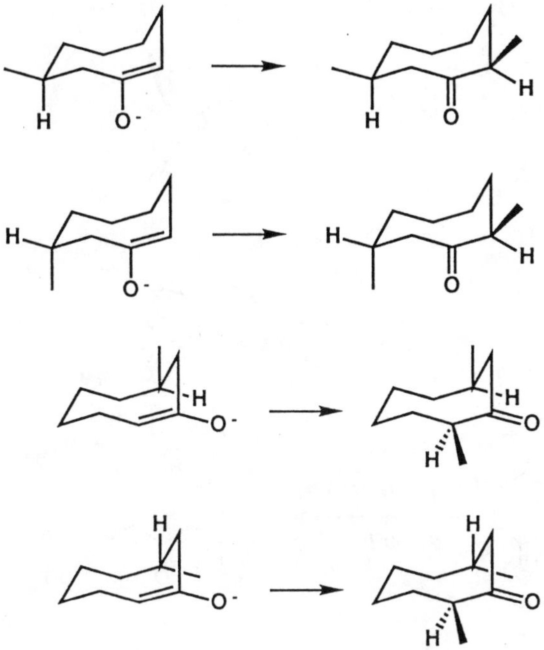

35-A
0.0 kcal/mol

37-A
0.0 kcal/mol

35-B
1.5 kcal/mol

37-B
1.7 kcal/mol

35-C
0.2 kcal/mol

37-C
2.9 kcal/mol

35-D
0.5 kcal/mol

37-D
2.3 kcal/mol

Figure 4.18 ■

molecular modeling also accurately forecast product stereochemistry and stereoselectivity. In dimethylcuprate addition or hydrogenation, conformational distribution of product (late transition state) or starting material (early transition state), respectively, leads to prediction of diastereoselectivity.

This macrocyclically stereocontrolled remote asymmetric induction has been applied effectively to a number of total syntheses of complex natural products.[31, 18a] For example, total synthesis of 3-deoxyrosaranolide (**54**), a 16-member macrolid, which makes extensive use of remote asymmetric induction, was reported by Still and Novack[31c] in 1984 (Figure 4.20). In this total synthesis, stereocontrol at the C-4, C-5, C-6, C-8, C-12, and C-13 asymmetric centers was performed successfully in a highly stereoselective manner by the influence of C-14 and C-15 chiral centers, which existed originally in the macrocyclic 16-membered ester (**53**), the starting material, using the new method for remote asymmetric induction.

As reviewed in this chapter, macrocyclic stereocontrolled reactions have been investigated extensively in the field of natural product chemistry, with successful application of molecular mechanics calculations to quantitative predictions of stereochemical course as a turning point. These quantitative studies seem to establish a general aspect of the macrocyclic reactions. This is to say, both of intramolecular and intermolecular reactions proceed via least-motion pathways connecting a starting material conformation with a closely related product conformation without significant alteration in three-dimensional arrangement in the course of the reactions. Further-

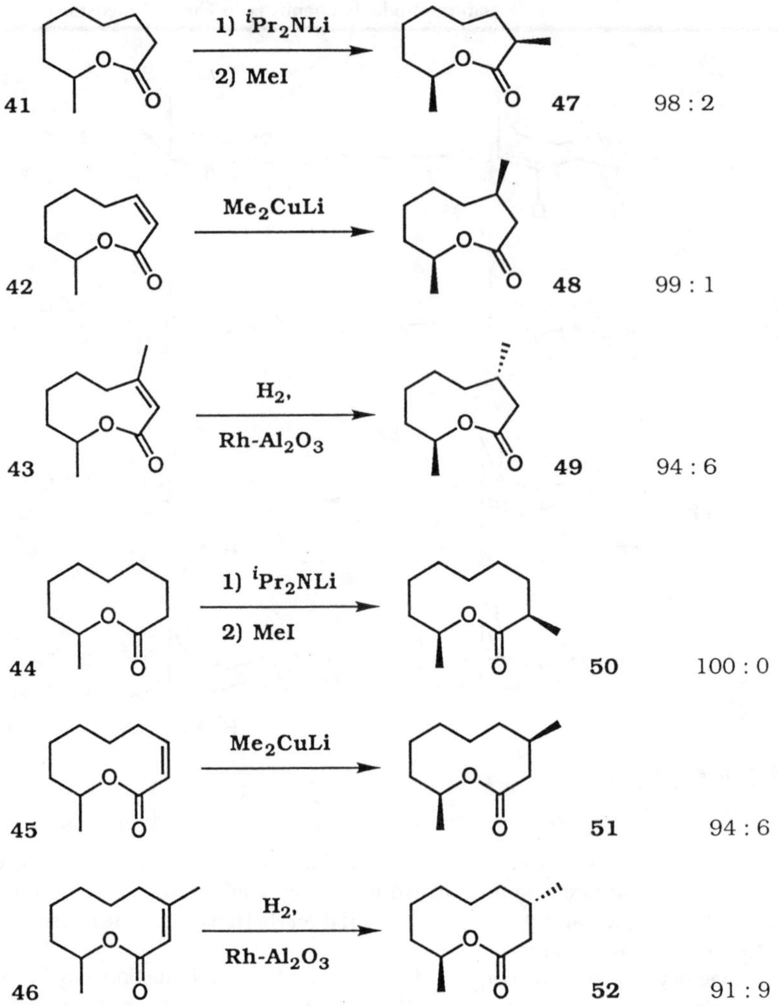

1) $^{i}Pr_2NLi$

2) MeI

41 → 47 98 : 2

Me_2CuLi

42 → 48 99 : 1

H_2,

Rh-Al_2O_3

43 → 49 94 : 6

1) $^{i}Pr_2NLi$

2) MeI

44 → 50 100 : 0

Me_2CuLi

45 → 51 94 : 6

H_2,

Rh-Al_2O_3

46 → 52 91 : 9

Figure 4.19 ■

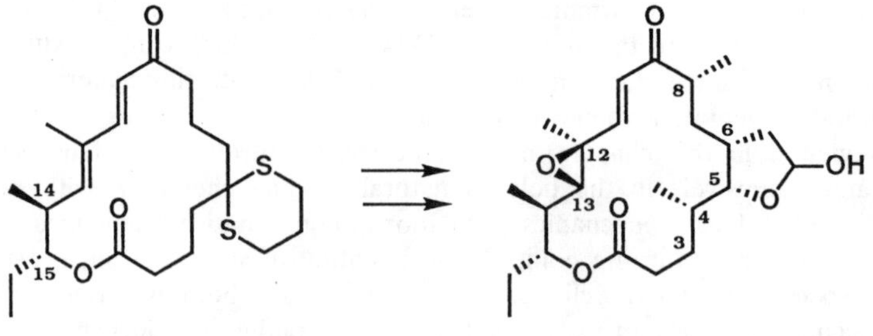

53 → 3-Deoxyrosaranolide (**54**)

Figure 4.20 ■

more, among the possible pathways, those leading smoothly from relatively low-energy conformations of starting materials to relatively low-energy product geometries are preferred. These characteristics seem to be a common feature not only in chemical reactions but also biogenetic transformations of macrocyclic molecules.

References

1. (a) Parker, W.; Roberts, J. S.; Ramage, R. *Quart. Rev., Chem. Soc.* **1967**, *21*, 331. (b) Herout, V. *Aspects of Terpenoid Chemistry and Biochemistry*; Goodwin, T. W., Ed.; Academic Press: New York, 1961; p. 53. (c) Roberts, J. S. *Chemistry of Terpenes and Terpenoids*; Newman, A. A., Ed.; Academic Press: New York, 1972; p. 88.
2. (a) Bhacca, N. S.; Fisher, N. H. *J. Chem. Soc., Chem. Commun.*, **1969**, 68. (b) Hikino, H.; Konno, C.; Takemoto, T.; Tori, K.; Ohtsuru, M.; Horibe, I. *J. Chem. Soc., Chem. Commun.* **1969**, 662. (c) Takeda, K.; Tori, K.; Horibe, I.; Ohtsuru, M.; Minato, H. *J. Chem. Soc. (C)* **1970**, 2697. (d) Tori, K.; Hirobe, I.; Yoshioka, H.; Mabry, T. J. *J. Chem. Soc. (B)* **1971**, 1084. (e) Tori, K.; Horibe, I.; Kuriyama, K.; Tada, H.; Takeda, K. *J. Chem. Soc., Chem. Commun.* **1971**, 1393. (f) Tori, K.; Horibe, I.; Minato, H.; Takeda, K. *Tetrahedron Lett.* **1971**, 4335. (g) Nishimura, K.; Horibe, I.; Tori, K. *Tetrahedron* **1973**, *29*, 271. (h) Horibe, I.; Tori, K.; Takeda, K. *Tetrahedron Lett.* **1973**, 735. (i) Takeda, K. *Tetrahedron* **1974**, *30*, 1525. (j) Borgen, G.; Dale, J. *J. Chem. Soc., Chem. Commun.* **1970**, 1105. (k) Anet, F. A. L.; Wagner, J. J. *J. Am. Chem. Soc.* **1972**, *94*, 9250. (l) Anet, F. A. L.; Cheng, A. K.; Krane, J. *J. Am. Chem. Soc.*, **1973**, *95*, 7877. (m) Anet, F. A. L. *Fortschr. Chem. Forsch.* **1974**, *45*, 169. (n) Anet, F. A. L.; Degen, P. J.; Yavari, I. *J. Org. Chem.* **1978**, *43*, 3021. (o) Anet, F. A. L.; Rawdah, T. N. *J. Am. Chem. Soc.* **1978**, *100*, 7166.
3. (a) McPhail, A. T.; Sim, G. A. *J. Chem. Soc. (B)* **1966**, 112. (b) Cradwich, M. E.; Cradwich, P. D.; Sim, G. A. *J. Chem. Soc., Perkin Trans. 2* **1973**, 404. (c) Allen, F. H.; Rogers, D. *J. Chem. Soc., Chem. Commun.* **1966**, 582. (d) Robertson, J. M.; Todd, G. *J. Chem. Soc.* **1955**, 1254. (e) Rogers, D.; Mazhar-ul-Haqae *Proc. Chem. Soc.* **1963**, 371. (f) Allen, F. H.; Rogers, D. *J. Chem. Soc., Chem. Commun.* **1967**, 588. (g) McClure, R. J.; Sim, G. A.; Coggon, P.; McPhail, A. T. *J. Chem. Soc., Chem. Commun.* **1970**, 128. (h) Messerotti, W.; Pagnoni, U. M.; Trave, R.; Zanasi, R.; Andreetti, G. D.; Bocelli, G.; Sgarabotto, P. *J. Chem. Soc., Perkin Trans. 2* **1978**, 217. (i) Paul, I. C.; Sim, G. A.; Hamor, T. A.; Robertson, J. M. *J. Chem. Soc* **1963**, 5502. (j) Ferguson, G.; Sim, G. A.; Robertson, J. M. *Proc. Chem. Soc.* **1962**, 385. (k) Dunitz, J. D. *Perspectives in Structural Chemistry*, Dunitz, J. D.; Ibers, J. A., Eds.; Wiley: New York, 1968; Vol. 2, p. 1. (l) Groth, P. *Acta Chem. Scand.* **1974**, *28A*, 294.
4. (a) Greenwood, J. M.; Sutherland, J. K.; Torre, A. *J. Chem. Soc., Chem. Commun.* **1965**, 410. (b) Greenwood, J. M.; Solomon, M. D.; Sutherland, J. K.; Torre, A. *J. Chem. Soc. (C)* **1968**, 3004. (c) McKervey, M. A.; Wright, J. R. *J. Chem. Soc., Chem. Commun.* **1970**, 117. (d) Brown, E. D.; Solomon, M. D.; Sutherland, J. K.; Torre, A. *J. Chem. Soc., Chem. Commun.* **1967**, 111. (e) Brown, E. D.; Sutherland, J. K. *J. Chem. Soc., Chem. Commun.* **1968**, 1060. (f) Sam, T. W.; Sutherland, J. K. *J. Chem. Soc., Chem. Commun.* **1971**, 970. (g) Jain, T. C.; McCloskey, J. E. *Tetrahedron Lett.* **1969**, 2917. (h) Jain, T. C.; McCloskey, J. E. *Tetrahedron Lett.* **1969**, 4525. (i) Jain, T. C.; McCloskey, J. E. *Tetrahedron Lett.* **1971**, 1415. (j) Tada, H.; Takeda, K. *J. Chem. Soc., Chem. Commun.* **1971**, 1391. (k) Iguchi, M.; Nishiyama, A.; Yamamura, S.; Hirata, Y. *Tetrahedron Lett.* **1970**, 855. (l) Niwa, M.; Iguchi, M.; Yamamura, S. *Bull. Chem. Soc. Jpn.* **1976**, *49*, 3148. (m) Doskotch, R. W.; Keely, S. L.; Hufford, C. D. *J. Chem. Soc., Chem. Commun.* **1972**, 1137. (n) Takeda, K.; Horibe, I.; Teraoka, M.; Minato, H. *J. Chem. Soc. (C)* **1969**, 1491. (o) Takeda, K.; Horibe, I.; Minato, H. *J. Chem. Soc. (C)* **1970**, 1142. (p) Takeda, K.; Horibe, I.; Minato, H. *J. Chem. Soc. (C)* **1970**, 2704. (q) Takeda, K.; Horibe, I.; Minato, H. *J. Chem. Soc., Perkin Trans. 1* **1973**, 2212. (r) Dowbenko, R.

Tetrahedron **1964**, *20*, 1843. (s) Gipson, R. M.; Guin, H. W.; Simonsen, S. H.; Skinner, C. G.; Shive, W. *J. Am. Chem. Soc.* **1966**, *88*, 5366. (t) Duffin, D.; Sutherland, J. K. *J. Chem. Soc., Chem. Commun.* **1970**, 627. (u) Duffin, D.; Sutherland, J. K. *J. Chem. Soc., Chem. Commun.* **1970**, 626.

5. (a) Allen, F. H.; Brown, E. D.; Rogers, D.; Sutherland, J. K. *J. Chem. Soc., Chem. Commun.* **1967**, 1116. (b) Sutherland, J. K. *Tetrahedron* **1974**, *30*, 1651. (c) Wharton, P. S.; Poon, Y.-C.; Kluender, H. C. *J. Org. Chem.* **1973**, *38*, 735.
6. (a) Allinger, N. L. *Adv. Phys. Org. Chem.* **1979**, *14*, 1. (b) Burkert, U.; Allinger, N. L. *Molecular Mechanics*, ACS Monograph 117; American Chemical Society: Washington, DC, 1982.
7. (a) Shirahama, H.; Osawa, E.; Matsumoto, T. *Tetrahedron Lett.* **1978**, 1987. (b) Shirahama, H.; Osawa, E.; Matsumoto, T. *J. Am. Chem. Soc.* **1980**, *102*, 3208.
8. (a) Ohfune, Y.; Shirahama, H.; Matsumoto, T. *Tetrahedron Lett.* **1975**, 4377. (b) Ohfune, Y.; Shirahama, H.; Matsumoto, T. *Tetrahedron Lett.* **1976**, 2795. (c) Ohfune, Y.; Shirahama, H.; Matsumoto, T. *Tetrahedron Lett.* **1976**, 2869. (d) Misumi, S.; Ohfune, Y.; Furusaki, A.; Shirahama, H.; Matsumoto, T. *Tetrahedron Lett.* **1976**, 2865. (e) Misumi, S.; Ohfune, Y.; Furusaki, A.; Shirahama, H.; Matsumoto, T. *Tetrahedron Lett.* **1977**, 279. (f) Hayano, K.; Ohfune, Y.; Shirahama, H.; Matsumoto, T. *Chem. Lett.* **1978**, 1301. (g) Hayano, K.; Ohfune, Y.; Shirahama, H.; Matsumoto, T. *Tetrahedron Lett.* **1978**, 1991. (h) Misumi, S.; Ohtsuka, T.; Ohfune, Y.; Sigita, K.; Shirahama, H.; Matsumoto, T. *Tetrahedron Lett.* **1979**, 31. (i) Misumi, S.; Ohtsuka, T.; Hashimoto, H.; Ohfune, Y.; Shirahama, H.; Matsumoto, T. *Tetrahedron Lett.* **1979**, 35. (j) Shirahama, H.; Ohfune, Y.; Misumi, S.; Matsumoto, T. *J. Synth. Org. Chem. Jpn.* **1978**, *36*, 569.
9. (a) Allinger, N. L.; Sprague, J. T.; Liljefors, J. *J. Am. Chem. Soc.* **1974**, *96*, 5100; *QCPE* 318. (b) Wertz, D. H.; Allinger, N. L. *Tetrahedron* **1974**, *30*, 1579.
10. Cradwick, P. D.; Sim, G. A. *J. Chem. Soc., Chem. Commun.* **1971**, 431.
11. Cane, D. E.; Nachbar, R. B. *J. Am. Chem. Soc.* **1978**, *100*, 3208; errata **1979**, *101*, 1908.
12. (a) Tursch, B.; Braekman, J. C.; Daloxe, D.; Fritz, P.; Kelecom, A.; Karlsson, R.; Losman, D. *Tetrahedron Lett.* **1974**, 747. (b) Karlsson, R. *Acta Crystallogr., Sect. B* **1976**, *32*, 2709.
13. Matsuo, A.; Nozaki, H.; Nakayama, M.; Kushi, Y.; Hayashi, S.; Komori, T.; Kamijo, N. *J. Chem. Soc., Chem. Commun.* **1979**, 174.
14. Shirahama, H.; Hayano, K.; Kanemoto, Y.; Misumi, S.; Ohtsuka, T.; Hashida, N.; Furusaki, A.; Murata, S.; Noyori, R.; Matsumoto, T. *Tetrahedron Lett.* **1980**, *21*, 4835.
15. Shirahama, H.; Hayano, K.; Ohtsuka, T.; Osawa, E.; Matsumoto, T. *Chem. Lett.* **1981**, 351.
16. Terada, Y.; Yamamura, S. *Tetrahedron Lett.* **1979**, 1623.
17. Shizuri, Y.; Ohtsuka, J.; Kosemura, S.; Terada, Y.; Yamamura, S. *Tetrahedron Lett.* **1984**, *25*, 5547.
18. (a) Takahashi, T.; Shimizu, K.; Doi, T.; Tsuji, J.; Fukazawa, Y. *J. Am. Chem. Soc.* **1988**, *110*, 2674. (b) Takahashi, T.; Doi, T.; Nemoto, H. *J. Synth. Org. Chem. Jpn.* **1989**, *47*, 135.
19. (a) Fukazawa, Y.; Usui, S.; Uchio, Y.; Shiobara, Y.; Kodama, M. *Tetrahedron Lett.* **1986**, *27*, 1825. Compare ref. 31e.
20. (a) Allinger, N. L. *J. Am. Chem. Soc.* **1977**, *99*, 8127; *QCPE* 395. (b) Jaime, C.; Osawa, E. *Tetrahedron* **1983**, *39*, 2769.
21. (a) Houk, K. N.; Paddon-Row, M. N.; Rondan, N. G.; Wu, Y.; Brown, F. K.; Spellmeyer, D. C.; Metz, J. T.; Li, Y.; Loncharich, R. *Science (Washington, D.C.)* **1986**, *231*, 1108. (b) Houk, K. N.; Rondan, N. G.; Wu, Y.; Metz, J. T.; Paddon-Row, M. N. *Tetrahedron* **1984**, *40*, 2257.
22. (a) Brown, F. K.; Houk, K. N. *Tetrahedron Lett.* **1984**, *25*, 4609. (b) Brown, F. K.; Houk, K. N. *Tetrahedron Lett.* **1985**, *26*, 2297.
23. Kirk, D. N.; Leonard, D. R. A. *J. Chem. Soc., Perkin Trans. 1* **1973**, 1836.

24. (a) Still, W. C. *J. Am. Chem. Soc.* **1979**, *101*, 2493. (b) Adams, M. A.; Nakanishi, K.; Still, W. C.; Arnold, E. V.; Clardy, J.; Persoons, C. J. *J. Am. Chem. Soc.* **1979**, *101*, 2495.
25. (a) Persoons, C. J.; Verwiel, P. E. J.; Ritter, F. J.; Talman, E.; Nooijen, P. F. J.; Nooijen, W. J. *Tetrahedron Lett.* **1976**, 2055. (b) Talman, E.; Verwiel, P. E. J.; Ritter, F. J.; Persoons, C. J. *Isr. J. Chem.* **1978**, *17*, 227. (c) Persoons, C. J.; Verwiel, P. E. J.; Talman, E.; Ritter, F. J. *J. Chem. Ecol.* **1979**, *5*, 219.
26. Karabatsos, G. J.; Fenoglio, D. J. *Topics Stereochem.* **1970**, *5*, 167.
27. Schreiber, S. L.; Santini, C. *J. Am. Chem. Soc.* **1984**, *106*, 4038.
28. Takahashi, T.; Kanda, Y.; Nemoto, H.; Kitamura, K.; Tsuji, J.; Fukazawa, Y. *J. Org. Chem.* **1986**, *51*, 3393.
29. Still, W. C.; Galynker, I. *Tetrahedron* **1981**, *37*, 3981.
30. Bare, T. M.; Herskey, N. D.; House, H. O.; Swain, C. G. *J. Org. Chem.* **1972**, *37*, 997.
31. (a) Still, W. C.; Murata, S.; Revial, G.; Yoshihara, K. *J. Am. Chem. Soc.* **1983**, *105*, 625. (b) Still, W. C.; Gennari, C.; Noguez, J. A.; Pearson, D. A. *J. Am. Chem. Soc.* **1984**, *106*, 260. (c) Still, W. C.; Novack, V. J. *J. Am. Chem. Soc.* **1984**, *106*, 1148. (d) Still, W. C.; Romero, A. G. *J. Am. Chem. Soc.* **1986**, *108*, 2105. (e) Still, W. C. *Current Trends in Organic Synthesis*; Nozaki, H., Ed.; Pergamon Press: Oxford, 1983; p. 223. (f) Still, W. C. In *Selectivity—A Goal for Synthetic Efficiency*; Bartmann, W.; Trost, B. M., Eds.; Verlag Chemie: Basel, 1983; p. 263. (g) Schreiber, S. L.; Sammakia, T.; Hulin, B.; Schulte, G. *J. Am. Chem. Soc.* **1986**, *108*, 2106. (h) Takahashi, T.; Nemoto, H.; Kanda, Y.; Tsuji, J.; Fujise, Y. *J. Org. Chem.* **1986**, *51*, 4315. (i) Takahashi, T.; Nemoto, H.; Kanda, Y.; Tsuji, J.; Fukazawa, Y.; Okajima, T. *Tetrahedron* **1987**, *43*, 5499.

5

Propellanes

Yoshito Tobe

Osaka University, Suita, Osaka, Japan

Contents

5.1. Introduction
5.2. Small-Ring Propellanes
  5.2A. Synthesis of Small-Ring Propellanes
    5.2A(1). Intramolecular Displacement
    5.2A(2). Intramolecular Carbene Addition
    5.2A(3). Cycloaddition of Bridgehead Alkenes
  5.2B. Theoretical and Structural Studies of Small-Ring Propellanes
    5.2B(1). [1.1.1]Propellane
    5.2B(2). [2.2.2]Propellane
  5.2C. Reactions of Small-Ring Propellanes
    5.2C(1). [*m*.*n*.1]Propellanes
    5.2C(2). [*m*.2.2]Propellanes
  5.2D. Silapropellanes and Stannapropellanes
5.3. Propellanes as Stereochemical Models
5.4. Synthetic Application of Propellanes
  5.4A. [1.1.1]Propellane
  5.4B. [*m*.*n*.1]Propellanes
  5.4C. [*m*.*n*.2]Propellanes
  5.4D. [*m*.3.3]Propellanes
5.5. Naturally Occurring Propellanes
References

5.1. Introduction

The tricyclic molecule (**1**) having three carbocyclic rings, which are conjoined along a common carbon–carbon bond between bridgeheads, was coined as *propellane* by Ginsburg in 1966,[1] from its propeller-like three-dimensional shape. The chemistry of propellanes has been studied extensively from many points of view and is already the subject of several reviews.[2–6]

As the size of the constituent rings of **1** decreases, the bridgehead carbons come to hold the "inverted" geometry of a normal tetrahedron. Much interest has been focused, therefore, on the bonding character of the central interbridgehead bond of **1** from both theoretical and synthetic aspects.[3,5,6] In 1982, Wiberg marked a breakthrough in this field: He succeeded in synthesizing [1.1.1]propellane (**2**), the ultimate member of this class of molecules,[7] Remarkable progress has been achieved since then that has served not only to better our understanding of chemical bonding under extremely deformed circumstances but also to provide an entry to a new synthetic chemistry starting from **2**.

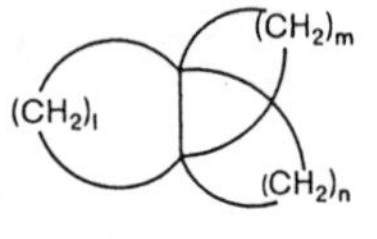

$(CH_2)_m$ $(CH_2)_l$ $(CH_2)_n$

1  2 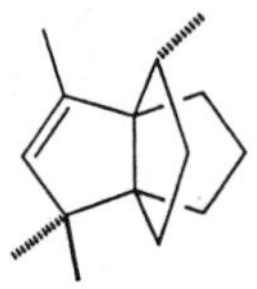 3

One of the major original interests in propellanes focused on those that serve as good stereochemical models because of their particular three-dimensional geometry.[2,4] It continues to be of special interest from this aspect, and many topologically as well as stereochemically intriguing molecules have been synthesized.

Ingenious utilization of the structural and stereochemical characteristics of propellanes has allowed selective chemical transformation in particular systems. In this respect, propellanes have received increasing recognition as useful and versatile synthetic intermediates for the synthesis of polycyclic natural and nonnatural products. Moreover, discovery of a naturally occurring carbocyclic propellane derivative, modhephene (**3**),[8] stimulated synthetic chemists to develop novel and efficient methods to construct the propellane skeleton.

This chapter reviews chemistry of propellanes developed mainly during the last decade from the aspects just described:

1. small-ring propellanes;
2. propellanes as stereochemical models;
3. synthetic application of propellanes;
4. naturally occurring propellanes.

5.2. Small-Ring Propellanes

5.2A. Synthesis of Small-Ring Propellanes

A variety of efficient methods have been developed to prepare small-ring propellanes with inverted tetrahedral geometry. On the whole, they are classified into three categories by the type of reactions:

1. intramolecular displacement;
2. intramolecular carbene addition;
3. cycloaddition with (or between) strained bridgehead alkenes.

5.2A(1). Intramolecular Displacement. Two modes of intramolecular substitution have been utilized successfully to construct small-ring propellanes: (i) bridgehead-to-bridgehead substitution of dihalo-substituted bridged bicyclic precursor (route A); (ii) bridgehead–to–side-chain displacement from fused bicyclic substrate with a leaving group in the side chain (route B).

The former method was successfully used to prepare [3.2.1]propellane (**4**) and [3.1.1]propellane (**5**) but was not successful for [2.2.2]propellane (**6**) and [2.2.1]propellane (**7**) because of higher reactivity of **6** and **7** toward reduction and reaction with nucleophiles.[5] Similarly, reactions of dihalobicyclo[2.2.2]octane (**9**, X = Y = Br or I) and dihalobicyclo[2.2.1]heptane (**10**, X = Y = Cl, Br, or I) with (trimethylstannyl)lithium was shown to proceed through an $S_{RN}1$ mechanism in the former case but probably through **7** in the latter case.[9] However, [1.1.1]propellane (**2**), the ultimate member of this class of molecules, was synthesized for the first time by Wiberg via this route.[7]

It is worth noting that **2** [which is smaller than **6**, **7**, and [2.1.1]propellane (**8**)] is much more stable to attack of a nucleophile than **6–8**, which immediately react under the conditions of their production. This remarkable stability of **2** relative to **6–8** toward a nucleophile was explained in terms of considerably lower calculated energy of hydrogenation of **2** to bicyclo[1.1.1]pentane **11** (X = Y = H, $\Delta H = -39$ kcal mol^{-1}) relative to those of **6–8** to the corresponding bicyclic alkanes **9**, **10**, and **12** (X = Y = H; **6**, $\Delta H = -93$ kcal mol^{-1}; **7**, $\Delta H = -99$ kcal mol^{-1}; **8**, $\Delta H = -73$ kcal mol^{-1}).[10] [2.2.1]Propellane (**7**)[11] and [2.1.1]propellane (**8**)[12] were finally synthesized by gas-phase reduction of bicyclic diiodides **10** and **12** (X = Y = I) with potassium vapor and were trapped in a nitrogen matrix at 30 K. Again, marked instability of **7** and **8** compared with that of **2** was interpreted on the basis of energies for homolytic cleavage of the central bond leading to bridgehead diradicals **10**, **12**, and **11** (X = Y = ·) (**7**, 35 kcal mol^{-1}; **8**, 30 kcal mol^{-1}; **2**, 65 kcal mol^{-1}).[10]

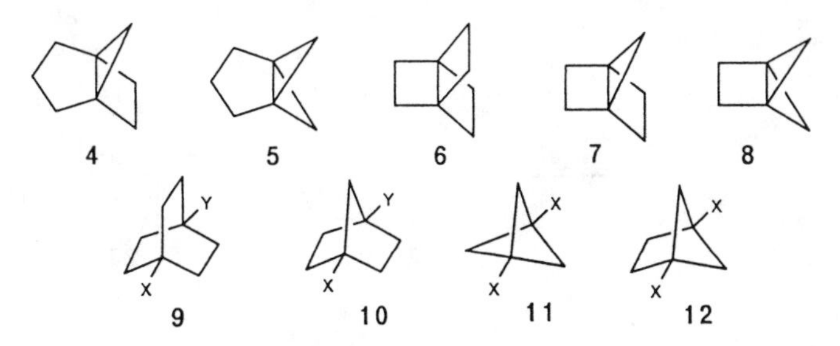

The second method, bridgehead–to–side-chain displacement developed by Szeimies, is a rather traditional type of reaction but turned out to be remarkably efficient for small-propellane synthesis. Thus, treatment of dibromo(bischloromethyl)cyclopropane (**13**), readily accessible from commercially available chloromethallyl chloride, with methyllithium yielded [1.1.1]propellane (**2**) by twofold intramolecular substitution.[13] Yield of the reaction was improved to 70%. Similarly, this method was used for the preparation of bridged [1.1.1]propellanes **14** and **15** and [3.1.1]propellane **16**.[14] However, [2.1.1]propellane **17** prepared by similar reaction eluded isolation.

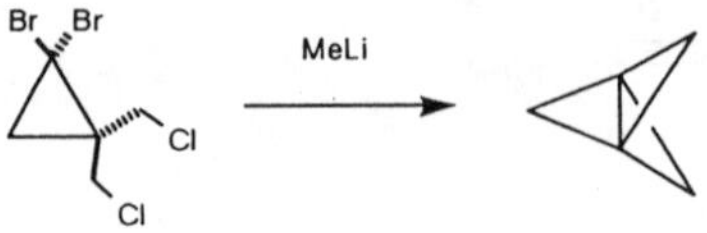

Br Br
MeLi
Cl
Cl
13
2

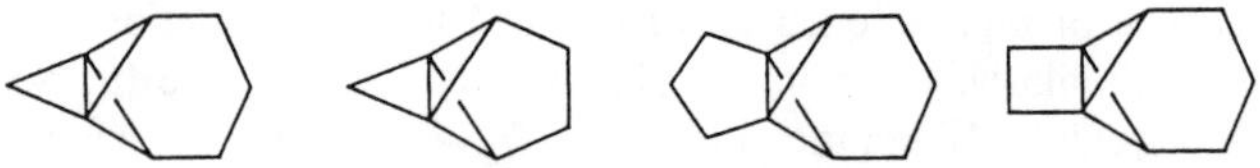

14 15 16 17

5.2A(2). Intramolecular Carbene Addition. Previously, intermolecular carbene addition to bicyclic olefins was used to prepare small-ring propellanes.[2,5] Majerski developed an intramolecular version of this reaction to prepare bridged and caged [3.1.1]propellanes **18**[15] and **19**[16] and [4.1.1]propellane **20**.[17] In addition, Szeimies reported that carbenoids generated by treatment of dihalomethylenebicyclo-[3.1.1]heptane (**21**) with methyllithium or LDA afforded bridged [1.1.1]propellane **22** with substituent(s) on the cyclopropane methylene.[14] Similarly, ethano bridged [1.1.1]propellane **23** and bispropellanes **24** and **25** were prepared utilizing this procedure.

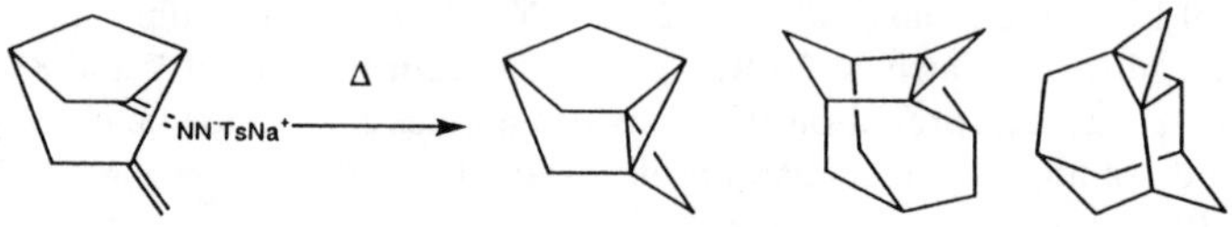

Δ
$^{-}$NN^{-}TsNa^{+}
18 19 20

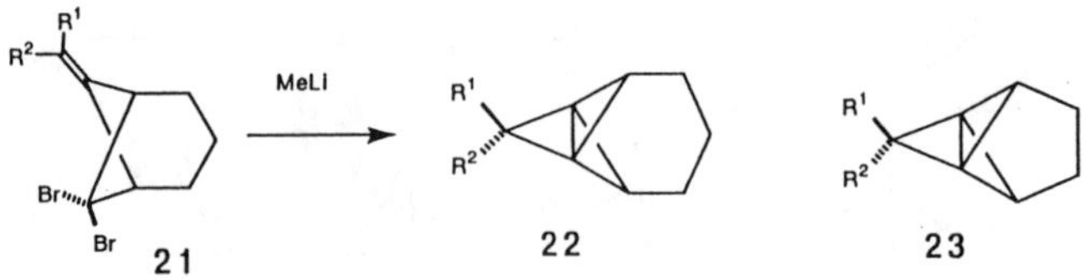

R^1
R^2
MeLi
Br
Br
21
R^1
R^2
22
R^1
R^2
23

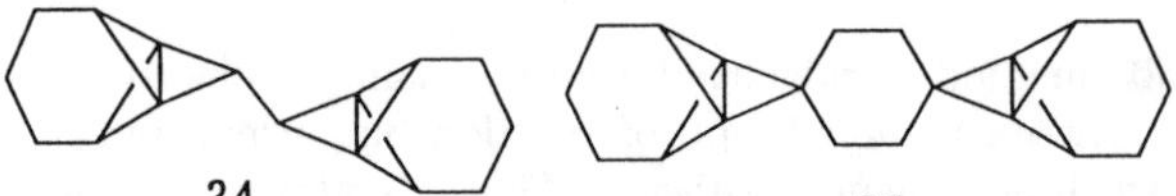

24 25

An expedient method for preparation of [3.3.1]propellanedione (**27**) was developed by Reingold, utilizing an intramolecular carbene addition in an eight-membered ring substrate **26**.[18] Intramolecular carbene insertion from dichloromethyldodecahedrane (**28**) produced [3.3.1]propellane **29** conjoined in a dodecahedrane framework.[19]

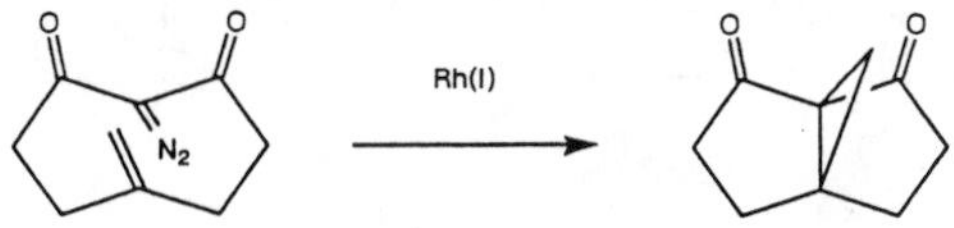

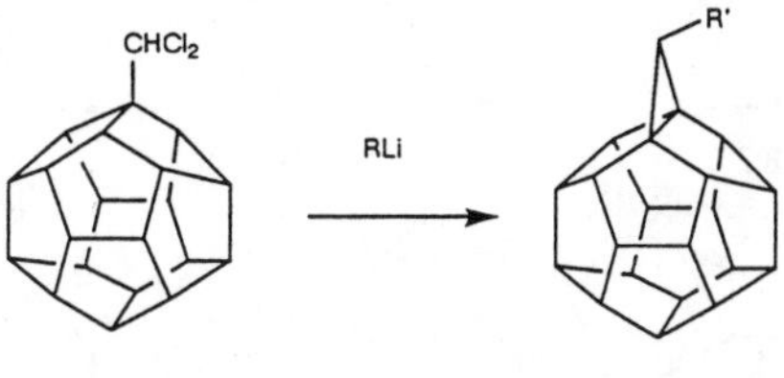

5.2A(3). Cycloaddition of Bridgehead Alkenes. Bridgehead olefins of fused small-ring bicyclic system with a central double bond and those having highly pyramidalized p orbitals have been known to be very reactive toward cyclodimerization and Diels–Alder reaction with dienes.[20] These reactions provided a new entry to small-ring propellanes.

Dimerization of small-ring bicyclic bridgehead olefin **30** generated from bromochlorobicyclo[2.2.0]hexane by electrolysis gave [4.2.2]propellane derivative **32** through bis[2.2.2]propellane (**31**), which was not isolated.[21] Similarly, treatment of dibromobicyclo[3.1.0]hexane with *tert*-butyllithium yielded anti bis[3.2.1]propellane (**34**) via dimerization of olefin **33**.[22]

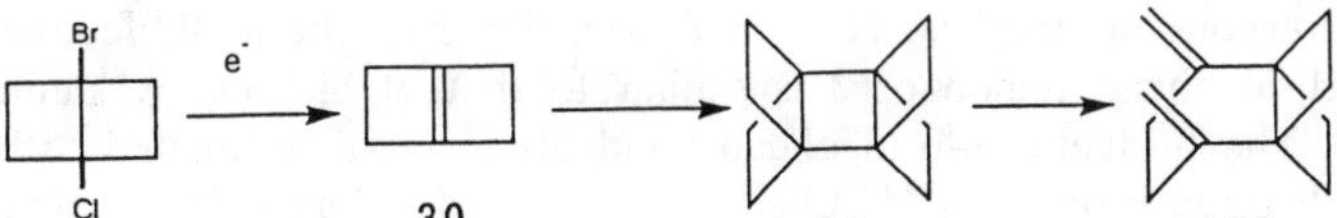

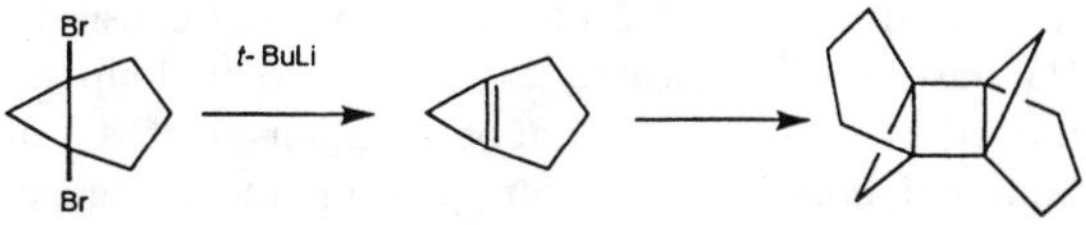

Diels–Alder reaction of bridged bridgehead olefins with pyramidalized p orbitals is exemplified by the reaction of cubene (**35**), which gave caged [4.2.2]propellane derivative **36**.[23] Similarly, bridged bicyclo[1.1.0]butene derivatives **37** and **38** and quadricyclene **39** was trapped with reactive dienes to afford [4.1.1]-propellanes and [3.1.1]propellanes such as **40**, **41**, and **42**, respectively.[24–26]

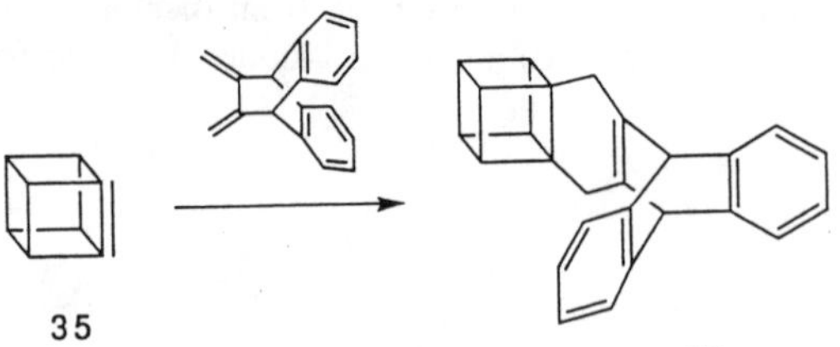

35 36 37 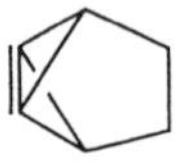 38 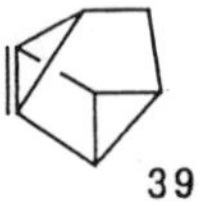 39 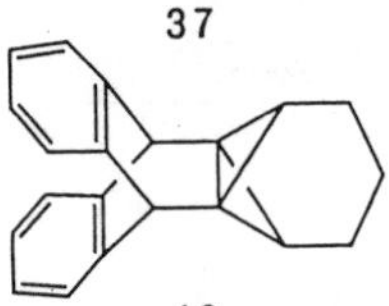 40 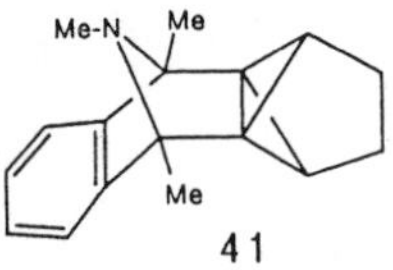 41 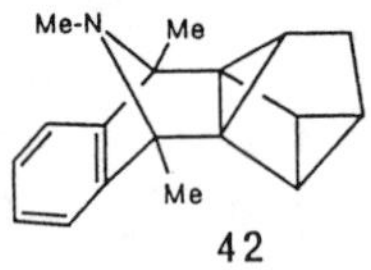 42

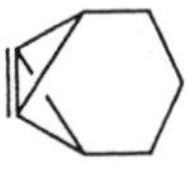

5.2B. Theoretical and Structural Studies of Small-Ring Propellanes

Theoretical studies on small-ring propellanes have mostly concentrated on [1.1.1]propellane (**2**) and [2.2.2]propellane (**6**), but from different points of view. Central to the interest in the former molecule has been the bonding character between bridgeheads that have totally inverted sp^3 geometry. On the other hand, because theoretical studies on **6** have shown that the central bond of **6** is of significant bonding character, interest in **6** stems from its behavior upon stretching the bridgehead–bridgehead bond.

5.2B(1). [1.1.1]Propellane. Early studies using ab initio methods (4-31 level)[27] and using extended Hückel calculations[28] concluded that **2** should be more stable than the corresponding diradical **11** (X = Y = ·). However, the total electron density of the interbridgehead region of **2** is similar to that of bicyclo[1.1.1]pentane (**11**, X = Y = H). The bridgehead–bridgehead bond distance was estimated to be about 1.6 Å with hybridization of $sp^{4.1}$. Like the case of [3.1.1]propellane derivative **41**, which was analyzed by high-resolution X-ray crystallography,[29] the calculated deformation density for the central bond of **2** obtained from self-consistent field (SCF) calculations of 6-31G basis set[30] indicated essentially no buildup of deformation density near the center of the molecule. These observations invited a question about the presence of traditional bonding character between the bridgeheads of **2**.[10,28] Despite the lack of traditional bonding character, calculated bond dissociation energy of **2** to diradical **11** (X = Y = ·) was estimated to be 65 kcal mol^{-1}, indicating the presence of considerable bonding energy![7]

The interbridgehead bonding character of **2** was described by a novel nonaxial orbital arrangement termed σ-bridged π that holds together (1.61 Å) the bridgehead atoms of inverted configuration.[31] Generalized valence bond description also supported considerable bonding character between bridgeheads with an enhanced p character ($sp^{4.5}$) of the orbitals connecting these atoms.[32] Moreover, the difference between the calculated (6-31G* UHF/MP2) one C—H bond dissociation energy of bicyclo[1.1.1]pentane **11** (X = Y = H) and the second C—H bond dissociation energy from radical **11** (X = H, Y = ·) was estimated to be 59 kcal mol^{-1}.[33] This considerable amount of stabilizing interaction (70% of normal C—C bond strength) might be regarded as an evidence of interbridgehead bonding of **2**. Finally, the arguments about the presence of bonding based on deformation density was pointed out to be misleading because it would be largely affected by the choice of spherical atom reference state.[34] Thus, analysis of the charge density distribution of **2** indicated that there is a total electronic charge density that is about four-fifths of a normal C—C single bond and that the charge density has a very broad maximum, suggesting the presence of "fat bond."

Analysis of through-space bond interaction of orbitals obtained from SCF calculations (STO-3G functions) showed that the energy level of the highest occupied molecular orbital (HOMO) of **2** is not much affected by the interaction with peripheral orbitals due to large energy gap between HOMO and the interacting orbitals.[35]

Infrared and Raman spectra of **2** and its d_6 derivative were measured, which led to an experimental determination of bridgehead–bridgehead distance of 1.60 ± 0.02 Å.[30] In addition, electron diffraction study of **2** provided an experimental bond distance of 1.596 ± 0.005 Å in accord with the value from infrared spectra.[36] ^{13}C–^{13}C coupling constant of **2** was determined [$J(C(1)–C(2)) = 9.9 \pm 0.1$ Hz], from which hybridization of interbridgehead bond was estimated to be $sp^{0.5}$.[37] The photoelectron spectrum of **2** was measured, which showed a sharp first ionization band at 9.74 eV, indicating a small geometry change during ionization.[38] The lack of structural change was attributed to nonbonding (or slightly antibonding) character of the HOMO of **2** and to its tight cage structure.

Photoelectron spectra of bridged [4.1.1]propellane (**43**), [3.1.1]propellanes **16** and **44**, and [1.1.1]propellanes **14** and **15** were recorded and the bands were assigned by MINDO/3 calculations.[39] Shift by about 1 eV toward lower energy was observed for the bands of **16** and **43** relative to a bicyclobutane **45** as well as for that of **44** compared with **46**, which were ascribed to the inductive effect of the bridge. In contrast, the first bands of **14** and **15** showed shifts toward higher energy relative to **45** and **46**, respectively, as a result of the inductive effect of the CH_2 group and the higher s character of the HOMO of the former compared with the latter.

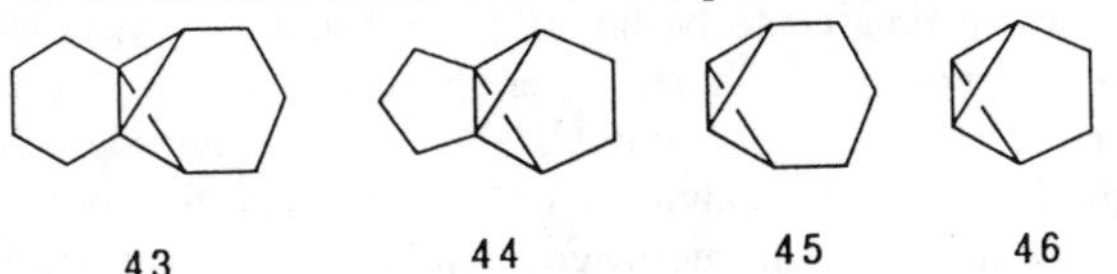

Photoelectron spectrum of caged [3.1.1]propellane **19** was measured and the bands were assigned by semiempirical calculations (MNDO and HAM/3).[40] The first ionization band of **19** (7.9 eV) shifted to lower energy relative to **2**. This was attributed to the inductive effect of the larger σ framework of **19** and lower s character of the HOMO of **19**.

A considerable amount of structural data has accumulated that were determined by X-ray crystallographic studies of crystalline derivatives of [1.1.1]propellanes **14**,[41]

Table 5.1 ▪ Observed and Calculated Bond Lengths of [1.1.1]Propellanes

Compound	Method	Central Bond (Å)	Side Bond (Å)	Ref.
2	4-31G	1.60	1.53	27
	6-31G*/MP2	1.594	1.515	30
	6-31G*/MP3	1.572	1.514	30
	IR/Raman	1.60	1.522	30
	Electron diffraction	1.596	1.525	36
14	X-ray	1.585	1.519–1.538	41
15	X-ray	1.587	1.517–1.529	41
23 ($R^1 = R^2 = Ph$)	X-ray	1.586, 1.592	1.517–1.534	14
24	X-ray	1.577	1.513–1.529	42
25	X-ray	1.587, 1.601	1.499–1.560	14

15,[41] and **23** ($R^1 = R^2 = Ph$),[14] and bispropellanes **24**[42] and **25**.[14] The bond length of the central and side bonds are summarized in Table 5.1 along with the calculated values and those determined by other methods. The bond length of the central and peripheral bonds of the propellane core differ considerably, varying from 1.577 to 1.601 Å for the central bond. The shortest observed length is not significantly longer than that of the corresponding bond length of [3.1.1]propellane derivative **41**.[29] On the other hand, the longest observed value is remarkably larger than that of **41**. Taking into account the substantial variation of the bond length due to rather small structural change, however, it is not very meaningful to compare them to each other. As in the case of [3.1.1]propellane derivative **41**,[29] deformation densities between the bridgeheads of **14** and **15** are slightly negative whereas those outside the bridgeheads are positive and diffuse.[41]

5.2B(2). [2.2.2]Propellane. Early theoretical studies based on extended Hückel,[28] ab initio (STO-3G),[43] and INDO calculations[44] already pointed out that there are two potential minima along the C(1)–C(4) coordinate, one corresponding to the propellane (**6**) structure with bond length of 1.52 (ref. 43) or 1.515 Å (ref. 44) and the other corresponding to bridgehead diradical **9** (X = Y = ·) with interatomic distance of 2.54 (ref. 43) or 2.290 Å (ref. 44). Moreover, the barrier at 1.95 Å distance between the two potential wells was estimated to be about 29 kcal mol^{-1}, suggesting a short lifetime for **6** at room temperature.[43] Shortly after these theoretical studies, a carboxamide derivative of **6**, **47** (the first and only [2.2.2]propellane thus far characterized), was synthesized by Eaton and Temme.[45] In accord with the prediction, **47** readily underwent cycloreversion to bismethylenecyclohexane derivatives **48** and **49**, with a half-life of only 1 h at 20° C.

Ab initio calculations at 6-31G* level found two energy minima at interbridgehead distances of 1.58 and 2.53 Å with comparable energy and a barrier of 14 kcal mol^{-1} at 2.00 Å.[33] When correction for electron correlation was involved, the outer minimum became considerably more stable and the barrier nearly zero. These results invited serious doubt about the existence of the parent hydrocarbon **6** as a gas-phase molecule.

Lowest energy transition densities were used to describe the change in the charge density for a displacement of nuclei of small-ring propellanes such as **2**, **6**, **7**, and **8**.[34] It was found that, for [2.2.2]propellane (**6**), the symmetric stretch is the most facile of the carbon framework motions. This indicates that although the bridgehead–bridgehead bond of **6** has higher bond order at its static geometry than [1.1.1]propellane (**2**) it is dramatically more susceptible to thermal rupture than **2**.

Through-space bond analysis of **6** showed that the central bond is much more delocalized than that of **2** through interactions with side bonds and the methylene orbitals.[35] Therefore, it was suggested that the HOMO level of **6** is more susceptible to the effect of the peripheral substituent than **2**.

5.2C. Reactions of Small-Ring Propellanes

5.2C(1). [*m.n.*1]Propellanes. Thermolysis of [1.1.1]propellane (**2**) in gas phase at 114°C yielded 3-methylenecyclobutane (**50**) with an E_a of approximately 30 kcal mol^{-1}.[7] On the contrary, pyrolysis of **2** in a flow system at 430°C gave bismethylenecyclopropane (**51**) as a single product.[46] In addition, bridged [1.1.1]propellanes **14** and **15** yielded products derived from the respective bismethylenecyclopropane derivatives.

50  51 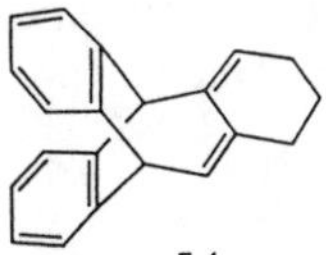 54 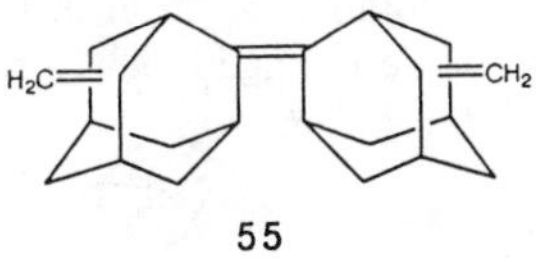 55

The steric course of thermal cycloreversion of a set of dimethyl[3.2.1]propellanes (**52**) to 1,3-bispropylidenecyclohexanes (**53**) was investigated.[47] Because only thermolysis of endo–endo **52** proceeded in a stereospecific manner giving trans–trans **53**, the mechanistic pathway was assumed to be down-disrotatory, a symmetry-forbidden $[2\sigma_s + 2\sigma_s]$ cleavage.

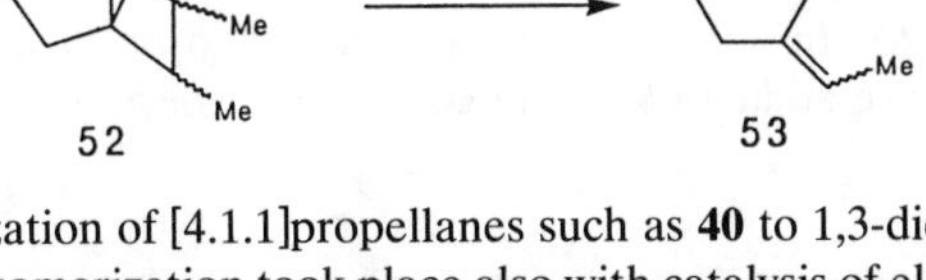

52 → 53 (Δ)

Thermal isomerization of [4.1.1]propellanes such as **40** to 1,3-dienes such as **54** was reported.[48] Similar isomerization took place also with catalysis of electrophilic reagents such as Ag(I) ion and Lewis acids. Silver(I)-catalyzed isomerization of caged [3.1.1]propellane **19** gave a 3:1 mixture of two bismethyleneadamantylidene dimers **55** as the major products.[49] The formation of **55** was interpreted in terms of dimerization of a silver(I)–carbene complex that was derived by metalocation-to-carbene rearrangement.

Addition of acetic acid to [1.1.1]propellane (**2**) gave methylenecyclobutyl acetate (**56**) with a heat of reaction of -35.2 ± 0.4 kcal mol^{-1}.[7,30] Estimation of ΔH_f° of **56** lead to an estimate of ΔH_f° of **2** to be 84 ± 1 kcal mol^{-1}, which was well in accord with a calculated value (89 kcal mol^{-1}).

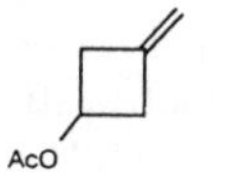

AcO
56 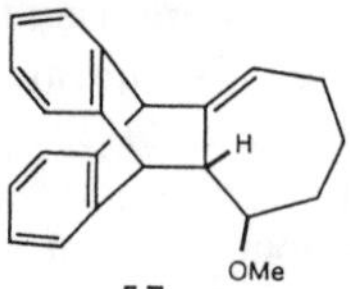
H
OMe
57 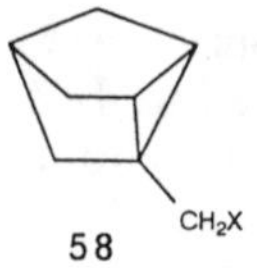
CH_2X
58 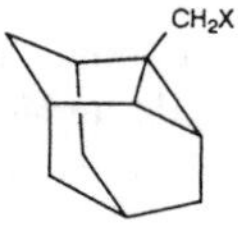
CH_2X
59 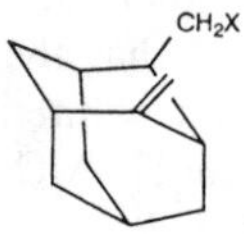
CH_2X
60 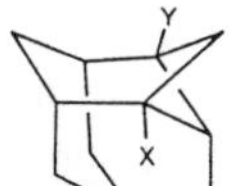
Y
X
61 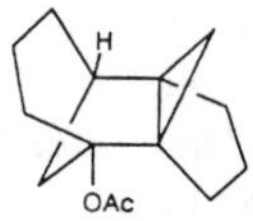
H
OAc
62

Acid-catalyzed addition of methanol to a [4.1.1]propellane **40** took place with skeletal rearrangement to afford ether **57**.[48] Addition of HCl to bridged [3.1.1]propellane **18** gave chloride **58** selectively derived by cleavage of one of the side bonds.[16] However, reaction of caged propellanes **19** and **20** with acids produced mixtures of products such as **59–61** through cleavage of a central bond, a side bond, or both, respectively.[15,17] Addition of acetic acid took place readily across the central bond of bis[3.2.1]propellane **34** to afford monoacetate **62**, which still retained one cyclopropane ring.[22] Similar addition of CH- and NH-acids such as diethyl malonate and *p*-toluenesulfamide to caged [3.3.1]propellane **63** gave adducts **64**.[50]

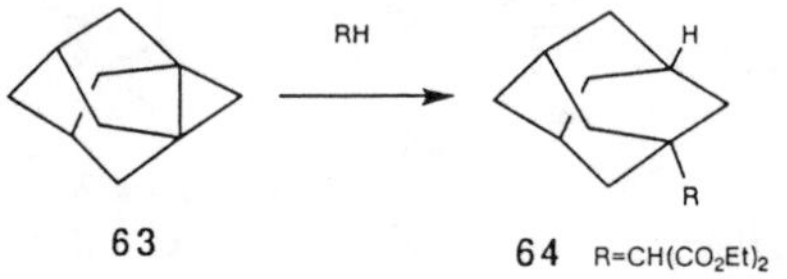

RH
63
H
R
64 R=$CH(CO_2Et)_2$
$NHSO_2Tol$

In contrast to addition of acid to [1.1.1]propellane (**2**), **2** underwent radical addition reaction across the central bond with many reagents without skeletal rearrangement. These include, for example, halomethanes,[51,52] thiophenol,[13] disulfides,[51,53,54] acetaldehyde,[51] methyl formate,[55] and biacetyl[56] in the absence or presence of an initiator or under photoirradiation. Depending on the reaction condition, the reaction gave not only monomeric product **11** but also oligomeric and polymeric products **65**. In addition, radical addition to bridged [1.1.1]propellane **14** also took place, to give adduct **66** and oligomers **67**.[57] Section 5.4 gives details about these telomers.

X Y n
65 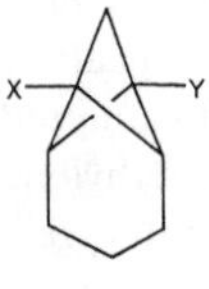
X Y
66 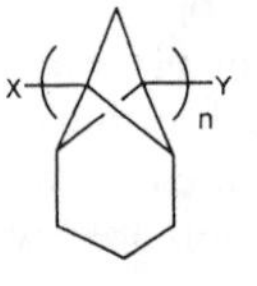
X Y n
67 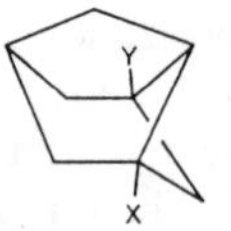
Y
X
68 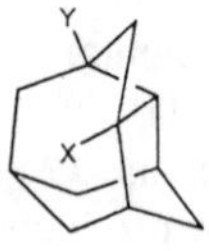
Y
X
69

Rates of addition of alkoxy and thiophenoxy radicals to **2** were measured by use of laser flash photolysis techniques.[58] It was found that the rate constant for thiophenyl addition of **2** was comparable to that of styrene, and even *tert*-butoxy radical, which rarely adds efficiently to unsaturated systems, reacted very readily with **2**.

Radical addition to [3.1.1]propellanes and [4.1.1]propellanes **18–20** took place across the central bond, yielding adducts **68**, **59**, and **69**, respectively.[16,17,59] An intermediate radical **59** ($X = CCl_3$, $Y = \cdot$) was detected by ESR spectrum.

[1.1.1]Propellane (**2**) underwent reaction with electron-deficient olefin and acetylene, presumably through a radical mechanism.[60] Reaction of **2** with tetracyanoethylene (TCNE) gave adduct **70** via addition of TCNE to the central bond followed by cleavage of a side bond. Similarly, dimethyl acetylenedicarboxylate (DMAD) gave 1 : 1 adduct **71** and two 2 : 1 adducts **72** and **73**, the latter being derived by addition of **71** to **2**.

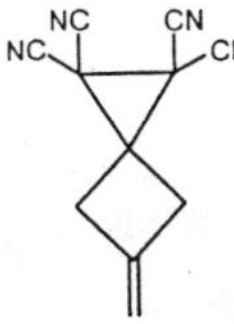

NC NC CN CN MeO2C CO2Me CO2Me CO2Me MeO2C CO2Me

70 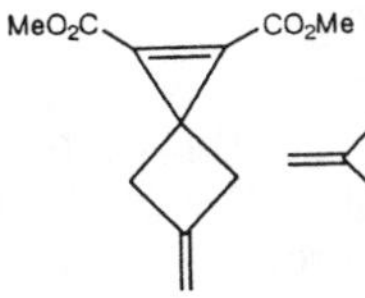 71 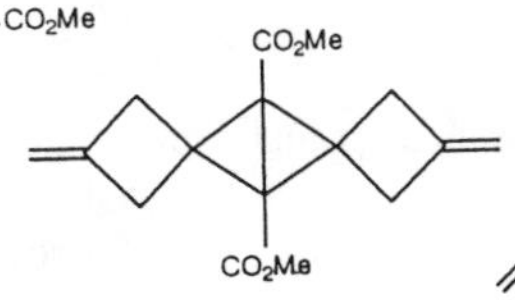 72 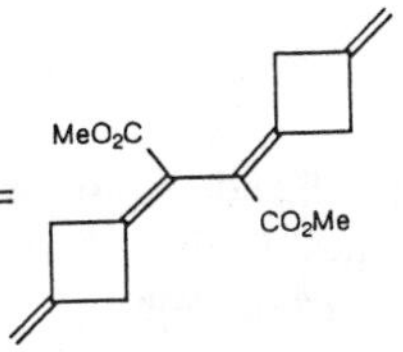 73

O O Me-N Me Me MeO2C CO2Me Ph Ph

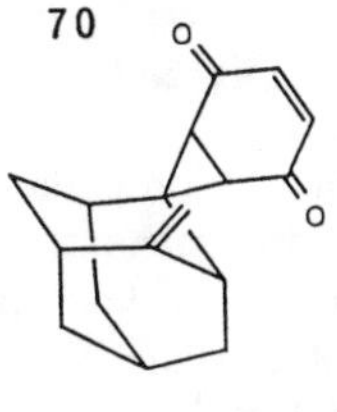

74 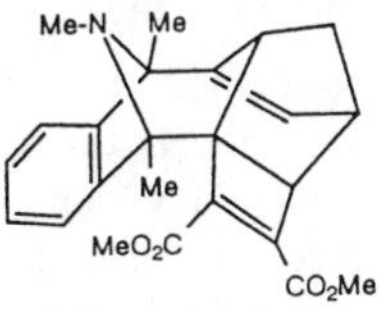 75 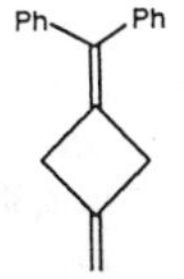 76

Reaction of caged [3.1.1]propellane **19** with *p*-benzoquinone also took place through a similar mechanism, giving adduct **74**.[59] However, reaction of DMAD with [3.2.1]propellane derivative **42** yielded cyclobutene **75** via cleavage of a remote cyclopropane ring rather than one of propellane bonds.[26]

[1.1.1]Propellane (**2**) reacted with diphenyl carbene to give bismethylene cyclobutane **76** through a mechanism similar to that of the reactions with electron-deficient olefins.[61] The rate of addition of the carbene to **2** was much faster than those of typical carbene additions to olefins.

Reaction of **2** and **14** with alkyllithium produced polymers of propellanes through anionic polymerization. These are described in Section 5.4.

5.2C(2). [*m*.2.2]Propellanes. Thermolysis of [4.2.2]propellane (**77**) leading to bismethylenecyclooctane (**78**) was investigated.[63] Low thermal reactivity of **77** ($E_a = 41$ kcal mol^{-1}) relative to bicyclo[2.2.0]hexane was attributed to the bridge, which is short enough to prevent the formation of chair cyclohexane-1,4-diyl. Remarkably higher reactivity of [2.2.2]propellane (**6**) ($E_a = 22$ kcal mol^{-1}) compared with **77** was ascribed to higher strain in **6**. Similarly, pyrolysis of [4.2.2]propellane (**79**) afforded triene **80**, presumably through a bridgehead diene as a transient intermediate or directly through a diradical intermediate.[63] Activation energy (35.8 kcal mol^{-1}) of this cycloreversion is lower than that of **77**, probably due to strain effect. Similarly, photolysis of [4.2.2]propellanones and [3.2.2]propellanones **81** and **82** with a cyclobutanone moiety in methanol gave products formed by β-fission involving cleavage of the central bond.[64]

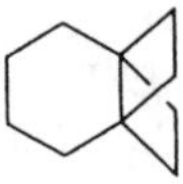

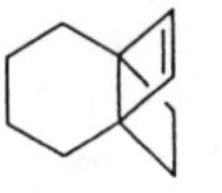

77 78 79 80

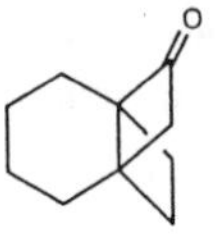

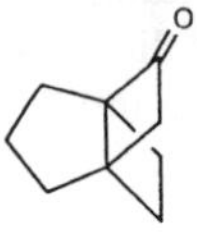

81 82

Thermal and photochemical transformation of [*m*.2.2]propelladienes to [*n*]paracyclophanes is described in Section 5.4, as well as bridgehead alkene synthesis utilizing carbocation rearrangement of [*m*.2.2]propellane derivatives.

5.2D. Silapropellanes and Stannapropellanes

Ab initio calculations were performed for pentasila[1.1.1]propellane (**83**) and octasila[2.2.2]propellane (**84**).[65–67] In the latter molecule, the calculated bond length of the central bond is 2.382 Å (STO-3G), which is slightly longer than normal Si—Si bond length but is not much different from that of cyclotetrasilane.[66] The estimated strain energy of **84** is 55.7 kcal mol^{-1}, being 40.3 kcal mol^{-1} less than that of carbocyclic homolog **6**. On the contrary, the calculated bridgehead–bridgehead bond length of **83** ranges from 2.686 to 2.885 Å, depending on the basis set employed, which is remarkably longer than normal. Also, calculations suggested substantial singlet diradical character of **83** with strain energy of 70.2 (STO-3G) or 62.2 kcal mol^{-1} (pseudopotential MC-SCF-21G*), being much smaller than carbocyclic analog **2**. The calculations predicted that **83** and **84** should be capable of existence, affording an interesting experimental challenge.

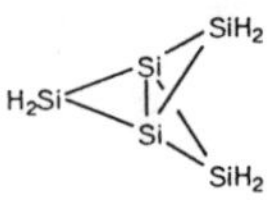

83 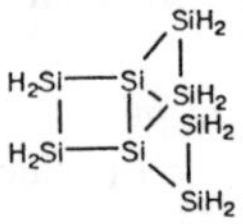 84 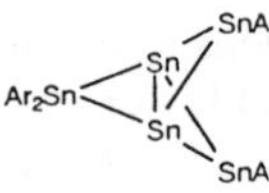 85 Ar=2,6-diethylphenyl

Although pentasila[1.1.1]propellane (**83**) has not yet been prepared, corresponding pentastannapropellane derivative **85** was synthesized by Sita and Bickerstaff.[68] X-ray structure analysis of **85** was done that revealed remarkably long Sn—Sn interbridgehead bond distance (3.367 Å), being the longest known tin–tin bond. That the bond distance is out of the normal range of tin–tin bond length supported the significant singlet diradical character in **85**.

5.3. Propellanes as Stereochemical Models

The unique three-dimensional geometry of propellanes provides a novel topology of organic molecules. Thus, conformational behavior of trisspiro[3.3.3]propellane **86** was investigated by temperature-dependent NMR studies.[69] Line-shape analysis for the interconversion of two chiral conformers of **86** gave activation parameters of $\Delta H^{\ddagger} = 11.3 \pm 0.6$ kcal mol^{-1} and $\Delta S^{\ddagger} = -6.1 \pm 2.2$ eu.

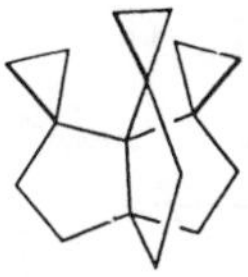

86 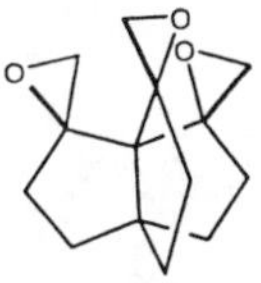 87 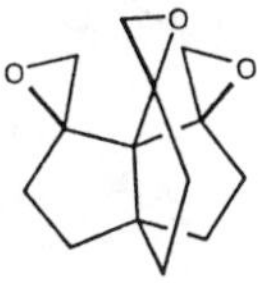 88

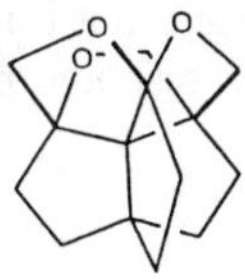

89 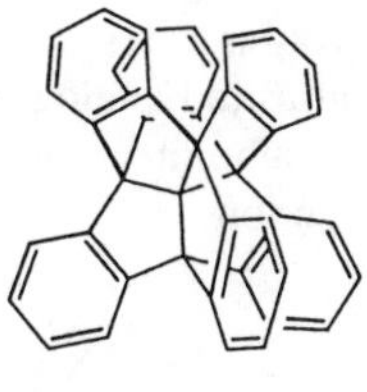 90

Thermal or acid-catalyzed isomerization of triepoxy[3.3.3]propellane **87** and **88** afforded triether **89**, the first topologically nonplanar molecule.[70] The mechanism of this rearrangement was investigated in detail using mono ^{13}C- and partly ^{18}O-labeled **87**.[71] The hexabenzo carbocyclic derivative **90**, coined as centrohexaindan, was also synthesized.[72]

Extensively unsaturated propellanes have been regarded as prototypes of that class of molecules that serve as excellent stereochemical models because of their three-dimensional array of π orbitals.[2,4] Polyunsaturated [4.4.4]propellanes, tetraene **91** and pentaene **92**, were synthesized by Paquette and co-workers.[73] Although **91** was thermally stable, **92** underwent smooth Diels–Alder fragmentation [$k(95°\text{C}) = 1.6 \times 10^{-4}$ s^{-1}] to give naphthalene and 1,3-butadiene. During efforts to manipulate **92**, several interesting transformations such as inter-ring bonding and skeletal rearrangement were observed. For example, upon treatment of tetrabromo enone **93** with potassium *tert*-butoxide, tetracyclic propellanone **94** was obtained via intramolecular substitution. [4.4.4]Propellapentaenyl cation, generated either from tetracyclic alcohol **95** or propellapentaenyl alcohol **96**, was transformed into dihydropleiadiene **(97)** through threefold Wagner–Meerwein rearrangement cascade.[74] A fully unsaturated congener of this class, hexaene **98**, was finally prepared and was shown to be more stable toward retro-Diels–Alder cleavage than pentaene **92**.[75]

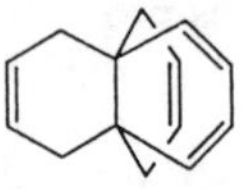

91 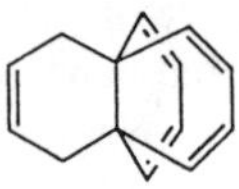 92 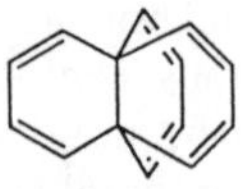 98

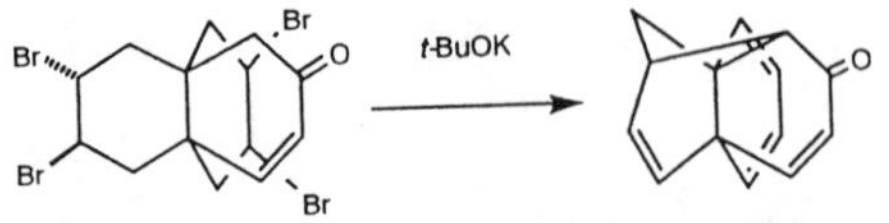

93 *t*-BuOK 94

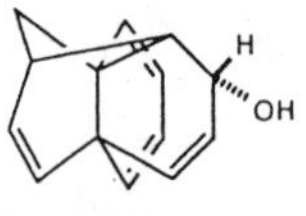

95 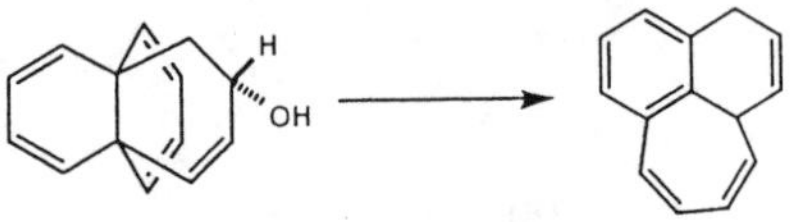 96 97

In the [4.4.3]propellane series, highly unsaturated propellanol **99a** was prepared for the purpose of investigating longicyclic stabilization in carbocations derived from it.[76] Comparison of solvolysis rates and products derived from dinitrobenzoate **99b** and those from less unsaturated analogs suggested that it ionizes with participation of only one cyclohexadiene subunit to produce a bishomotropylium ion rather than a purported bicycloaromatic ion.

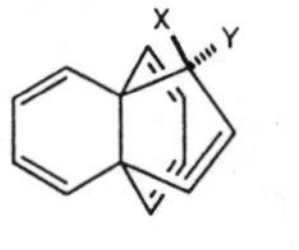

99a X=H, Y=OH
99b X=H, Y=ODNB
100 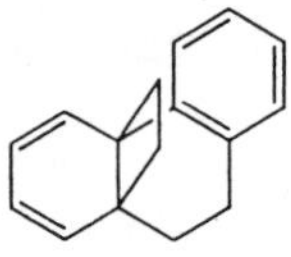

Optically active benzo[4.4.2]propellatrienes such as **100** were prepared and their absolute configurations were determined.[77]

Fully unsaturated [4.2.2]propellatetraenes **101a** and **101b** were synthesized by Tsuji and Nishida.[78] Photoelectron spectra and molecular orbital calculations of **101a** showed that the electronic interaction between the constituent rings is not important.[79] Ester **101b** underwent thermal rearrangement in solution at 200°C to the positional isomer **102**.[80] A mechanism was proposed involving a cyclobutadiene intermediate that was intercepted by methanol to afford bicyclic tetraene **103**.

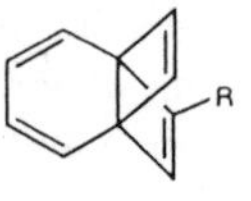

101a R=H
101b R=CO_2Me
102 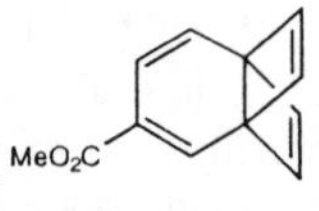
103 

Pagodane (**104**), a strained undecacyclic cage molecule of unique architecture, itself can be envisioned as caged bis[3.3.2]propellane or bis[4.4.2]propellane. Synthesis of this molecule was achieved by photochemical intramolecular [6 + 6] cycloaddition of dibenzosesquinorbornadiene (**105**) having faced aromatic rings to the "bird-cage" bispropellane **106** and the subsequent domino Diels–Alder reaction as pivotal steps of the synthesis.[81] Gas-phase isomerization of **104** produced dodecahedrane (**107**)[82] (8% at optimized condition), which marked the second total synthesis of this aesthetically fascinating molecule.[83] [3.3.1]Propellanes **108** and **109** were prepared, which afforded methyldodecahedranes and dimethyldodecahedranes more efficiently than **104**.

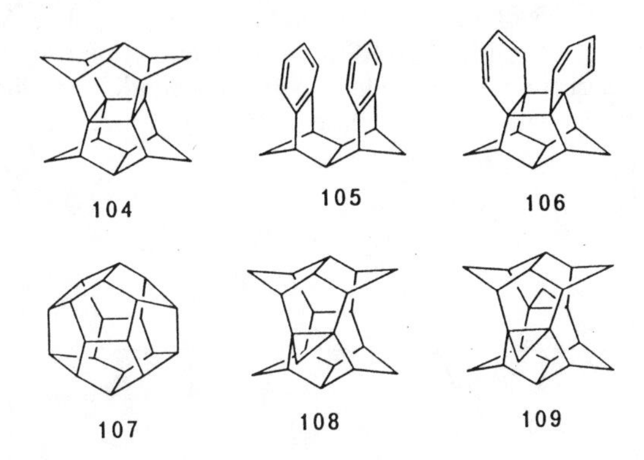

Stereoisomeric chloro[4.3.1]propellanes **110** and **111** were used for a mechanistic study on S_N2 substitution at cyclopropyl carbon.[84] Treatment of **110** with *tert*-BuOK gave tricyclic ether **112**, whereas **111** was unchanged under the reaction conditions. Reaction of monodeuterated **110** produced **112**, which retained deuterium with inversion of stereochemistry. These observations are in accord with S_N2 reaction with inversion at a cyclopropane carbon atom.

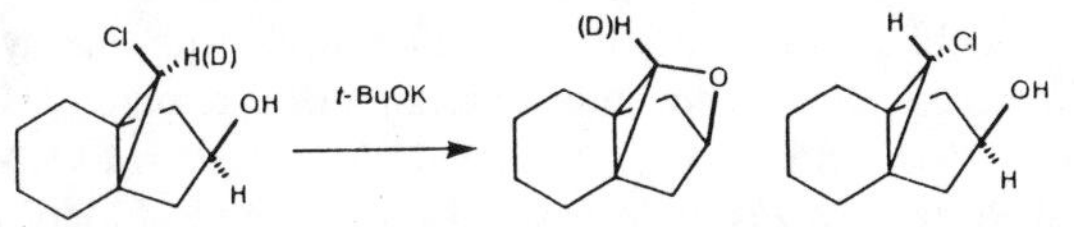

Cl, H(D), OH, H — *t*-BuOK → (D)H, O — H, Cl, OH, H

110 112 111

[3.3.1]Propellanes **113** (X = O, NPh, CH_2, CHMe) having sesquinorbornene structure were prepared and dyotropic hydrogen migration of their endo protons to the proximal double bond leading to equilibrium between **113** and **114** was examined.[85] A good correlation was observed between the reaction rate (over 10^4 range) and the intracavity distance (0.3 Å range) determined by X-ray crystallographic analyses.

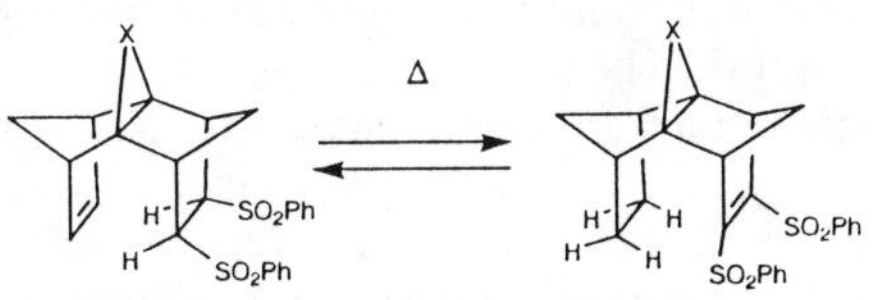

X, H, H, SO_2Ph, SO_2Ph ⇌ (Δ) X, H, H, H, H, SO_2Ph, SO_2Ph

113 114

5.4. Synthetic Application of Propellanes

5.4A. [1.1.1]Propellane

As described in the previous section, because highly reactive [1.1.1]propellane (**2**) and its bridged derivative **14** have now become readily available in quantity, new synthetic chemistry starting from these compounds has been evolved. Oligomers and polymers of **2**, which were coined by Michl as [*n*]staffanes,[55] have a rigid rodlike structure with a van der Waals radius of 2.3 Å and a length increment of 3.35 Å for each element. These telomers were obtained by radical or anionic polymerization of **2**; the product distribution was susceptible to reagent and reaction conditions. Dimeric, trimeric, and up to hexameric compounds were isolated and characterized, as well as polymers.

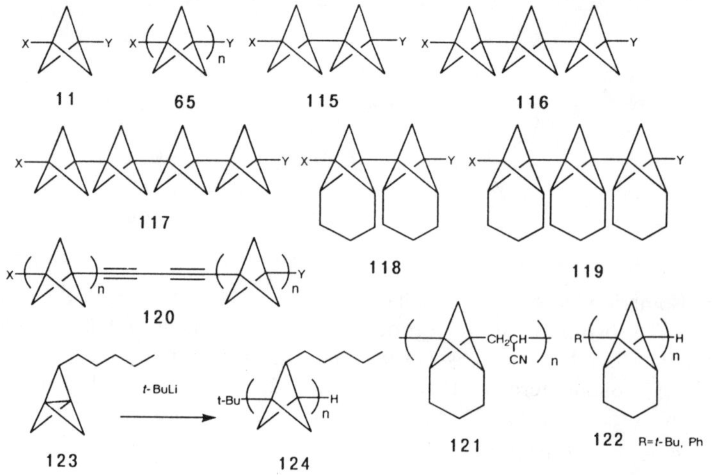

11 65 115 116 117 118 119 120 121 122 R=*t*-Bu, Ph 123 *t*-BuLi 124

For example, Michl reported that irradiation of **2** with methyl formate in the presence of an initiator gave mono endo-functionalized telomers **11** (3%), **115** (3%), **116** (1.5%), **117** (1.5%), and **65** ($n = 5$, 0.8%) (X = H, Y = CO_2Me).[55] Ultraviolet irradiation of **2** with diacetyl disulfide gave **11** and bis endo-functionalized oligomers **65** ($n = 2, \ldots, 5$, X = Y = SCOMe, 1–20%).[54] Similarly, Szeimies found that heating **2** with dimethyl disulfide with AIBN produced **11** (49%), **115** (21%), **116** (7%), and **117** (4%) (X = Y = SMe).[53] Dimer **118** and trimer **119** of bridged propellane **14** was also prepared (X = H, Y = CHMeOEt).[57] Moreover, parent hydrocarbons of [*n*]staffane, **115** (13%), **116** (9%), **117** (6%), and **65** ($n = 5$, 2%; $n = 6$, 0.3%), were prepared by Michl by addition of **2** across the H—H bond.[86] X-ray structure analyses of dimers **115** (X = H, Y = CO_2Me),[55] (X = Y = SMe, X = Y = SO_2Me),[53] trimers **116** (X = Y = H),[86] (X = Y = SCOMe),[54] and a tetramer **117** (X = Y = H)[86] showed that each constituent unit has interbridgehead distance 1.86–1.89 Å, and the bonds connecting each bicyclo[1.1.1]pentane unit are uniformly short (1.457–1.498 Å) for an sp^3–sp^3 bond.

Monomers **11** and dimers **115** (X = Cl, Y = CCl_2Me; X = H, Y = CCl_2Me) obtained by radical addition of trichloroethane and dichloroethane to **2** were converted to acetylenes **11** and **115** (X = Cl, Y = CCH; X = H, Y = CCH) connected with bicyclo[1.1.1]pentane units.[52] Coupling of the latter gave linear biacetylene **120** (X = Cl; X = H, $n = 1, 2$) endo-capped with bicyclopentane units.

Copolymerization of bridged propellane **14** and acrylonitrile took place when they were mixed at temperature ranging from $-25°C$ to room temperature, giving an alternating copolymer **121**.[87] Detailed analysis of CP-MAS ^{13}C NMR of **121** revealed that it was clearly alternating with an atactic, head-to-tail assembly.

Schlüter found that alkyllithium induced anionic polymerization of **14** to afford polymers **122** whose degree of polymerization was greater than 20.[88] In addition, soluble polymer **124** with greater polymerization degree was also prepared by anionic polymerization of pentyl[1.1.1]propellane (**123**).[89]

5.4B. [*m.n.*1]Propellanes

1,5-Methano[10]annulene (homoazulene) (**127a**) was synthesized by Scott via oxidative cleavage of the central bond of [5.3.1]propellatriene (**125**), giving diacetate **126**, followed by twofold elimination.[90] Dimethoxy derivative **127b** was also prepared through silica-gel-catalyzed isomerization of propelladienone **128** to **129**.[91] Moreover, solvolysis of [5.3.1]propelladienol derivative **130** afforded bridgehead alcohol **131** through cyclopropylcarbinyl–homoallyl rearrangement, which was converted to 1,5-bridged tropylium ion **132**.[92]

$Pb(OAc)_4$ AcO OAc

125 126 127a

SiO_2 OMe OMe

128 129 127b

PNBO H_2O OH

130 131 132

A considerable amount of work has been done on bridgehead alkene synthesis,[93] taking advantage of the propensity of [*m.n.*1]propellanes having a halogenated cyclopropane ring toward ring-opening through cyclopropyl–allyl transformation. [5]Metacyclophane (**135a**), the smallest-bridged [*n*]metacyclophane isolated to date, was successfully synthesized by Bickelhaupt and co-workers via twofold elimination of HCl from dichloro[5.3.1]propellane (**134**), which was obtained by partial reduction of tetrachloropropellane **133**.[94] Silver-catalyzed isomerization of **133** gave 8,11-dichloro[5]paracyclophane (**135b**).

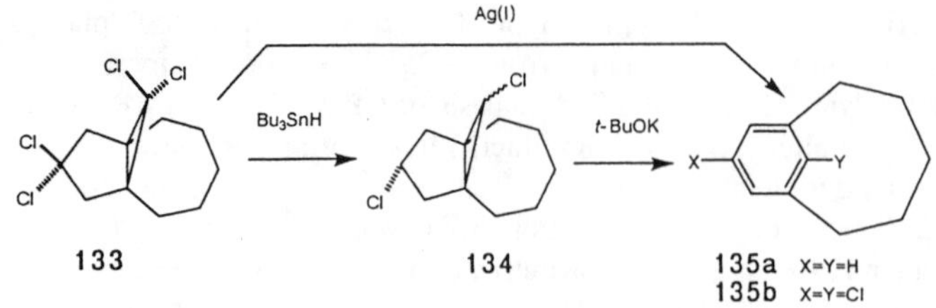
Ag(I)
Bu3SnH
t-BuOK
133
134
135a X=Y=H
135b X=Y=Cl

Solvolytic rearrangement, promoted by silver ion, of bromochloro[4.4.1]propellenes **136a** and **136b** gave various products, which were derived from initially formed noninterconverting bridgehead olefins **137a** and **137b**.[95]

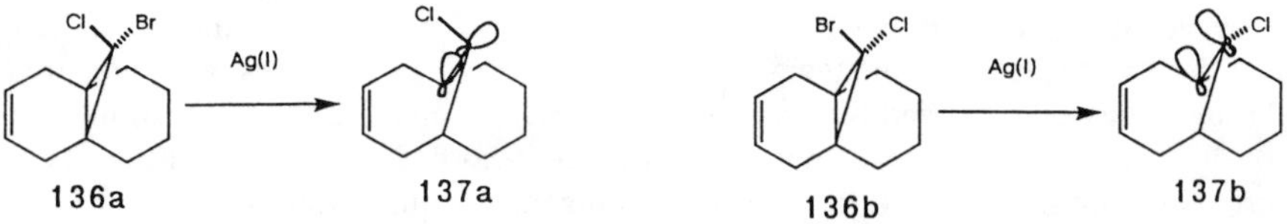
Ag(I)
Ag(I)
136a
137a
136b
137b

Warner reported that cyclopropylidenes generated from dibromo[4.3.1]propellene (**138**) produced bridgehead diene **139**, which was trapped with diphenylisobenzofuran (DPIBF) to give an adduct **140**.[96] More remarkable is the formation of bridgehead vinyllithium intermediate **142** from dichloro[4.2.1]propellene (**141**).[97] Generation of **142** was indicated by the structure and stereochemistry of DPIBF adduct **143**.

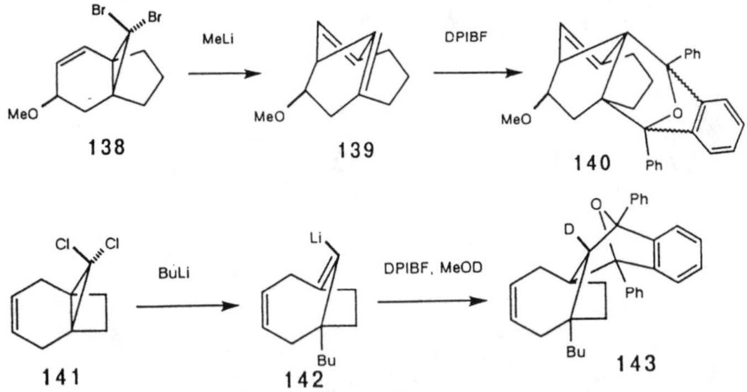
MeLi
DPIBF
MeO
138
139
140
BuLi
DPIBF, MeOD
141
142
143

An interesting transformation of [4.3.1]propellanone **144** was reported by Warner, Chu, and Boulanger[98] that gave bridgehead enone **145** through ring opening of a copper enolate intermediate. Elimination of HCl from **145** afforded bridgehead dienone **146**.

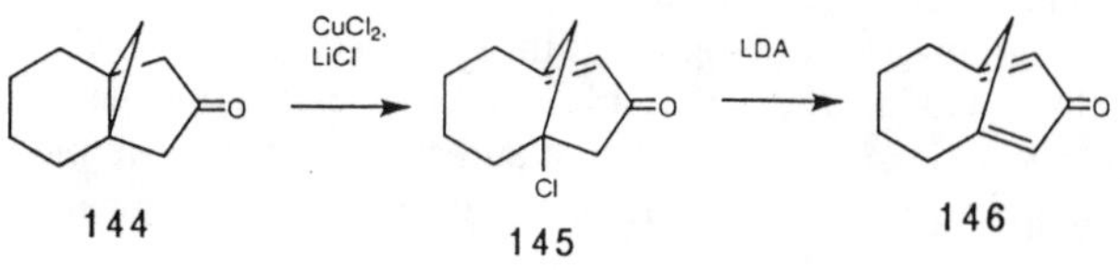
CuCl2, LiCl
LDA
144
145
146

A totally different synthetic use of [*m.n*.1]propellanes developed by Jones and co-workers[99] is a novel method of photochemical generation of carbenes. The reaction is formally the reversal of synthesis of the starting materials, that is, [*m.n*.1]propellanes. Thus, irradiation of dibromo[4.4.1]propellatriene **(147)** or dibromo[4.3.1]propelladiene **(148)** produced dibromocarbene and dihydronaphthalene or indane. More remarkable is the generation of homocubylidene from homocubane-conjoined [4.3.1]propelladiene **149**.

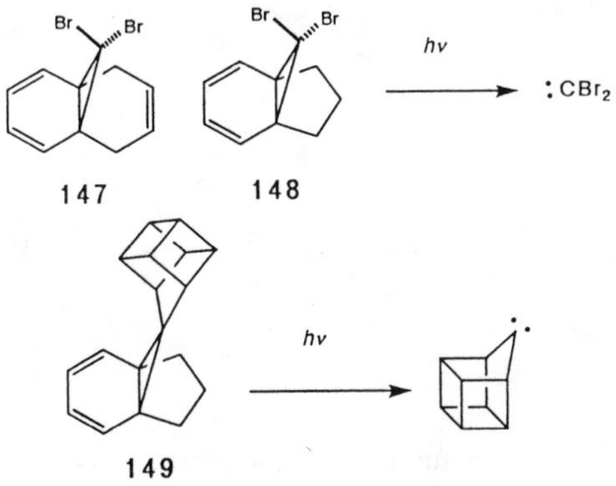

147 148 149

5.4C. [*m.n*.2]Propellanes

[*m.n*.2]Propellanes having cyclobutane ring(s) are in most cases prepared expediently by photochemical enone–olefin [2 + 2] cycloaddition. In this respect, rearrangement of these compounds leading to other polycarbocyclic systems has been studied extensively as a versatile method of synthesis.

Acid-catalyzed rearrangement of [*m.n*.2]propellanones **(150)** having a cyclohexanone ring in the absence of a nucleophile has been known as the Cargill rearrangement.[100] The course of the reaction is explained by an initial migration of external cyclobutane bond followed by a second migration leading to propellanones **151** with two cyclopentane rings.

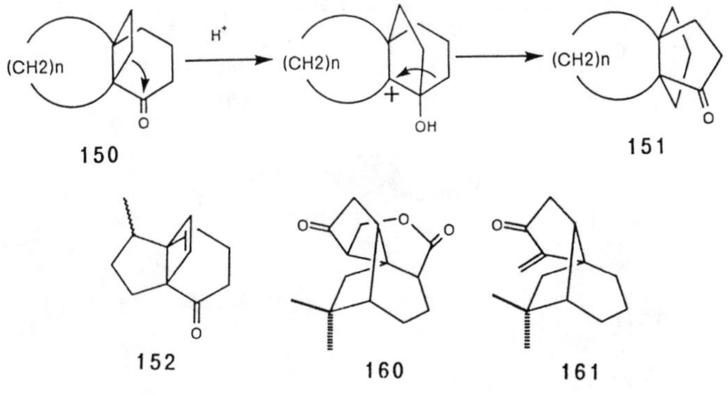

150 151 152 160 161

An ingenious utilization of this reaction is exemplified by the synthesis of (±)-modhephene (**3**), a unique sesquiterpene with [3.3.3]propellane skeleton, from [4.3.2]propellenone **152**.[101] On the other hand, in the presence of a nucleophile, [4.2.2]propellanone (**153**)[102] and [4.3.2]propellanone (**155**)[103] underwent selective migration of the central bond to give tricyclic diols **154** and **156**, respectively.

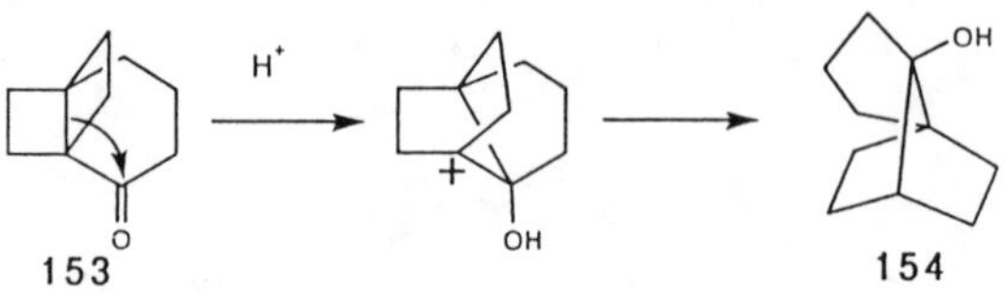

153 154

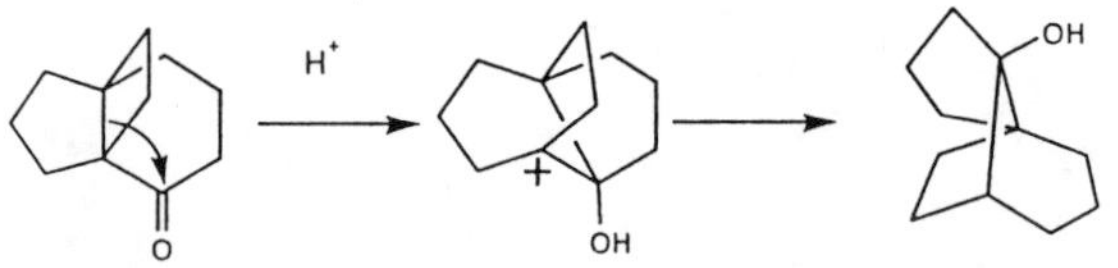

155 156

Acid-catalyzed rearrangements of [4.3.2]propellanols (**157a** and **157b**) and their derivatives were investigated[104] in detail by Smith, Odaira, Kakiuchi, and co-workers. *exo*-Propellanol **157a** gave tricyclic alcohol **158a** through migration of a central bond, whereas endo **157b** yielded alcohol **158b** of the same skeleton through external bond migration, along with triquinane **159**, a product of further rearrangement of **158b**. This remarkable difference in migration selectivity was explained in terms of stereoelectronic effects arising from the alignment of the leaving group and migrating cyclobutane bond. Selective central bond migration of the [4.3.2]propellane system was applied to the construction of the core tricyclic ring system of (+)-quadrone (**160**)[105] and (±)-descarboxyquadrone (**161**),[106] antibiotics with antitumor activities, in their stereoselective total syntheses.

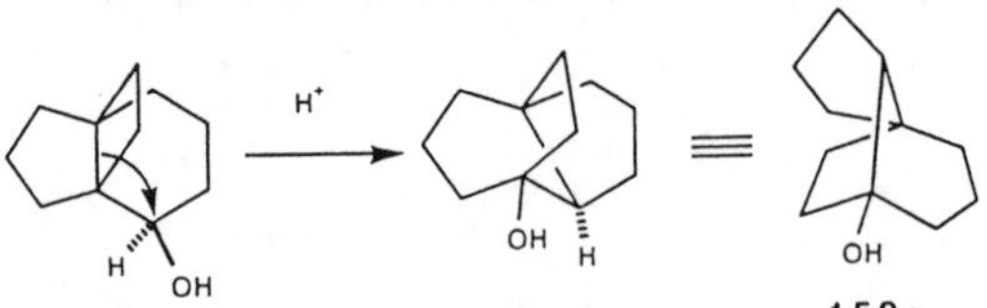

157a 158a

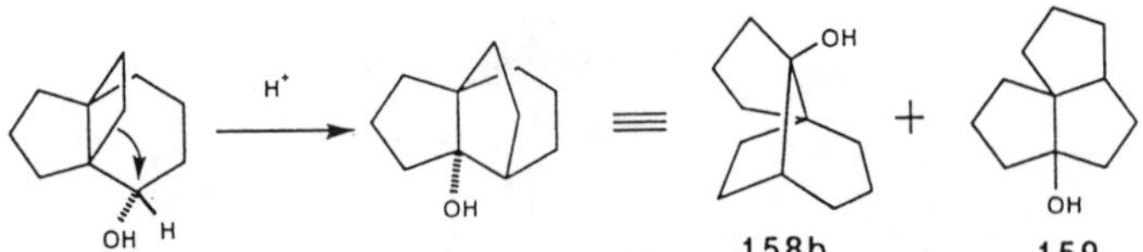

157b 158b 159

A similar type of transformation was observed in carbanion rearrangement of propellanes. Thus Grob-type fragmentation of [4.4.2]propellanedione (**162**) produced tricyclic hydroxyketone **164** through a dianion **163**.[107]

Li

162 163 164

[n.3.2]Propellanes (**165a** and **165b**) ($n = 3, \ldots, 6$) with a leaving group on the five-membered ring underwent only peripheral bond migration, to afford tricyclic olefins **166** or ester **167** ($n = 3$) as major rearrangement products.[108] Only in the case of the smallest, *exo*-[3.3.2]propellanyl tosylate (**165a**, $n = 3$), diacetate **169** was obtained as a minor product (15% yield) that was derived by cleavage of central bond and the subsequent addition of acetic acid to bridgehead olefin **168**.[109]

RCO_2H

165a X=OTs, Y=H
165b H=H, Y=OTs

166 167

AcOH AcOH

165a 168 169

In the case of [n.2.2]propellanes with a leaving group on the four-membered ring, the situation is much different because the cleavage and/or migration of central bond would be more facilitated than the larger propellanes. Solvolysis of *exo*-[4.2.2]propellanyl dinitrobenzoate (**170a**) gave alcohols **171a** and **172a** derived from 1,2-shift and cleavage of a central bond, respectively.[110] The endo tosylate **170b** yielded acetate **173** through external bond migration and the subsequent double Wagner–Meerwein rearrangement. The relative rate of the solvolysis [k(**170a**)/k(**170b**) $= 4.7 \times 10^8$] indicated strong σ participation of a central bond in solvolysis of **170a**. Moreover, oxidative decarboxylation of [n.2.2]propellanecarboxylic acids (**174**) ($n = 4, \ldots, 6$) with lead tetraacetate gave cyclopropylcarbinyl-type acetates **171b** and/or allylcarbinyl-type acetates **172b** in good yields.[111] Acid-catalyzed isomerization of **172b** ($n = 4, 5$) took place smoothly to afford **172b**, providing an efficient method for synthesis of bicyclo[n.2.2]-type bridgehead alkenes. Oxidation of **174** ($n = 3, 4$) with an excess lead tetraacetate yielded bicyclic alkenes **175** doubly substituted at bridgeheads through further oxidation of **171b** ($n = 3, 4$).[112] Similarly, oxidation of [n.2.1]propellanecarboxylic acids (**176**) ($n = 4, \ldots, 6$) gave bridgehead alkenes **177** ($n = 5, 6$) of bicyclo[n.2.1] type and/or diacetate **178** ($n = 4, 5$).[113]

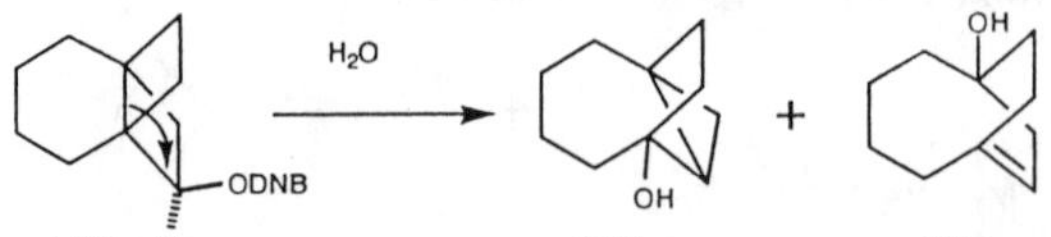

170a 171a 172a

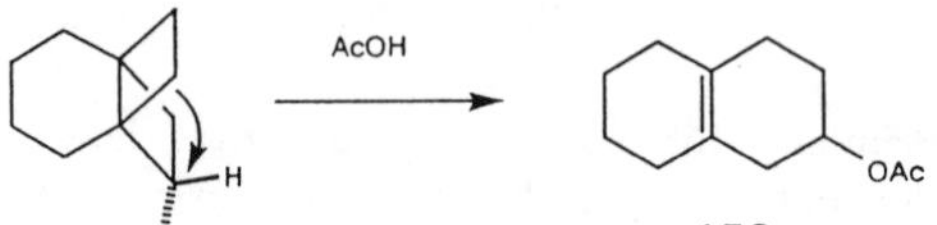

170b 173

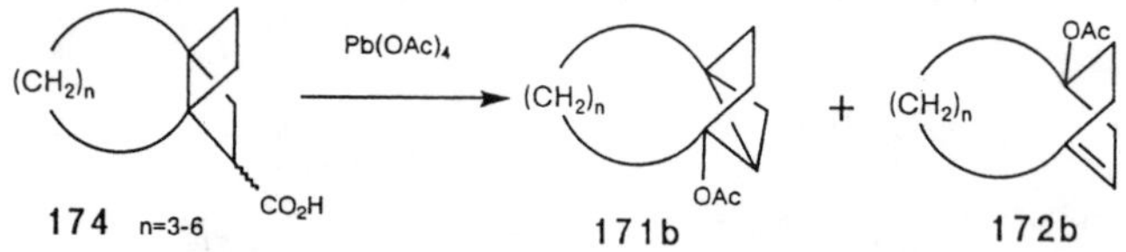

174 n=3-6 171b 172b

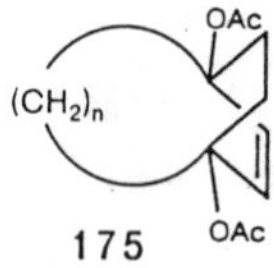

175

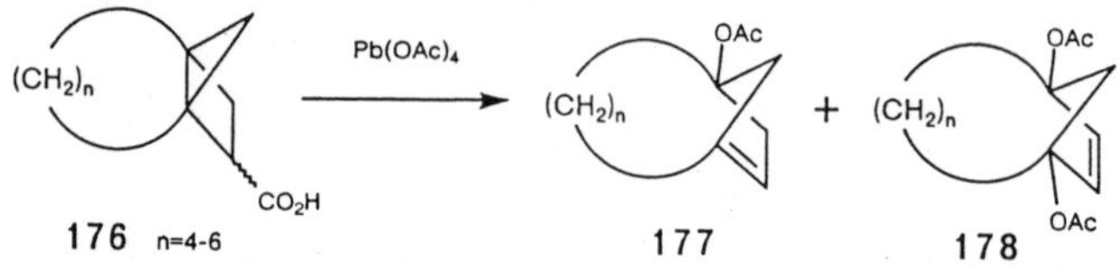

176 n=4-6 177 178

By utilizing this technique, an efficient method for preparation of [6]paracyclophane (**180a**), the smallest-bridged [n]paracyclophane so far isolated, was developed by Tobe et al.[114] On oxidation of [6.2.2]propellenecarboxylic acid (**179**) with lead tetraacetate, cyclophane **180a** (39%) and acetate **181** (21%) were obtained. The latter was converted to **180a** quantitatively by treatment with a base.

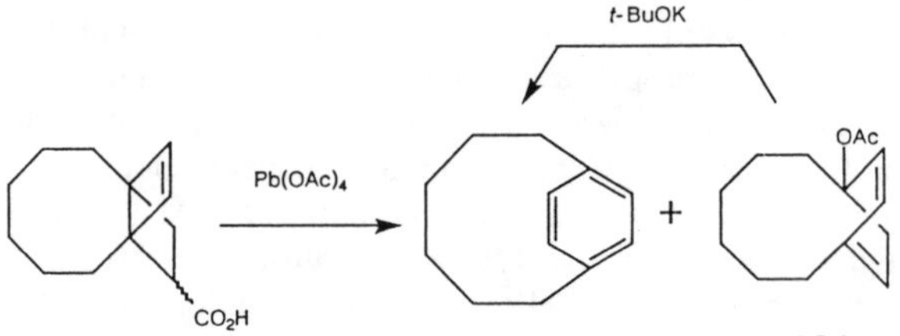

179 180a 181

Valence bond isomerization of unsaturated [*n*.2.2]propellanes and [*n*.4.2]propellanes provided a novel and general synthetic route for strained [*n*]paracyclophanes and [*n*](1,4)cyclooctatetraenophanes, respectively. Previously, thermal isomerization of [6.2.2]propelladiene (**182a**), a Dewar benzene type valence isomer of [6]paracyclophane (**180a**), was reported by Jones and Bickelhaupt to take place smoothly ($E_a = 19.9$ kcal mol^{-1}).[115] This transformation was applied by Tobe to monoester and diester derivatives **182b** ($E_a = 24.9$ kcal mol^{-1}) and **182c** ($E_a = 29.0$ kcal mol^{-1}) for the synthesis of substituted [6]paracyclophanes **180b** and **180c**, for which X-ray structure analyses were undertaken.[114,116] Moreover, [6]paracyclophene (**183**), the most highly deformed paracyclophane that has been isolated, was prepared by similar transformations.[117]

182a R^1=R^2=H
182b R^1=H, R=CO_2Me
182c R^1=R^2=CO_2Me

180a,b,c

183

184

185

186

187

Thermolysis of the next lower homolog, [5.2.2]propelladiene (**184**), however, gave spiro triene **185** and benzocycloheptene (**186**) although intermediacy of [5]paracyclophane (**187**) was suggested.[118] On the other hand, Bickelhaupt and Tobe found that irradiation of **184** and the derivatives with electron-withdrawing group(s) at low temperature (−60°C) gave [5]paracyclophane (**187**) and its derivatives despite low yields (2–15%).[119] These [5]paracyclophanes were characterized by their ^{1}H NMR and UV spectra but have eluded isolation due to their lability.

Irradiation of the lower homolog, [4.2.2]propelladiene (**188**), in methanol was performed by Bickelhaupt, Tsuji, and Nishida, affording [4]paracyclophane (**189**) as a reactive intermediate that was trapped by methanol to give a 1,4-adduct **190**.[120] Cyclophane **189** was characterized by UV spectra taken at 77 K. Moreover, similar irradiation of tetraene **101a** produced [4]paracyclophadiene (**191**), which was trapped as bis-Diels–Alder adduct **192**.[121]

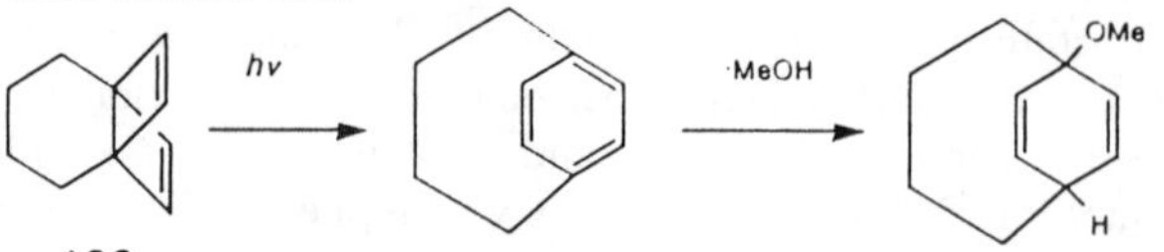

hν MeOH OMe H

188 189 190

hν

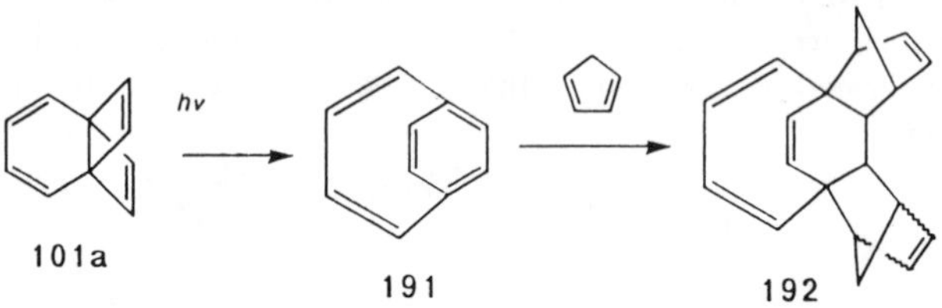

101a 191 192

Optically active [5](1,4)cyclooctatetraenophane (**194**) was prepared by Paquette and Trova[122] from [5.4.2]propellatriene (**193**) (although not isolated) using valence isomerization of [*n*.4.2]propellatrienes developed previously. The chiral cyclooctatetraene **194** did not undergo racemization because of steric effects induced by the relatively short bridge.

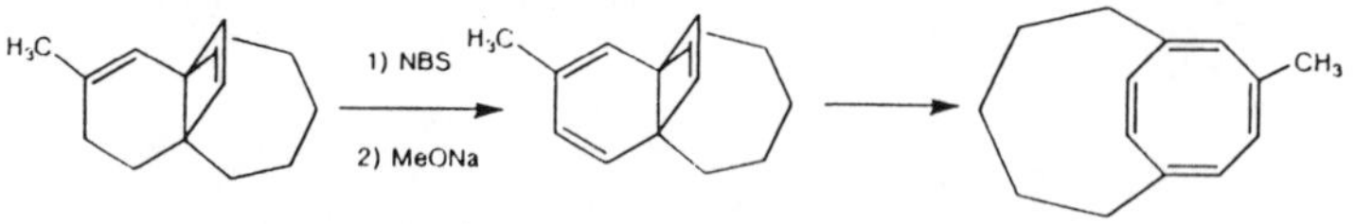

H_3C 1) NBS 2) MeONa H_3C CH_3

193 194

5.4D. [*m*.3.3]Propellanes

Mehta reported that acid-catalyzed rearrangement of a [3.3.3]propellanol **196**, which was derived from readily available propellanone **195**, gave tricyclic enone **197a**.[123] Enone **197a** was transformed straightforwardly into descarboxyquadrone (**161**). Rearrangement of **195** and its methyl-substituted derivatives afforded ketone **197b** and triquinane **198**; the ratio being critically dependent on the substitution pattern. Similarly, mutual isomerization between [3.3.3]propellane (**199**), tricyclo[4.3.2.0^{1,5}]undecane (**200**), and angularly fused triquinane **201** was investigated by Kakiuchi and Fitjer in connection with biogenesis of sesquiterpenes such as modhephene (**3**), quadrone (**160**), and isocomene (**202**) as well as from the viewpoint of homoadamantane rearrangement of tricycloundecanes.[124]

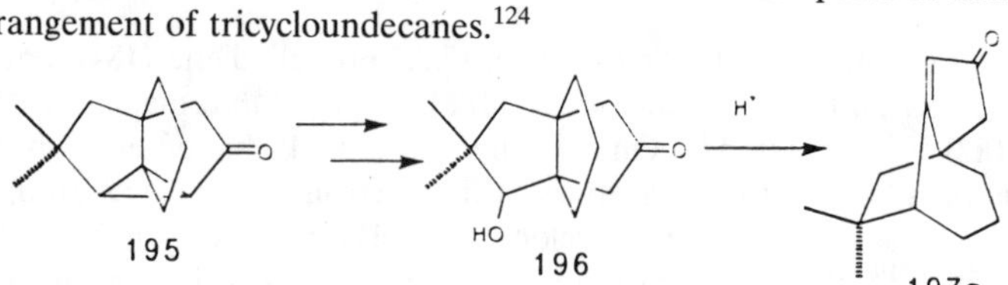

O HO H^+ O

195 196 197a

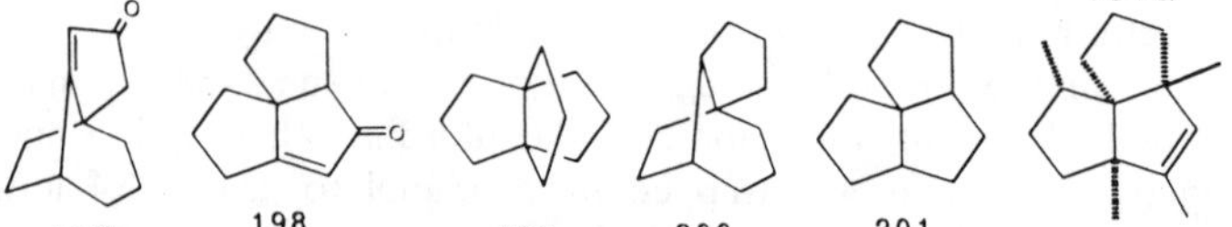

197b 198 199 200 201 202

A novel and convenient way to [*n*](1,5)cyclooctatetraenophanes (**204**) ($n = 5, \ldots, 8, 10$) was developed by Paquette et al.[125] based on valence isomerization by FVP of cycloalkane-fused semibullvalenes **203**, [*n*.3.3]propelladiene derivatives. [2.2](1,5)Cyclooctatetraenophane **205** was also prepared by this method.[126]

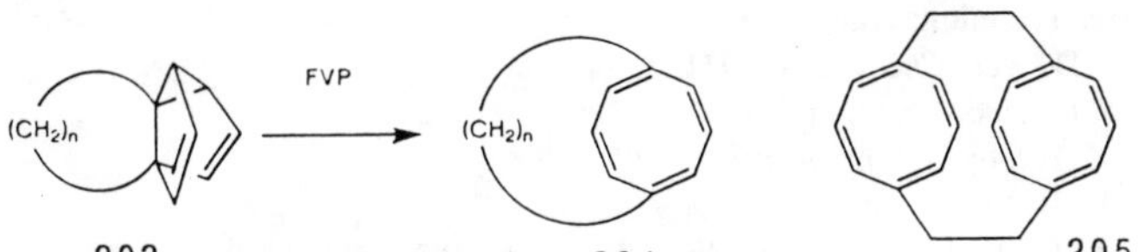

FVP

$(CH_2)_n$

203 204 205

5.5. Naturally Occurring Propellanes

Modhephene (**3**)[8] and 13-acetoxymodhephene (**206**)[127] are the only two carbocyclic propellanes that have been discovered in nature. **3** attracted a lot of interest as an intriguing target because of its unique structure, and a number of total syntheses of **3** have been accomplished.[101,128] Moreover, a variety of novel synthetic methods were developed during the stereocontrolled synthesis of **3**. Because some of them have already been the subject of review articles,[129] only references to the literature are listed here.

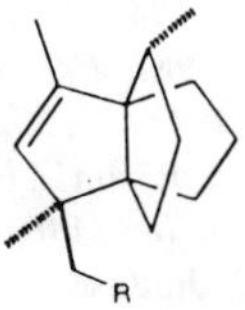

3 R=H
206 R=OAc

In connection with the natural products, some novel methods for the construction of [3.3.3]propellane skeleton have been developed. For example, transannular cyclization of eight-membered bisolefin **207** produced propellenes **208** via an intramolecular version of metal-catalyzed cycloaddition of methylenecyclopropane with olefins.[130] Acid-catalyzed rearrangement of bisspirane such as **209** was evolved to prepare propellanones such as **210**.[131]

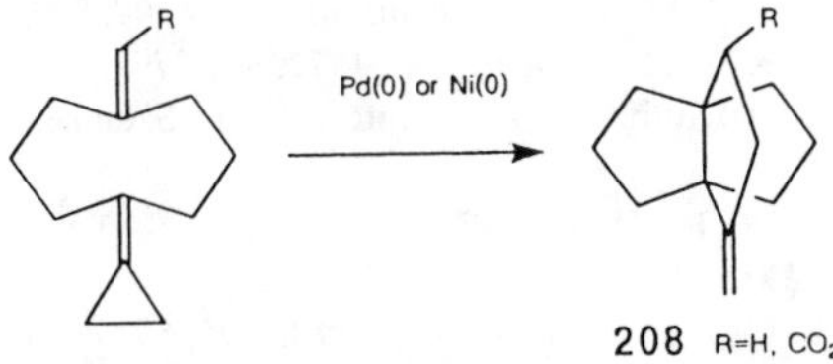

Pd(0) or Ni(0)

207

208 R=H, CO_2Me

H^+

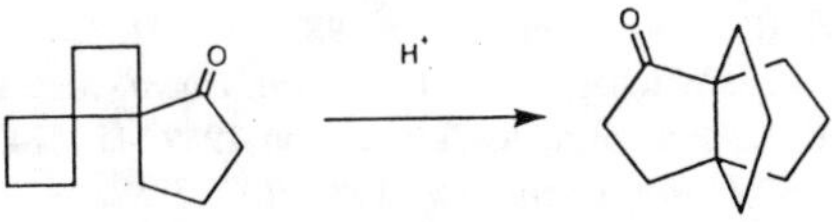

209 210

References

1. Altman, J.; Babad, E.; Itzchaki, J.; Ginsburg, D. *Tetrahedron Suppl.* **1966**, *8* (Part 1), 279.
2. Ginsburg, D. *Propellanes—Structure and Reactions*; Verlag Chemie: Weinheim, 1975; *Sequel I*, Department of Chemistry, Technion: Haifa, 1980; *Sequel II*, Department of Chemistry, Technion: Haifa, 1985.
3. Ginsburg, D. *Acc. Chem. Res.* **1972**, *5*, 249.
4. Ginsburg, D. *Acc. Chem. Res.* **1974**, *7*, 286.
5. Wiberg, K. B. *Acc. Chem. Res.* **1984**, *17*, 379.
6. Wiberg, K. B. *Chem. Rev.* **1989**, *89*, 975.
7. Wiberg, K. B.; Walker, F. H. *J. Am. Chem. Soc.* **1982**, *104*, 5239.
8. Zalkow, L. H.; Harris, R. N., III; van Derveer, D. *J. Chem. Soc., Chem. Commun.* **1978**, 420.
9. Adcock, W.; Iyer, V. S.; Kitching, W.; Young, D. *J. Org. Chem.* **1985**, *50*, 3706. Adcock, W.; Gangodawila, H. *J. Org. Chem.* **1989**, *54*, 6040.
10. Wiberg, K. B. *J. Am. Chem. Soc.* **1983**, *105*, 1227.
11. Walker, F. H.; Wiberg, K. B.; Michl, J. *J. Am. Chem. Soc.* **1982**, *104*, 2056.
12. Wiberg, K. B.; Walker, F. H.; Pratt, W. E.; Michl, J. *J. Am. Chem. Soc.* **1983**, *105*, 3638.
13. Semmler, K.; Szeimies, G.; Belzner, J. *J. Am. Chem. Soc.* **1985**, *107*, 6410. Belzner, J.; Bunz, U.; Semmler, K.; Szeimies, G.; Opitz, K.; Schlüter, A.-D. *Chem. Ber.* **1989**, *122*, 397.
14. Belzner, J.; Gareiß, B.; Polborn, K.; Schmid, W.; Semmler, K.; Szeimies, G. *Chem. Ber.* **1989**, *122*, 1509.
15. Mlinarić-Majerski, K.; Majerski, Z. *J. Am. Chem. Soc.* **1983**, *105*, 7389.
16. Vinković, V.; Majerski, Z. *J. Am. Chem. Soc.* **1982**, *104*, 4027.
17. Majerski, Z.; Zuanić, M. *J. Am. Chem. Soc.* **1987**, *109*, 3496.
18. Reingold, I. D.; Drake, J. *Tetrahedron Lett.* **1989**, *30*, 1921.
19. Paquette, L. A.; Kobayashi, T.; Gallucci, J. C. *J. Am. Chem. Soc.* **1988**, *110*, 1305.
20. For a recent review, see Borden, W. T. *Chem. Rev.* **1989**, *89*, 1095.
21. Wiberg, K. B.; Matturo, M. C.; Okarma, P. J.; Jason, M. E.; Dailey, W. P.; Burgmaier, G. J.; Bailey, W. F.; Warner, P. *Tetrahedron* **1986**, *42*, 1895.
22. Wiberg, K. B.; Bonneville, G. *Tetrahedron Lett.* **1982**, *23*, 5385.
23. Eaton, P. E.; Maggini, M. *J. Am. Chem. Soc.* **1988**, *110*, 7230.
24. Szeimies-Seebach, U.; Schoffer, A.; Romer, R.; Szeimies, G. *Chem. Ber.* **1981**, *114*, 1767.
25. Schlüter, A.-D.; Harnish, H.; Harnish, J.; Szeimies-Seebach, U.; Szeimies, G. *Chem. Ber.* **1985**, *118*, 3513.
26. Baumgartel, O.; Harnish, J.; Szeimies, G.; Meerssche, M. V.; Germain, G.; Declerq, J.-P. *Chem. Ber.* **1983**, *116*, 2205.
27. Newton, M. D.; Schulman, J. M. *J. Am. Chem. Soc.* **1972**, *94*, 773.
28. Stohrer, W.; Hoffmann, R. *J. Am. Chem. Soc.* **1972**, *94*, 779.
29. Chakrabarti, P.; Seiler, P.; Dunitz, J. D.; Schlüter, A.-D.; Szeimies, G. *J. Am. Chem. Soc.* **1981**, *103*, 7378.
30. Wiberg, K. B.; Dailey, W. P.; Walker, F. H.; Waddell, S. T.; Crocker, L. S.; Newton, M. *J. Am. Chem. Soc.* **1985**, *107*, 7247.
31. Jackson, J. E.; Allen, L. C. *J. Am. Chem. Soc.* **1984**, *105*, 591.
32. Messmer, R. P.; Schultz, P. A. *J. Am. Chem. Soc.* **1986**, *108*, 7407.
33. Feller, D.; Davidson, E. R. *J. Am. Chem. Soc.* **1987**, *109*, 4133.
34. Wiberg, K. B.; Bader, R. F. W.; Lau, C. D. H. *J. Am. Chem. Soc.* **1987**, *109*, 985.
35. Ushio, T.; Kato, T.; Ye, K.; Imamura, A. *Tetrahedron* **1989**, *45*, 7743.
36. Hedberg, L.; Hedberg, K. *J. Am. Chem. Soc.* **1985**, *107*, 7257.
37. Jarret, R. M.; Cusumano, L. *Tetrahedron Lett.* **1990**, *31*, 171.

38. Honogger, E.; Huber, H.; Heilbronner, E.; Deiley, W. P.; Wiberg, K. B. *J. Am. Chem. Soc.* **1985**, *107*, 7172.
39. Gleiter, R.; Pfeifer, K.-H.; Szeimies, G.; Belzner, J.; Lehne, K. *J. Org. Chem.* **1990**, *55*, 636.
40. Eckert-Maksić, M.; Mlinarić-Majerski, K.; Majerski, Z. *J. Org. Chem.* **1987**, *52*, 2098.
41. Seiler, P.; Belzner, J.; Bunz, U.; Szeimies, G. *Helv. Chim. Acta* **1988**, *71*, 2100.
42. Kottirsch, G.; Polborn, K.; Szeimies, G. *J. Am. Chem. Soc.* **1988**, *110*, 5588.
43. Newton, M. D.; Schulman, J. M. *J. Am. Chem. Soc.* **1972**, *94*, 4391.
44. Danneberg, J. J.; Prociv, T. M. *J. Chem. Soc., Chem. Commun.* **1973**, 291.
45. Eaton, P. E.; Temme, G. H. *J. Am. Chem. Soc.* **1973**, *95*, 7508.
46. Belzner, J.; Szeimies, G. *Tetrahedron Lett.* **1986**, *27*, 5839.
47. Blaustein, M. A.; Berson, J. A. *J. Am. Chem. Soc.* **1983**, *105*, 6337.
48. Baumgart, K.-D.; Harnish, H.; Szeimies-Seebach, U.; Szeimies, G. *Chem. Ber.* **1985**, *118*, 2883.
49. Majerski, Z.; Mlinarić-Majerski, K. *J. Org. Chem.* **1986**, *51*, 3219.
50. Kogay, B. E.; Sokolenko, W. A. *Tetrahedron Lett.* **1983**, *24*, 613.
51. Wiberg, K. B.; Waddell, S. T.; Laidig, K. *Tetrahedron Lett.* **1986**, *27*, 1553.
52. Bunz, U.; Szeimies, G. *Tetrahedron Lett.* **1989**, *30*, 2087.
53. Bunz, U.; Polborn, K.; Wagner, H.-U.; Szeimies, G. *Chem. Ber.* **1988**, *121*, 1785.
54. Friedli, A. C.; Kaszynski, P.; Michl, J. *Tetrahedron Lett.* **1989**, *30*, 455.
55. Kaszynski, P.; Michl, J. *J. Am. Chem. Soc.* **1988**, *110*, 5225.
56. Kaszynski, P.; Michl, J. *J. Org. Chem.* **1988**, *53*, 4593.
57. Belzner, J.; Szeimies, G. *Tetrahedron Lett.* **1987**, *28*, 3099.
58. McGarry, P. F.; Johnston, L. J.; Scaiano, J. C. *J. Org. Chem.* **1989**, *54*, 6133.
59. Mlinarić-Majerski, K.; Majerski, Z.; Rakvin, B.; Veksli, Z. *J. Org. Chem.* **1989**, *54*, 545.
60. Wiberg, K. B.; Waddell, S. T. *Tetrahedron Lett.* **1987**, *28*, 151.
61. McGarry, P. F.; Johnston, L. J.; Scaiano, J. C. *J. Am. Chem. Soc.* **1989**, *111*, 3750.
62. Wiberg, K. B.; Matturro, M. G. *Tetrahedron Lett.* **1981**, *22*, 3481.
63. Baldwin, J. E.; Cheng, G. E. C. *J. Org. Chem.* **1982**, *47*, 848.
64. Tobe, Y.; Kanazawa, Y.; Kakaiuchi, K.; Odaira, Y. *Chem. Lett.* **1982**, 1177.
65. Schleyer, P. v. R.; Janoschek, R. *Angew. Chem., Int. Ed. Engl.* **1987**, *26*, 1267.
66. Nagase, S.; Kudo, T. *Organometallics* **1987**, *6*, 2456.
67. Schoeller, W. W.; Dabisch, T.; Busch, T. *Inorg. Chem.* **1987**, *26*, 4383.
68. Sita, L. R.; Bickerstaff, R. D. *J. Am. Chem. Soc.* **1989**, *111*, 6454.
69. Maggio, J. E.; Simmons, H. E., III; Kouba, J. K. *J. Am. Chem. Soc.* **1981**, *102*, 1579.
70. Simmons, H. E., III; Maggio, J. E. *Tetrahedron Lett.* **1981**, *22*, 287. Paquette, L. A.; Vazeux, M. *Tetrahedron Lett.* **1981**, *22*, 291.
71. Benner, S. A.; Maggio, J. E.; Simmons, H. E., III *J. Am. Chem. Soc.* **1981**, *103*, 1581.
72. Kuck, D.; Schuster, A. *Angew. Chem., Int. Ed. Engl.* **1988**, *27*, 1192.
73. Jendralia, H.; Jelick, K.; DeLucca, G.; Paquette, L. A. *J. Am. Chem. Soc.* **1986**, *108*, 3731.
74. Paquette, L. A.; Waykole, L.; Jendralla, H.; Cottrell, C. E. *J. Am. Chem. Soc.* **1986**, *108*, 3739.
75. Waykole, L.; Paquette, L. A. *J. Am. Chem. Soc.* **1987**, *109*, 3174.
76. Paquette, L. A.; Ohkata, K.; Jelich, K.; Kitching, W. *J. Am. Chem. Soc.* **1983**, *105*, 2800.
77. Klobucar, W. D.; Paquette, L. A.; Blount, J. F. *J. Org. Chem.* **1981**, *46*, 4021.
78. Tsuji, T.; Komiya, Z.; Nishida, S. *Tetrahedron Lett.* **1980**, *21*, 3583. Tsuji, T.; Nishida, S. *Tetrahedron Lett.* **1983**, *24*, 3361.
79. Gleiter, R.; Krennich, G.; Bischop, P.; Tsuji, T.; Nishida, S. *Helv. Chim. Acta* **1986**, *69*, 962.
80. Tsuji, T.; Nishida, S. *Tetrahedron Lett.* **1983**, *24*, 1269. Tsuji, T.; Nishida, S. *Chem. Lett.* **1986**, 1389.

81. Fessner, W.-D.; Sedelmaier, G.; Spurr, P. R.; Rihs, G.; Prinzbach, H. *J. Am. Chem. Soc.* **1987**, *109*, 4626.
82. Fessner, W.-D.; Murty, B. A. R. C.; Worth, J.; Hunkler, D.; Fritz, H.; Prinzbach, H.; Roth, W. D.; Schleyer, P. v. R.; McEwen, A. B.; Maier, W. F. *Angew. Chem., Int. Ed. Engl.* **1987**, *26*, 452. Spurr, P. R.; Murty, B. A. R. C.; Fessner, W.-D.; Fritz, H.; Prinzbach, H. *Angew. Chem., Int. Ed. Engl.* **1987**, *26*, 455.
83. For a review, see Paquette, L. A. *Chem. Rev.* **1989**, *89*, 1051.
84. Turkenburg, L. A. M.; de Wolf, W. H.; Bickelhaupt, F.; Stam, C. H.; Konijn, M. *J. Am. Chem. Soc.* **1982**, *104*, 3471.
85. Paquette, L. A.; Kesselmager, M. A.; Rogers, R. D. *J. Am. Chem. Soc.* **1990**, *112*, 284.
86. Murthy, G. S.; Hassenruck, K.; Lynch, U. M.; Michl, J. *J. Am. Chem. Soc.* **1989**, *111*, 7262.
87. Bothe, H.; Schlüter, A.-D. *Makromol. Chem., Rapid Commun.* **1988**, *9*, 529.
88. Schlüter, A.-D. *Macromolecules* **1988**, *21*, 1208.
89. Opitz, K.; Schlüter, A.-D. *Angew. Chem., Int. Ed. Engl.* **1989**, *28*, 456.
90. Scott, L. T.; Brunsvold, W. R.; Kirms, M. A.; Erden, I. *J. Am. Chem. Soc.* **1981**, *103*, 5216.
91. Scott, L. T.; Oda, M.; Erden, I. *J. Am. Chem. Soc.* **1985**, *107*, 7213.
92. Scott, L. T.; Hashemi, M. M. *Tetrahedron* **1986**, *42*, 1830.
93. For recent reviews, see Shea, K. J. *Tetrahedron* **1980**, *36*, 1683, and Warner, P. M. *Chem. Rev.* **1989**, *89*, 1067.
94. Jenneskens, L. W.; de Kanter, F. J. J.; Turkenburg, L. A. M.; de Boer, H. J. R.; de Wolf, W. H.; Bickelhaupt, F. *Tetrahedron* **1984**, *40*, 4401.
95. Warner, P. M.; Ah-King, M.; Palmer, R. F. *J. Am. Chem. Soc.* **1982**, *104*, 7166.
96. Warner, P. M.; Chang, S.-C.; Powell, D. R.; Jacobson, R. A. *J. Am. Chem. Soc.* **1980**, *102*, 5125.
97. Warner, P. M.; Chang, S.-C.; Koszewski, N. J. *J. Org. Chem.* **1985**, *50*, 2605.
98. Warner, P. M.; Chu, I.-S.; Boulanger, W. *Tetrahedron Lett.* **1983**, *24*, 4165.
99. Hartwig, J. F.; Jones, M., Jr.; Moss, R. A.; Lawrynowicz, W. *Tetrahedron Lett.* **1986**, *27*, 5907. Warner, P. M.; Lu, S.-L.; Gurumurthy, R. *J. Phys. Org. Chem.* **1988**, *1*, 281. Le, N. A.; Jones, M., Jr.; Bickelhaupt, F.; de Wolf, W. H. *J. Am. Chem. Soc.* **1989**, *111*, 8491. Chen, N.; Jones, M., Jr. *Tetrahedron Lett.* **1989**, *30*, 6969.
100. Cargill, R. L.; Jackson, T. E.; Peet, N. P.; Pond, D. M. *Acc. Chem. Res.* **1974**, *7*, 106, and references cited therein.
101. Smith, A. B., III; Jerris, P. J. *J. Org. Chem.* **1982**, *47*, 1845.
102. Eaton, P. E.; Jobe, P. G. *J. Am. Chem. Soc.* **1980**, *102*, 6636.
103. Kakiuchi, K.; Itoga, K.; Tsugaru, T.; Hato, Y.; Tobe, Y.; Odaira, Y. *J. Org. Chem.* **1984**, *49*, 659. Cargill, R. L.; Bushey, D. F.; Dalton, J. R.; Prasad, R. S.; Dyer, R. D.; Border, J. *J. Org. Chem.* **1981**, *46*, 3389.
104. Kakiuchi, K.; Tsugaru, T.; Takeda, M.; Wakaki, I.; Tobe, Y.; Odaira, Y. *J. Org. Chem.* **1985**, *50*, 488. Smith, A. B., III; Wexlaer, B. A.; Tu, C.-Y.; Konopelski, J. P. *J. Am. Chem. Soc.* **1985**, *107*, 1308.
105. Smith, A. B., III; Konopelski, J. P. *J. Org. Chem.* **1984**, *49*, 4094.
106. Kakiuchi, K.; Nakao, T.; Takeda, M.; Tobe, Y.; Odaira, Y. *Tetrahedron Lett.* **1984**, *25*, 557. Kakiuchi, K.; Ue, M.; Tadaki, T.; Tobe, Y.; Odaira, Y. *Chem. Lett.* **1986**, 507.
107. Jeffrey, D. A.; Cogen, J. M.; Maier, W. F. *J. Org. Chem.* **1986**, *51*, 3026.
108. Tobe, Y.; Hayauchi, Y.; Sakai, Y.; Odaira, Y. *J. Org. Chem.* **1980**, *45*, 637. Tobe, Y.; Terashima, K.; Sakai, Y.; Odaira, Y. *J. Am. Chem. Soc.* **1981**, *103*, 2307. Eaton, P. E.; Jobe, P. G.; Reingold, I. D. *J. Am. Chem. Soc.* **1984**, *106*, 6437.
109. Tobe, Y.; Odaira, Y. Unpublished results.
110. Tobe, Y.; Ohtani, M.; Kakiuchi, K.; Odaira, Y. *J. Org. Chem.* **1983**, *48*, 5114.
111. Sakai, Y.; Toyotani, S.; Ohtani, M.; Matsumoto, M.; Tobe, Y.; Odaira, Y. *Bull. Chem. Soc. Jpn.* **1981**, *54*, 1474.

112. Sakai, Y.; Terashima, K.; Tobe, Y.; Odaira, Y. *Bull. Chem. Soc. Jpn.* **1981**, *54*, 2229.
113. Tobe, Y.; Fukuda, Y.; Kakiuchi, K.; Odaira, Y. *J. Org. Chem.* **1984**, *49*, 2012.
114. Tobe, Y.; Ueda, K.-I.; Kakiuchi, K.; Odaira, Y.; Kai, Y.; Kasai, N. *Tetrahedron* **1986**, *42*, 1851.
115. Kammula, S. L.; Iroff, L. D.; Jones, M., Jr.; van Straten, J. W.; de Wolf, W. H.; Bickelhaupt, F. *J. Am. Chem. Soc.* **1977**, *99*, 5815.
116. Tobe, Y.; Nakayama, A.; Kakiuchi, K.; Odaira, Y.; Kai, Y.; Kasai, N. *J. Org. Chem.* **1987**, *52*, 2639.
117. Tobe, Y.; Ueda, K.-I.; Kaneda, T.; Kakiuchi, K.; Odaira, Y.; Kai, Y.; Kasai, N. *J. Am. Chem. Soc.* **1987**, *109*, 1136.
118. van Straten, J. W.; Turkenburg, L. A. M.; de Wolf, W. H.; Bickelhaupt, F. *Recl. Trav. Chim. Pays-Bas* **1985**, *104*, 89.
119. Jenneskens, L. W.; de Kanter, F. J. J.; Kraakman, P. A.; Turkenburg, L. A. M.; Koolhaas, W. E.; de Wolf, W. H.; Bickelhaupt, F.; Tobe, Y.; Kakiuchi, K.; Odaira, Y. *J. Am. Chem. Soc.* **1985**, *107*, 3716. Tobe, Y.; Kaneda, T.; Kakiuchi, K.; Odaira, Y. *Chem. Lett.* **1985**, 1301. Kostermans, G. B. M.; de Wolf, W. H.; Bickelhaupt, F. *Tetrahedron* **1987**, *43*, 2955.
120. Tsuji, T.; Nishida, S. *J. Am. Chem. Soc.* **1988**, *110*, 2157. Kostermans, G. B. M.; Bobeldijk, M.; de Wolf, W. H.; Bickelhaupt, F. *J. Am. Chem. Soc.* **1987**, *109*, 2471.
121. Tsuji, T.; Nishida, S. *J. Am. Chem. Soc.* **1989**, *111*, 368.
122. Paquette, L. A.; Trova, M. P. *J. Am. Chem. Soc.* **1988**, *110*, 8197.
123. Mehta, G.; Pramod, K.; Subrahmanyam, D. *J. Chem. Soc., Chem. Commun.* **1986**, 247. Mehta, G.; Subrahmanyam, D. *Tetrahedron Lett.* **1989**, *30*, 2709.
124. Kakiuchi, K.; Kumanoya, S.; Ue, M.; Tobe, Y.; Odaira, Y. *Chem. Lett.* **1985**, 989. Kakiuchi, K.; Ue, M.; Wakaki, I.; Tobe, Y.; Odaira, Y.; Yasuda, M.; Shima, K. *J. Org. Chem.* **1986**, *51*, 281. Fitjer, L.; Kanschik, A.; Majewski, M. *Tetrahedron Lett.* **1985**, *26*, 5277. Fitjer, L.; Kanschik, A.; Majewski, M. *Tetrahedron Lett.* **1988**, *29*, 5525.
125. Paquette, L. A.; Trova, M. P.; Luo, J.: Clough, A. E.; Anderson, L. B. *J. Am. Chem. Soc.* **1990**, *112*, 228.
126. Paquette, L. A.; Kesselmayer, M. A. *J. Am. Chem. Soc.* **1990**, *112*, 1258.
127. Bohlman, F.; Zdero, C.; Bohlman, R.; King, R. M.; Robinson, H. *Phytochemistry* **1980**, *19*, 579.
128. Karpf, M.; Dreiding, A. S. *Helv. Chim. Acta* **1981**, *64*, 1123. Schostarez, H.; Paquette, L. A. *Tetrahedron* **1981**, *37*, 4431. Oppolzer, W.; Marazza, F. *Helv. Chim. Acta* **1981**, *64*, 1575. Oppolzer, W.; Battig, K. *Helv. Chim. Acta* **1981**, *64*, 2489. Wender, P. A.; Dreyer, G. B. *J. Am. Chem. Soc.* **1982**, *104*, 5805. Wrobel, J.; Takahashi, K.; Honkan, V.; Lannoye, G.; Cook, J. M.; Bertz, S. H. *J. Org. Chem.* **1983**, *48*, 141. Tobe, Y.; Yamashita, S.; Yamashita, T.; Kakiuchi, K.; Odaira, Y. *J. Chem. Soc., Chem. Commun.* **1984**, 1259. Wilkening, D.; Mundy, B. P. *Tetrahedron Lett.* **1984**, *25*, 4619. Mash, E. A.; Math, S. K.; Flann, C. J. *Tetrahedron Lett.* **1988**, *29*, 2147.
129. Paquette, L. A. *Topics Curr. Chem.* **1987**, *119*, 1. Paquette, L. A.; Doherty, A. M. *Polyquinane Chemistry*; Springer-Verlag: Berlin, 1987.
130. Yamago, S.; Nakamura, E. *J. Chem. Soc., Chem. Commun.* **1988**, 1112.
131. Fitjer, L.; Majewski, M.; Kanschik, A. *Tetrahedron Lett.* **1988**, *29*, 1263. Fitjer, L.; Quabeck, U. *Angew. Chem., Int. Ed. Engl.* **1989**, *28*, 94.

6

Cyclophanes from Vinylarenes

Jun Nishimura

Gunma University, Kiryu, Japan

Contents

6.1. Introduction
6.2. Intramolecular [2 + 2] Photocycloaddition of Styrene Derivatives toward a New Class of Cyclophanes
6.2A. [2 + 2] Photocyclization of Styrene
6.2B. Intramolecular [2 + 2] Photocycloaddition of Styrene and Vinylnaphthalene Derivatives
6.2C. Birch Reduction of (1,2)Ethano[2.*n*]cyclophanes
6.2D. Multibridged Cyclophanes
6.2E. Crown Ethers
6.2F. Scope and Limitation
6.3. Cationic Cyclocodimerization
6.3A. Cationic Cyclocodimerization: Conditions and Requirements
6.3B. Cationic Cyclocodimerization: Scope and Limitation
6.3C. Cationic Cyclization of Styrene Derivatives
Acknowledgments
References

6.1. Introduction

This chapter contains work on the intramolecular reactions of vinylarenes and their applications toward cyclophane syntheses that has been done in 10 years by this author's group.[1] Because the subjects herein are not arranged by the date when they were carried out, a certain amount of explanation is necessary to look over the contents.

Vinylarenes like styrene are attractive compounds in both organic chemistry and polymer science. They are highly reactive and polymerizable, as is well known. Paradoxically, their reactivity, or instability, is very attractive and seems to be the main property to be utilized in synthetic chemistry. One can choose almost all kinds of cationic, neutral, and anionic species to react with these vinylarenes and may obtain products in all cases.

Their *intramolecular* reactions seemed to be a bonanza a decade ago. For such a purpose, one needs some well-designed high-molecular-weight vinylarenes. Sufficient synthetic methods, however, especially for high-boiling-point vinylarenes had not been provided. Therefore the first tough task was to develop a general and facile method for these vinylarenes and to obtain them in reasonable yields. This subject is discussed in a few of the articles reported previously.[2]

In 1980, intramolecular reactions of styrene derivatives like 1,3-bis(*p*-vinylphenyl)propane were investigated in order to make cyclophanes,[3] whose general structures and properties are attractive in the field of physical and organic chemistry.[4]

First of all, I would like to emphasize how the so-called cationic cyclocodimerization[3] (see Section 6.3), one of the new cyclophane synthetic methods, was found. Thus, it is a quite common phenomenon that a cation can form a donor–acceptor complex with a π donor and stabilize itself. Hence, if the interaction is woven into an intramolecular system dexterously, a macrocyclic system or even a strained macrocyclic system is envisioned to be produced by cationic methods. On the other hand, when a molecule has donating systems at both ends that are going to be connected intramolecularly, any anionic C—C bond formation is scarcely suited because of the electrostatic repulsion between them. One of the typical cases, where the anionic C—C bond formation has such a difficulty, is the synthesis of cyclophanes that have two donating π systems, namely, arene moieties, in a face-to-face arrangement. For such cases, however, the cationic method may form a C—C bond quite readily, because an attractive force can work between a donating π system and a cationic center, which may be derived from one of the π systems, and can bring the two groups close to each other for the intramolecular bond formation. In fact, from such an expectation the cationic cyclocodimerization was found to make [3.n]cyclophanes.[3]

It is interesting to survey the synthetic methods of [3.3]paracyclophane because it is quite unique within a large family of cyclophanes. The strain (12 kcal mol^{-1})[5] is not that large compared with those of other homologs, but its synthesis was rather difficult and has experienced an intriguing progress. The first stage of its history was just in the struggle with the repulsive interaction between two benzene rings (π donors) that appeared in Cram's ingenious pioneer works;[6] that is, Cram and his associates reduced 1,3-diphenylpropane derivative to 1,3-dicyclohexylpropane derivative in order to avoid the repulsive interaction during the successive cyclization, namely, intramolecular acyloin condensation (a radical anion as the intermediate).

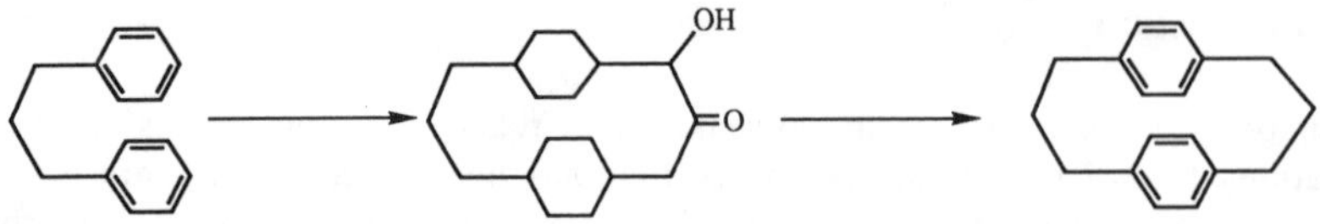

OH

=O

(6.1)

At the second stage again Cram and his associates made major contributions.[7] They used [2.2]paracyclophane as the starting material and modified linkages to make [*n.m*]paracyclophanes, including [3.3]paracyclophane, because [2.2]paracyclophane is easily made from xylylene and is even commercially available.[8]

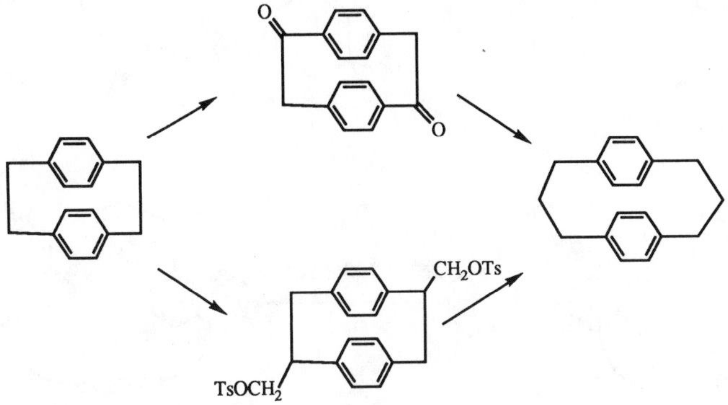

(6.2)

At the third stage several groups reported utilizing the sulfur extrusion method in the synthesis of cyclophanes, especially [3.3]cyclophanes.[9] They used, namely, the entropy advantage. Of course, dithiacyclophanes have two atoms more than the final target cyclophanes so that it is more easily prepared.[8]

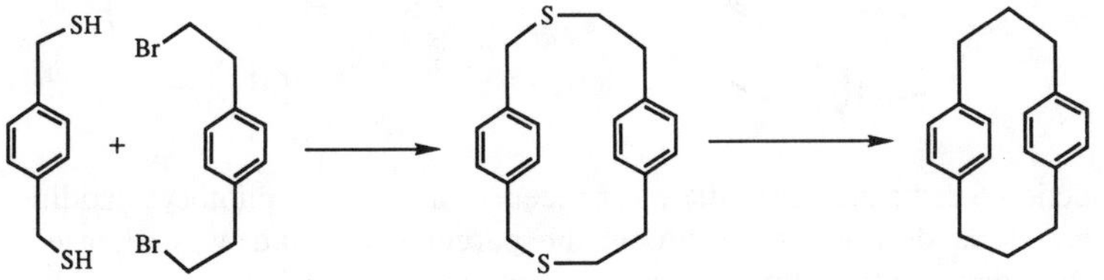

(6.3)

At the fourth stage of the history, people started to weave the intramolecular attractive interaction at the cyclization step. The pioneer work at this stage was reported by Tsuji and Nishida, who successfully cyclized both tetracyanoquinodimethane (acceptor) and spiro[2.2.2.2]deca-4,9-diene (donor) into the [3.3]paracyclophane skeleton in a surprisingly high 60% yield and suggested the donor–acceptor interaction between the starting materials.[10] Their evidence strengthened our working hypothesis mentioned previously. Quite recently, Mitchell and co-workers reported the synthesis of long-sought *syn*-[2.2]metacyclophane, using the donor–acceptor interaction working at the sulfur extrusion.

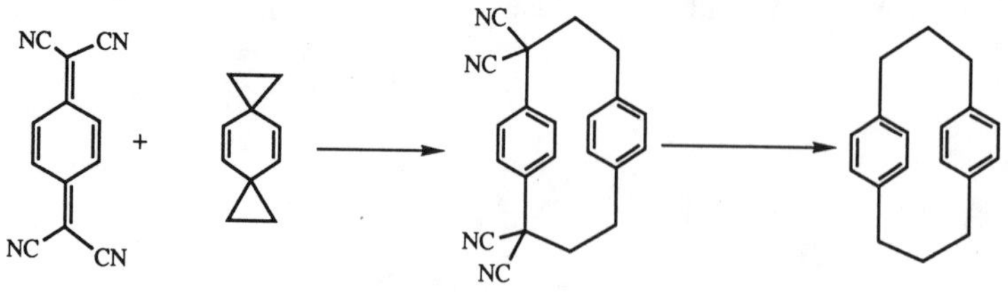

(6.4)

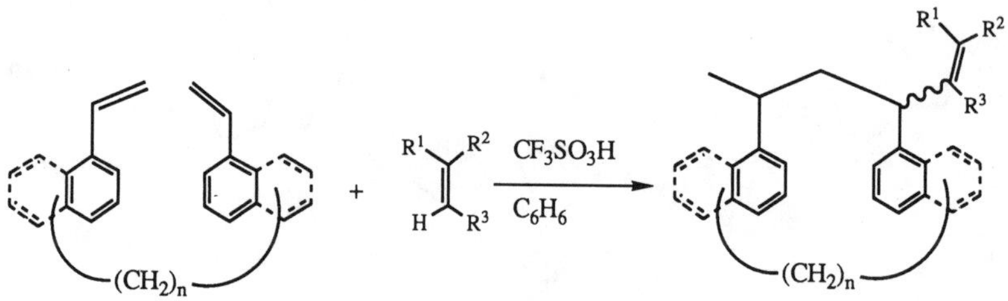

(6.5)

Actually, cationic cyclocodimerization (6.5) led to the successful synthesis of [3.3]cyclophane skeletons (see Section 6.3). So this cyclocodimerization can be added as another example at the fourth stage of the synthetic approach.

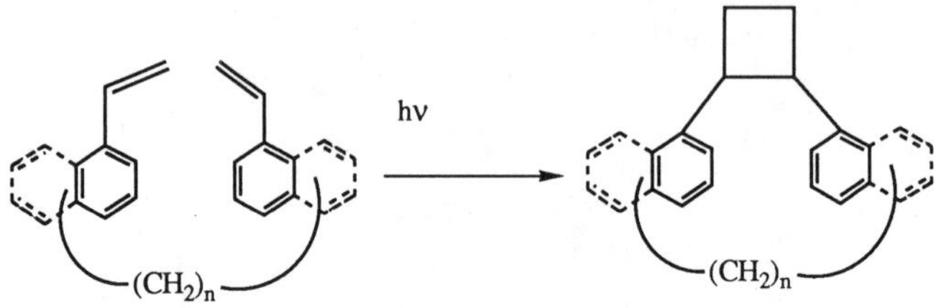

(6.6)

In Section 6.2, I summarize the most recent works on the photocycloaddition (6.6) of these styrene derivatives.[12] Styrene, the parent compound, was reported to undergo photodimerization, but very poorly, and to afford 1,2-diphenylcyclobutane in meager yields,[13] so that nobody had dreamed of the success in the intramolecular version of the reaction toward the synthesis of cyclophanes. Even so, why was the reaction tried, using styrene derivatives like 1,3-bis(*p*-vinylphenyl)propane? The reasons are that the secret of the poor results in the photoreaction of styrene was noticed as mentioned later and that, moreover, the desire for making symmetrical or structurally sophisticated cyclophanes was so intense. Now, the number of cyclophanes, accumulated by the photoreaction, are larger than those previously made by cationic cyclocodimerization, despite the short research period. Additionally, by this fact you can understand how easy this photochemical reaction is. The following section contains the details.

6.2. Intramolecular [2+2] Photocycloaddition of Styrene Derivatives Toward a New Class of Cyclophanes

6.2A. [2+2] Photocyclization of Styrene

Photocyclodimerization of styrene was found two decades ago and was investigated extensively.[13] The formation of diphenylcyclobutanes by this reaction has opposing demands on the styrene concentration in that the first dimerization step needs a high concentration of styrene in order to react with the short-lived excited styrene species, but for the second cyclization step a dilute condition is preferable in order to avoid intermolecular side reactions of the highly reactive species with this easily polymerizable monomer. Therefore, there had not yet been related synthetic work with styrene and alkyl-substituted styrene in high yields.

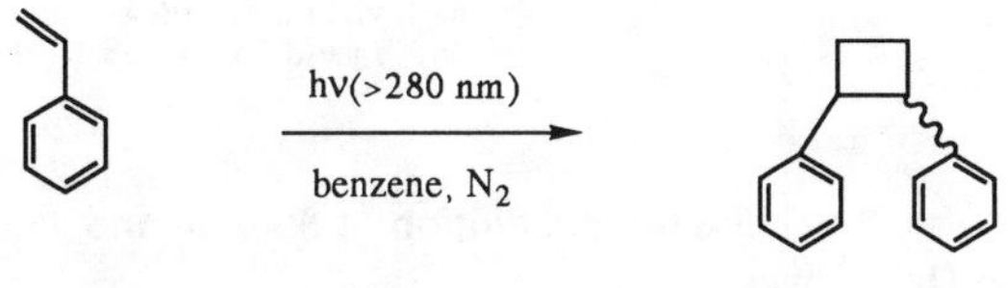

hν(>280 nm)

benzene, N_2

(6.7)

I may be obliged to mention the reported activities on the synthesis of cyclophanes by other $[2+2]$ photocyclization of related olefins. Using stilbene derivatives,[14] several interesting, yet odd-looking, cyclophanes have been prepared. The most striking one among these examples is the synthesis of a triethano$[2_3](1,3,5)$cyclophane derivative.[14c] By only one step, they made complexed cage-compounds in reasonable yields. There were, however, no examples of the synthesis of their parent cyclophanes by photoreactions until 1991.

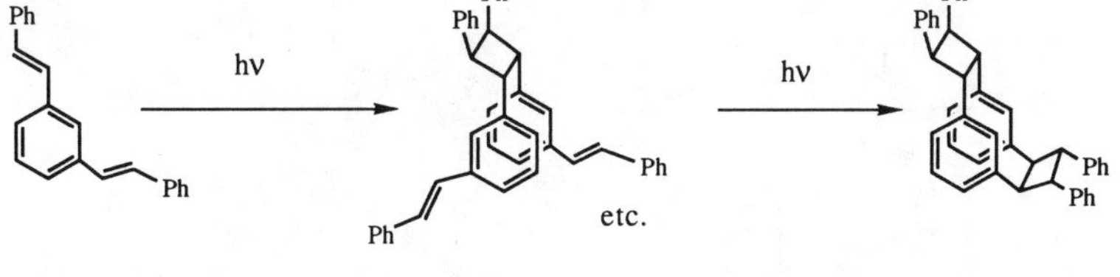

Ph Ph hν Ph Ph Ph etc. hν Ph Ph Ph Ph

(6.8)[14a]

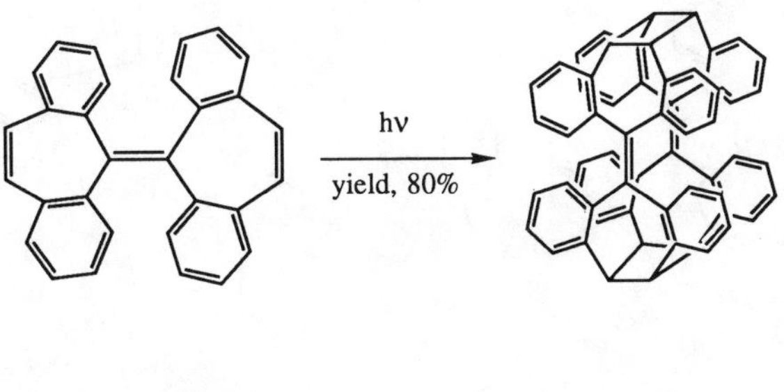

hν

yield, 80%

(6.9)[14b]

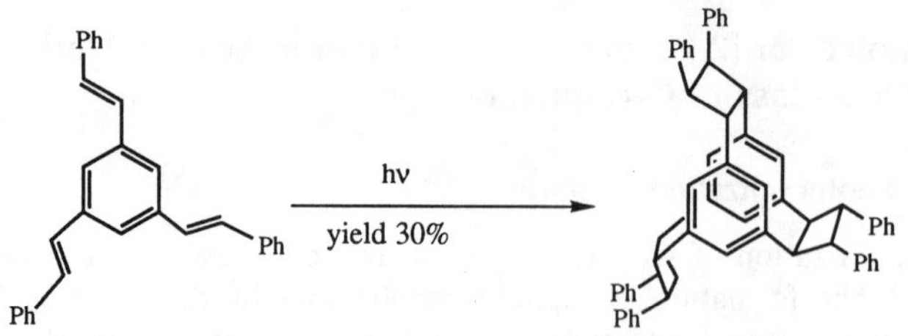
Ph
hν
yield 30%
Ph
Ph
Ph
Ph
Ph
Ph
Ph
Ph

(6.10)[14c]

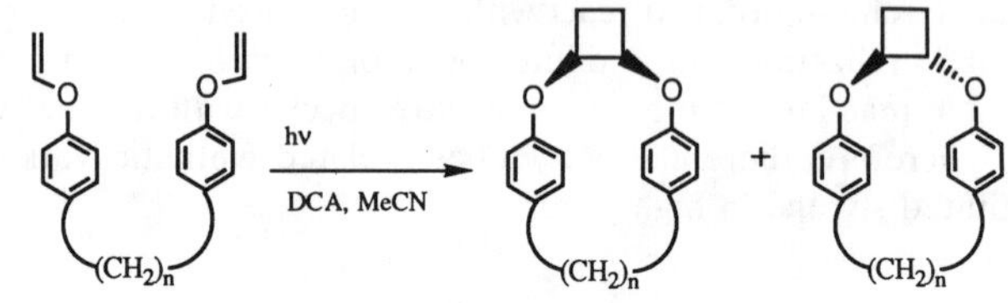
O
hν
DCA, MeCN
$(CH_2)_n$
+

(6.11)[14d]

n=4; yield 25% 96:4
n=20; yield 66% 42:58

6.2B. Intramolecular [2+2] Photocycloaddition of Styrene and Vinylnaphthalene Derivatives

If one weaves the preceding styrene photodimerization within a molecule, the difficulty due to the monomer concentration effects can be easily overcome and then the deeper inside part of the reaction can be revealed. So α,ω-bis(vinylaryl)alkanes were treated under photoirradiation. This successfully gave a new class of cyclophanes in reasonable and sometimes excellent yields.[12]

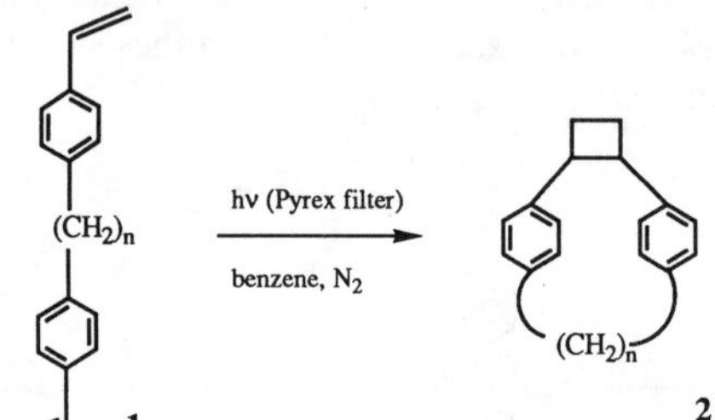
$(CH_2)_n$
hν (Pyrex filter)
benzene, N_2

(6.12)

1 2

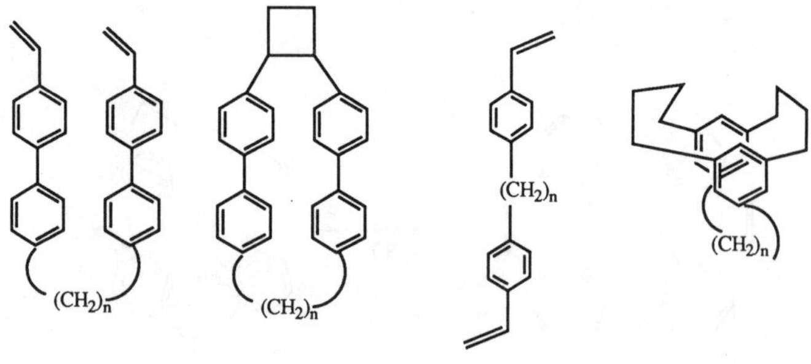
$(CH_2)_n$

3 4 5 6

Table 6.1 ▪ Preparation of (1,2)Ethano[2.*n*]cyclophanes[a]

Entry	Monomer	Product[b]	Isolated Yield (%)[c]
1	**1a** ($n = 3$)	**2a**	(13.8)
2	**1b** ($n = 4$)	**2b**	68.9
3	**1c** ($n = 3$)	**2c**	(48.5)
4	**1d** ($n = 4$)	**2d**	(12.2)
5	**3a** ($n = 3$)	**4a**	35.1
6	**3b** ($n = 4$)	**4b**	32.9
7	**5a** ($n = 3$)	**6a**[d]	95.2
8	**5b** ($n = 4$)	**6b**[d]	73.0

[a]Reaction conditions: in benzene through Pyrex filter at 30°C.
[b]Molecular models showed that only cis configurations of cyclobutane linkages are possible because of the short linkages.
[c]When yields were determined by gas chromatography (QF-1, 3%, 2m, 170–250°C) and corrected on the basis of conversion, they are given in parentheses.
[d]See (6.21) for details.

Some of the results are summarized in Table 6.1. Direct irradiation through a Pyrex filter gave normally satisfactory results, although 1,2-bis(*p*-vinylphenyl)ethane did not give any cyclophanes.

Note that the yields of cyclophanes **2** from α,ω-bis(*p*-vinylphenyl)alkanes **1** depend on the methylene chain length (n). The n number of **1** must be greater than 2 for the cyclization, due to high strain loaded in transition states. The cyclophane yield becomes maximum when $n = 4$ and decreases as n increases, because of decreasing entropy advantage. The feature was proved by photophysical investigation.[15] As shown in Table 6.2, the product formation process (Φ_d, k_d) is accelerated by the entropy advantage (small n), but decelerated by the strain loaded (parallel to ΔE_{st}).

α,ω-Bis(*m*-vinylphenyl)alkanes **7** also give cyclophanes with a cyclobutane ring as one of linkages.[16] The general synthetic method is depicted in Scheme 6.1, with that of olefins. In contrast to *p*-isomers **1**, they gave the desired products even when $n = 2$.

Table 6.2 ▪ Kinetic Parameters of the Photocycloaddition

Olefin	n	Φ_f	τ_f (ns)[a]	Φ_d	k_f (10^9 s^{-1})	k_n^{intra} (10^9 s^{-1})[b]	k_d (10^9 s^{-1})	ΔE_{st} (kcal mol^{-1})[c]
1a	3	0.048	2.30	0.031	0.021	0.34	0.014	52.6
1b	4	0.047	2.46	0.38	0.019	0.19	0.15	43.5
1c	5	0.040	2.15	0.39	0.019	0.21	0.18	37.4
1d	6	0.085	4.39	0.25	0.019	0.095	0.057	34.9
—[d]	∞	0.20	13.37	—	0.015	—	—	28.7[e]

[a]Excitation at 294 nm; emission at 310 nm.
[b]The rate constant of nonradiative process k_n is assumed to be equal to the rate constant for *p*-methylstyrene: $k_n = 0.060 \times 10^9$ s^{-1}.
[c]Strain energies calculated by MM2 for *cis*-(1,2)ethano[2.*n*]paracyclophanes **2**.
[d]*p*-Methylstyrene as a model of the olefin possessing the infinitely long linkage.
[e]For *cis*-diphenylcyclobutane; the calculated strain energy for the trans isomer is 27.6 kcal mol^{-1}.

a) CuBr/THF-HMPA. b)HCl/DOX-H_2O. c)LAH/ether. d)$ZnCl_2$-CCl_3COOH-DMSO, 170 °C.

Scheme 6.1 ■

Moreover, when $n = 3$ and 4, it gave cis- and trans-isomerized products (**8** and **9**). The other nature of the reaction is the same as that found in a series of *p*-isomers. The results are listed in Table 6.3.

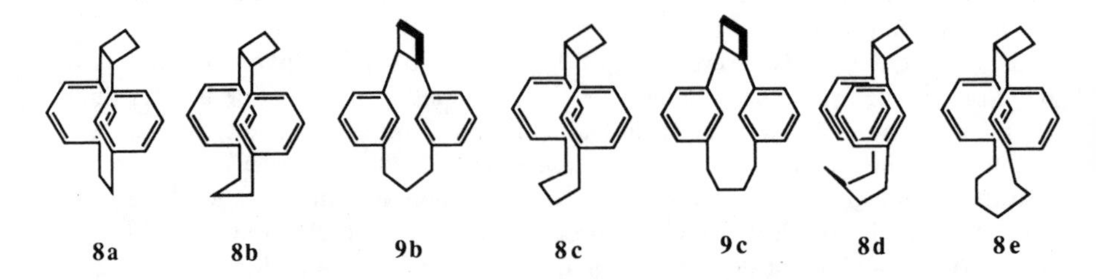

For cis metacyclophanes **8**, the major conformations were studied by ^1H NMR spectroscopy. Those determined by means of Lehner's $\Delta\delta$ value,[17] are the same as the reported results, which are summarized also in Table 6.3.

Table 6.3 ■ **Synthesis of Metacyclophanes**[a]

Monomer	Product	Yield (%)[b]	Isomer Ratio	syn/anti Conformation For *cis*-**8**	lit.[c]	For *trans*-**9**
7a	**8a**	60.9	—	anti	anti	
7b	**8b/9b**	80.4	4.0	anti	anti	anti
7c	**8c/9c**	70.2	3.9	anti	anti	anti
7d	**8d**	58.1	—	syn	syn	
7e	**8e**	31.4	—	anti	anti	

[a]Reaction conditions: 400-W high-pressure mercury lamp; Pyrex filter; in dry benzene under an N_2 atmosphere at 30°C.
[b]Obtained by GLC, based on conversion.
[c]Ref. 17.

Table 6.4 ▪ Preparation of (1,2)Ethano[2.*n*]naphthalenophanes[a]

Entry	Monomer	Product[b]	Isolated Yield (%)
1	**10a** ($n = 3$)	**11a**	8
2	**12a** ($n = 3$)	**13a**[c]	35
3	**12b** ($n = 4$)	**13b**	33
4	**12c** ($n = 5$)	**13c**	46
5	**14a** ($n = 3$)	**15a**[c]	29
6	**14b** ($n = 4$)	**15b**	28
7	**16a** ($n = 3$)	**17a**	49
8	**16b** ($n = 4$)	**17b**	53

[a]Reaction conditions: in benzene through Pyrex filter at 30°C.
[b]Compounds **11**, **13**, **15**, and **17** do not show any aromatic proton signals at $\delta = 5.9$–6.1 that are characteristic of anti naphthalenophanes, so that their syn configurations are apparent. The endo,exo configuration of cyclobutane ring in naphthalenophanes was determined by nuclear Overhauser effect (NOE) experiments.
[c]Structures were confirmed by NOE experiments. Compounds **13a** and **15a** show 18 and 22% NOE enhancements between a peri-hydrogen and a cyclobutane-methine hydrogen, respectively.

Intramolecular [2 + 2] photocycloaddition of vinylnaphthalene derivatives **10**, **12**, **14**, and **16** successfully gave corresponding naphthalenophanes **11**, **13**, **15**, and **17**[12] (Table 6.4). Before going into the details, let us summarize the previous synthetic work on these particular phanes.

Several methods for the synthesis of naphthalenophanes were already reported, including 1,6-Hoffman degradation,[18] solvolysis of appropriate tosylate,[19] malonate synthesis[20] and tosmic synthesis,[21] and the use of dithia-intermediates as the most general precursor.[22] Until now the following were disclosed: *syn*- and *anti*-[2.2](1,4)-naphthalenophane[18]; chiral and achiral [2.2](2,6)napthalenophane[22]; [2.2](2,7)naphthalenophane[23]; [2.2](2,6)(2,7)naphthalenophane[24]; [2.2]paracyclo(1,4) and (2,6)naphthalenophane[25]; *syn*- and *anti*-[3.3](1,4)naphthalenophane[20]; and chiral and achiral [3.3](2,6)naphthalenophane.[18b]

hν(Pyrex filter)

Benzene, N_2

10 → **11** (6.13)

By this photochemical method *syn*-[2.*n*]paracyclo-, *syn*-[2.*n*](1,4)-, *syn*-[2.n](1,5)-, and [2.*n*](2,6)naphthalenophanes were prepared.[26] An example is illustrated in (6.13). The reaction gives exclusively syn isomer for **11**, **13**, and **15** because the vinyl group has only exo conformation (conformational rigidity of vinyl group) and never attains an *endo,exo*-divinyl combination for the formation of anti isomer. This selectivity is quite important, because previously reported synthetic methods listed before generally give conformationally stable *anti*-naphthalenophanes exclusively (*vide infra*).

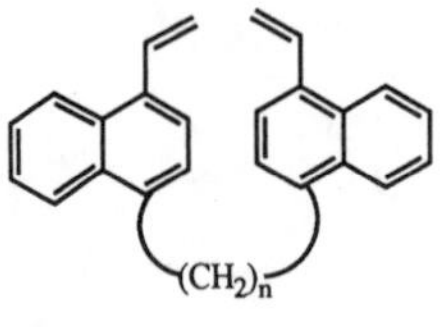

$(CH_2)_n$ 12

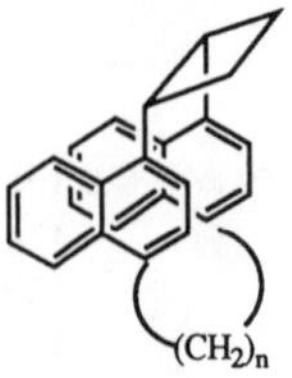

$(CH_2)_n$ 13

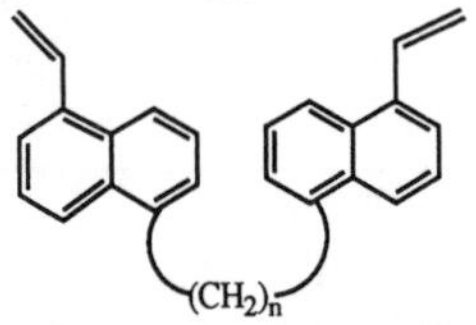

$(CH_2)_n$ 14

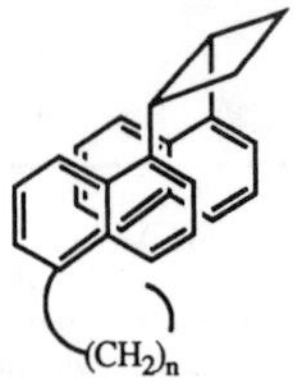

$(CH_2)_n$ 15

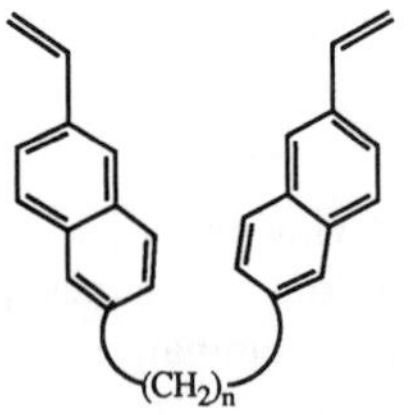

$(CH_2)_n$ 16

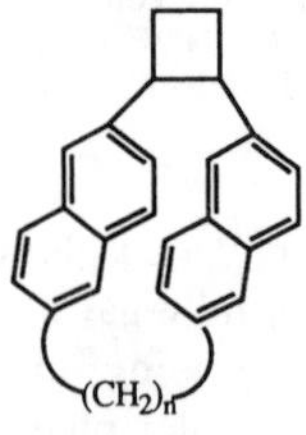

$(CH_2)_n$ 17

6.2C. Birch Reduction of (1,2)Ethano[2.*n*]cyclophanes

The cyclobutane ring, produced in the cyclophane, is readily cleaved by Birch reduction, to afford a tetramethylene unit.[27] This reaction and the preparation of [4.*n*]cyclophane are discussed in detail in this section.

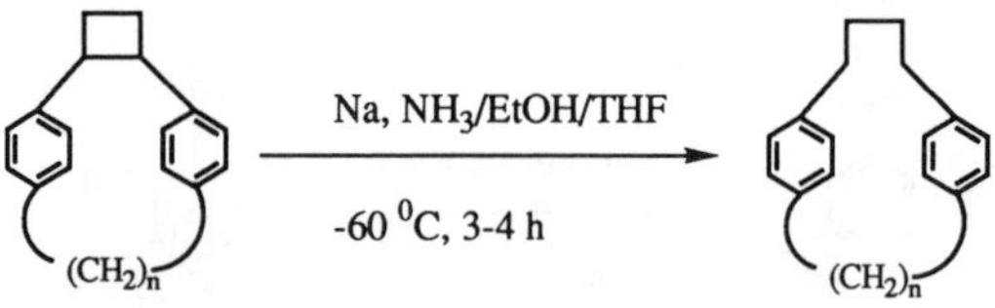

2 $(CH_2)_n$ → Na, NH_3/EtOH/THF, -60 ^{0}C, 3-4 h → 18 $(CH_2)_n$ (6.14)

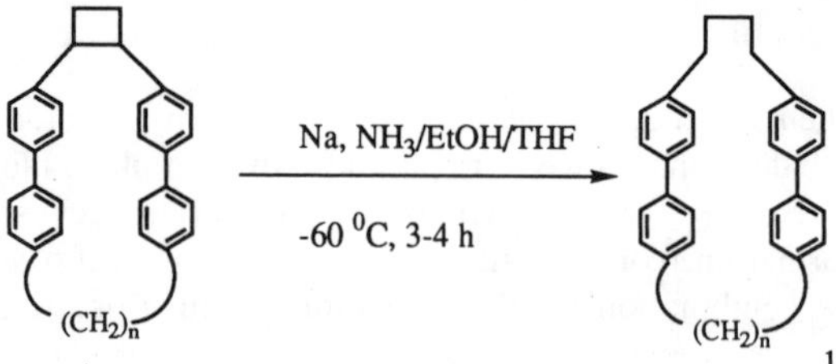

4 $(CH_2)_n$ → Na, NH_3/EtOH/THF, -60 ^{0}C, 3-4 h → 19 $(CH_2)_n$ (6.15)

Photocycloadducts **2** reacted very readily under Birch reaction conditions and gave [4.*n*]paracyclophanes **18** in around 90% yields. Biphenylophanes **4** also gave simple [4.*n*]biphenylophanes **19** in 63–76% yields.

Metacyclophanes **8** showed the same behavior to the reaction as before, if the linkages are long enough (**8c**, **8d**, and **8e**). But strained cyclophanes like **8a** and **8b** gave a considerable amount of open-chain compounds **21a** and **21b**.

$$\mathbf{8} \xrightarrow[-60\,^{\circ}\mathrm{C},\ 4\ \mathrm{h}]{\mathrm{M,\ NH_3/ROH/THF}} \mathbf{20} + \mathbf{21} \qquad (6.16)$$

There are many unknown factors involved in discussing the mechanism. Yet we observed very interesting effects of strain on the fate of the intermediate anion radical. The strained cyclophanes cleaved themselves to α,ω-bis(ethylphenyl)alkanes after accepting electrons. Actually, once it opens, the reduction may occur quite readily, because model compound 1,5-bis(*p*-vinylphenyl)pentane afforded 1,5-bis(*p*-ethylphenyl)pentane in a quantitative yield under the same reaction conditions as those for cyclophanes. The strain release is considered to be one of the driving forces of the reaction (see Table 6.5). Generally one can observe only the products from 1,4-tetramethylene anion radical species, but not from divinyl anion radical species, in the case of the reduction of cyclophanes. Hence, this Birch reduction with strained cyclophanes is very unusual and, thus, intriguing. (Scheme 6.2.)

Table 6.5 ▪ Yields and Strain Energy of Cyclophanes[a]

Compound	Yield (%)	Fragmentation	Strain Energy (kcal mol^{-1})	$\Delta\Delta E_{st}$[b] (kcal mol^{-1})
8a	17	Yes	61.6	32.9
2a	99	No	52.6	23.9
9a	—	Yes	50.8	22.1
8b	82	Yes	45.6	16.9
2b	88	No	43.5	14.8
9b	—	No	42.8	14.1
8c	90	No	38.9	10.2
8d	98	No	35.4	6.7
8e	98	No	34.5	5.8
—[b]	—	—	28.7	0

[a]Strain energies were calculated by MM2.
[b]Strain energy differences based on 28.7 kcal mol^{-1} of *cis*-diphenylcyclobutane.

As described in the reaction of three-bridged photoadducts in Section 6.2E, when photoproducts are structurally complicated, this reductive method is useful because of the simplification of structures, an extremely clean and mild reaction, and remarkably high yields.[12] Yet naphthalenophanes obtained by photocyclization did not form straightforward products. They produced overreduced products, whose structures are not yet fully examined. In order to obtain desired naphthalenophanes from them, the reduced materials were oxidized in decalin with 2,3-dichloro-5,6-dicyano-1,4-benzoquinone (DDQ) at 170°C.[26]

11 → 1)Na, NH_3/EtOH/THF; 2)DDQ, decalin → **22** (6.17)

13 $(CH_2)_n$ → 1)Na, NH_3/EtOH/THF; 2)DDQ, decalin → **23** $(CH_2)_n$ (6.18)

15 $(CH_2)_n$ → 1)Na, NH_3/EtOH/THF; 2)DDQ, decalin → **24** $(CH_2)_n$ (6.19)

17 $(CH_2)_n$ → 1)Na, NH_3/EtOH/THF; 2)DDQ, decalin → **25** $(CH_2)_n$ + **26** $(CH_2)_n$ (6.20)

From naphthalenophanes **15**, products **24** were obtained without conformational alteration on naphthalene nuclei, whereas **13** gave conformationally stable *anti*-naphthalenophanes **23** exclusively. Compound **17a** (chiral/achiral isomer ratio = 4.2) gave two isomeric [3.4](2,6)naphthalenophanes **25a** and **26a** [(chiral **26a**)/(achiral **25a**) isomer ratio = 4.2].

The syn selectivity of the methods mentioned previously has a great advantage, because syn isomers are generally conformationally unstable and given as minor products by methods reported previously. By using this advantage of the syn selectivity, the room required for the intraannular rotation of naphthalene nuclei was determined. By the enlargement of the ring system from [2.3]- to [4.4]-one, the

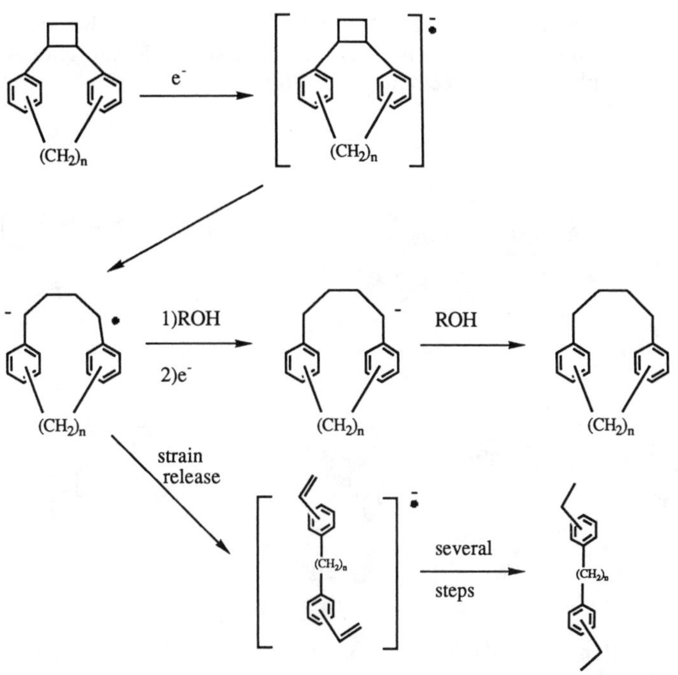

Scheme 6.2 ■

(1,5)naphthalenophanes always kept syn conformation under the experimental conditions of −60 to 180°C, because quite a large volume is necessary if the nucleus rotates around the axis through C-1 and C-5 of the naphthalene ring.

Interestingly the (1,4)naphthalenophanes showed the limiting point for the nuclei to start the rotation. It was determined between [3.3] and [3.4] systems. Namely, [3.4](1,4)naphthalenophane takes only anti conformation after the transformation of *cis,exo*-1,2-ethano-*syn*-[2.3](1,4)naphthalenophane **13a**, whereas *syn*-[3.3]- and [2.4]-naphthalenophane take only the original syn conformation at least up to their melting points. The room requirement for the rotation of naphthalene rings is just similar to that previously reported for [*n.m*]paracyclophanes[28]: Activation free energies for [3.4]paracyclophanes and [4.4]paracyclophanes are reported as 33 kcal mol^{-1} at 160°C and 15 kcal mol^{-1} at ca. 15°C, respectively. From the coalescence temperatures, activation free energies for **23a** and **23b** were calculated to be ≥ 25 kcal mol^{-1} (≥ 200°C) and 13 kcal mol^{-1} (10°C), respectively.

6.2D. Multibridged Cyclophanes

Using reaction (6.21), multibridged cyclophanes were also synthesized.[12] Monomer **5** gave an isomer mixture of three-bridged cyclophanes **27** as a beautiful fine needle in surprisingly high yields (see Table 6.1). The mixture was treated under Birch reduction conditions (Na in liquid NH_3–EtOH–THF) to cleave the cyclobutane rings to the tetramethylene chains. The reaction proceeded very smoothly and gave isomer-free [4.4.*n*](1,3,5)cyclophanes **6** in excellent yields (86.1–95.4%).

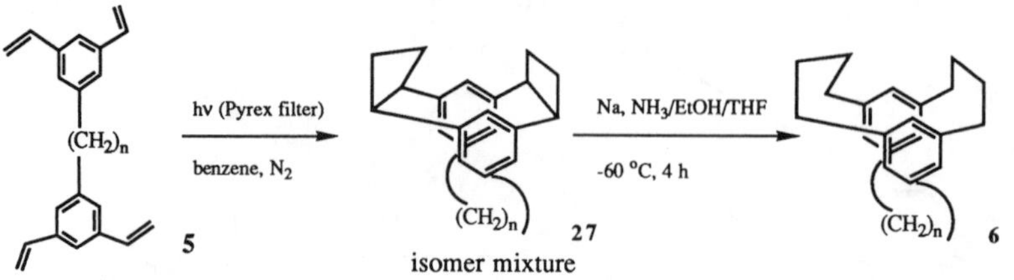

hv (Pyrex filter), benzene, N_2 → isomer mixture **27**; Na, NH_3/EtOH/THF, -60 °C, 4 h → **6**

(6.21)

28 **29** **30** **31**

When one intends to make a kind of paddlanes (**28–31**) by the sequence including photoreaction, one needs to modify linkages after the cyclization. In order to modify cyclophanes at the linkages, heterocyclophanes are valuable intermediates. Generally such cyclophanes are afforded by the cyclization using facile carbon–heteroatom bond formations, whose examples are found in many syntheses of dithiacyclophanes, etc.[8] There is, however, another one with the C—C bond formation using a material that already has heteroatoms at the linkage between aromatic nuclei. The formation of oxacyclophanes was achieved along with the latter strategy, namely, the photochemical C—C bond formation of appropriate ethers.[29]

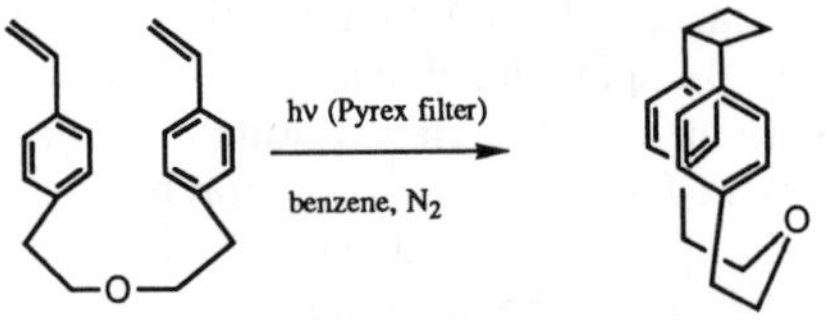

32 —hv (Pyrex filter), benzene, N_2→ **33**

(6.22)

Olefins with ether linkages like **32** recorded better yields (40–85%) than those having pentamethylene linkage like **1e**. The relatively stable gauche conformation of C—O bond is believed to be favorable for the cyclization.[30] Olefin **34** gave cis and trans isomeric cyclophanes **35** and **36** whose ratio was 4.1 : 1.0 (see Scheme 6.3). Cis and trans isomers show characteristic cyclobutane methine proton peaks at $\delta = 3.98$ (normal) and 3.56 (shielded), respectively, which indicate clearly their structures.

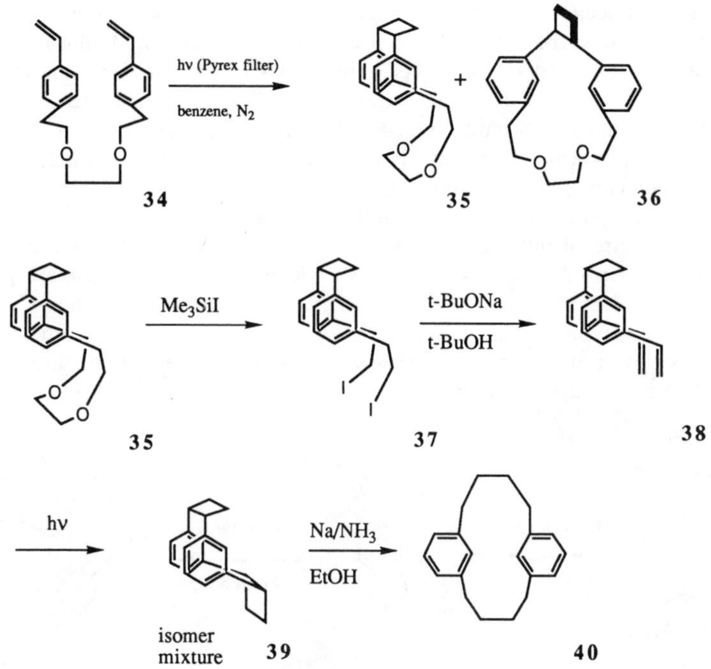

hν (Pyrex filter)
benzene, N_2

34 35 + 36

Me_3SiI t-BuONa t-BuOH

35 37 38

hν Na/NH_3 EtOH

isomer mixture 39 40

Scheme 6.3 ■

Trimethylsilyl iodide was used in order to cleave the ether linkage of the cyclophanes like **32**, but it resulted in the recovery of the starting materials. The reactivity of the ether linkage in these compounds is considered to be usual, so that the failure might be due to a steric reason that their oxygen lone pairs are hidden and not available for the attack of silyl iodide. Accordingly, the stable conformations of the cyclophanes were calculated by the MM2 method,[31] which suggests the tail-in structures like that depicted for cyclophane **33**. The structure is also independently proved by the ^1H NMR spectra of the parent cyclophanes,[16] *cis*-(1,2)ethano[2.5]paracyclophane and *cis*-(1,2)-ethano-*syn*-[2.5]metacyclophane, which show their central methylene groups in the linkages as the very high-field-shifted peaks at $\delta = 0.40$ and 0.58, respectively: that is, the upfield shift is ascribed to their stable tail-in conformations. On the other hand, **35** and **36** show relatively sharp singlets for methylene groups of oxyethylene units to suggest that these structural units rotate quite rapidly to expose the oxygen lone pairs for the attack of trimethylsilyl iodide. In fact, as shown in Scheme 6.3, compounds **35** and **36** reacted smoothly with this reagent to afford diiodides **37** exclusively. Diiodides **37** underwent E2 reaction with sodium *tert*-butoxide in *tert*-butanol to give unstable diolefins **38**, which were purified by column chromatography (SiO_2, cyclohexane).

Olefin **38** gave desired cyclobutane-linked cyclophane **39** in a 20% yield after photoirradiation (> 280 nm) for 8 h. Cyclophane **39** was separated as an isomer mixture and reduced by Na/NH_3–EtOH to afford the known compound **40**, *anti*-[4.4]metacyclophane, in a 98% yield. Hence the photocyclization of **38** was confirmed.

By this strategy, photocycloadducts were always composed in complexed isomer mixtures that got into trouble for separation. Accordingly, a new method has been searched for to make such multibridged cyclophanes as **39** not only stereoselectively, but also from easily available olefins. Recently it was found that an *o*-methoxyl group to a vinyl group can effectively control the conformation of the vinyl group and makes the cyclobutane ring direct exclusively anti to the methoxyl group.[32]

In fact, we carried out the photoreaction of olefin **41** successfully and obtained cyclophanes **42** ($n = 2$, 3, and 4) in 11–78% yields. Cyclophanes **42** were gradually decomposed under irradiation, so that their yields began to decrease after the maximum reached in ca. 12 h under the conditions applied. No other isomeric cyclophanes except an intermediate **43** were found in the reaction mixture by careful chromatographic analysis (*vide infra*).

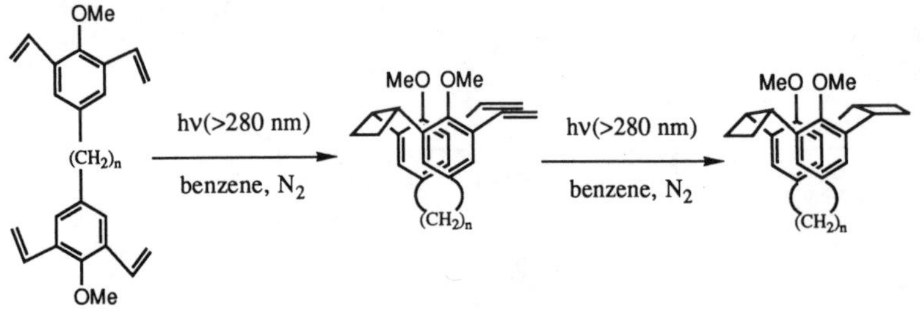

hv(>280 nm), benzene, N_2 → hv(>280 nm), benzene, N_2 (6.23)

				Yield (%)
41a	$n = 2$	**43a**	**42a**	11
41b	$n = 3$	**43b**	**42b**	78
41c	$n = 4$	**43c**	**42c**	48

The configuration of the cyclobutane ring is determined to be cis from the chemical shift of cyclobutane methine protons which appear at $\delta = 4.44$–4.49. Moreover its configuration to the methoxyl group was concluded to be anti by the NOESY experiment: Their methylene protons show the NOE interaction with aromatic protons, but the methine protons of the cyclobutane ring show it with the methoxyl protons. The ^1H NMR spectra show only a set of proton peaks for a half-molecule, due to C_{2v} symmetry of the molecule. The ^1H NMR chemical shift of the methoxyl groups does not show any downfield shift caused by the compression. On the other hand, ^1H NMR chemical shift of methoxyl groups considerably shifts to the high-field region by 0.3–0.4 ppm in comparison with those of anisole derivatives, due to the shielding of the aromatic ring. Accordingly, they are concluded to be directed outside in order to avoid the steric repulsion with each other. Consequently, the structure of cyclophanes **42** is determined as depicted in (6.23).

Molecular mechanics calculation (MM2) denotes that the most stable conformer of **42** has the methoxyl groups perpendicular to benzene rings and pointing outside the cage. This supports the structure of **42** deduced previously.

The MM2 calculation also indicates that olefin **41** already has its vinyl groups of anti conformation against the methoxyl groups. Therefore the selectivity shown in this cyclization is concluded to be due to the vinyl conformation.

The mechanism has not been thoroughly examined, but an experimental run gave intermediate cyclophane **43b** ($n = 3$), whose structure is depicted in (6.23). After it was irradiated through a Pyrex filter, it was completely converted to cyclophane **42b** ($n = 3$), so that the present photocycloaddition is believed to proceed stepwise.

The anisole moiety can be converted to vinylphenyl moiety through phenyl triflate moiety,[33] so that the latter method seems to provide a reasonable and relatively simple approach to the structurally interesting paddlanes (**28–31**).

6.2E. Crown Ethers

This photocycloaddition can be applied to prepare a new kind of crown ethers,[34] possessing face-to-face oriented aromatic ring systems, which have some possibilities to modify their properties. Crown ethers of this sort do not have many examples. According to our literature survey, Cram and co-workers[35a] and Misumi and co-workers[35b] designed and prepared such crown ethers.

hν (Pyrex filter), MeOH, N_2 (6.24)

			Yield (%)		Yield (%)
44a	$n = 3$	**45a**	91	**46a**	2
44b	$n = 4$	**45b**	95	**46b**	5

cis-Crown ethers **45** ($n = 3$ and 4) were formed predominantly and isolated from the reaction mixture by column chromatography (SiO_2, acetone–benzene; isolated yields were more than 75%). They showed a typical multiplet signal of *cis*-cyclobutane methine protons at $\delta = 3.92$ (**45a**) and 4.03 (**45b**) and high-field-shifted aromatic protons caused by the layered aromatic nuclei.

As shown in (6.24) the crown ethers were obtained in considerably high yields, even if they are compared with those of Okahara cyclization, which is one of the most efficient synthetic methods of crown compounds from oligoethylene glycols.[36]

According to the data obtained, small template and solvent effects on the yield are recognized. The addition of an alkali metal ion resulted in the increase of the yield, due to the template effect. Solvents affected the reaction to some extent, although the origin is not clear at this moment.

Crown ethers **45** having a cyclobutane moiety are not stable under photoirradiation. Eventually the yield of **45** reaches a maximum and then decreases. Even though every experiment was carried out after bubbling nitrogen gas into the reaction mixture for at least 30 min, photooxidation took place and gave cyclohexenones **46**, whose yields were less than 10%. Products **46** seem to be formed by the oxidation of a radical intermediate with oxygen that remained in the reaction system.

The crown ethers prepared can bind alkali metal ions as well as conventional crown ethers. The binding ability was determined from the nitrogen content of the organic layer, after prolonged contact (24 h) of **45** with finely ground thiocyanate salts (MSCN) in methylene chloride.[37] Crown ethers **45** effectively extracted the salts into methylene chloride and showed high selectivity for LiSCN ($Li^+/Na^+ = 3$, $Li^+/K^+ = 10$) under the conditions applied. The results are notable, because lithium metal is very attractive in many fields, such as the future electric power generation by nuclear fusion, etc.[38]

The new kind of crown ethers have several structural advantages for further modification, that is, the cyclobutane ring to be enlarged, aromatic moieties for hydrophobicity and asymmetrical substitution, etc.

6.2F. Scope and Limitation

The scope of this reaction can be easily realized by the summaries shown in Tables 6.1–6.4. The application of this reaction, of course, is limited by the preparation of styrene derivatives.[2] Most of the monomers that have been prepared could give cyclophanes in reasonable yields, so several cyclophanes, naphthalenophanes, biphenylophanes, and even large-membered paracyclophanes can be safely said to be synthesized by this method.

One of the linkages in a cyclophane skeleton is inevitably *cis*- or *trans*-1,2-disubstituted cyclobutane ring, but one can cleave the ring by Birch reduction, so that one can also design [4.*n*]cyclophane by the reaction, as discussed in Section 6.2C. Note that the cyclophane skeleton itself can be cleaved if its strain is high enough. We experienced the cases at the reductions of (1,2)ethano[2.2]- and -[2.3]-metacyclophane. They gave α,ω-bis(m-ethylphenyl)alkanes in considerable yields.

For the smallest ring system, [2.3]paracyclophanes can be made, whereas [2.2]metacyclophane is still possible to be made. [1.2]Orthophane, therefore, could be prepared by the reaction. Generally speaking, the photocyclization can afford the molecule of which strain energy is less than 31 kcal mol^{-1} or that of [2.2]paracyclophane.

What kind of functional groups can be introduced in the cyclophanes? There is not so much information on this question, but at least it can be said that many functional groups are not harmful in this photoreaction as already appeared in many examples, compared with the cationic cyclocodimerization summarized in Section 6.3.

6.3. Cationic Cyclocodimerization

Cationic C—C bond formation is not always clean and straightforward, because the highly unstable carbocationic species undergo many undesirable reactions such as rearrangements. Common rings, however, often are prepared by cationic bond formation, because *intra*molecular reactions are faster and cleaner than *inter*molecular reactions. However, the preparation of macrocyclic systems by this method is not so usual. Our idea leading to the preparative cationic method for macrocyclic systems, namely, cationic cyclocodimerization, was already mentioned in the introduction.[3] The outline of the reaction is shown in Scheme 6.4, that is, the sequence of

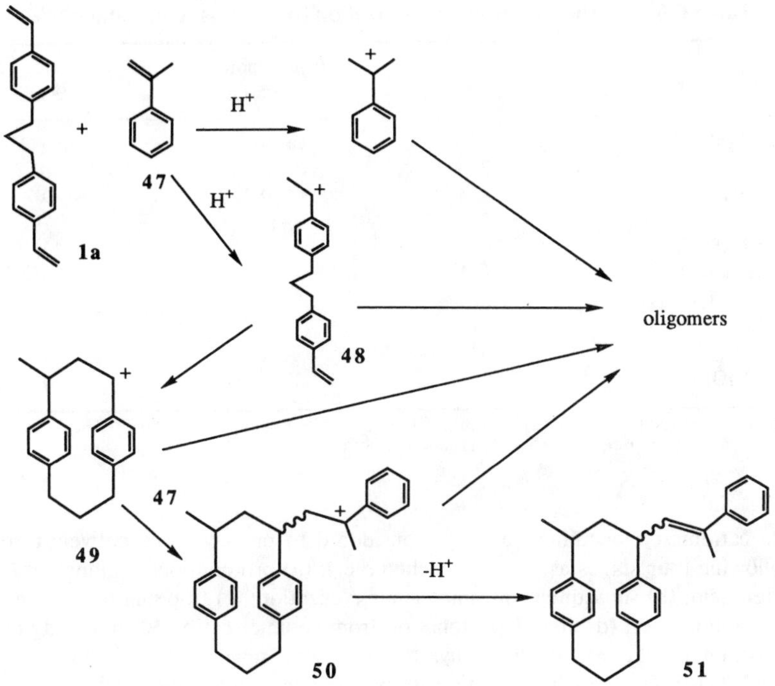

Scheme 6.4 ■

intramolecular cationic C—C bond formation, intermolecular C—C bond formation, and finally deprotonation to the neutral compound.

6.3A. Cationic Cyclocodimerization: Conditions and Requirements

Conditions were optimized, using the pair of monomer **1a** and comonomer styrene. Among the catalysts examined, CF_3SO_3H was the most effective and easily available, as shown in Table 6.6. The concentration of the acid should be lower than 0.2 mM (ca. $\frac{1}{100}$ mole ratio to the monomer), when the monomer concentration is 24 mM. Reaction temperatures from 40 to 80°C were practical.

The effects of concentrations of styrene and monomer **1a** on the yield were studied. At concentrations of monomer **1a** lower than 0.1 M, desired cyclocodimers were obtained in 20–38% yields. Note that the cyclocodimer yield did not decrease so significantly by increasing the concentration of the monomer up to 0.1 M. This is because the stability of cyclic cation, which is attributed to the intramolecular donor–acceptor interaction between styryl group and styryl cation. This tendency seems to indicate a very sharp contrast to other preparation methods of [3.3]cyclophanes in which almost always high-dilution technique is indispensable for the cyclization.[8]

Table 6.6 ▪ Cationic Cyclocodimerization by Various Acid Catalysts[a]

Acid	mM	Conversion (%)	Yield (%)
CF_3SO_3H	0.2	ca. 100	33
$AcClO_4$	2	ca. 100	28
FSO_3H	2	88	9
$ClSO_3H$	2	72	7
CH_3SO_3H	30	59	12
Nafion-H	[b]	100	28
$HClO_4$ (70%)	30	30	0
H_2SO_4 (95%)	30	44	0
$SnCl_4$	10	84	0
BF_3OEt_2	10	92	0

[a]Reaction conditions: [monomer **1a**] = 24 mM; [styrene]/[**1a**] = 5; temperature 50°C; time 3 h.
[b]The w/w ratio of the monomer/Nafion-H was 3; at 80°C for 16 h.

The cationic cyclocodimerization is considered to proceed consecutively through the following four stages as shown in Scheme 6.4: (a) protonation of monomers by a Brønsted acid; (b) subsequent intramolecular cyclization; (c) trapping of cyclic cation **49** by a comonomer; (d) final deprotonation from codimer cation **50** to afford neutral cyclic codimers like **51**. Additionally, the following major side reactions can be expected: (e) trapping of linear cation **48** by the comonomer and (f) when a highly nucleophilic comonomer is employed, protonation taking place predominantly to it and the corresponding carbocation, thus formed, causing the formation of higher oligomers and/or linear codimers with monomers.

Steps (a) and (d) occur as long as acid catalysts like CF_3SO_3H, $AcClO_4$, FSO_3H, $ClSO_3H$, CH_3SO_3H, and Nafion-H were used. Also, the competition between steps (b) and (e) was not serious as long as the reaction was carried out at less than 0.06 M of monomer concentration above 0°C. Therefore, the competitive protonation between a monomer and a comonomer olefin should be examined together with the competitive trapping [step (c)] of the cyclic cation between the olefin and the remaining monomer.

The ionization potential (IP) can be used as the reactivity parameter of olefins for the protonation and the addition of some cationic species, because IP normally represents the HOMO level and olefins are considered to interact first with cationic species through this molecular orbital, although the discussion should be limited to kinetically controlled reactions. In the presence of an olefin whose IP is higher than the 8.12-eV value of *p*-methylstyrene (a model of monomer **1a**), the protonation is supposed to occur mainly to monomer **1a**, and the cyclocodimerization goes well. The results of the cyclocodimerization with ring-substituted styrenes support this expectation; that is, *p*-methoxystyrene no cyclocodimers and *p*-methylstyrene gave rather lower product yield than styrene. According to the data summarized in Table 6.7, olefins with IPs higher than 8.12 eV can be used as comonomers. This is the upper limitation of the reactivity of olefins toward cationic species for the cyclocodimerization. The reaction with 1,1-diphenylethylene (IP = 8.0 eV) is an exception:

Table 6.7 ▪ Reactivities to Olefins

Entry	Olefin	IP (eV)	Total Delocalization[a] (eV)	Remarks[b]
1	Anethole	7.68		No
2	*p*-Methoxystyrene	7.92	0.6216	No
3	Acenaphthylene	8.02		No
4	*trans*-1-Phenylpropene	8.17	0.5653	No
5	1,1-Diphenylethylene	8.00		Yes
6	*p*-Methylstyrene	8.12		Yes
7	Monomer **1a**	—		Yes
8	Indene	8.13		Yes
9	2-Phenylpropene	8.34	0.5828	Yes
10	*p*-Chlorostyrene	8.45		Yes
11	Styrene	8.49	0.6443	Yes
12	*m*-CF_3-styrene	—		Yes
13	*p*-Nitrostyrene	—		No
14	1,3-Cyclohexadiene	8.25	0.9992	Yes
15	*trans*-1,3-Pentadiene	8.56	0.7551 (tail)[c] 0.3545 (head)[d] 0.5366 (tail)[e] 0.3354 (head)[f]	Yes
16	Cyclopentadiene	8.61	0.6314	No
17	1,3-Cyclooctadiene	8.68		No
18	2,3-Dimethyl-2-butene	8.40	0.4685	No
19	Norbornene	8.97		No
20	Cyclohexene	9.12		No
21	1-Octene	9.52	0.1940	No
22	2-Chloroethyl vinyl ether	9.97[g]	0.2597[g]	

[a]CNDO method; with benzyl cation.
[b]Availability for the cyclocodimerization with monomer **1a**.
[c]For *S*-cis-tail attack.
[d]For *S*-cis-head attack.
[e]For *S*-trans-tail attack.
[f]For *S*-trans-head attack.
[g]For methyl vinyl ether as a model.

Diphenylethyl cation from this comonomer is stabilized and slowly propagates. Therefore, the trapping of the cyclic cation with this olefin seems to be observed.

The reactivities of olefins toward the cyclic cation were estimated by the total delocalization energy (DE) calculated by using a benzyl cation–olefin pair as a model for the transition state: Both molecules are arranged face-to-face with the distance of 0.3 nm between the cationic site and the reacting site of the olefin.

The experimental results are reasonably interpreted by the calculated DE values. Ineffective comonomers like 2,3-dimethyl-2-butene and 2-chloroethyl vinyl ether (methyl vinyl ether as a model) gave rather small total delocalization energies (0.19–0.47 eV), whereas pertinent styrene derivatives (entries 1–13 in Table 6.7) and dienes (entries 14–17) like 2-phenylpropene and 1,3-cyclohexadiene gave large values of 0.6–1.00 eV, although these values may not be compared directly beyond each group.

Among the three easily cationically polymerizable cyclic dienes examined, 1,3-cyclohexadiene has an almost planar 1,3-diene moiety and gives the largest DE value. It accounts well for the experimental results that only this cyclic diene was able to cyclocodimerize with monomer **1**. The calculation can also predict the attacking mode of *trans*-1,3-pentadiene to the cyclic cation. Although several attacking modes are conceivable, the following four models were chosen for the total DE calculation as the most probable cases: *S*-cis-head, *S*-cis-tail, *S*-trans-head, and *S*-trans-tail attacks. The *S*-cis-head attack gives the highest total DE value, so that the contribution of this attacking mode may be significant in this cyclocodimerization.

Table 6.8 ▪ Cyclocodimers[a] Obtained by Cationic Cyclocodimerization

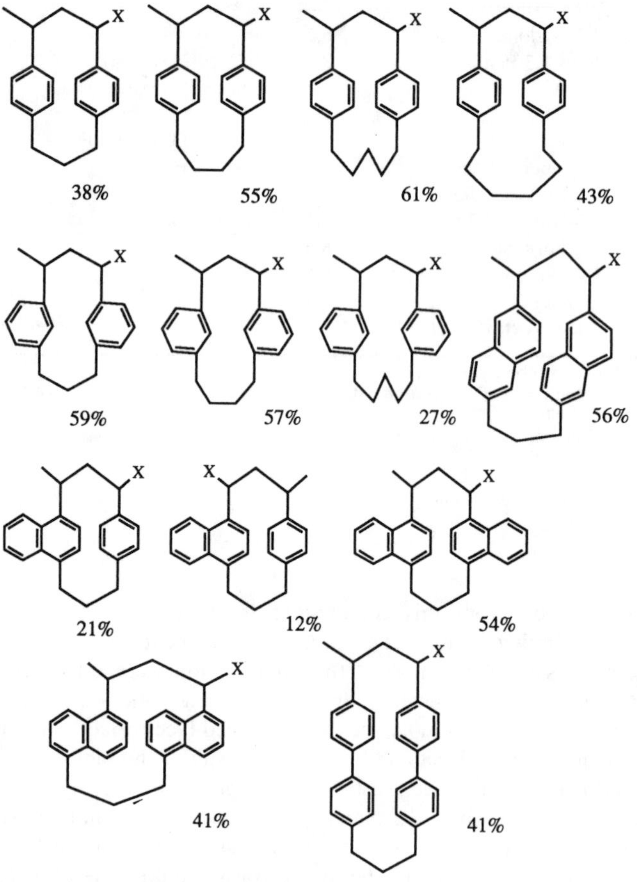

X X X X

38% 55% 61% 43%

X X X X

59% 57% 27% 56%

X X X

21% 12% 54%

X X

41% 41%

[a]X = comonomer residue.

Because this total DE depends largely on the HOMO level of an olefin, the other end of the limitation can also be provided by IP, ca. 8.6 eV, of *trans*-1,3-pentadiene and cyclopentadiene. Hence it can be said qualitatively that styrene derivatives and dienes having their IPs in the range of 8.1–8.6 eV are suitable for the reaction.

Although *trans*-1-phenylpropene has a pertinent property with regard to the reactivity based on IP and DE, it did not cyclocodimerize. This could be due to the steric hindrance caused by the methyl group substituted at the double bond, when this olefin attacks the cyclic cation, because its homolog, indene, successfully gave the desired cyclocodimers.

The highest yield of cyclocodimers was obtained in the reaction with 2-phenylpropene. This result is ascribed to its ceiling temperature, in addition to its ideal properties that are promised by IP and DE. Near the ceiling temperature, it propagates little. It has the same advantage as 1,1-diphenylethylene.

6.3B. Cationic Cyclocodimerization: Scope and Limitation

Table 6.8 summarizes the cyclophanes obtained by cationic cyclocodimerization.[3] The scope of this reaction can be well understood by this summary. The application of this

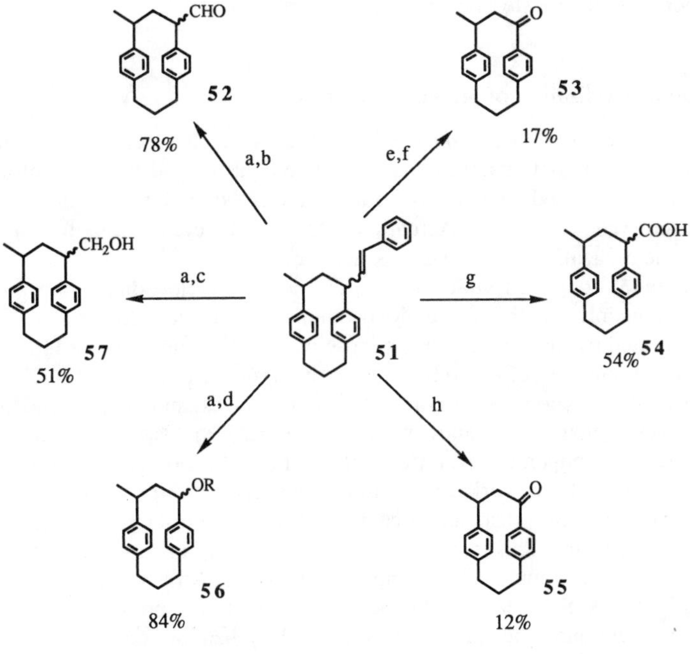

CHO 52 78% a,b
O 53 17% e,f
CH$_2$OH 57 51% a,c
51
g COOH 54 54%
a,d OR 56 84%
h O 55 12%

a)Ozone/CCl_4. b)Zn. c)LAH/ether. d)70%$HClO_4$/ROH. e)mCPBA. f)$Pb(OAc)_4$. g)OsO_4/$NaIO_4$. h)O_2, 80 °C.

Scheme 6.5 ■

reaction is again limited by the preparation of styrene derivatives and the number of isomers produced.

One of the linkages is inevitably a disubstituted three-carbon as shown in the codimer structure. This chain has a double bond, so that one can prepare some intriguing substances, applying some reactions at the double bond (see Scheme 6.5). Aromatic parts right now should be benzene, naphthalene, and biphenyl. Several functional groups like —O—, >N—, etc., which behave as proton acceptors, could not be incorporated into the structure. This is the major limitation of this reaction. The linkage should be longer than an ethano-bridge, but the longer the linkage, the lower the yield became. Very large cyclophanes that can accommodate a trans double bond as a bridging unit, however, could be made by cationic cyclization mentioned later. These large ring systems are attractive in host–guest chemistry.

One thing to be emphasized is the selectivity of this reaction. If any stereoselectivity does not work, one can obtain many kinds of isomers as products. Generally speaking the reaction would be useless, unless some stereoselectivity works. In the formation of (1,4)- and (2,6)naphthalenophanes, and metacyclophanes, conformational rigidity or preference worked well, so that there were only two products, that is, α-vinylnaphthalene moiety has only exo conformation,[39,40] and β-vinylnaphthalene moiety[26] and *m*-vinylalkylbenzene[41] moiety have preferable exo conformation. Moreover, the comonomer residue always gave *trans*-styryl moiety. Therefore, one does not need to worry about the stereochemistry of the residue.

6.3C. Cationic Cyclization of Styrene Derivatives

So far cationic *cyclocodimerization* has been discussed, where an intermediate cyclic cation cannot deprotonate itself to generate endo double bond for the neutralization. Yet it is easily expected that a large-membered cyclic cation does deprotonate to afford a macrocyclic compound. Actually we could make cyclic compounds like **59** by direct cationic cyclization, as shown in Scheme 6.6.[42]

Monomers (**58**) depicted in Scheme 6.6 did not give cyclocodimers but only linear cooligomers, mainly by the usual cyclocodimerization procedure. That is, linear cations generated by the protonation of those monomers readily reacted intermolecularly with a comonomer before cyclization, so that they were treated with CF_3SO_3H in the absence of comonomers. Among the monomers examined, **58b** and **58c** gave desired paracyclophanes **59b** and **59c** in 80 and 62% yields, respectively, whereas others formed only oligomers even under the same cyclization conditions.

On the basis of framework examination, two styryl groups in both **58a** and **60** cannot approach within a distance of ca. 0.3 nm in a face-to-face manner, due to the severe steric constraint. This means that intermediate styryl cations are not stabilized sufficiently by the intramolecular complexation with a styrene moiety or electron donor, and this is the reason why both monomers did not afford any desired macrocyclic compounds, because their statistical cyclization seems to be sterically possible.

The results draw two important aspects of the cationic cyclization of those styrene derivatives. First, the stabilization due to the intramolecular face-to-face complexation between the styrene moiety and the styrene cation is essential. Second, if a

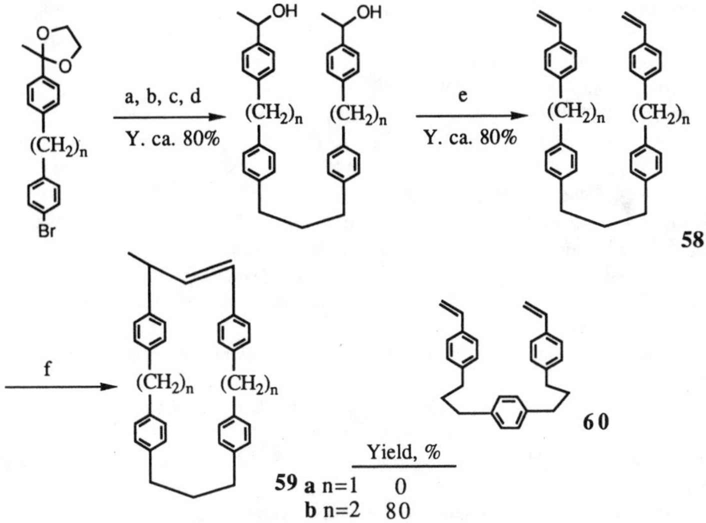

58 **a** n=1
b n=2
c n=3

60

	Yield, %
59 **a** n=1	0
b n=2	80
c n=3	62

a)Mg/THF. b)$Br(CH_2)_3Br$, CuBr-HMPA/THF. c)H^+, DOX/H_2O. d)LAH/ether.
e)CCl_3COOH-$ZnCl_2$/DMSO, 170 °C. f)CF_3SO_3H/dry benzene, 50 °C.

Scheme 6.6 ■

desired product like **59b** and **59c** seems to accommodate a trans double bond easily in the three-carbon linkage, then the present cationic cyclization will proceed successfully just as observed in the case of **58b** and **58c**.

Acknowledgments. I would like to express sincere thanks to Professor Dr. Junji Furukawa, Professor Dr. Shinzo Yamashita, Professor Dr. Akira Oku, Dr. Hideo Takahashi, Dr. Yokihiro Okada, and Dr. Sei-ichi Inokuma, for their valuable discussions, and cooperation. Also, I would like to express my heartfelt thanks to many co-workers and co-thinkers who are responsible for the evolution of this research program and whose names appear in literature cited.

Support of this work through grants from the Ministry of Education, Science, and Culture, Japan, and from Torey Science Foundation is gratefully acknowledged.

References

1. For related reviews, see the following: (a) Nishimura, J.; Yamashita, S. In *Cyclopolymerization and Polymers with Chain-Ring Structures*; Butler, G.; Kresta, J. E., Eds.; American Chemical Society Symposium Series, No. 195; American Chemical Society: Washington, DC, 1977; p. 177. (b) Nishimura, J. *Yuki Gosei Kagaku Kyokaisi* **1988**, *46*, 574.

2. (a) Nishimura, J.; Ishida, Y.; Hashimoto, K.; Shimizu, Y.; Oku, A.; Yamasita, S. *Polym. J.* **1981**, *13*, 635. (b) Nishimura, J.; Yamada, N.; Horiuchi, Y.; Ueda, E.; Ohbayashi, A.; Oku, A. *Bull. Chem. Soc. Jpn.* **1986**, *59*, 2035.
3. Nishimura, J.; Hashimoto, K.; Okuda, T.; Hayami, H.; Mukai, Y.; Oku, A. *J. Am. Chem. Soc.* **1983**, *105*, 4758.
4. For reviews, see the following: (a) Cram, D. J.; Cram, J. M. *Acc. Chem. Res.* **1971**, *4*, 204. (b) Ferguson, J. *Chem. Rev.* **1986**, *86*, 957. (c) Misumi, S. *Pure Appl. Chem.* **1987**, *59*, 1627. (d) Wong, H. N. C. *Acc. Chem. Res.* **1989**, *22*, 145.
5. (a) Boyd, R. H. *Tetrahedron* **1966**, *22*, 119. (b) Boyd, R. H. *J. Chem. Phys.* **1968**, *49*, 2574. (c) Shieh, C.-F.; McNally, D.; Boyd, R. H. *Tetrahedron* **1969**, *25*, 3653.
6. Cram, D. J.; Allinger, N. L.; Steinberg, H. *J. Am. Chem. Soc.* **1954**, *76*, 6132.
7. (a) Cram, D. J.; Helgeson, R. C. *J. Am. Chem. Soc.* **1966**, *88*, 3515. (b) Hedaya, E.; Kyle, L. M. *J. Org. Chem.* **1967**, *32*, 187.
8. As a review, see Keehn, P. M.; Rosenfeld, S. M. *Cyclophanes, I and II*; Academic Press: New York, 1983.
9. (a) Otsubo, T.; Kitasawa, M.; Misumi, S. *Chem. Lett.* **1977**, 977. (b) Longone, D. T.; Küsefoglu, S. H.; Gladysz, J. A. *J. Org. Chem.* **1977**, *42*, 2787. (c) Haenel, M. W.; Flatow, A.; Taglieber, V.; Staab, H. A. *Tetrahedron Lett.* **1977**, 1733. (d) Haenel, M. W.; Flatow, A. *Chem. Ber.* **1979**, *112*, 249.
10. For a review, see Tsuji, T.; Nisida, S. *Acc. Chem. Res.* **1984**, *17*, 56.
11. Mitchell, R. H.; Vinod, T. K.; Bushnell, G. W. *J. Am. Chem. Soc.* **1985**, *107*, 3340.
12. Nishimura, J.; Doi, H.; Ueda, E.; Ohbayashi, A.; Oku, A. *J. Am. Chem. Soc.* **1987**, *109*, 5293.
13. (a) Mayo, F. R. *J. Am. Chem. Soc.* **1968**, *90*, 1289. (b) Brown, W. G. *J. Am. Chem. Soc.* **1968**, *90*, 1916.
14. (a) Schönberg, A.; Sodtke, U.; Praefcke, K. *Tetrahedron Lett.* **1968**, 3669. (b) Müller, E.; Meier, H.; Sauerbier, M. *Chem. Ber.* **1970**, *103*, 1356. (c) Juriew, J.; Skorochodowa, T.; Merkuschew, J.; Winter, W.; Meier, H. *Angew. Chem.* **1981**, *93*, 285. (d) Mizuno, K.; Kagano, H.; Otsuji, Y. *Tetrahedron Lett.* **1983**, *24*, 3849.
15. Nishimura, J.; Ohbayashi, A.; Wada, Y.; Oku, A.; Ito, S.; Tsuchida, A.; Yamamoto, M.; Nishijima, Y. *Tetrahedron Lett.* **1988**, *29*, 5375.
16. Nishimura, J.; Ohbayashi, A.; Doi, H.; Nishimura, K.; Oku, A. *Chem. Ber.* **1988**, *121*, 2019.
17. Krois, D.; Lehner, H. *Tetrahedron* **1982**, *38*, 3319.
18. (a) Wasserman, H. H.; Keehn, P. M. *J. Am. Chem. Soc.* **1969**, *91*, 2374. (b) Blank, N. E.; Haenel, M. W. *Chem. Ber.* **1981**, *114*, 1531.
19. Brown, G. W.; Sondheimer, F. *J. Am. Chem. Soc.* **1967**, *89*, 7116.
20. Kawabata, T.; Shinmyozu, T.; Inazu, T.; Yoshino, T. *Chem. Lett.* **1979**, 315.
21. Kurosawa, K.; Suenaga, M.; Inazu, T. *Tetrahedron Lett.* **1982**, 5335.
22. Grivens, R. S.; Olsen, R. J.; Wylie, P. L. *J. Org. Chem.* **1979**, *44*, 1608. Hereafter the prefixes "chiral" and "achiral" are used for the configuration of naphthalene rings, even if the products have other asymmetric centers.
23. Jessup, P. J.; Reiss, J. A. *Aust. J. Chem.* **1977**, *30*, 843.
24. Boekelheide, V.; Tsai, C. H. *Tetrahedron* **1976**, *32*, 423.
25. Otsubo, T.; Kitasawa, M.; Misumi, S. *Bull. Chem. Soc. Jpn.* **1979**, *52*, 1515.
26. Nishimura, J.; Takeuchi, M.; Takahashi, H.; Ueda, E.; Matsuda, Y.; Oku, A. *Bull. Chem. Soc. Jpn.* **1989**, *62*, 3161.
27. Nishimura, J.; Ohbayashi, A.; Ueda, E.; Oku, A. *Chem. Ber.* **1988**, *121*, 2025.
28. (a) Cram, D. J.; Allinger, N. L. *J. Am. Chem. Soc.* **1955**, *77*, 6289. (b) Cram, D. J.; Wechter, W. J.; Kierstead, R. W. *J. Am. Chem. Soc.* **1958**, *80*, 3126. (c) Reich, H. J.; Cram, D. J. *J. Am. Chem. Soc.* **1969**, *91*, 3517.
29. Nishimura, J.; Horikoshi, Y.; Takahashi, H.; Machino, S.; Oku, A. *Tetrahedron Lett.* **1989**, *30*, 5439.

30. Deslongchamps, P. *Stereoelectronic Effects in Organic Chemistry*; Pergamon Press: Oxford, 1983.
31. We are indebted to Prof. Eiji Osawa, Toyohashi University of Technology, for providing us with the MM2 program.
32. Okada, Y.; Sugiyama, K.; Wada, Y.; Nishimura, J. *Tetrahedron Lett*. **1990**, *31*, 107.
33. Chen, Q.-Y.; Yang, Z.-Y. *Tetrahedron Lett*. **1986**, *27*, 1171.
34. Inokuma, S.; Yamamoto, T.; Nishimura, J. *Tetrahedron Lett*. **1990**, *31*, 97.
35. (a) Helgeson, R. C.; Tarnowski, T. L.; Timko, J. M.; Cram, D. J. *J. Am. Chem. Soc*. **1977**, *99*, 6411. (b) Kawashima, N.; Kawashima, T.; Otsubo, T.; Misumi, S. *Tetrahedron Lett*. **1978**, 5025.
36. Kuo, P.-L.; Miki, M.; Okahara, M. *J. Chem. Soc., Chem. Commun*. **1978**, 504.
37. Hiratani, K.; Aiba, S. *Bull. Chem. Soc. Jpn*. **1984**, *57*, 2657.
38. Hiratani, K.; Taguchi, K.; Sugihara, H.; Iio, K. *Bull. Chem. Soc. Jpn*. **1984**, *57*, 1976.
39. Nishimura, J.; Okuda, T.; Mukai, Y.; Hashiba, H.; Oku, A. *Tetrahedron Lett*. **1984**, *25*, 1495.
40. Nishimura, J.; Yamada, N.; Okuda, T.; Mukai, Y.; Hashiba, H.; Oku, A. *J. Org. Chem*. **1985**, *50*, 836.
41. Nishimura, J.; Ohbayashi, A.; Horiuchi, Y.; Okada, Y.; Yamanaka, S.-H.; Oku, A. *J. Org. Chem*. **1987**, *52*, 1409.
42. Nishimura, J.; Yamada, N.; Ueda, E.; Ohbayashi, A.; Oku, A. *Tetrahedron Lett*. **1986**, *27*, 4331.

7

Syntheses of Prismanes

Goverdhan Mehta[a] and S. Padma[a,b]

[a]University of Hyderabad, Hyderabad, India
[b]J. Nehru Centre for Advanced Scientific Research, Bangalor, India

Contents

7.1. Introduction
7.2. Synthetic Challenge and Tactics
7.3. [3]-Prismane
7.4. [4]-Prismane
7.5. [5]-Prismane
7.6. Toward [6]-Prismane
7.7. Toward [7]-Prismane
7.8. Outlook
Acknowledgments
References

7.1. Introduction

The convex polyhedral solids endowed with multitudinous shapes and natural splendor have fascinated people since time immemorial.[1] Some of these exquisite, highly symmetrical, fundamental polyhedral shapes can be realized among organic molecules by replacing the vertices with CH units and edges with C—C bonds. The resulting hydrocarbon frameworks (polyhedranes) are compact, rigid, aesthetically pleasing carbocyclic molecules of high symmetry.[2] These polyhedranes, comprising rings of multifarious shapes and sizes, possess many interesting structural features such as unusual C—C bond lengths and bond angles, steric compression and overcrowding, and high strain energy. Such structural deviations are readily manifest in novel and

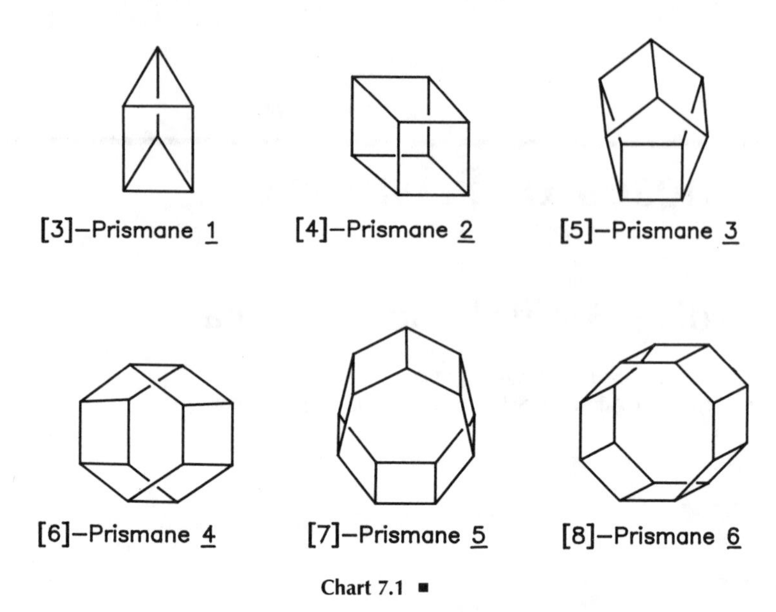

Chart 7.1 ■

diverse patterns of chemical reactivity and spectroscopic properties. Thus, polyhedranes constitute excellent substrates for the testing of various concepts of structure and bonding and consequently hold considerable appeal to theoretical, physical–organic, and synthetic chemists.[2,3] However, it is the synthetic pursuit of these carbocyclic molecules that is of paramount importance, as it offers the exciting prospect of creating real-world molecules to enable physicochemical investigations. To the synthetic chemist, creation of the intricate network of rings present in polyhedranes is a challenge of exceptional magnitude and many practitioners of synthetic craft, around the world, have been enticed into this exciting arena.

Among the polyhedranes, the [n]-prismanes constitute an enthralling class of saturated polycyclic hydrocarbons. The lowest member of this family, [3]-prismane, was conceived over a century ago by Ladenberg as a formulation for benzene.[4] Prismanes are made up of an even number of methine units of general formula $(CH)_{2n}$, arranged at the corners of a regular prism, and they belong to D_{nh} symmetry point group. Structurally and in organic chemists' parlance, they can be considered as composed of two identical n-membered rings cojoined by n four-membered rings, all fused in a cis,syn manner. The value of n represents the order of the prismane and in principle can vary from 3 to ∞. However, prismanes with $n = 3, \ldots, 12$ only, have been predicted to be planar with D_{nh} symmetry.[5] Prismanes **1–6** with $n = 3, \ldots, 8$, respectively, are displayed in Chart 7.1.

There has been a spate of papers in recent years dealing with the structure and geometry of prismanes.[5] Different methods ranging from molecular mechanics to ab initio molecular orbital (MO) calculations have been applied to elucidate the detailed structural parameters of prismanes and are summarized in Table 7.1. Prismanes as a family are recognized among the most strained organic molecules. The

Table 7.1 ■ Geometrical and Structural Parameters of [*n*]-Prismanes

Prismane	Formula	Symmetry	Number of Faces	Face Angles	H_f^a	Steric Energy[b]	SE[c]	C—C[d] MM2′ (Ref. 5v)	C—C[d] Ab Initio (Ref. 5u)	r(C—C)[e] MM2′ (Ref. 5v)	r(C—C)[e] Ab Initio (Ref. 5u)
[3]-Prismane	C_6H_6	D_{3h}	Triangle, 2	60°	136.4	319.6	148.7	1.535	1.507	1.516	1.549
			Square, 3	90°							
[4]-Prismane	C_8H_8	O_h	Square, 6	90°	148.5	171.5	164.9	1.562	1.559	1.562	1.559
[5]-Prismane	$C_{10}H_{10}$	D_{5h}	Pentagon, 2	108°	119.6	143.6	140.1	1.571	1.552	1.562	1.558
			Square, 5	90°							
[6]-Prismane	$C_{12}H_{12}$	D_{6h}	Hexagon, 2	120°	153.1	164.4	177.7	1.574	1.551	1.537	1.553
			Square, 6	90°							
[7]-Prismane	$C_{14}H_{14}$	D_{7h}	Heptagon, 2	128.6°	212.0	215.3	240.7	1.576	1.553	1.530	1.553
			Square, 7	90°							
[8]-Prismane	$C_{16}H_{16}$	D_{8h}	Octagon, 2	135°	283.0	290.7	315.9	1.579	1.556	1.532	1.551
			Square, 8	90°							
[9]-Prismane	$C_{18}H_{18}$	D_{9h}	Nonagon, 2	140°	368.7		405.6	1.583	1.561	1.537	1.550
			Square, 9	90°							

[a] Heat of formation calculated by ab initio method using STO 6-31G* (RMP 2) level, ref. 5u.
[b] Steric energy (in kilocalories per mole), ref. 5o.
[c] Strain energy (in kilocalories per mole), ref. 5u.
[d] Single C—C bond distance ring to ring.
[e] Single C—C bond distance within *n*-membered ring.

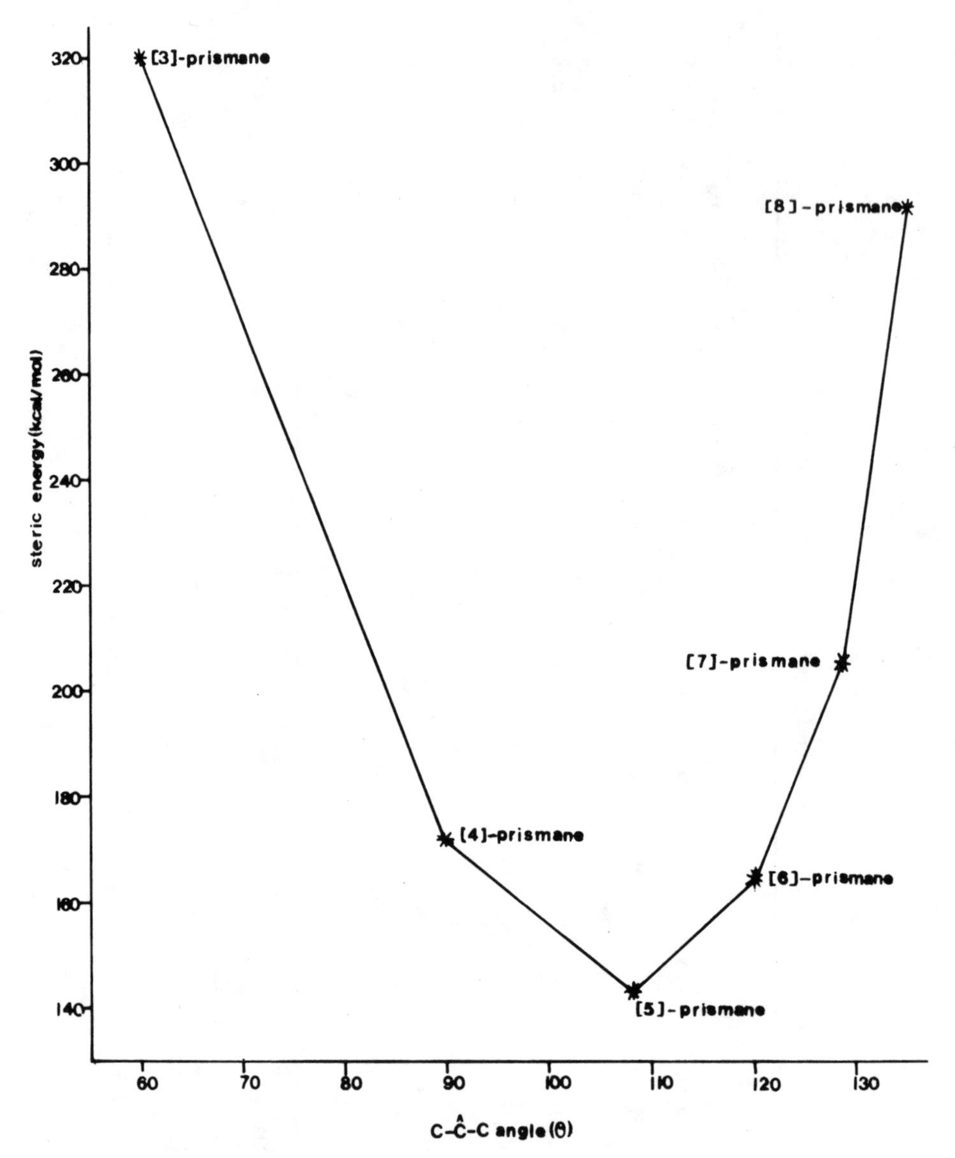

Figure 7.1 ■ Steric energy calculated by MM2 versus C—Ĉ—C angle (θ) of the n-membered ring of [n]-prismanes.[5o]

variation of steric energy computed by MM2[5o] of [3]-prismane (320 kcal mol^{-1}) to [8]-prismane (290 kcal mol^{-1}) follows a parabolic curve with a minimum corresponding to [5]-prismane (143 kcal mol^{-1}) and is depicted in Figure 7.1. Despite the high strain energy, the prismanic frameworks are blessed with unusual kinetic stability due to the symmetry-imposed barrier toward ring opening.[6] For example, [4]-prismane **2** remains essentially unchanged even at temperatures up to ~200°C[3f] and [3]-prismane **1**, at 90°C decomposes with a half-life of 11 h.[7a] In fact, Woodward and Hoffmann[6a] have likened the stability of [3]-prismane to "an angry tiger unable to break out of a paper cage."

Theoretical calculations indicate that higher $(CH)_{2n}$ homologs ($n \geq 12$) exist as puckered structures that are the more stable isomeric forms, as they have reduced angle strain and less-pronounced nonbonded H interactions. These "prismanes" have been designated as Helvetanes and Israelanes depending upon the number of cis- and trans-ring fusions.[8]

7.2. Synthetic Challenge and Tactics

Although prismanes were conceptualized over a century ago, progress toward their attainment remained dormant for a long period and the first practical realization of a prismanic framework was only accomplished in the early 1960s. Because prismanes are composed of many cyclobutane rings, fused in cagelike structures, the main task in their assembly resides in adopting strategies that would rapidly generate multiple cyclobutane rings with rigorous stereochemical control. As conventional ring forming strategies are not suitable for this purpose, the emergence of photochemical [2 + 2] cycloaddition as a general and versatile reaction in the 1950s and 1960s in both intermolecular and intramolecular versions, was an important development that promoted synthetic efforts toward prismanes. Indeed, by employing this strategy, straightforward solutions to [3]-**1** and [4]-prismane **2** (cubane) were conceived as shown in Scheme 7.1. However, in reality, neither 1,1′-bicyclopropenyl **7** nor tricyclo[4.2.0.0^{2,5}]octa-2,7-diene **8** could be cyclized to **1** and **2**, respectively, perhaps

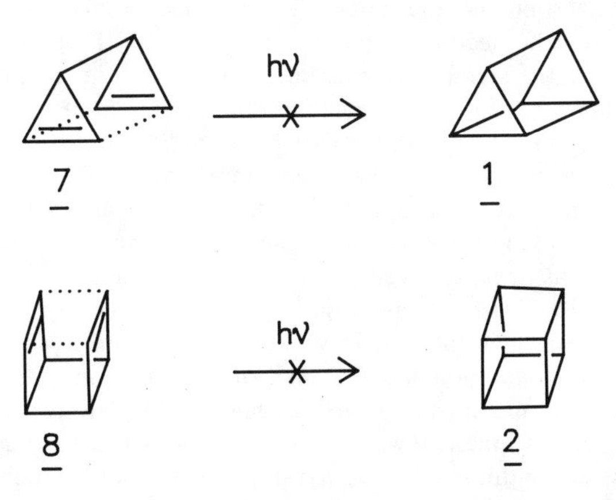

Scheme 7.1 ■

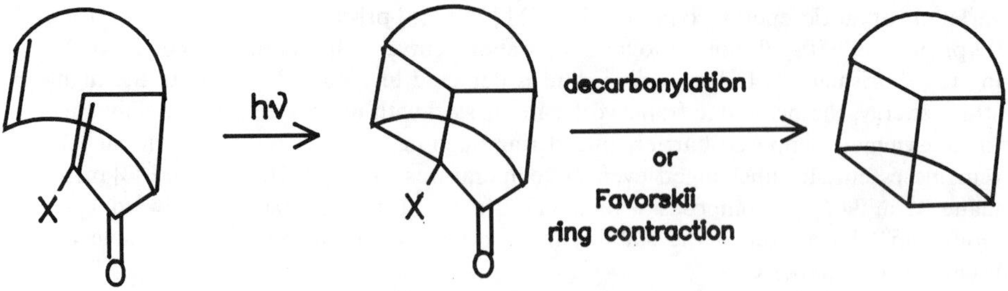

decarbonylation or Favorskii ring contraction

Scheme 7.2 ■

due to the lack of cooperative stereoelectronic factors. An early report on the photolysis of hexaphenyl compound **9** to hexaphenyl benzene **10** did invoke the intermediacy of the prismane **11**[9] but subsequent work discounted such an involvement.[10] However, more embellished derivatives of **8** have been shown to photocyclize to the corresponding cubanes.[11]

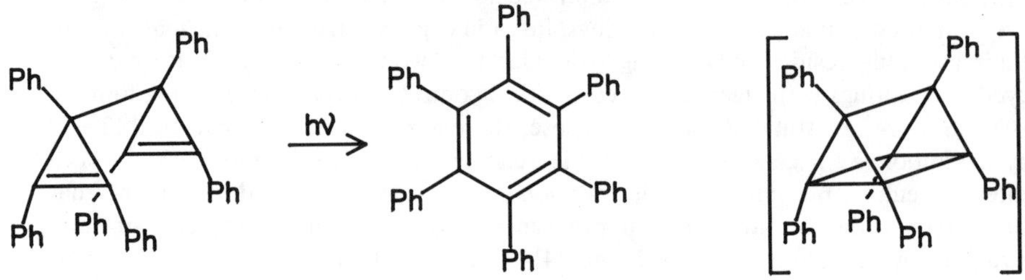

9 10 11

The failure of some of the direct [2 + 2] cycloadditions to yield the target prismanes indicated the need for suitable modification of the precursor substrates to provide more favorable stereoelectronic environment for the crucial ring closure. In this regard, deployment of the more facile intramolecular enone–olefin cycloadditions in homologous systems in tandem with ring contraction protocols appeared to be a more viable strategy for access to the target frameworks as shown in Scheme 7.2. It will be seen (*vide infra*) that this approach indeed proved successful in the realization of [4]- and [5]-prismane and has also been applied towards the [6]-prismane problem.

Some progress has been achieved in the quest for prismanes during the past three decades. Parent [3]-**1**,[7] [4]-**2**,[12,13] and [5]-**3**[14] prismanes and many of their derivatives have yielded to synthesis in the past 25 years. Indeed, **2** and **3** have been synthesized employing [2 + 2] cycloaddition and Favorskii ring contraction as the prime reactions. However, [6]-**4**, [7]-**5**, and higher prismanes have stubbornly resisted synthetic conquest, but some advancement toward their attainment has been made and further synthetic efforts are continuing. In this account, developments toward the synthesis of each member of this hydrocarbon family are briefly described.

7.3. [3]-Prismane

Triprismane **1** (tetracyclo[2.2.0.0^{2,6}.0^{3,5}]hexane), more often referred to as prismane, is one of the important valence isomers of benzene. The most logical and direct method to attain this C_6H_6 hydrocarbon is obviously through photochemical valence isomerization of benzene. Indeed, a series of papers has appeared since the mid-1960s on the photochemistry of various substituted benzenes leading to the formation of prismanic frameworks[15,16,18–22]; the results are summarized in chronological order in Table 7.2. These reactions work well with sterically encumbered benzene derivatives because the enforced deviation from planarity facilitates the formation of Dewar benzene valence isomers, which in turn undergo facile intramolecular [2 + 2] photocycloaddition to deliver substituted prismanes. This approach has met with considerable success in preparing substituted prismanes in spite of the fact that sometimes complex mixtures of benzene valence isomers are obtained through the attainment of a photostationary state. However, the parent prismane **1** has not been accessible through direct irradiation of benzene.

Several Dewar benzene derivatives, synthesized through nonbenzenoid routes, have also been shown to photocyclize to the corresponding prismanes[7b,17,23–25] (Table 7.2). Heteroprismanes, particularly azaprismanes, have also been reported through the photochemical valence isomerization of substituted pyridines.[26]

The first synthesis of parent triprismane was achieved by Katz and Acton[7a] in 1972 through an elegant strategy[27] in which nitrogen extrusion from a polycyclic azo compound served as the pivotal step (Scheme 7.3). Benzvalene **12**, a valence isomer of benzene accessible from 1,3-cyclopentadienide anion,[28] was reacted with the powerful dienophile *N*-phenyltriazoline dione to furnish the 1:1 cycloadduct **13**. Hydrolysis and oxidation led to the azo compound **14**, which on irradiation furnished a complex mixture of products from which **1** could be isolated in 1.8% yield as a colorless explosive liquid! A modified photochemical protocol has been shown to furnish **1** in slightly improved yield from **14**.[29]

As indicated before, attempts to prepare [3]-prismane directly through photochemical valence isomerization of benzene have not been very successful and fulvene and benzvalene were encountered as the major products of the reaction.[30] However, Turro, Ramamurthy, and Katz[7b] have succeeded in preparing **1** in 15% yield through the direct excitation of Dewar benzene, (entry 10, Table 7.2). Despite the successful attainment of **1**, it is still far from conveniently accessible for the exploration of its interesting chemistry. Clearly, newer and more efficient approaches need to be devised.

7.4. [4]-Prismane

Tetraprismane **2** (pentacyclo[4.2.0.0^{2,5}.0^{3,8}.0^{4,7}]octane), commonly referred to as cubane is a regular polyhedron of O_h symmetry and, therefore, one of the Platonic hydrocarbons.[3f] The synthesis of this C_8H_8 molecule in 1964 by Eaton and Cole was a pioneering achievement and demonstrated that complex, highly strained hydrocarbons can be made and are stable. The original Eaton synthesis is summarized in

Table 7.2 ■ [3]-Prismane Derivatives from the Corresponding Benzene and / or Dewar Benzene Precursors

S·No·	Precursor	[3]–Prismane	Other Products	Ref·
1·	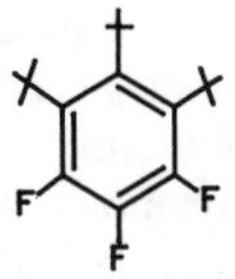 F F F	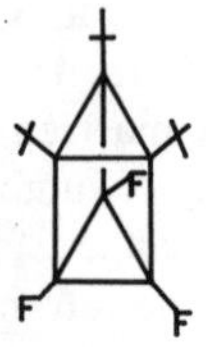 F F F	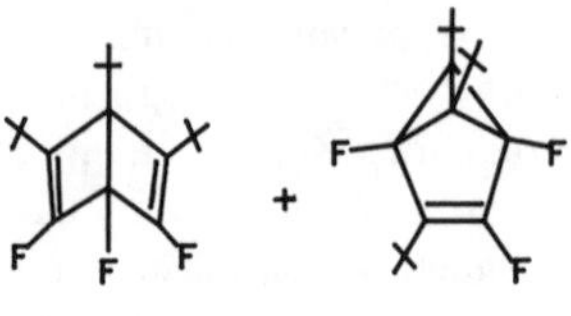 F F F + F F F	15
2·	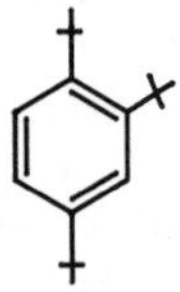	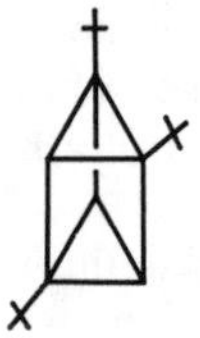	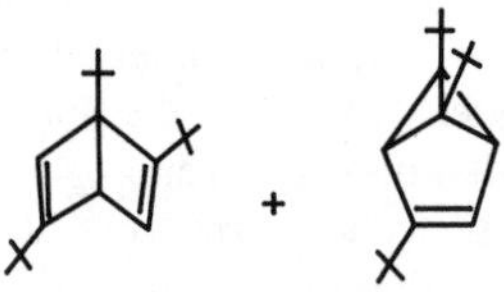 +	16
3·	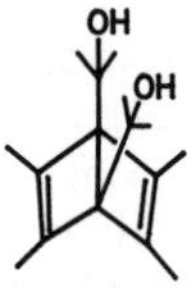 OH OH	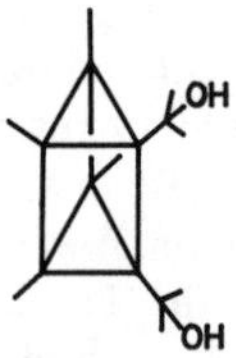 OH OH	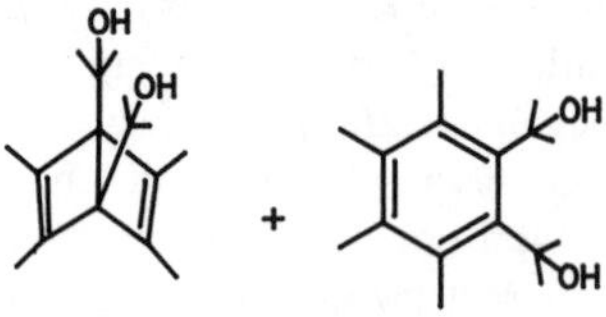 OH OH + OH OH	17
4·	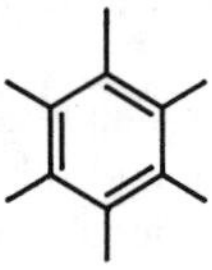		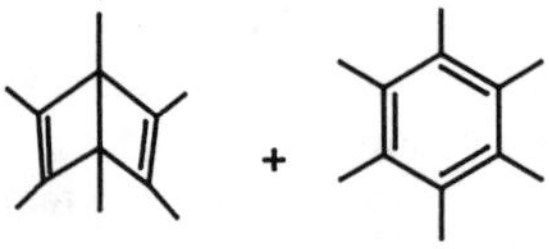 +	18,19
5·	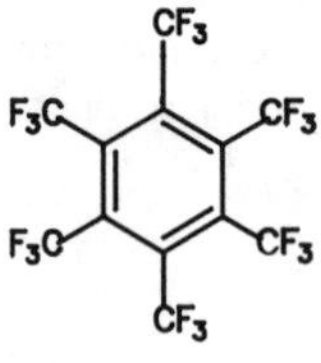 CF_3 F_3C CF_3 F_3C CF_3 CF_3	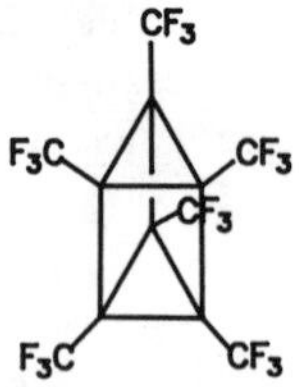 CF_3 F_3C CF_3 CF_3 F_3C CF_3	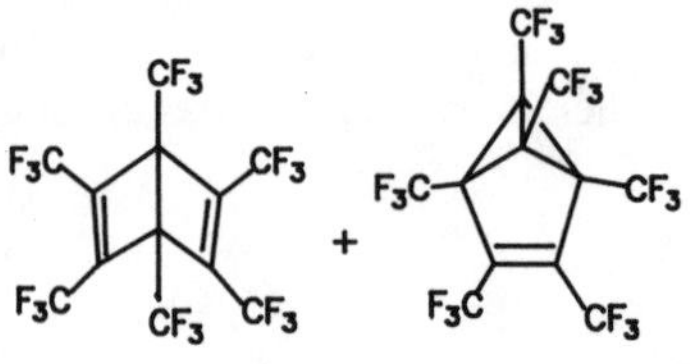 CF_3 F_3C CF_3 F_3C CF_3 CF_3 + CF_3 CF_3 F_3C CF_3 F_3C CF_3	20 21a,b
6·	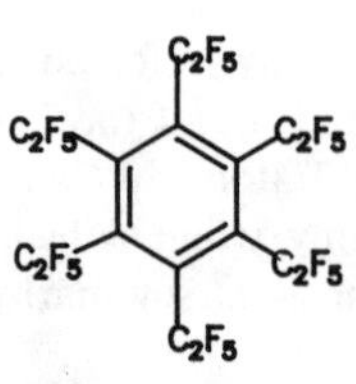 C_2F_5 C_2F_5 C_2F_5 C_2F_5 C_2F_5 C_2F_5	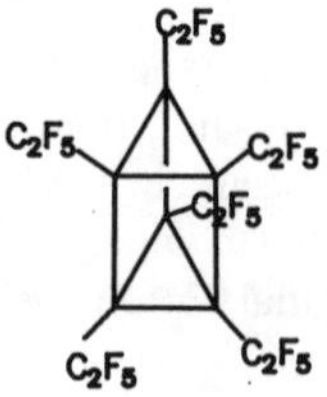 C_2F_5 C_2F_5 C_2F_5 C_2F_5 C_2F_5 C_2F_5	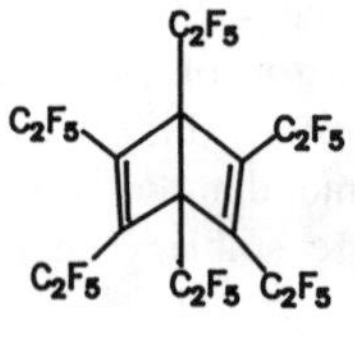 C_2F_5 C_2F_5 C_2F_5 C_2F_5 C_2F_5 C_2F_5	21b,c

Table 7.2 (*Continued*)

S·No·	Precursor	[3]–Prismane	Other Products		Ref·
7·	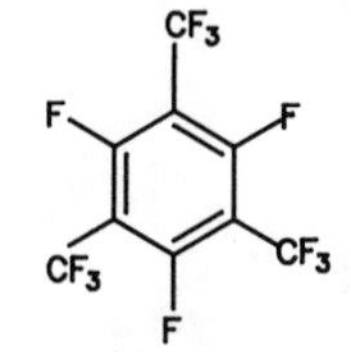	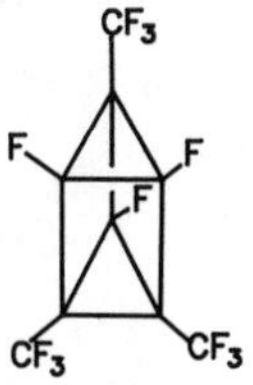	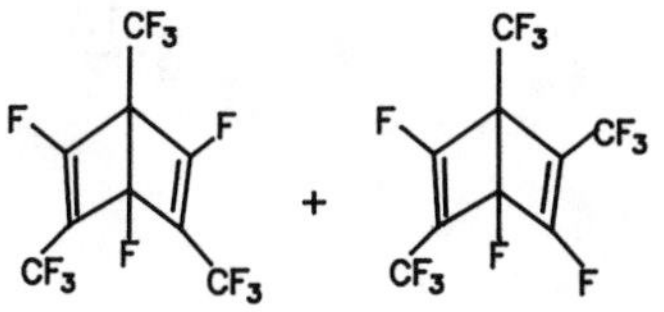 +		22
8·	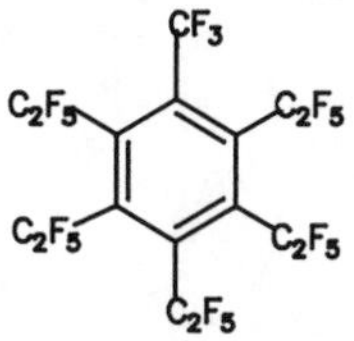	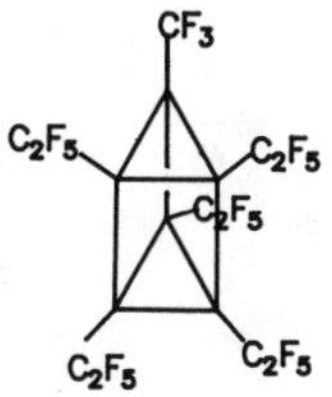	—	—	21c
9·	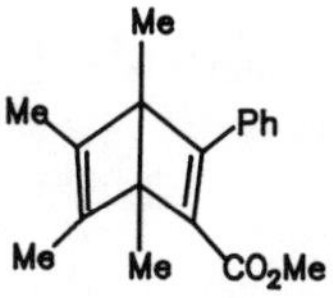	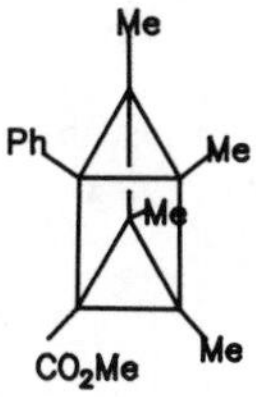	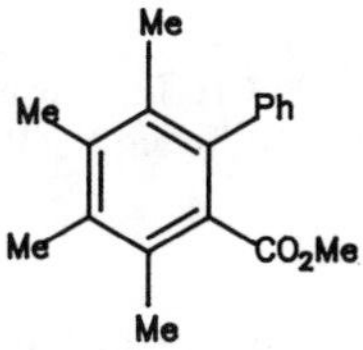	—	23
10·	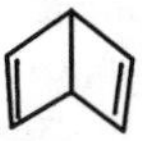	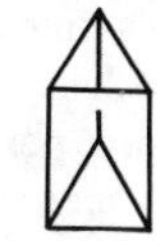			7b
11·	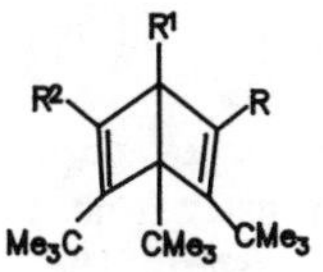	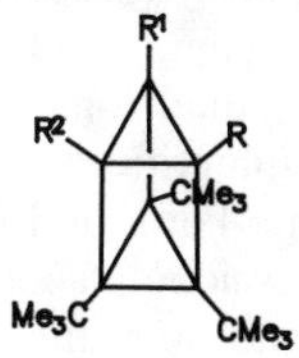	—		24
12·	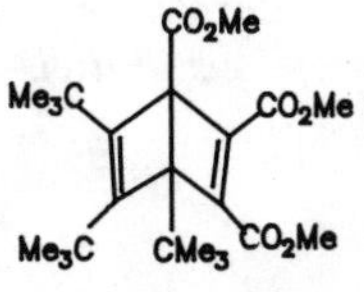	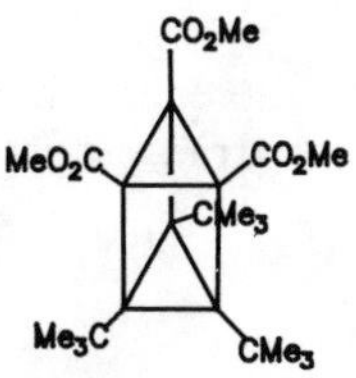	—		25

Structure labels:

7. Precursor: CF_3, F, F, CF_3, CF_3, F; [3]-Prismane: CF_3, F, F, F, CF_3, CF_3; Other products: CF_3, F, F, CF_3, F, CF_3 + CF_3, F, CF_3, CF_3, F, F

8. Precursor: CF_3, C_2F_5, C_2F_5, C_2F_5, C_2F_5, C_2F_5; [3]-Prismane: CF_3, C_2F_5, C_2F_5, C_2F_5, C_2F_5, C_2F_5

9. Precursor: Me, Me, Ph, Me, Me, CO_2Me; [3]-Prismane: Me, Ph, Me, Me, CO_2Me, Me; Other products: Me, Me, Ph, Me, CO_2Me, Me

11. Precursor: R^1, R^2, R, Me_3C, CMe_3, CMe_3; [3]-Prismane: R^1, R^2, R, CMe_3, Me_3C, CMe_3

$R=R^1=CO_2Me, CO_2CMe_3, CN, CF_3$; $R^2=CO_2CMe_3$

12. Precursor: CO_2Me, Me_3C, CO_2Me, Me_3C, CMe_3, CO_2Me; [3]-Prismane: CO_2Me, MeO_2C, CO_2Me, CMe_3, Me_3C, CMe_3

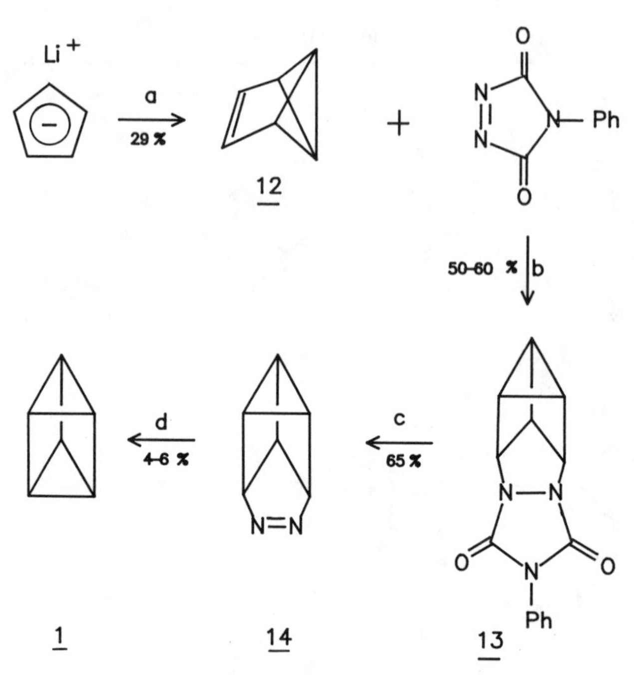

Scheme 7.3 ▪ *Reagents:* (a) CH_3Li, dichloromethane; (b) ether–dioxane; (c) KOH, CH_3OH–H_2O (85 : 15), acidic $CuCl_2$, aq. NaOH; (d) $h\nu$, Pyrex.

Scheme 7.4.[12] An intramolecular [2 + 2] photocycloaddition in ketal-enone **16** derived from the dimer **15** of 2-bromocyclopentadienone and two Favorskii-like rearrangements either stepwise (**17** → **18** and **20** → **21**) or as a single step (**22** → **23**) were the two prime multiple cyclobutane generating reactions. The resulting cubane carboxylic acids **21** and **23** were reductively decarboxylated using perester decomposition technology. The original Eaton cubane synthesis has been subjected to considerable refinement with respect to several key steps, namely, preparation of dicyclopentadienone ketal precursor,[12b,31] Favorskii rearrangement,[32] and decarboxylation.[33] It is our understanding that, following the modified methodologies, cubane and its immediate precursors are being prepared in kilogram quantities.

Immediately after the first synthesis of cubane, Pettit and Barborak[13] designed a short and no less elegant synthesis of **2** wherein cyclobutadiene transfer from its iron tricarbonyl complex to 2,5-dibromobenzoquinone was the key step. The endo Diels–Alder adduct **24** on 2 + 2 photocycloaddition followed by single-step double Favorskii ring contraction furnished 1,3-cubanedicarboxylic acid **25**, which was transformed to the hydrocarbon **2** (Scheme 7.5), via the decarboxylation of its *tert*-butylperester.

Cubane 1,3-dicarboxylic acid **25** has also been prepared by an alternate approach by Eaton through the photochemical rearrangement of *cis,anti,cis*-4,9-

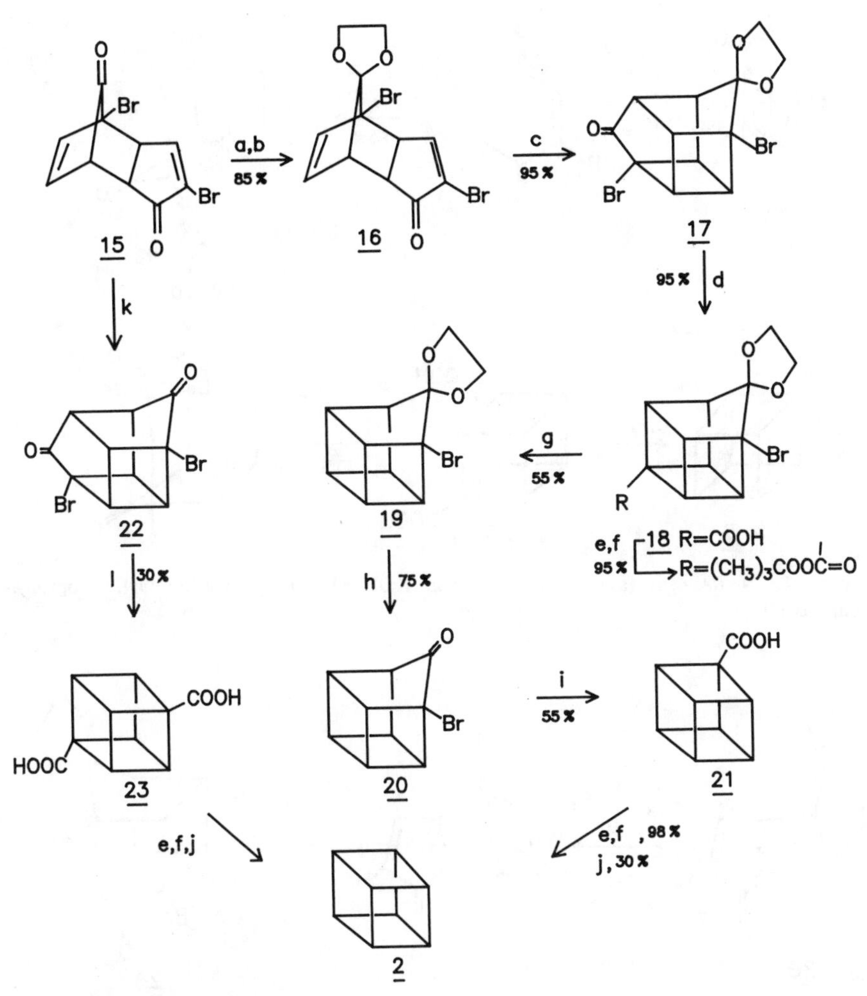

Scheme 7.4 ■ *Reagents:* (a) $(CH_2OH)_2$, H^+; (b) aq. HCl; (c) $h\nu$; (d) 10% KOH; (e) $SOCl_2$; (f) $(CH_3)_3COOH$–py.; (g) cumene, 152°C; (h) 75% H_2SO_4; (i) 25% KOH; (j) diisopropylbenzene, 100°C; (k) $h\nu$, MeOH, HCl; (l) 50% aq. KOH.

dibromotricyclo[5.3.0.0^{2,6}]deca-4,9-diene-3,8-dione **26** (Scheme 7.6).[34] Photochemical 1,3-sigmatropic process in **26** furnished **27**, which was further subjected to an intramolecular [2 + 2] cycloaddition to the pentacyclic compound **28**. Steps analogous to the earlier synthesis of cubane 1,4-dicarboxylic acid[12] led to **25**,[13] thus constituting an alternate formal synthesis of [4]-prismane **2**.

Masamune and co-workers have reported a formal synthesis of cubane from basketene **29**, involving oxidative cleavage of double bond followed by Dieckman's cyclization to substituted homocubanone **30** (Scheme 7.7).[35] The ester functionality in

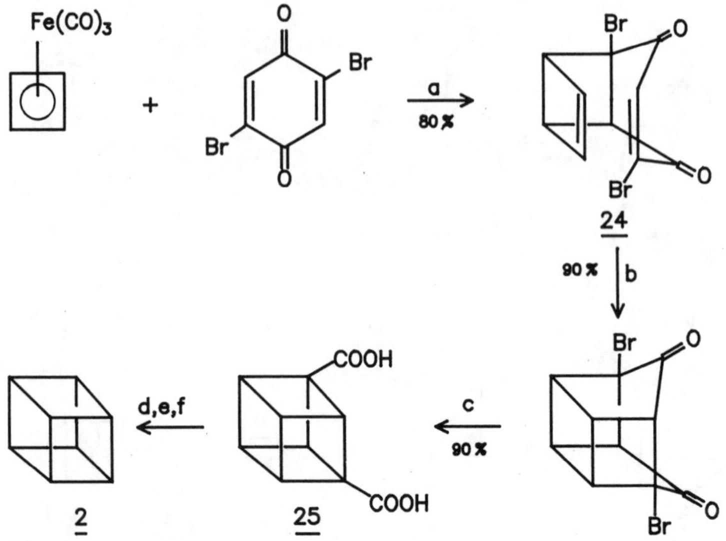

Scheme 7.5 ■ *Reagents*: (a) Ce^{4+}; (b) $h\nu$; (c) aq · KOH; (d) $SOCl_2$; (e) *t*-BuOOH, diisopropylbenzene; (f) Δ.

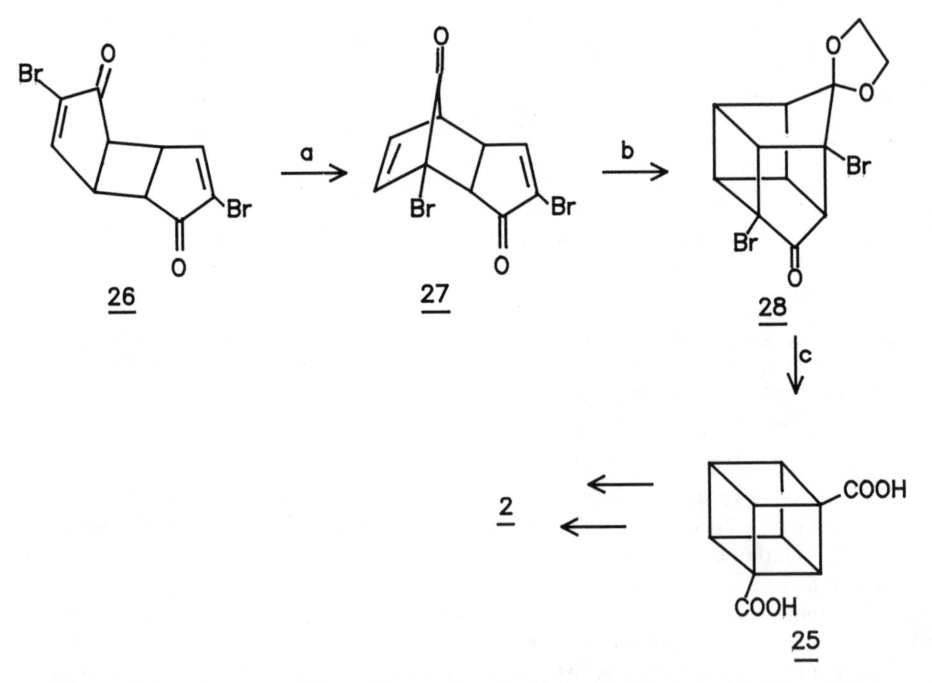

Scheme 7.6 ■ *Reagents*: (a) $h\nu$, trifluroacetic acid; (b) $h\nu$, MeOH, HCl; (c) 50% aq · KOH.

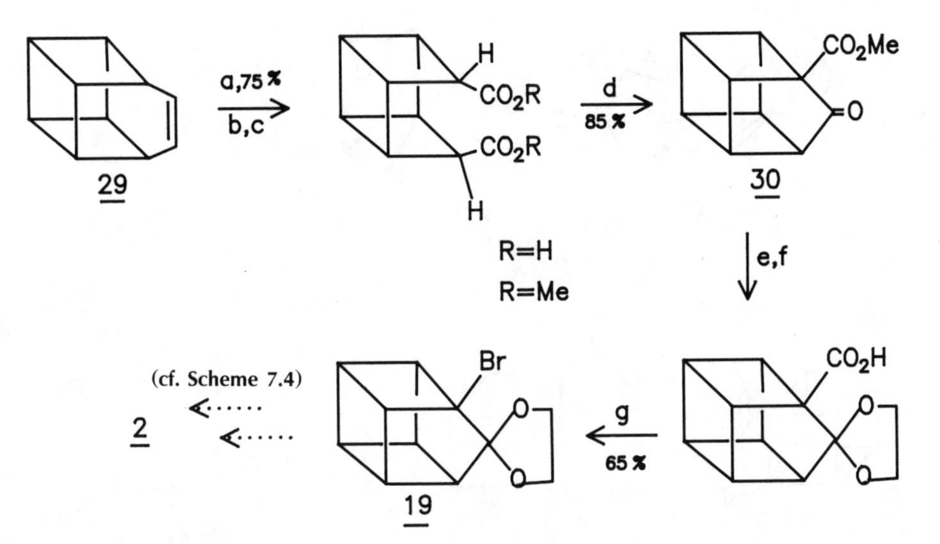

Scheme 7.7 ■ *Reagents*: (a) OsO_4; (b) Jones reagent; (c) diazomethane; (d) methylsulfinylcarbanion, 0°C; (e) ethylene glycol. PTSA; (f) aq · KOH; (g) HgO, Br_2.

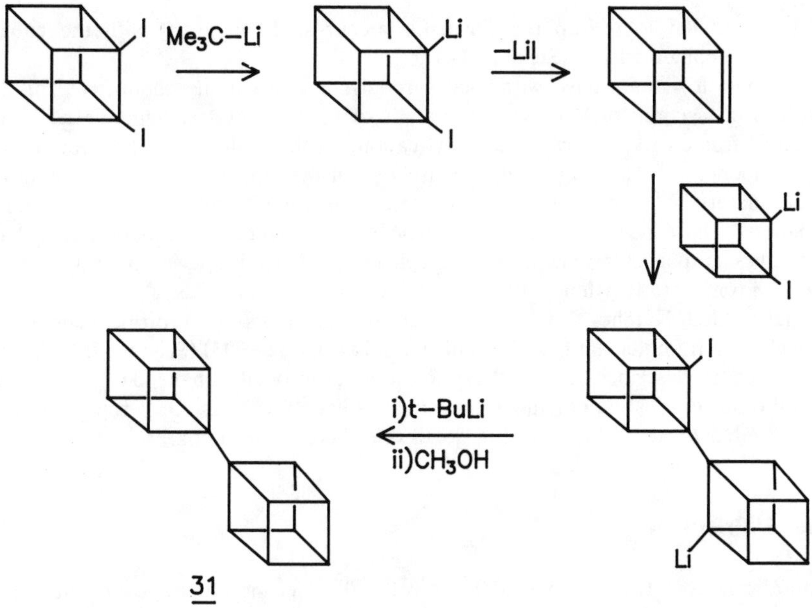

I
I
$Me_3C{-}Li$
Li
I
−LiI
Li
I
I
Li
i)t−BuLi
ii)CH_3OH
31

Scheme 7.8 ■

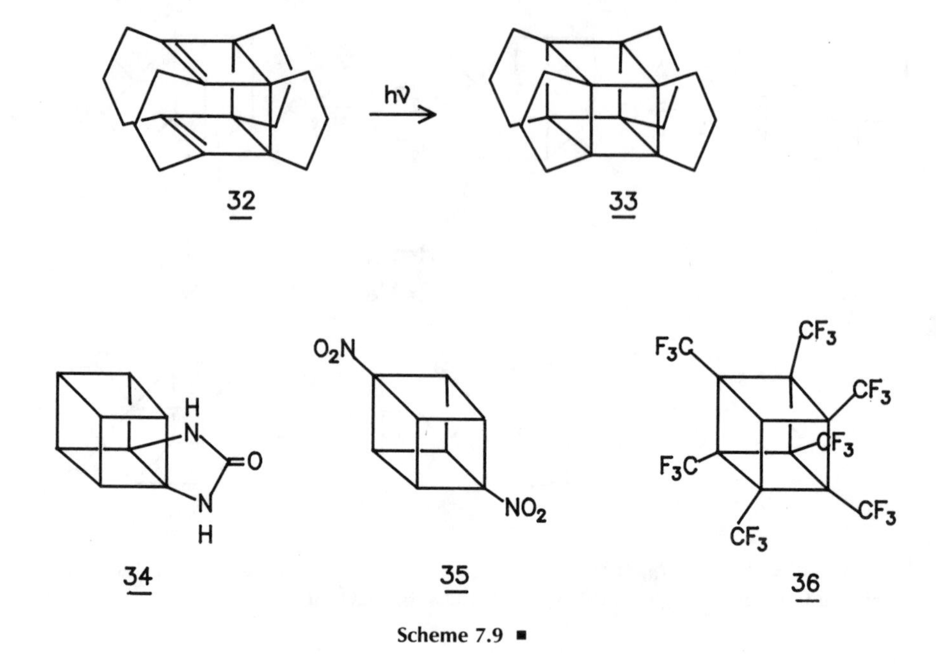

Scheme 7.9 ▪

30 was modified to set up the Favorskii precursor **19** identical with the cubane precursor obtained earlier (Scheme 7.4).[12]

The past few years have witnessed a renaissance in cubane chemistry. This has occurred on account of interest in the study of reactive intermediates derived from the cubyl framework, for example, cubyl cation,[36] cubyl radical,[33,37] 1,4-cubyl diradical,[38] cubene,[38c,39] etc., and the promise of polyfunctional cubanes as potentially useful materials.[40] Application of directed metalations[41] and hypervalent iodine oxidations[42] have been remarkably efficacious in the synthesis of many new cubane derivatives. In particular, synthesis of cubylcubane **31** needs special mention because it holds promise for further synthesis of polycubanes, Scheme 7.8.[39a]

Gleiter and Karcher[43] have described the synthesis of a propellacubane **33** through an intramolecular [2 + 2] photocycloaddition (**32** → **33**) (Scheme 7.9). A more recent report describes the synthesis of a heteropropellacubane **34**.[44] There are several other interesting polyfunctional cubanes like **35**[40a,45] and **36**[11] reported in the literature but these are discussed in detail elsewhere in this book.

7.5. [5]-Prismane

Pentaprismane **3** (hexacyclo[5.3.0.0.^{2,6}0.^{3,10}0.^{4,9}0.^{5,8}]decane, housane), is the least-strained member of the prismane family (Table 7.1). Taking cognizance of the pivotal role of intramolecular [2 + 2] photocycloaddition in the synthesis of prismanes, hypostrophene **37** was considered as the most logical precursor, which on [2 + 2] cycloaddition would deliver [5]-prismane **3**. However, various attempts to effect intramolecular photocycloaddition in **37** have been singularly unsuccessful.[46] Also unsuccessful was the approach based on pinacolic coupling in **38** to furnish the diol **39**.[47]

The extrusion of dinitrogen from appropriately constituted cyclic azo compounds has been gainfully employed for making small ring polycyclic compounds (e.g., triprismane). Consequently, pentacyclic cyclic azo compounds **40** and **41** were assembled as shown in Scheme 7.10. However, nitrogen extrusion reactions on **40** and **41** took a wayward course and pentaprismane **3** could not be realized.[48]

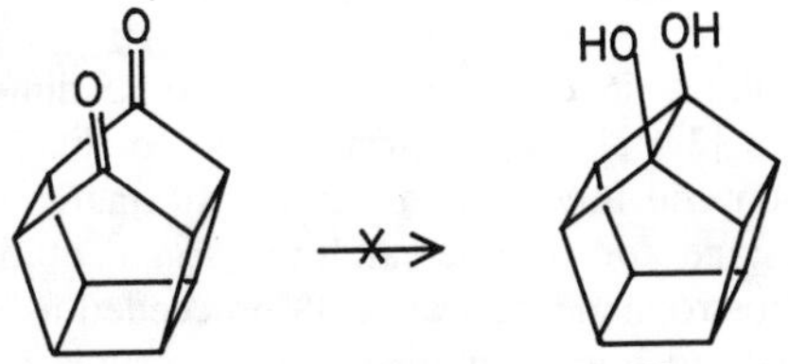

It had been observed earlier that, unlike hypostrophene **37**, homohypostrophene **42** readily underwent intramolecular [2 + 2] photocycloaddition to the homopen-

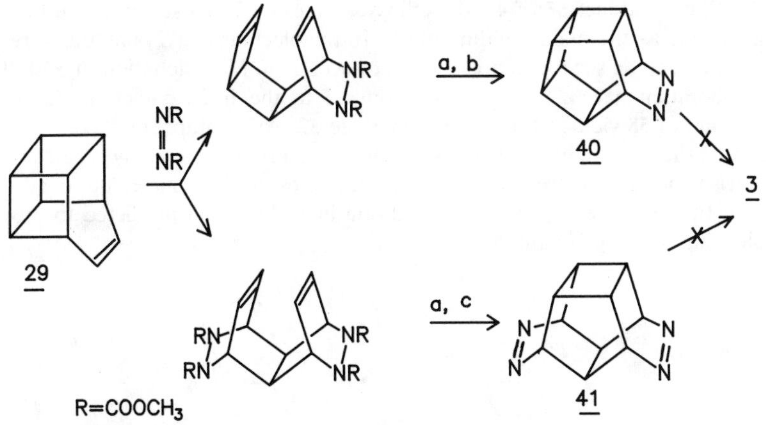

Scheme 7.10 ▪ *Reagents:* (a) *hν*, acetone; (b) (1) KOH, $CH_3OH–H_2O$, (2) HgO; (c) (1) KOH, CH_3OH, $CuCl_2$, (2) NaOH.

taprismane **43**.[49] Eaton, therefore, opted for the homopentaprismane route to pentaprismane, which required incorporation of elements of a ring contraction in an appropriately substituted homopentaprismane **44** to furnish a pentaprismane derivative **45**. Successful realization of this approach was announced in 1981 and the sequence is summarized in Scheme 7.11.[14]

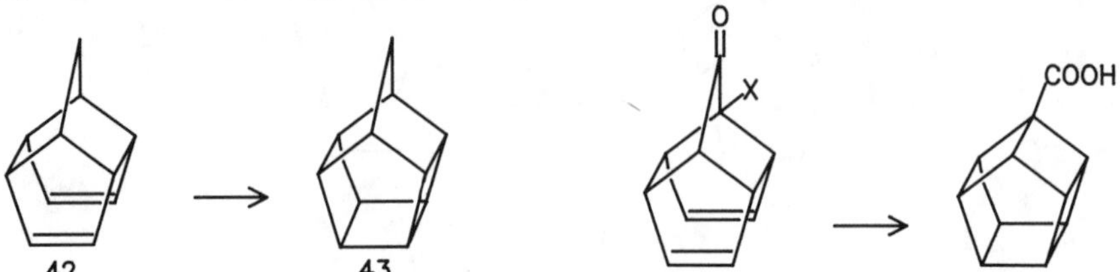

42 → 43   44 → 45

The Diels–Alder adduct **46** of 1,2,3,4-tetrachloro-5,5-dimethoxycyclopentadiene and *p*-benzoquinone on [2 + 2] cage cyclization led to the pentacyclic dione **47**.[50] Reductive dechlorination and functional group transformations led to the iodo-tosylate **48** and set the stage for a base-catalyzed fragmentation to the diene **49**. Irradiation of homohypostrophene derivative **49** proceeded as planned and deacetalization of the functionalized homopentaprismane derivative **50** furnished the corresponding ketone **51**. Suitable bridgehead functionalization in **51** was now expected to set the stage for the key Favorskii ring contraction to pentaprismane skeleton. However, at this stage direct methods of bridgehead substitution via enolate chemistry were discounted due to violation of Bredt's rule and a more circuitous route via the keto-ester **52** was devised. Acyloin-type coupling reaction in **52** was employed to obtain the bridgehead substituted homopentaprismane **53**. Functional group manipulations, Favorskii ring contraction, and decarboxylation furnished pentaprismane **3** (Scheme 7.11).

A formal synthesis of pentaprismane has been realized by Dauben and Cunningham[51] through Eaton's advanced intermediate keto-ester **52** (Scheme 7.12). The [4 + 2]-cycloadduct **55** of 4,4-dimethoxycyclohexa-2,5-dien-1-one **54** and cyclobutadiene served as the main building block. Intramolecular [2 + 2] photoclosure in **55** gave the pentacyclic secopentaprismane keto-acetal **56**. Methylenation of **56** followed by hydroboration gave **57**, which was oxidized to the hydroxy-lactone **58**. Finally, esterification of **58** yielded the key intermediate **52**[14] of pentaprismane **3**.

To date, the homopentaprismane ring contraction route is the only access available to [5]-prismane and therefore alternate strategies to this molecule need to be explored. In this context, [2 + 2] cycloaddition in **59** has been predicted to provide a possible route to [5]-prismane.[52]

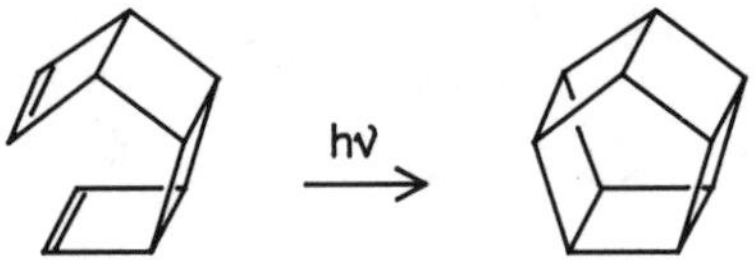

59 —hν→ 3

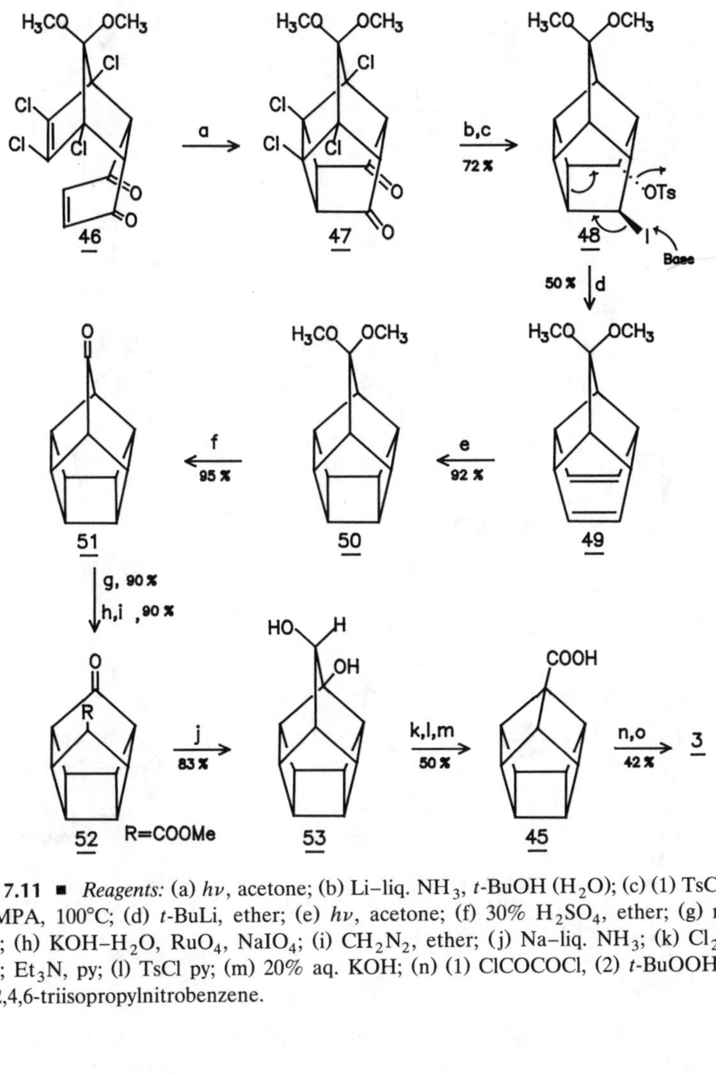

Scheme 7.11 ■ *Reagents:* (a) $h\nu$, acetone; (b) Li–liq. NH_3, *t*-BuOH (H_2O); (c) (1) TsCl, py, (2) NaI, HMPA, 100°C; (d) *t*-BuLi, ether; (e) $h\nu$, acetone; (f) 30% H_2SO_4, ether; (g) mCPBA, CH_2Cl_2; (h) KOH–H_2O, RuO_4, $NaIO_4$; (i) CH_2N_2, ether; (j) Na–liq. NH_3; (k) Cl_2, Me_2S, CH_2Cl_2; Et_3N, py; (l) TsCl py; (m) 20% aq. KOH; (n) (1) ClCOCOCl, (2) *t*-BuOOH, py; (o) 150°C, 2,4,6-triisopropylnitrobenzene.

7.6. Toward [6]-Prismane

Hexaprismane **4** (heptacyclo[$6.4.0.0^{2,7}.0^{3,12}.0^{4,11}.0^{5,10}.0^{6,9}$]dodecane), a $C_{12}H_{12}$ hydrocarbon is formally a face-to-face dimer of benzene and to date remains unconquered. It is composed of 12 identical methine units $(CH)_{12}$ arranged at the corners of a regular hexagonal prism and thus the two parallel six-membered rings are cojoined by six four-membered rings. Furthermore, **4** is the first [*n*]-prismane in the series in which the carbon atoms have C—C—C bond angles, both less (90°) and greater

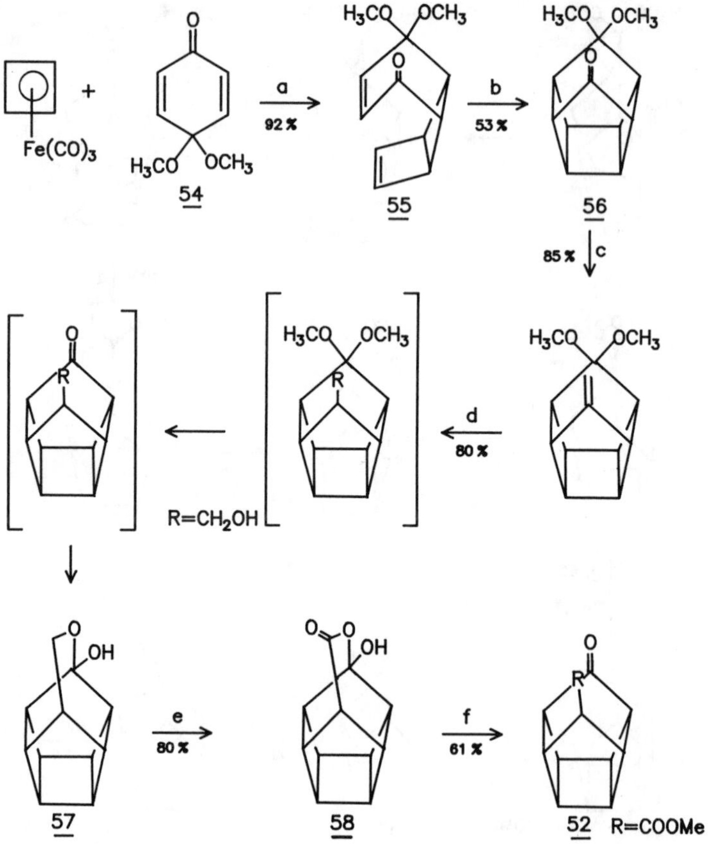

Scheme 7.12 ■ *Reagents:* (a) $Pb(OAc)_4$–py; (b) $h\nu$, ether, uranium filter; (c) PPh_3CH_3Br, $C_5H_{11}ONa$, benzene; (d) (1) BH_3–THF, THF, (2) 3M H_2SO_4, (3) 3M NaOH, (4) H_2O_2; (e) RuO_4, CCl_4, NaOH, $NaIO_4$; (f) KF, MeI, DMF.

(120°) than the normal tetrahedral angle.[5q] The strain energy in [6]-prismane is only marginally higher than that of [5]-prismane (Table 7.1), but synthetically it has proved to be quite a formidable proposition and despite many efforts has not yielded to synthesis. Therefore, only various approaches that have been or are being pursued toward **4** and the progress achieved so far will be discussed here. Synthesis of various homologs and secologs of hexaprismane reported so far have also been included.

Keeping in view the strategic options revealed by earlier successful prismane syntheses, two types of approaches have been pursued toward [6]-prismane. These are summarized in Scheme 7.13. The first set of approaches is based on multiple intramolecular [2 + 2] photocycloadditions leading to two or more cyclobutane rings and eventuating directly to the target framework. The other set of approaches are based on assembling homologous systems ([6]-asterane derivatives) with a minimum

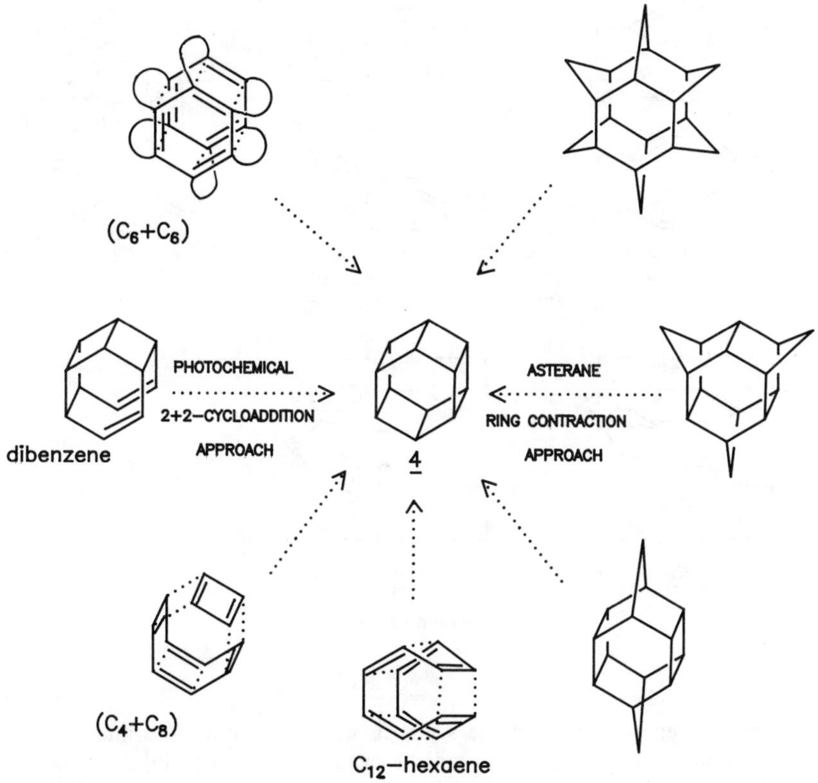

Scheme 7.13 ■

number of or no cyclobutane rings and then subjecting them to the ring contraction protocols (Scheme 7.13).

In the [2 + 2] photocycloaddition approaches, the most direct way to **4** would be through the union of two benzene rings held face-to-face for threefold intramolecular [2 + 2] ring closure. In general, cyclophanes ranging from superphane to less intricate multilayered derivatives, with disposable bridges and face-to-face held benzene rings, can be considered as promising precursors for direct access to the hexaprismane framework. Although the distance between the two aromatic rings in many of the cyclophanes is favorable for ring closure, the presence of bridges imposes excessive strain on the product of photocycloaddition process. Thus, this approach has received only limited success despite the elegant efforts of Misumi's group[53] with layered heteroatom-bridged cyclophanes. The oxa-, thia-, and selenium-bridged quadruple-layered cyclophanes **60a–60c** and **63** on irradiation undergo two facile intramolecular $\pi_2 + \pi_2$ cycloadditions to furnish the corresponding bridged dibenzene derivatives **61a–61c** and **64**, respectively (Scheme 7.14). However, efforts to convert **61** to the hexaprismane derivative **62** have not met with success.

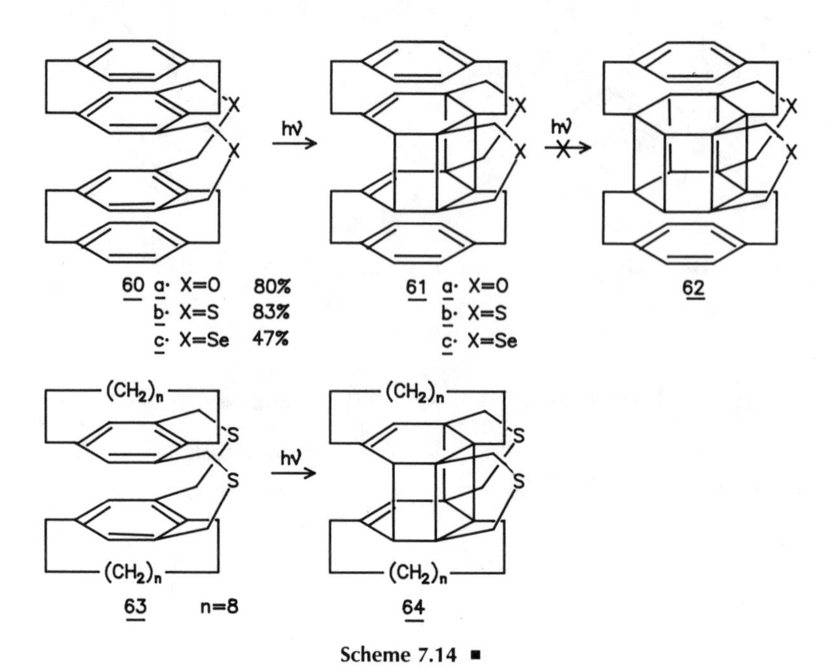

Scheme 7.14 ■

Musso's group has reported the synthesis of several novel and interesting polycyclic compounds (formally [4]-asterane derivatives) related to [6]-prismane framework. The underlying theme in this work is to unite two aromatic (or dihydro aromatic) moieties in face-to-face arrangement. Building on the early observations of Cookson, Fox, and Hudec[54] that quinones can be photochemically dimerized to bis-seco-[6]-prismane tetraones,[54,55] Musso synthesized several face-to-face photocyclized dimers **66a–66c** of quinones **65a–65c** and transformed them into many interesting derivatives (Scheme 7.15).[56,57a]

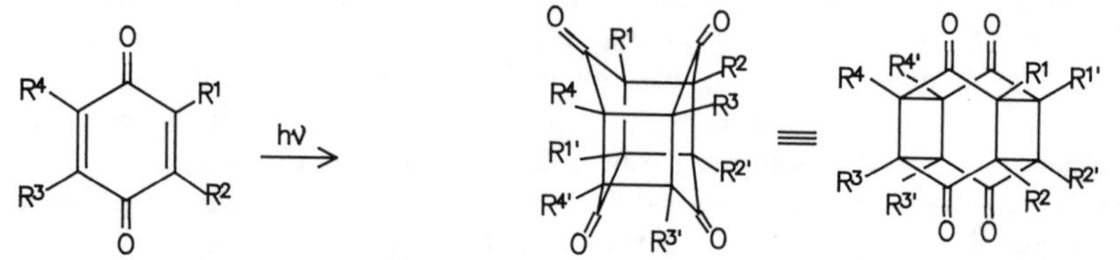

65a $R^1 = R^4 = CH_3$; $R^2 = R^3 = H$
65b $R^1 = R^2 = CH_3 = R^3 = R^4 = H$
65c $R^1 = R^3 = CH_3$; $R^2 = R^4 = H$

66a $R^1 = R^4 = R^{1'} = R^{4'} = CH_3$; $R^2 = R^3 = R^{2'} = R^{3'} = H$
66b $R^1 = R^2 = R^{1'} = R^{2'} = CH_3$; $R^3 = R^4 = R^{3'} = R^{4'} = H$
66c $R^1 = R^3 = R^{1'} = R^{3'} = CH_3$; $R^2 = R^4 = R^{2'} = R^{4'} = H$

Scheme 7.15 ■

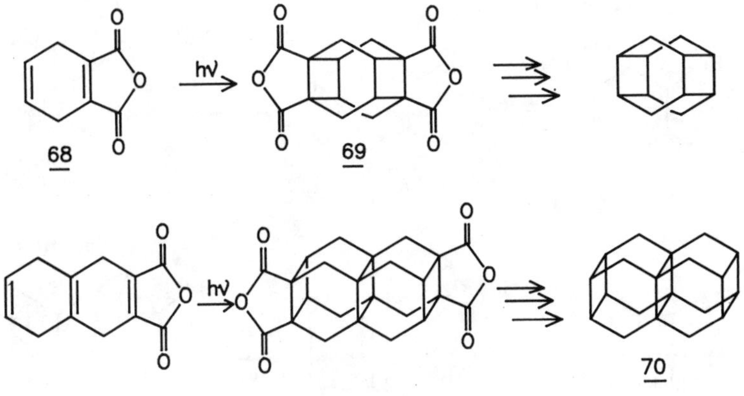

Scheme 7.16 ■

In principle, quinone dimers **66** can be converted to [6]-prismane derivatives **67** through pinacolic coupling or related reaction but such possibilities have not succeeded due to stereoelectronically unfavorable alignment of carbonyl group.

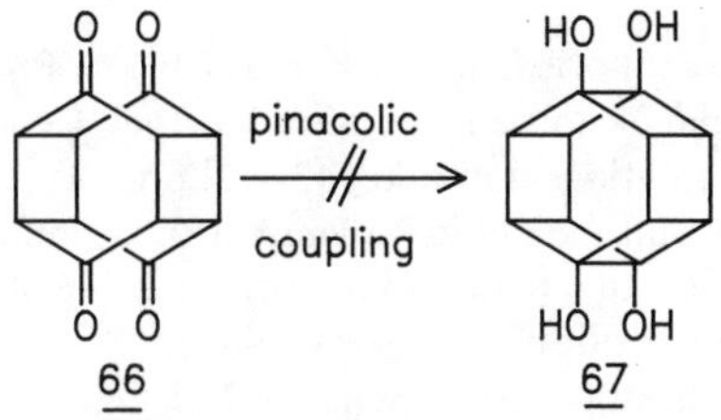

In a related study, Musso and co-workers studied the photochemical dimerization of dihydrophthalic anhydride **68** and obtained **69** among the complex mixture of products.[57] More recently, this study was extended to the synthesis of a double [4]-asterane **70**,[58] which can be considered as a double bis-seco-[6]-prismane, Scheme 7.16.

A conceptually pleasing approach to [6]-prismane through a single-shot photocyclization of an *all cis*-cyclododecahexaene **71**, held in appropriate conformation through metal complexation, has been conceived. Although the synthesis of related *all cis* tetraene **72** has been successfully achieved,[59] further work in the area is awaited.

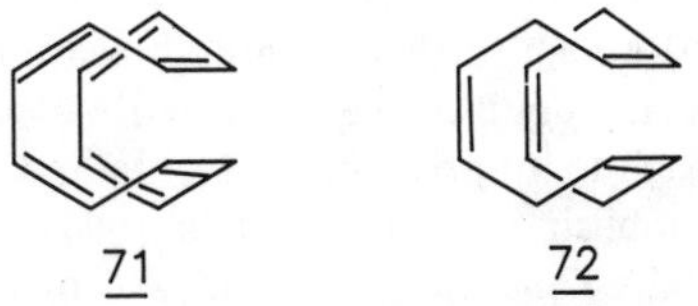

Yang and Horner have sought to unite two benzene rings through a *p,p'*-benzene dimer equivalent and have succeeded in the synthesis of a pentacyclic dimer of

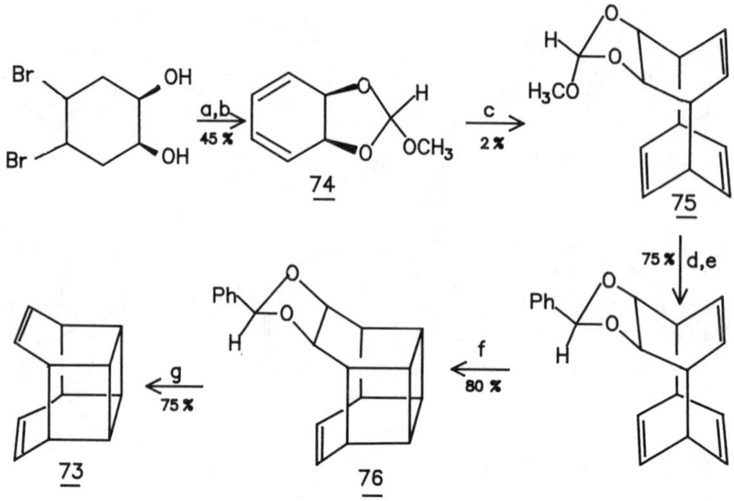

Scheme 7.17 ■ *Reagents:* (a) $CH(OCH_3)_3$, TsOH; (b) DBU, benzene; (c) *hν*, benzene; (d) HCl, H_2O, THF; (e) $PhCH(OCH_3)_2$, TsOH; (f) *hν*, xanthone; (g) *t*-BuLi, THF.

benzene **73** en route to [6]-prismane (Scheme 7.17).[60] Photochemical addition of substituted cyclohexadiene **74** to benzene afforded the [4 + 4] adduct **75**, which after functional group transformations and a key [2 + 2] photocycloaddition step furnished **76**. Elimination of the acetal moiety in **76** yielded the pentacyclic benzene dimer **73**. Although the synthesis of the penultimate precursor of **4** has been achieved, its further photocyclization has not been reported so far.

The $C_4 + C_8$ approach indicated in Scheme 7.13 constitutes an attractive way to [6]-prismane **4** provided the two partners, cyclooctatetraene (COT) and cyclobutadiene (CB) can be coaxed into proper spatial alignment with complementary cycloaddition reactivities. It is known that both COT and CB react preferentially as 4π components in cycloadditions. Therefore, their utilization in the context of [6]-prismane would require deployment of carefully chosen equivalents with altered preferences toward cycloaddition chemistry. Eaton was the first to exploit this strategy, employing 1,2,3,4-tetrachloro-5,5-dimethoxycyclopentadiene **77** as CB equivalent and 1,5-cyclooctadiene **78** as a model for COT.[61,62] Diels–Alder reaction between them furnished the known adduct **79**.[63] Intramolecular [2 + 2] cycloaddition proceeded smoothly to **80**, in which the acetal moiety was deprotected to set up the Favorskii ring contraction (**81** → **82**). Reductive decarboxylation in **82** gave a bis-seco-[6]-prismane **83** (Scheme 7.18). Although further progress toward [6]-prismane has not been reported along these lines, the equivalence of **77** with CB was demonstrated.

In their $C_4 + C_8$ approach to [6]-prismane **4**, Mehta and Padma[64] have successfully developed protocols to establish equivalency of both **77** and **78** with CB and COT, respectively (Scheme 7.19).[64] Thus, the functionalized pentacyclic dimer of benzene **86** was assembled through the intermediate **85**, which incorporates two new olefinic bonds and thus established complete equivalence with COT. Although the penultimate precursor of [6]-prismane framework was realized, attempts to photocyclize **86**

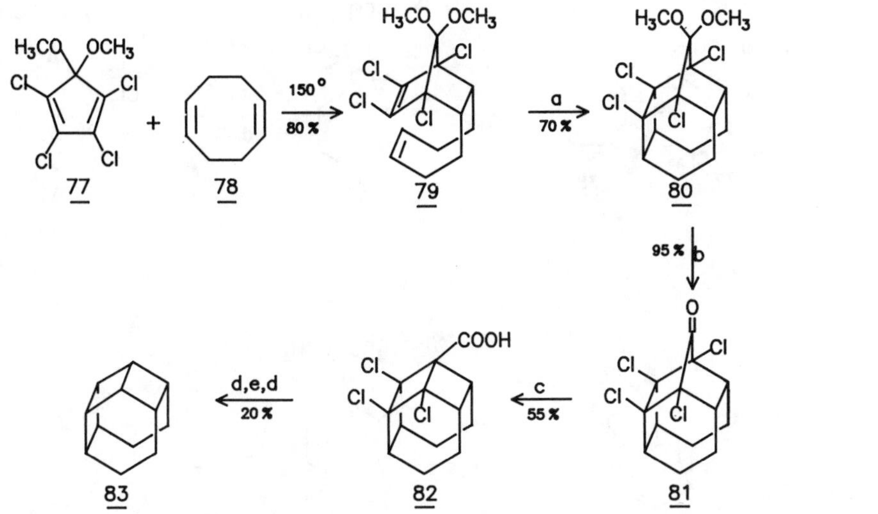

Scheme 7.18 ▪ *Reagents:* (a) $h\nu$, acetone; (b) H_2SO_4, H_2O; (c) NaOH, toluene; (d) Li, *t*-BuOH, THF; (e) Br_2, HgO, CCl_4.

to **87** have remained unsuccessful.[65] Even the homologous diene **85** resisted any attempts to photocyclize to **88** despite the proximity of the two double bonds (~ 2.8 Å) in it.

With the failure of **85** and **86** to photocyclize, access to the [6]-prismane framework was denied via this approach and, therefore, it was decided to attempt the synthesis of relatively less strained secohexaprismane **89**, a one-bond-less, but closest possible isomer of **4**. The successful attainment of **89** from the earlier synthesized dimesylate **84** (Scheme 7.19) is depicted in Scheme 7.20.[64] The key operation in this venture was the preparation of tetracyclic diene **91** from **90** through a boron-mediated fragmentation reaction. The secohexaprismane framework was secured through Favorskii ring contraction (**92** → **93**) and defunctionalization led to the pentacyclic $C_{12}H_{14}$ hydrocarbon **89**. Attempts at dehydrogenative ring closure in **89** to **4** have not been successful.

The consistent failure of the intramolecular [2 + 2] cycloaddition strategies to deliver [6]-prismane stressed the need to explore the alternative asterane ring contraction route to this hydrocarbon (Scheme 7.13). Among the various asterane precursors, those having fewer bridges, well separated to minimize steric interactions between the methylenes, like **94** and **95** appeared to be more serviceable.

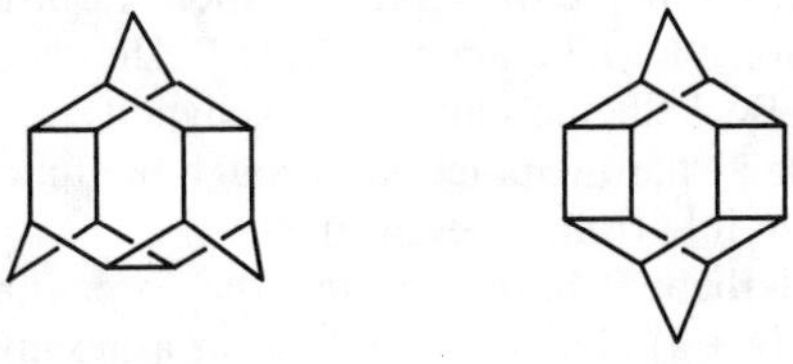

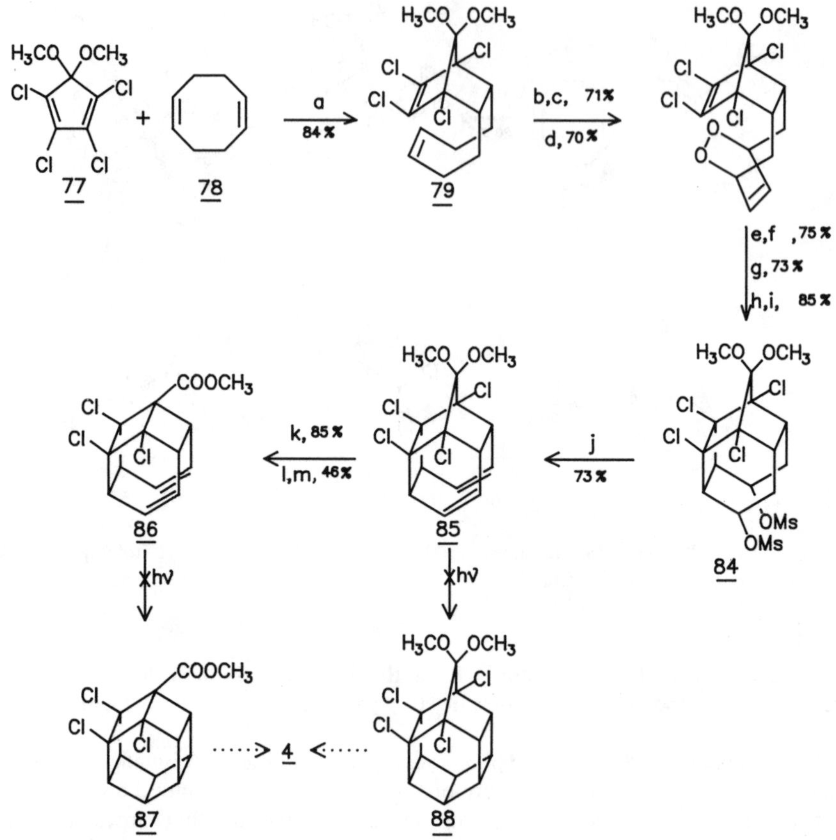

Scheme 7.19 ▪ *Reagents:* (a) reflux; (b) *N*-bromosuccinimide, AIBN, carbon tetrachloride; (c) DBU, DMSO; (d) O_2, methylene blue, dichloromethane, 500-W tungsten lamp; (e) $LiAlH_4$, ether; (f) Ac_2O, py; (g) $h\nu$, acetophenone, benzene, Pyrex; (h) aq. KOH, methanol; (i) CH_3SO_2Cl, py; (j) NaI, HMPA, 130°C; (k) 90% H_2SO_4, dichloromethane; (l) powdered NaOH, toluene; (m) CH_2N_2–ether, MeOH.

The first asterane-like precursor **97** of [6]-prismane was prepared quite accidentally in 1973 by Boekelheide and Hollins. They observed that [2.2.2](1,3,5)-cyclophane **96** on exposure to $AlCl_3$–HCl milieu underwent transannular cyclization to polycyclic chloro derivatives that on reductive dechlorination furnished tris-1,3,5-bishomohexaprismane **97** in reasonable yield (Scheme 7.21).[66] The sequence leading to **97** was recently repeated[67] and Boekelheide's intuitively assigned structure has been found to be correct. The polycycle **97** has the potential through the functionalization of bridges for further elaboration to [6]-prismane derivatives.

Srikrishna and Sunderbabu[68] have reported the formation of a novel diene **99** through intramolecular [4 + 4] photocycloaddition in a hexahydroanthracene precur-

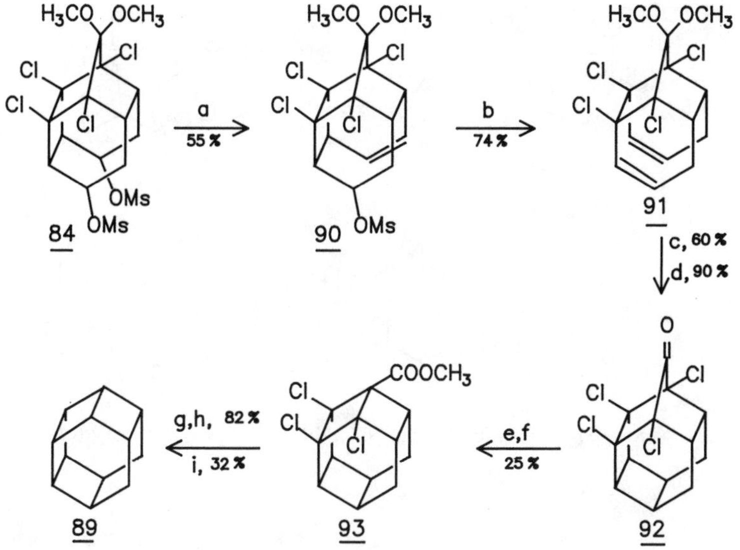

Scheme 7.20 ▪ *Reagents:* (a) NaI, HMPA, 100°C; (b) B_2H_6, THF, aq. NaOH; (c) $h\nu$, Pyrex, acetone; (d) 90% H_2SO_4, dichloromethane; (e) powdered NaOH, toluene; (f) CH_2N_2–ether, MeOH; (g) aq. KOH, MeOH; (h) HgO, CH_2Br_2, Br_2; (i) Li, *t*-BuOH, THF.

sor **98** (Scheme 7.22).[68] The pentacyclic diene **99** could serve as a precursor of 1,2-bishomo-[6]-prismane **100**, and further efforts in this direction are warranted.

Among the asterane ring contraction based approaches to [6]-prismane, notable success has been achieved by Mehta and Padma[69] through their synthesis of 1,4-bishomo-[6]-prismane **95** ("garudane") (Scheme 7.23).[69] The heptacyclic hydrocarbon **95** is formally a face-to-face dimer of norbornadiene and has been painstakingly sought for nearly a quarter of a century among the products of catalyzed dimerization of norbornadiene.[70] Indeed, structure **95** was invoked in earlier studies for one of the dimers of norbornadiene[70a,b,k] but was later revised to the octacyclic formulation **101** ("isogarudane").[70f,h,j,k] Because the norbornadiene dimerization approach to **95** had

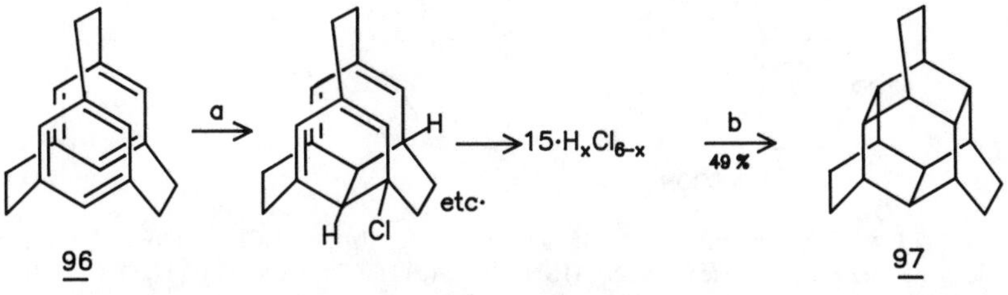

Scheme 7.21 ▪ *Reagents:* (a) (1) $AlCl_3$, CH_2Cl_2, (2) HCl; (b) Li, *t*-BuOH, THF.

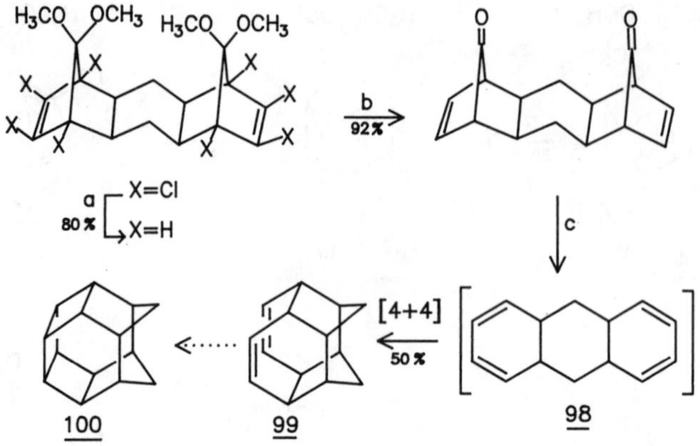

H_3CO OCH_3 H_3CO OCH_3 X X X X X X X X
b 92%
a 80% X=Cl → X=H
c
[4+4] 50%
100 99 98

Scheme 7.22 ■ *Reagents:* (a) Na–EtOH, liq. NH_3; (b) 20% H_2SO_4, CH_2Cl_2; (c) $h\nu$, benzene.

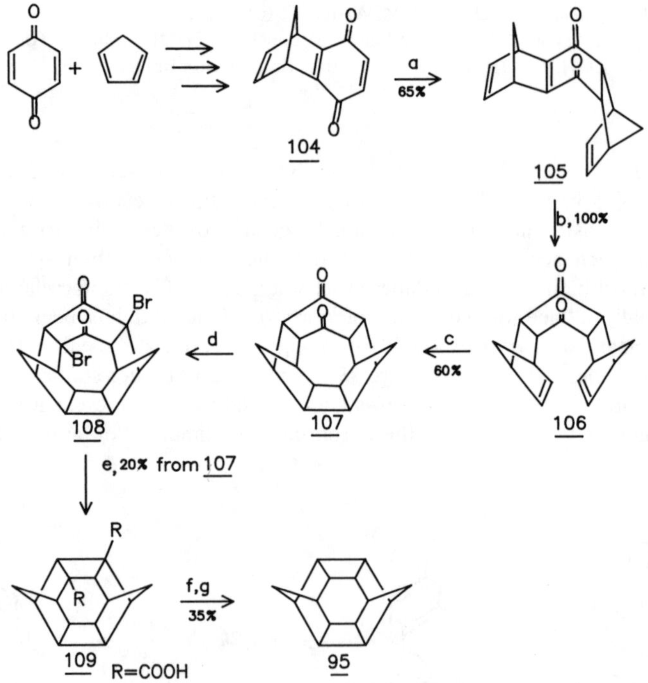

+
a 65%
104 105
b, 100%
Br Br
d
c 60%
108 107 106
e, 20% from 107
R R
f,g 35%
109 R=COOH 95

Scheme 7.23 ■ *Reagents:* (a) cyclopentadiene, benzene; (b) aq. $TiCl_3$, acetone; (c) $h\nu$, vycor, 10% acetone–benzene; (d) excess NBS, AIBN, CCl_4; (e) NaOH, toluene; (f) HgO, CH_2Br_2, Br_2; (g) Li, *t*-BuOH, THF.

repeatedly failed, a rational approach to **95** was devised as indicated in **102**. This required tandem cycloadditions between two cyclopentadienes and cyclobutadiene as shown in **102**. Here again, to induce CB to act twice as a 2π component, an equivalent was sought with pronounced dienophilic character. 2,5-Dibromobenzoquinone was chosen as CB equivalent and **102** to a more practically realizable picture, **103**. For the expression of this conceptual theme, norbornenobenzoquinone **104** was chosen as the abundantly accessible starting material.

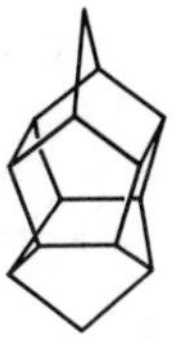

101

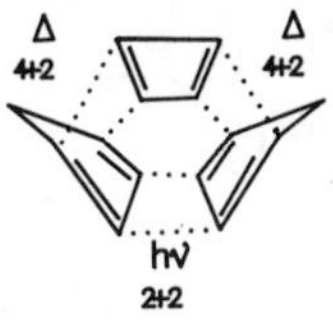

Δ 4+2 Δ 4+2 hν 2+2

102

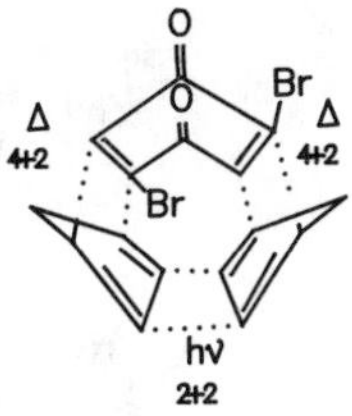

O O Br Br Δ 4+2 Δ 4+2 hν 2+2

103

Diels–Alder reaction between norbornenobenzoquinone **104**[71] and cyclopentadiene predominantly furnished the endo,syn adduct **105**. The enedione moiety in **105** was regioselectively and stereoselectively reduced with aq. $TiCl_3$ to furnish the endo,syn,endo adduct **106**. On irradiation **106** underwent smooth [2 + 2] intramolecular photocycloaddition and the heptacyclic compound, 1,4-bishomo-6-seco-[7]-prismane dione **107** was obtained. This molecule, in addition to being the key precursor of **95**, is potentially serviceable for elaboration to various interesting polycyclic compounds, including [7]- and [8]-prismane analogs. Further transformation of the dione **107** to the diacid **109** was achieved by two α-brominations to **108** (cf. **103**) followed by a single-shot double Favorskii ring contraction. Conversion to the target hydrocarbon **95** was effected through routine functional group manipulations (Scheme 7.23).

The successful attainment of garudane **95** indicates that this approach is particularly suited for the synthesis of [6]-prismane.[47] Efforts directed toward the preparation of suitably functionalized derivatives like **110** and **111** are expected to present less-formidable problems in view of the ample literature precedences. Efforts along these lines are currently underway.[72]

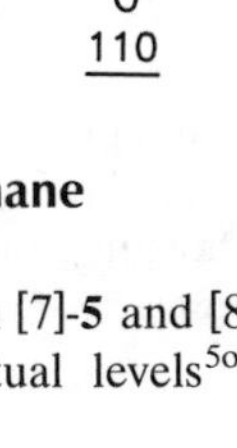

O X O

110

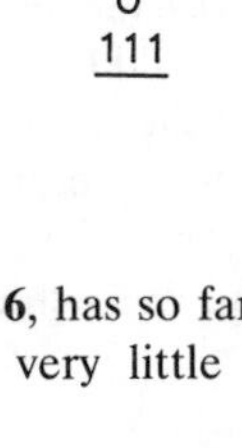

O X X O

111

7.7. Toward [7]-Prismane

The chemist's interest in [7]-**5** and [8]-prismane **6**, has so far been confined largely to theoretical and conceptual levels[5o,s,u,v,y] and very little synthetic effort directed

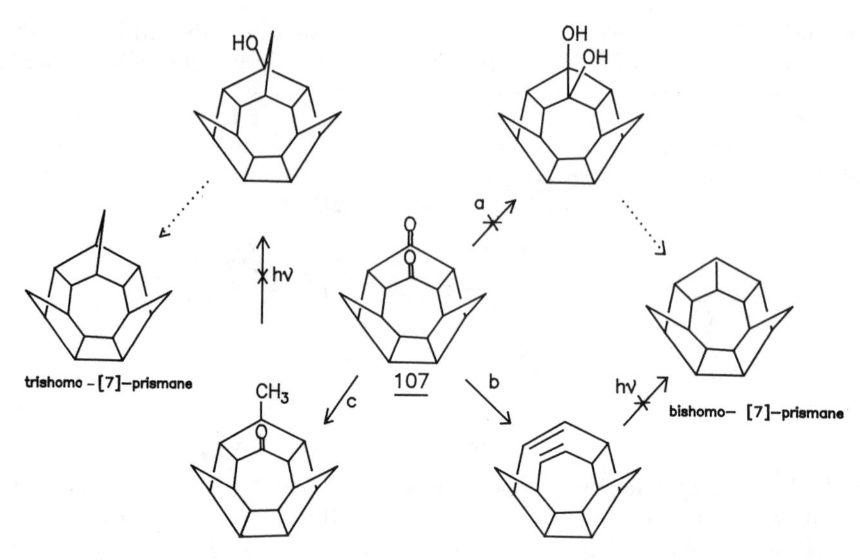

Scheme 7.24 ▪ *Reagents:* (a) Mg, HgCl, THF, $TiCl_4$; (b) (1) $NaBH_4$, MeOH, (2) MsCl, py, 81%, (3) NaI, HMPA, 63%, (4) Na–K alloy, THF, 70%; (c) (1) CH_3PPh_3Br, t-$C_5H_{11}ONa$, toluene, 80%, (2) H_2, 10% Pd/C, EtOAc.

toward them as appeared in literature. Attempts to photodimerize tropone and heptafulvene, in a manner analogous to benzoquinones, have not met with much success.[73] Photoirradiation of tetrabenzoheptafulvene **112** have led only to **113** and nothing beyond.[74]

The heptacyclic dione **107** (Scheme 7.23) possesses a bishomosecoheptaprismane framework and, therefore, could be a potential precursor for gaining access to **5** and **6** and their analogs. Some of the preliminary endeavors in this area, although not very successful, are schematically represented in Scheme 7.24.[75]

Thus far, **114** is the nearest to a [7]-prismane homolog that has been synthesized. Its unanticipated formation was achieved via an intramolecular Cannizzaro reaction (Scheme 7.25).[76]

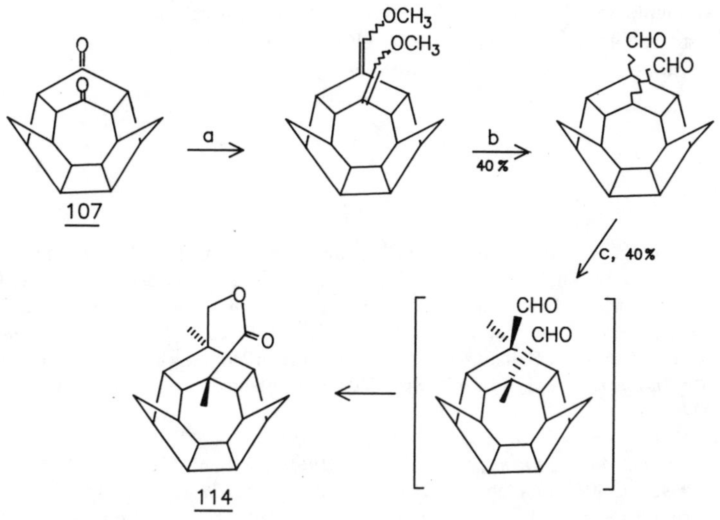

Scheme 7.25 ▪ *Reagents:* (a) $CH_3OCH_2PPh_3Cl$, $C_5H_{11}O^-Na^+$, ether–THF; (b) 35% $HClO_4$, ether; (c) KH, THF, MeI.

7.8. Outlook

Despite their almost irresistible synthetic appeal, progress toward the synthesis of [*n*]-prismanes has been slow and halting. For example, in each of the preceding three decades, only one [*n*]-prismane has been synthesized and it is already a decade since the last one (pentaprismane) was attained.[14] The next member, [6]-prismane is defying synthetic conquest[64,69] and the higher [7]- and [8]-prismanes are still on the synthetic chemists' drawing boards, a long distance from practical realization. This would imply that either prismanes are exceptionally difficult to make or the more enterprising among the synthetic community are too preoccupied in making molecules conceived by Nature, that is, natural products, and are far less excited about creating these aesthetically pleasing structures envisioned by Man, that is, "unnatural products." Perhaps, there is need to redress this imbalance, particularly as it is becoming abundantly clear that prismanes and related polyhedranes are no longer mere intellectual curiosities or physical–organic chemists' delight but hold great promise as important new and hitherto unexplored materials. The renaissance in cubane chemistry[40] and discovery of [*n*]-staffanes[77] are just two such examples.

The intramolecular [2 + 2] photocycloaddition and Favorskii ring contraction protocols have served well the cause of [4]- and [5]-prismane synthesis and provided some leeway toward [6]-prismane. But how far? For those already fascinated by the prismanic frameworks and active in this area, it is time to think anew and explore and develop new synthetic strategies. Hopefully, we will witness new strides in the prismane arena during this decade and perhaps more than one new higher prismane will be synthesized before the end of this century.

Acknowledgments. S. P. thanks CSIR for the award of a research associateship. We would like to record our appreciation to UGC for supporting our research through Special Assistance and COSIST Programmes.

References

1. (a) Lyusternik, L. A. *Convex Figures and Polyhedra*; Smith, T. J., Transl.; Dover Publications: New York, 1963. (b) Shubnikov, A. V.; Koptsik, V. A. *Symmetry in Science and Art*; Plenum Press: New York, 1974. (c) Rosen, J. *Symmetry Discovered Concepts and Applications in Nature and Science*; Cambridge University Press; Cambridge, 1975. (d) Ernst, B. *The Magic Mirror of M. C. Escher*; Ballantine Books: New York, 1976. (e) Hargittai, I.; Hargittai, M. *Symmetry Through the Eyes of a Chemist*; VCH: Weinheim, 1986. (f) Stoddart, F. *Nature*, **1988**, *334*, 10.
2. Schultz, H. P. *J. Org. Chem.* **1965**, *30*, 1361.
3. (a) Seebach, D. *Angew, Chem., Int. Ed. Engl.*, **1965**, *4*, 121. (b) Ferguson, L. N. *J. Chem. Educ.*, **1969**, *46*, 404. (c) Liebman, J. F.; Greenberg, A. *Chem. Rev.*, **1976**, *76*, 311. (d) Greenberg, A.; Liebman, J. F. *Strained Organic Molecules*; Academic Press: New York, 1978. (e) Mehta, G. *J. Sci. Ind. Res.* (*India*) **1978**, *37*, 256. (f) Eaton, P. E. *Tetrahedron* **1979**, *35*, 2189. (g) Paquette, L. A. *Topics in Current Chemistry*; Springer-Verlag: Berlin, 1979; Vol. 79, p. 43.
4. Ladenburg, A. *Chem. Ber.* **1869**, *2*, 140.
5. (a) Randic, M.; Majerski, Z. *J. Chem. Soc.* (*B*) **1968**, 1289. (b) Teijiro, Y.; Simizu, K.; Kato, H. *Bull. Chem. Soc. Jpn.* **1968**, *41*, 2336. (c) Baird, N. C.; Dewar, M. J. S. *J. Am. Chem. Soc.* **1969**, *91*, 352. (d) Engler, E. M.; Andose, J. D.; Schleyer, P. v. R. *J. Am. Chem. Soc.* **1973**, *95*, 8005. (e) Newton, M. D.; Schulman, J. M.; Manus, M. M. *J. Am. Chem. Soc.* **1974**, *96*, 17. (f) Minkin, V. I.; Minyaev, R. M. *Zh. Org. Khim.* **1981**, *17*, 221. (g) Allinger, N. L.; Eaton, P. E. *Tetrahedron Lett.*, **1983**, 3697. (h) Engelke, R.; Hay, P. J.; Kleier, D. A.; Wadt, W. R. *J. Am. Chem. Soc.* **1984**, *106*, 5439. (i) Dai, Y.; Dunn, K.; Boggs, J. E. *THEOCHEM*, **1984**, *18*, 127. (j) Wiberg, K. B. *J. Comput. Chem.* **1984**, *5*, 197. (k) Wiberg, K. B. *J. Org. Chem.* **1985**, *50*, 5285. (l) Disch, R. L.; Schulman, J. M.; Sabio, M. L. *J. Am. Chem. Soc.* **1985**, *107*, 1904.(m) Schulman, J. M.; Disch, R. L. *J. Am. Chem. Soc.* **1985**, *107*, 5059. (n) Schulman, J. M.; Disch, R. L. *Chem. Phys. Lett.* **1985**, *113*, 291. (o) Reddy, V. P.; Jemmis, E. D. *Tetrahedron Lett.* **1986**, *27*, 3771. (p) Mehta, G.; Padma, S.; Osawa, E.; Barbiric, D. A.; Mochizuki, Y. *Tetrahedron Lett.* **1989**, *28*, 1295. (q) Dailey, W. P. *Tetrahedron Lett.* **1987**, *28*, 5787. (r) Dodziuk, H.; Nowinski, K. *Bull. Pol. Acad. Sci. Chem.* **1987**, *35*, 195. (s) Mehta, G.; Padma, S.; Jemmis, E. D.; Leela, G.; Osawa, E.; Barbiric, D. A. *Tetrahedron Lett.* **1988**, *29*, 1613. (t) Miller, M. A.; Schulman, J. M. *J. Mol. Struct.* **1988**, *163*, 133. (u) Disch, R. L., Schulman, J. M. *J. Am. Chem. Soc.* **1988**, *110*, 2102. (v) Jemmis, E. D.; Rudzinski, J. M.; Osawa, E. *Chem. Express* **1988**, *3*, 109. (w) Miller, M. A.; Schulman, J. M. *THEOCHEM* **1988**, *40*, 133. (x) Politzer, P.; Seminario, J. M. *J. Phys. Chem.* **1989**, *93*, 588. (y) Martin, R. M. Unpublished results quoted in ref. 14b.
6. (a) Woodward, R. B.; Hoffmann, R. *The Conservation of Orbital Symmetry*; Academic Press; New York, 1971. (b) Woodward, R. B.; Hoffmann, R. *Angew, Chem., Int. Ed. Engl.* **1969**, *8*, 781.
7. (a) Katz, T. J.; Acton, N. *J. Am. Chem. Soc.* **1973**, *95*, 2738. (b) Turro, N. J.; Ramamurthy, V.; Katz, T. J. *Nouv. J. Chim.* **1977**, *1*, 363.
8. Osawa, E.; Rudzinski, J. M.; Barbiric, D. A.; Jemmis, E. D. In *Strain and Its Implications in Organic Chemistry*; de Meijere, A.; Blechert, S., Eds.; NATO ASI Series; Kluwer Academic Publishers: Dordrecht, 1989; Vol. 273, p. 259.

9. (a) Breslow, R.; Gal, P. *J. Am. Chem. Soc.* **1959**, *81*, 4747. (b) Breslow, R.; Gal, P.; Chang, H. W.; Altman, L. J. *J. Am. Chem. Soc.* **1965**, *87*, 5139.
10. Weiss, R.; Kölbl, H. *J. Am. Chem. Soc.* **1975**, *97*, 3222.
11. Pelosi, L. F.; Miller, W. T. *J. Am. Chem. Soc.* **1976**, *98*, 4311.
12. (a) Eaton, P. E.; Cole, T. W., Jr. *J. Am. Chem. Soc.* **1964**, *86*, 962. (b) Eaton, P. E.; Cole, T. W., Jr. *J. Am. Chem. Soc.* **1964**, *86*, 3157.
13. Barborak, J. C.; Watts, L.; Pettit, R. *J. Am. Chem. Soc.* **1966**, *88*, 1328.
14. (a) Eaton, P. E.; Or, Y. S.; Branca, S. J. *J. Am. Chem. Soc.* **1981**, *103*, 2134. (b) Eaton, P. E.; Or, Y. S.; Branca, S. J.; Shankar, B. K. R. *Tetrahedron* **1986**, *42*, 1621.
15. Viehe, H. G.; Merenyi, D.-I. R.; Oth, J. F. M.; Senders, J. R.; Valange, P. *Angew. Chem., Int. Ed. Engl.* **1964**, *3*, 755.
16. Wilzbach, K. E.; Kaplan, L. *J. Am. Chem. Soc.* **1965**, *87*, 4004.
17. Criegee, R.; Askani, R. *Angew. Chem., Int. Ed. Engl.* **1966**, *5*, 519.
18. Lemal, D. M.; Lokensgard, J. P. *J. Am. Chem. Soc.* **1966**, *88*, 5934.
19. Schäfer, W.; Criegee, R.; Askani, R.; Grüner, H. *Angew, Chem., Int. Ed. Engl.* **1967**, *6*, 78.
20. Lemal, D. M.; Staros, J. V.; Austel, V. *J. Am. Chem. Soc.* **1969**, *91*, 3373.
21. (a) Barlow, M. G.; Haszeldine, R. N.; Hubbard, R. *J. Chem. Soc., Chem. Commun.* **1969**, 202. (b) Barlow, M. G.; Haszeldine, R. N.; Hubbard, R. *J. Chem. Soc. (C)* **1970**, 1232. (c) Barlow, M. G.; Haszeldine, R. N.; Kershaw, M. J. *Tetrahedron* **1975**, *31*, 1649.
22. Barlow, M. G.; Haszeldine, R. N.: Kershaw, M. J. *J. Chem. Soc., Perkin Trans. 1* 1974, 1736.
23. Dopper, J. H.; Greijdanus, B.; Wynberg, H. *J. Am. Chem. Soc.* **1975**, *97*, 216.
24. Wingert, H.; Regitz, M. *Chem. Ber.* **1986**, *119*, 244.
25. Maier, G.; Bauer, I.; Huber-Patz, U.; Jahn, R.; Kallfaß, D.; Rodewald, H.; Irngartinger, H. *Chem. Ber.* **1986**, *119*, 1111.
26. (a) Barlow, M. G.; Dingwall, J. G.; Haszeldine, R. N. *J. Chem. Soc., Chem. Commun.* **1970**, 1580. (b) Barlow, M. G.; Haszeldine, R. N.; Dingwall, J. G. *J. Chem. Soc., Perkin Trans. 1* **1973**, 1542.
27. Trost, B. M.; Cory, R. M. *J. Am. Chem. Soc.* **1971**, *93*, 5572.
28. Katz, T. J.; Wang, E. J.; Acton, N. *J. Am. Chem. Soc.* **1971**, *93*, 3782.
29. (a) Turro, N. J.; Renner, C. A.; Waddell, W. H.; Katz, T. J. *J. Am. Chem. Soc.* **1976**, *98*, 4320. (b) Turro, N. J.; Ramamurthy, V. *Recl. Trav., Chim. Pays-Bas* **1979**, *98*, 173.
30. (a) Angus, H. J. F.; Blair, J. M.; Bryce-Smith, D. *J. Chem. Soc.* **1960**, 2003. (b) Ward, H. R.; Wishnok, J. S.; Sherman, P. D., Jr. *J. Am. Chem. Soc.* **1967**, *89*, 162. (c) Kaplan, L.; Wilzbach, K. E. *J. Am. Chem. Soc.* **1967**, *89*, 1030. (d) Wilzbach, K. E.; Ritsher, J. S.; Kaplan, L. *J. Am. Chem. Soc.* **1967**, *89*, 1031. (e) Ward, H. R.; Wishnok, J. S. *J. Am. Chem. Soc.* **1968**, *90*, 5353. (f) Kaplan, L.; Walch, S. P.; Wilzbach, K. E. *J. Am. Chem. Soc.* **1968**, *90*, 5646. (g) Bryce-Smith, D.; Gilbert, A.; Robinson, D. A. *Angew. Chem., Int. Ed. Engl.* **1971**, *10*, 745.
31. Chapman, N. B.; Key, J. M.; Toyne, K. J. *J. Org. Chem.* **1970**, *35*, 3860.
32. (a) Luh, T.-Y.; Stock, L. M. *J. Am. Chem. Soc.* **1974**, *96*, 3712. (b) Klunder, A. J. H.; Zwanenburg, B. *Tetrahedron* **1972**, *28*, 4131.
33. (a) Abeywickrema, R. S.; Della, E. W. *J. Org. Chem.* **1980**, *45*, 4226. (b) Della, E. W.; Tsanaktsidis, J. *Aust. J. Chem.* **1986**, *39*, 2061. (c) Della, E. W.; Tsanaktsidis, J. *Aust. J. Chem.* **1989**, *42*, 61.
34. Eaton, P. E.; Cole, T. W., Jr. *J. Chem. Soc., Chem. Commun.* **1970**, 1493.
35. Chin, C. G.; Cuts, H. W.; Masamune, S. *J. Chem. Soc., Chem. Commun.* **1966**, 880.
36. (a) Reddy, D. S.; Sollot, G. P.; Eaton, P. E. *J. Org. Chem.* **1989**, *54*, 722. (b) Eaton, P. E.; Yang, C.-X; Xiong, Y. *J. Am. Chem. Soc.* **1990**, *112*, 3225. (c) Hrovat, D. A.; Borden, W. T. *J. Am. Chem. Soc.* **1990**, *112*, 3227. (d) Moriarty, R. M.; Tuladhar, S. M.; Penmasta, P.; Awasthi, A. K. *J. Am. Chem. Soc.* **1990**, *112*, 3228. (e) Kevill, D. N.; D'Souza, M. J.; Moriarty, R. M.; Tuladhar, S. T.; Penmasta, R.; Awasthi, A. K. *J. Chem. Soc., Chem. Commun.* **1990**, 623.

37. (a) Knight, L. B., Jr.; Arrington, C. A.; Gregory, B. W.; Cobranchi, S. T.; Liang, S.; Paquette, L. A. *J. Am. Chem. Soc.* **1987**, *109*, 5521. (b) Reddy, D. S.; Maggini, M.; Tsanaktsidis, J.; Eaton, P. E. *Tetrahedron Lett.* **1990**, *31*, 805. (c) Della, E. W.; Elsey, G. M.; Head, N. J.; Walton, J. C. *J. Chem. Soc., Chem. Commun.* **1990**, 1589. (d) Eaton, P. E.; Maggini, M. *J. Am. Chem. Soc.* **1988**, *110*, 7230.
38. (a) Hassenrück, K.; Radziszewski, J. G.; Balaji, V.; Murthy, G. S.; McKinley, A. J.; David, D. E.; Lynch, V. M.; Martin, H.-D.; Michl, J. *J. Am. Chem. Soc.* **1990**, *112*, 873. (b) Hrovat, D. M.; Borden, W. T. *J. Am. Chem. Soc.* **1990**, *112*, 875. (c) Eaton, P. E.; Tsanaktisidis, J. *J. Am. Chem. Soc.* **1990**, *112*, 876. (d) Dagani, R. *Chem. Eng. News.* **1990**, (Feb 12), p. 24.
39. (a) Eaton, P. E.; Maggini, M. *J. Am. Chem. Soc.* **1988**, *110*, 7230. (b) Worthy, W. *Chem. Eng. News* **1988** (Nov. 14), p. 45.
40. (a) Marchand, A. P. *Tetrahedron* **1988**, *44*, 2377. (b) Griffin, G. W.; Marchand, A. P. *Chem. Rev.* **1989**, *89*, 997.
41. (a) Eaton, P. E.; Castaldi, G. *J. Am. Chem. Soc.* **1985**, *107*, 724. (b) Worthy, W. *Chem. Eng. News* **1985** (Feb. 25), p. 31. (c) Eaton, P. E.; Cunkle, G. T.; Marchioro, G.; Martin, R. M. *J. Am. Chem. Soc.* **1987**, *109*, 948. (d) Eaton, P. E.; Higuchi, H.; Millikan, R. *Tetrahedron Lett.* **1987**, 1055. (e) Jayasuriya, K.; Alster, J.; Politzer, P. *J. Org. Chem.* **1987**, *52*, 2306. (f) Della, E. W.; Tsanaktsidis, J. *J. Organometallics* **1988**, *7*, 1178. (g) Bashir-Hashemi, A. *J. Am. Chem. Soc.* **1988**, *110*, 7234.
42. (a) Eaton, P. E.; Cunkle, G. T. *Tetrahedron Lett.* **1986**, *27*, 6055. (b) Moriarty, R. M.; Khosrowshahi, J. S.; Dalecki, T. M. *J. Chem. Soc., Chem. Commun.* **1987**, 675. (c) Moriarty, R. M.; Khosrowshahi, J. S.; Awasthi, A. K.; Penmasta, R. *Synth. Commun.* **1988**, *18*, 1179. (d) Moriarty, R. M.; Khosrowshahi, J. S. *Synth. Commun.* **1989**, *19*, 1395.
43. Gleiter, R.; Karcher, M. *Angew. Chem., Int. Ed. Engl.* **1988**, *27*, 840.
44. Eaton, P. E.; Pramod, K.; Gilardi, R. *J. Org. Chem.* **1990**, *55*, 5746.
45. Eaton, P. E.; Wicks, G. E. *J. Org. Chem.* **1988**, *53*, 5353.
46. (a) Eaton, P. E.; Cerefice, S. A. Unpublished results. (b) McKennis, J. S.; Brener, L.; Ward, J. S.; Pettit, R. *J. Am. Chem. Soc.* **1971**, *93*, 4957. (c) Paquette, L. A.; Davis, R. F.; James, D. R. *Tetrahedron Lett.* **1974**, 1615.
47. Osawa, E.; Barbiric, D. A.; Lee, O. S.; Padma, S.; Mehta, G. *J. Chem. Soc., Perkin Trans. 2* **1989**, 1161.
48. (a) Shen, K.-W. *J. Am. Chem. Soc.* **1971**, *93*, 3064. (b) Allred, E. L.; Beck, B. R. *Tetrahedron Lett.* **1974**, 437. (c) Askane, R.; Schneider, W. *Chem. Ber.* **1983**, *116*, 2366.
49. (a) Eaton, P. E.; Cassor, L.; Hudson, R. A.; Hwang, D. R. *J. Org. Chem.* **1976**, *41*, 1445. (b) Smith, E. C.; Barborak, J. C. *J. Org. Chem.* **1976**, *41*, 1433. (c) Marchand, A. P.; Chou, T.-C.; Ekstrand, J. D.; van der Helm, D. *J. Org. Chem.* **1976**, *41*, 1438.
50. (a) Marchand, A. P.; Chou, T.-C. *J. Chem. Soc., Perkin Trans. 1* **1973**, 1948. (b) Marchand, A. P.; Chou, T.-C. *Tetrahedron* **1975**, *31*, 2655.
51. Dauben, W. G.; Cunningham, A. F., Jr. *J. Org. Chem.* **1983**, *48*, 2842.
52. Jemmis, E. D.; Leela, G. *Proc. Indian Acad. Sci. (Chem. Sci.)* **1987** *99*, 281.
53. (a) Higuchi, H.; Takatsu, K.; Otsubo, T.; Sakata, Y.; Misumi, S. *Tetrahedron Lett.* **1982**, *23*, 671. (b) Misumi, S. *Pure Appl. Chem.* **1987**, *59*, 1627.
54. (a) Cookson, R. C.; Fox, D. A.; Hudec, J. *J. Chem. Soc.* **1961**, 4499. (b) Cookson, R. C.; Frankel, J. J.; Hudec, J. *J. Chem. Soc., Chem. Commun.* **1965**, 16.
55. (a) Flaig, W.; Salfeld, J.-C.; Llanos, L. C. Q. A. *Angew. Chem.* **1960**, *72*, 110. (b) Rabinovich, D.; Schmidt, G. M. J. *J. Chem. Soc. (B)* **1967**, 144.
56. (a) Hopf, H.; Musso, H. *Chem. Ber.* **1973**, *106*, 143. (b) Chesick, J. P.; Dunitz, J. D.; Gizycki, U. V.; Musso, H. *Chem. Ber.* **1973**, *106*, 150.
57. (a) Hutmacher, H.-M.; Fritz, H.-G.; Musso, H. *Angew. Chem., Int. Ed. Engl.* **1975**, *14*, 180. (b) Fritz, H.-G.; Hutmacher, H.-M.; Musso, H.; Ahlgren, G.; Akermark, B.; Karlsson, R. *Chem. Ber.* **1976**, *109*, 3781. (c) Kaiser, G.; Musso, H. *Chem. Ber.* **1985**, *118*, 2266.
58. Hoffmann, V. T.; Musso, H. *Angew. Chem., Int. Ed. Engl.* **1987**, *26*, 1006.

59. Brudermüller, M.; Musso, H. *Angew. Chem., Int. Ed. Engl.* **1988**, *27*, 298.
60. Yang, N. C.; Horner, M. G. *Tetrahedron Lett.* **1986**, *27*, 543.
61. Eaton, P. E.; Chakraborty, U. R. *J. Am. Chem. Soc.* **1978**, *100*, 3634.
62. (a) Mehta, G.; Srikrishna, A.; Nair, M. S. *Ind. J. Chem.* **1983**, *22B*, 621. (b) Mehta, G.; Nair, M. S.; Srikrishna, A. *Ind. J. Chem.* **1983**, *22B*, 959.
63. Akhtar, I. A.; Fray, G. I.; Yarrow, J. M. *J. Chem. Soc. (C)* **1968**, 812.
64. (a) Mehta, G.; Padma, S. *J. Am. Chem. Soc.* **1987**, *109*, 2212. (b) Mehta, G.; Padma, S. *Tetrahedron* **1991**, *47*, 7783.
65. Mehta, G. In *Strain and Its Implications in Organic Chemistry*; de Meijere, A.; Blechert, S., Eds.; NATO ASI Series; Kluwer Academic Publishers: Dordrecht, 1989; Vol. 273, p. 269.
66. Boekelheide, V.; Hollins, R. A. *J. Am. Chem. Soc.* **1973**, *95*, 3201.
67. Mehta, G.; Shah, S. R. *Indian J. Chem.* **1990**, *29B*, 101.
68. Srikrishna, A.; Sunderbabu, G. *J. Org. Chem.* **1987**, *52*, 5037.
69. (a) Mehta, G.; Padma, S. *J. Am. Chem. Soc.* **1987**, *109*, 7230. (b) Mehta, G.; Padma, S. *Tetrahedron* **1991**, *47*, 7807.
70. (a) Lemal, D. M.; Shim, K. S. *Tetrahedron Lett.* **1961**, 368. (b) Bird, C. W.; Colinese, D. L.; Cookson, R. C.; Hudec, J.; Williams, R. O. *Tetrahedron Lett.* **1961**, 373. (c) Jolly, P. W.; Stone, F. G. A.; Mackenzie, K. *J. Chem. Soc.* **1965**, 6416. (d) Katz, T. J.; Acton, N. *Tetrahedron Lett.* **1967**, 2601. (e) Schrauzer, G. N.; Ho, R. K. Y.; Schlesinger, G. *Tetrahedron Lett.* **1970**, 543. (f) Acton, N.; Roth, R. J.; Katz, T. J.; Frank, J. K.; Maier, C. A.; Paul, I. C. *J. Am. Chem. Soc.* **1972**, *94*, 5446. (g) Ennis, M.; Manning, A. R. *J. Organomet. Chem.* **1976**, *116*, C31. (h) Neely, S. C.; van der Helm, D.; Marchand, A. P.; Hayes, B. R. *Acta Crystallogr. Sect. B. Struct. Crystallogr. Cryst. Chem.* **1976**, *B32*, 561. (i) Marchand, A. P.; Hayes, B. R. *Tetrahedron Lett.* **1977**, 1027. (j) Chow, T. J.; Liu, L.-K.; Chao, Y. S. *J. Chem. Soc., Chem. Commun.* **1985**, 700. (k) Hargittai, I.; Brunvoll, J.; Cyvin, S. J.; Marchand, A. P. *J. Mol. Struct.* **1986**, *140*, 219. (l) Chow, T. J.; Chao, Y.-S; Liu, L.-K. *J. Am. Chem. Soc.* **1987**, *109*, 797.
71. (a) Meinwald, J.; Wiley, G. A. *J. Am. Chem. Soc.* **1958**, *80*, 3667. (b) Cookson, R. C.; Hill, R. R.; Hudec, J. *J. Chem. Soc.* **1964**, 3043. (c) Mehta, G.; Padma, S.; Karra, S. R.; Gopidas, K. R.; Cyr, D. R.; Das. P. K.; George, M. V. *J. Org. Chem.* **1989**, *54*, 1342.
72. Mehta, G.; Reddy, S. H. K. Unpublished results.
73. Dauben, W. G.; Koch, K.; Smith, S. L.; Chapman, O. L. *J. Am. Chem. Soc.* **1963**, *85*, 2616, and references cited therein.
74. Schönberg, A.; Sodtke, U.; Praefcke, K. *Tetrahedron Lett.* **1968**, 3669.
75. Mehta, G.; Reddy, S. H. K.; Padma, S. *Tetrahedron* **1991**, *47*, 7821.
76. Mehta, G.; Padma, S. *J. Org. Chem.* **1991**, *56*, 1298.
77. (a) Bunz, U.; Polborn, K.; Wagner, H.-U.; Szeimies, G. *Chem. Ber.* **1988**, *121*, 1785. (b) Murthy, G. S.; Hassenrück, K.; Lynch, V. M.; Michl, J. *J. Am. Chem. Soc.* **1989**, *111*, 7262. (c) Hassenrück, K.; Murthy, G. S.; Lynch, V. M.; Michl, J. *J. Org. Chem.* **1990**, *55*, 1013.

8

Recent Developments in the Chemistry of Cubane

Hiroyuki Higuchi[1] and Ikuo Ueda[2]

[1]Toyama University, Toyama, Japan
[2]Osaka University, Osaka, Japan

Contents

8.1. Introduction
8.2. Methodology for Direct Functionalization of Cubane
8.2A. Metallocubanes
8.2A(1). Synthesis of Stable Mercurated Cubane
8.2A(2). Reactivities of Mercurated Cubane with Iodine
8.2A(3). Reverse Transmetalation of Mercurated Cubane
8.2A(4). Transmetalations of the Lithiated Cubane with Other Transition Metals (Mg, Zn, Cd)
8.2B. Synthetic Applications for Cubane Derivatives by the Metallocubane Methodology
8.2B(1). Synthesis of Cubane Tetracarboxylic Acids
8.2B(2). Transformation of the Directing Diisopropylamide Function
8.2C. Synthesis of Cubane Derivatives by Other Methods
8.2C(1). Photosensitized Substitutions of Iodinated Cubane
8.2C(2). Oxidative Substitutions of Iodinated Cubane
8.2C(3). New Decarboxylation Procedures of Cubyl Carboxylic Acid
8.3. Reactions of Cubanes with an Electron-Deficient Species
8.3A. Ring Enlargement Reactions of Cubylcarbinols under Acidic Conditions—Cubylmethyl Carbonium Cation
8.3B. Twisted Olefins in the Cubane System
8.3B(1). Photolysis and Thermolysis of Cubylazide—Cubylmethyl Nitrene
8.3B(2). Photolysis and Thermolysis of Cubylphenyldiazomethane—Cubylmethyl Carbene

8.4. Highly Pyramidalized Olefins in the Cubane System—Homocubene and Cubene
8.4A. Homocubene (4,5-Dehydrohomocubane)
8.4B. Cubene (1,2-Dehydrocubane)
8.5. New Type of Cubane Derivative—Combination of Cubane with π-Electronic Ring System
8.5A. Phenylcubanes
8.5B. Quinone Methide Cubanes
8.5B(1). Transformation of the Carboxylic Acid to the Quinone Methide π-Electronic System
8.5B(2). Electronic Absorption of Quinone Methide Cubanes
8.6. Developments of the Related Chemistry of Cubane
8.6A. Photocycloaddition of *syn*-Dienes
8.6A(1). Strategy for Intramolecular [2 + 2] Photocycloaddition in the Cage Molecules
8.6A(2). Propellane Type of Cubane
8.6B. Tetraphosphacubane
8.7. Future Chemistry of Cubane and Its Perspectives
8.7A. Synthesis of Polynitrocubanes—High Energy Density Materials
8.7B. Research for New Types of Pharmaceuticals Containing Cubane Skeleton
8.8. Conclusion
Acknowledgment
References

1. Introduction

Cubane (**1**; C_8H_8; pentacyclo[4.2.0.0^{2,5}.0^{3,8}.0^{4,7}]octane), the most typical three-dimensional polycyclic molecule with the simplest form (O_h point group) among nonnatural products, which are the fruits of the inventions of one's mind, was synthesized by Eaton in 1964.[1]

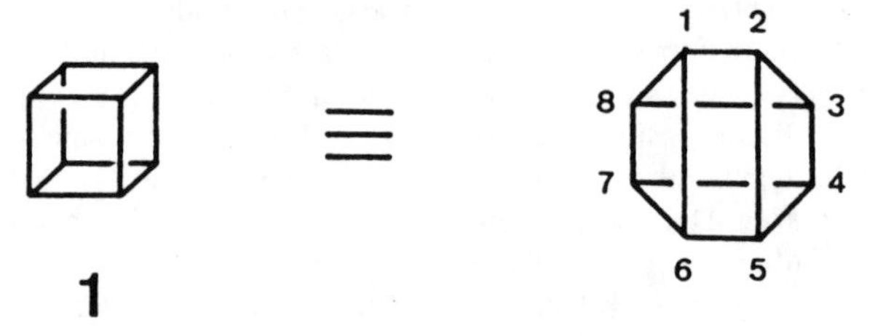

For the last 25 years, since the big success in a conclusive synthesis of this extremely strained structure, a great deal of effort has been concentrated mostly upon the study of its peculiar structure and bond by unveiling curious physicochemical properties that one could never explore (Table 8.1). It not only has brought about a construction of the present elegant and resolute field, "cubane chemistry," but also has con-

Table 8.1 ■ Physical Properties of Cubane 1[2]

Melting point	130–131°C
Boiling point	~130°C
IR	2992, 1235, 852 cm^{-1}
UV	Clear
^{1}H NMR	4.04 ppm
^{13}C NMR	47.3 ppm
$J_{^{13}C-H}$	155 Hz
C—H s character	31%
C—C length	1.550 ± 0.003 Å
C—H length	1.06 ± 0.05 Å
C—C—C angle	90.0 ± 0.5°
ΔH_f° (298°)	144.5 ± 1.3 kcal mol^{-1}
Strain	166 kcal mol^{-1}

tributed more or less to the drastic developments of many other chemistries of nonnatural products by encouraging chemists to keep challenging themselves toward the realization of dreams (unknown structures) produced in their scientific or nonscientific minds.

As Eaton says,[2] it would not be wise for chemists to divide the chemistry of nonnatural products from that of natural products. Rather these two worlds should be affecting each other intensively, to make themselves more developed and sublimated. Therefore it can be clearly pointed out and should be kept in mind that the chemistry of cubane itself has been also growing up on the basis of fundamental acts and phenomena of Nature; development of the chemistry of cubane would be impossible without contributions of the chemistry of natural products.

When we turn our eyes to the current chemistry of cubane, it seems that physicochemical studies of its structural and bond properties have almost come to an end and now have shifted to the next stage, where the chemistry for seeking applications or cultivating newer fields on the basis of a huge number of accumulated results must be required. In fact, from this viewpoint many distinguished papers have been appearing in journals and research activities are continuing.[2,3] Although, regrettably, all of them cannot be picked up here, this chapter will introduce particularly significant and interesting recent research developments that have been leading (and will certainly continue to lead) the field of cubane chemistry. Some prospective research in the future chemistry of cubane will also be introduced.

8.2. Methodology for Direct Functionalization of Cubane

Establishment of the general synthetic method for introducing desired functional groups into the parent skeleton to form a wide variety of derivatives, with which systematic studies of particular physicochemical properties can be performed, does enhance a drastic development of cubane chemistry, as is generally the case with any field of organic chemistry.

As we see the recent high level of research activity on the chemistry of cubane, it may be hard to realize that the study of functionalization of the cubane skeleton started only a few years ago. This can be ascribed to the difficulty of introducing a functional group directly into the cubane skeleton due to low reactivity (such as reactivities of the usual aliphatic hydrocarbons) and additionally, seemingly inconsistent, to the potentially high reactivity due to the strained skeleton, ready destruction of which takes place under various conditions.[24–26]

Therefore cubane derivatives have been prepared by the limited number of methods where functional groups that are to be introduced into cubane must be held, from the beginning, in the building blocks, sometimes in a different form convertible into the desired functions, and must be kept safe until the construction of cubane skeleton is completed.[4] Such weak points due to the particular reactivity of the cubane skeleton, however, have been cleared up recently and the field of cubane chemistry is about to step up again more widely and intensively.

8.2A. Metallocubanes

8.2A(1). Synthesis of Stable Mercurated Cubane. The extremely strained geometry of cubane[5] requires substantial rehybridization of the component carbons from normal sp^3 bonding. The $^{13}C—H$ nuclear coupling constant in cubane (155 Hz) translates into 31% s character in that bond[6] and an expectation of enhanced acidity.[7] Based on these bond properties, Eaton was successful in establishing the methodology for efficient formation of cubyl anion, which reacts with electrophiles to afford various types of substituted cubanes.

From the fact that in aromatic systems there are many directing groups, such as amides, esters, and ethers, that encourage lithiation at an adjacent ortho position,[8] a similar substituent effect using diisopropylamide (E), which is easily derived from carboxylic acid, was employed to generate the cubyl anion **3** in a similar manner with lithium tetramethylpiperidide (LiTMP) as the best base[9] (Schemes 8.1 and 8.2). Although the thermodynamical concentration of the formed lithiated cubyl anion **3** reached only 3%, mercury(II) chloride was found to be a useful trapping agent of the lithiated cubyl species. Consequently, addition of powdered $HgCl_2$ into the reaction mixture draws the equilibrium with the starting material **2** effectively to the right through the ready transmetalation process, affording an irreversible formation of the

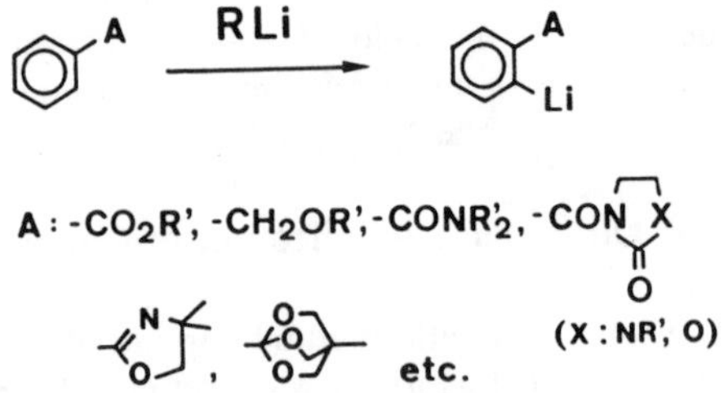

RLi

A : $-CO_2R'$, $-CH_2OR'$, $-CONR'_2$, -CON X (X : NR', O) etc.

Scheme 8.1 ■ Directed ortho-lithiation of aromatics.

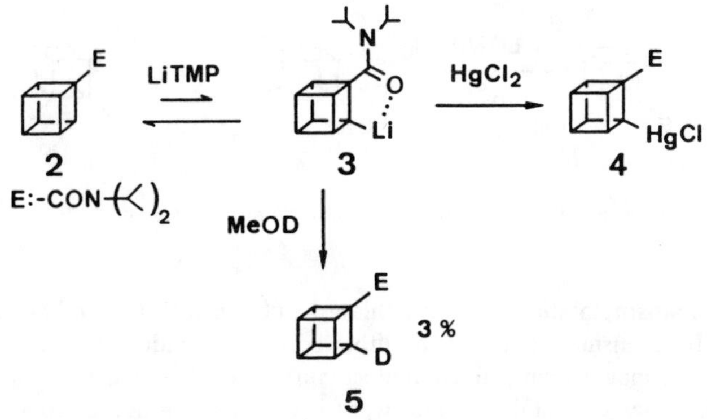

Scheme 8.2 ▪ Preparation of mercurated cubane **4**.

mercurated cubane diisopropylamide **4**, which is isolable and fairly stable to both air and water.

8.2A(2). Reactivities of Mercurated Cubane with Iodine. Although mercurated cubane **4** thus prepared is stable enough to be stored for a long time, it suffers from activation of C—Hg bond with elemental iodine I_2, followed by its ready cleavage to yield an iodinated cubane diisopropylamide derivative **6** (Scheme 8.3). Two diisopropylamide directing groups at 1 and 4 positions (**8**) strengthen the acidity of cubyl C—H bond and promote the lithiation process more rapidly and effectively. Similarly, treatment of the formed bismercurated cubane with I_2 also gives diiodide **9** in 70–90% yield along with a small amount of triiodide **10**[9] (Scheme 8.4).

8.2A(3). Reverse Transmetalation of Mercurated Cubane. These mercurated cubanes were found to exchange mercury for other metals such as lithium and magnesium from alkyllithiums and the Grignard reagents, to give the corresponding metallocubanes **7** in excellent yields.[10–13,15] Especially this reverse transmetalation procedure, through which the regenerated lithiated or magnesiated cubane is highly reactive to ordinary electrophilic reagents, is not only practically useful and convenient but also mechanistically significant, in that this methodology may provide a general synthetic method for a wide variety of polysubstituted cubanes and the like.

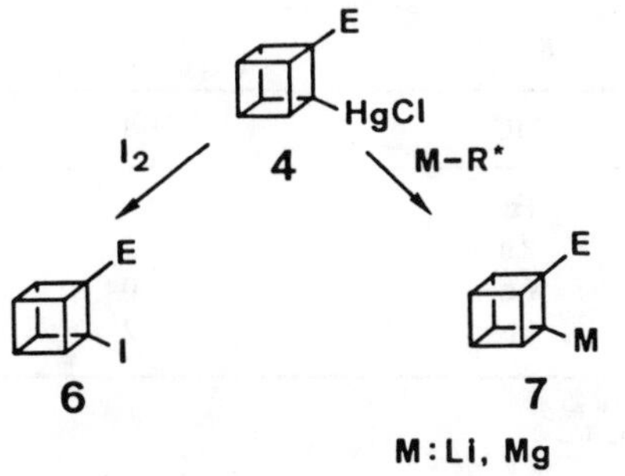

Scheme 8.3 ▪ Reactivity of mercurated cubane **4**.

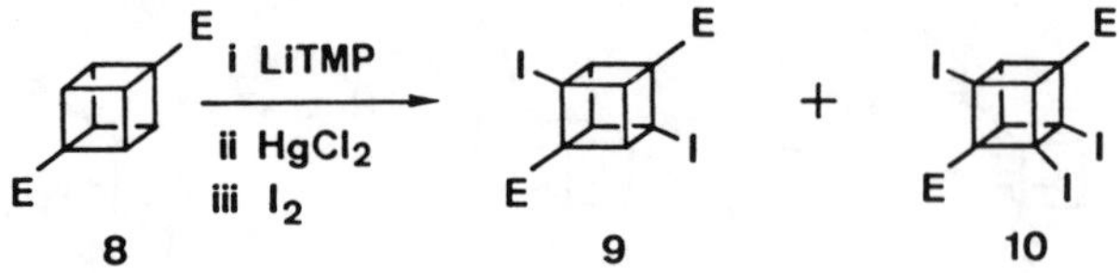

Scheme 8.4 ▪ Iodination of cubane via mercuration procedure.

8.2A(4). Transmetalations of the Lithiated Cubane with Other Transition Metals (Mg, Zn, Cd). Transmetalation of the lithiated cubane under an equilibrated condition with zinc-, magnesium-, or cadium-salt in place of mercuric chloride was also examined. In all cases metallocubanes were irreversibly formed in good to moderate yields similarly to the mercurated cubane as shown in Table 8.2.[11] Because these three metallocubanes are not so stable as to be isolated, unlike the mercurated cubane, but give the iodinated cubane **9** by adding I_2 into the reaction mixture in adequate yields, they may offer a utility for the one-pot synthesis of cubane derivatives alternative to the mercurated cubane procedure.

8.2B. Synthetic Applications for Cubane Derivatives by the Metallocubane Methodology

8.2B(1). Synthesis of Cubane Tetracarboxylic Acids. In one particular case, zincated cubane prepared in situ in the same manner as that for mercurated cubane reacts with benzoyl chloride in the presence of tetrakis(triphenylphosphine)palladium(0) to give benzoyl diisopropylamide cubane **11**, which can be regarded as the first synthetic example of cubane-1,2,4,7-tetracarboxylic acid derivative[12] (Scheme 8.5). In another case, a strategy of tandem orthomagnesiation–carboxylation was also successfully applied to the cyanoamide **12**, to obtain the corresponding

Table 8.2 ▪ **Yield of Cubane Diiodide 9**

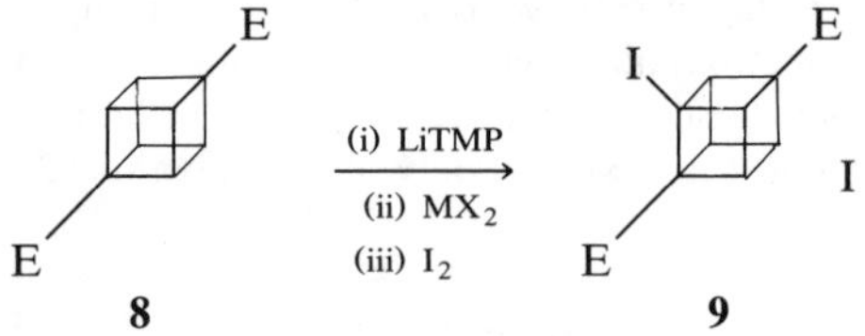

MCl_2	Yield (%)
Hg	90
Zn	75
Cd	70
Mg[a]	72[b]

[a] $MgBr_2$ used.
[b] Ref. 46.

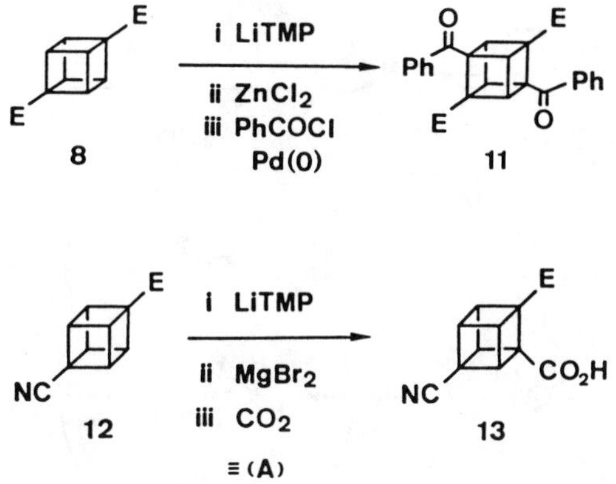

8 → 11: i LiTMP; ii $ZnCl_2$; iii PhCOCl, Pd(0)

12 → 13: i LiTMP; ii $MgBr_2$; iii CO_2 ≡ (A)

Scheme 8.5 ■ Benzoylation and carboxylation of cubanes via zincation and magnesiation procedures.

cubane carboxylic acid **13**.[13] Especially the latter example clearly indicates that carbonyl function in the amide group plays an important role for stabilizing the ortho-lithiated cubyl anion and that cyano group, the more electron-withdrawing group, assists just to increase the whole acidity of cubyl C—H bonds.

8.2B(2). Transformation of the Directing Diisopropylamide Function. Although the diisopropylamide group also displayed superior ability to encourage ortho-lithiation in the cubane system as a useful directing group, there was one crucial point to be resolved where the amide group in each product thus obtained outstandingly resists hydrolysis to the original carboxylic acid under both strongly acidic and basic conditions (**8**; no reaction even at reflux in various saturated solutions of MOH). After many trials, Higuchi was successful in converting it into the *tert*-amine **14** with $LiAlH_4$ (Scheme 8.6), followed by a combination of oxidation–reduction of the amine **14** according to Ferris's C—N bond cleavage procedure[14] to give the cubyl aldehyde derivative **15** in moderate yield without noticeable destruction of the cubane skeleton.[15] Consequently, cubane-1,2,4,7-tetracarboxylic acid **16** was for the first time prepared in this way. Recently, cubane-1,3,5,7-tetracarboxylic acid **22** of nominal tetrahedral symmetry (T_d)[13] could be obtained also by a subsequent removal of the amide group from the pentasubstituted cubane **18** by the skillful methodology as shown in Scheme 8.7.

8.2C. Synthesis of Cubane Derivatives by Other Methods

8.2C(1). Photosensitized Substitutions of Iodinated Cubane. Apart from the synthetic methodology with metallocubanes for functionalization of the cubane skeleton, studies of conversions of substituents on the cubane molecule into the other desired

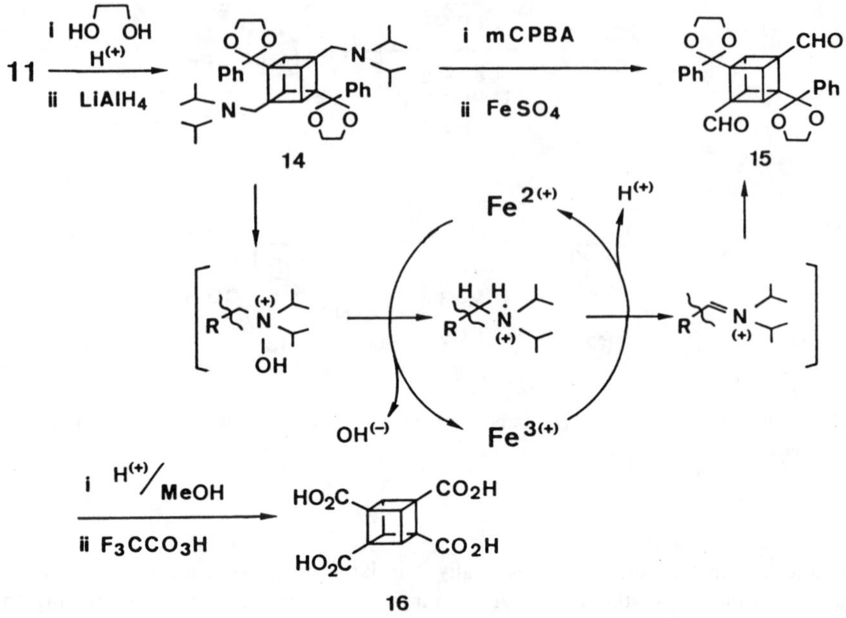

Scheme 8.6 ■ Synthesis of cubane-1,2,4,7-tetracarboxylic acid **16**.

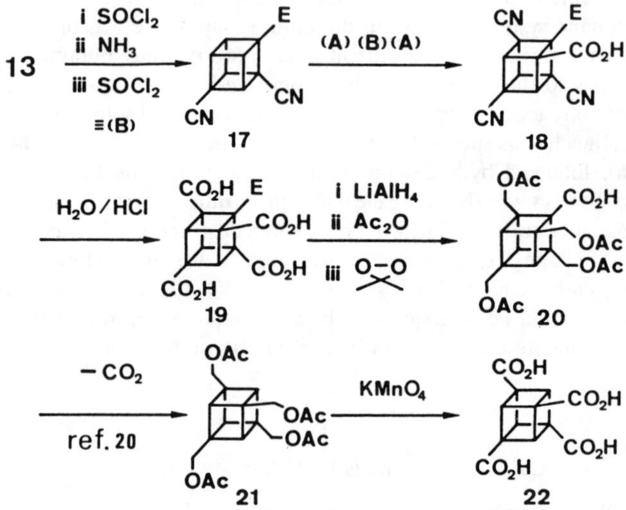

Scheme 8.7 ■ Synthesis of cubane-1,3,5,7-tetracarboxylic acid **22**.

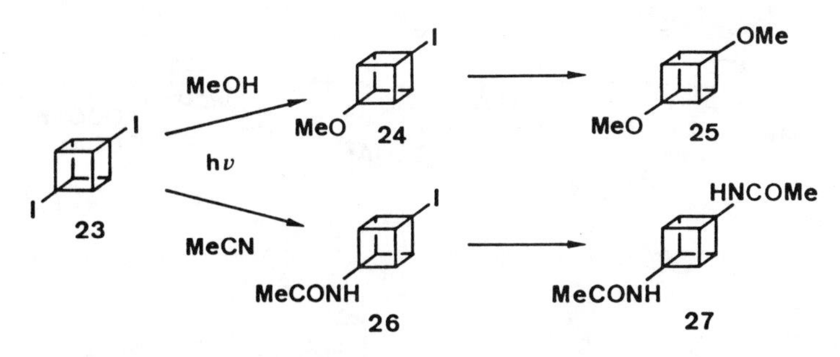

Scheme 8.8 ■ Photolysis of iodocubanes **23**.

function with the conventional and/or up-to-date procedures also have continued in many places. Among them, almost simultaneously, several unique and excellent procedures for the transformation of iodinated cubanes to other valuably functionalized cubanes have been reported recently. Reddy and co-workers, based on the high sensitivity of C—I bond to light, generated cubyl cation[16] by irradiation of iodinated cubane **23** and employed this reactive species for introducing another atom from solvent to cubane.[17] When the reaction was carried out at 40°C with light at 254 nm in methanol (Scheme 8.8), mono- and di-methoxy cubanes **24** and **25** were obtained, whereas in acetonitrile, mono- and di-acetamidocubanes **26** and **27**, respectively, were obtained in good yields.

8.2C(2). Oxidative Substitutions of Iodinated Cubane. On the other hand, Eaton and Cunkle[18] and Moriarty and co-workers[19] have shown that iodine replacement can occur on the cubane molecules **28** and **31** after oxidation of the iodine substituent to the hypervalent species,[16] which also leads to the formation of the cubyl cation (Scheme 8.9). In such conversion processes the oxidation of iodine weakens the C—I bond and activates it for cleavage. However, it must be noticed that the substituents incorporated into cubane molecule, in these cases, are not from solvents employed but those on the hypervalent iodine, because these reactions proceed via frontside 1,2-shift of the nucleophile from iodide to cubane carbon.

8.2C(3). New Decarboxylation Procedures of Cubyl Carboxylic Acid. The last step in the reaction sequence for cubane synthesis established first was to eliminate the carboxyl function that was left after the complete construction of cubane skeleton **35** via Favorskii ring contraction from 1,3-bishomocubanone dibromide **34**.[1] This decarboxylation procedure is a very important step in the cubane chemistry and might be experienced by many chemists before and/or after the particular reactions. Eaton led to the parent hydrocarbon, cubane **1**, utilizing thermolysis of the peracid ester **36** prepared from acid chloride and *tert*-butylhydroperoxide. However, this procedure itself is slightly inconvenient with respect to the general synthetic method for highly strained molecules without any damage to other remaining substituents, because it is necessary to employ higher reaction temperature and peroxy reagent in this procedure (Scheme 8.10). Barton, Crich, and Motherwell[20] recently have developed an elegant decarboxylation procedure to form the corresponding hydrocarbon com-

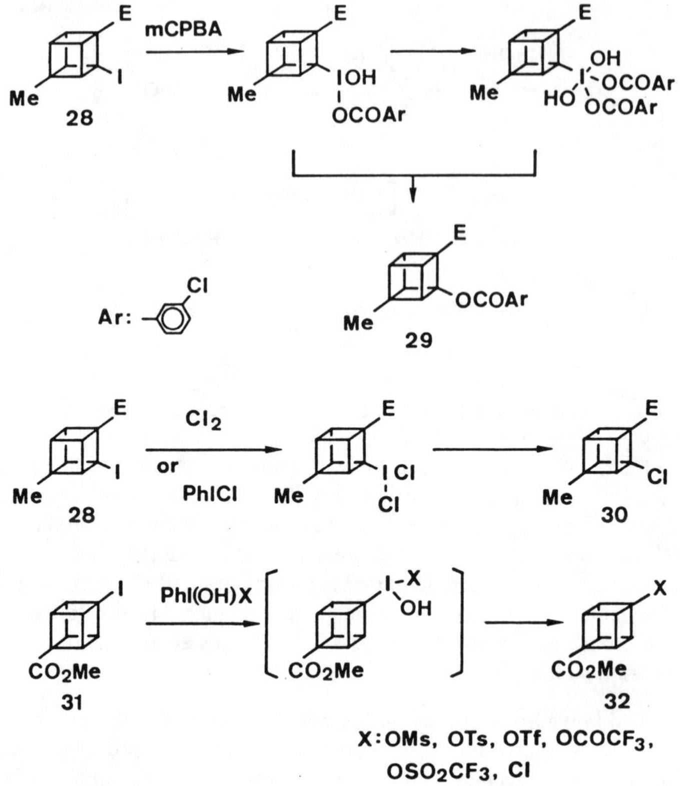

Scheme 8.9 ■ Oxidative deiodination of iodocubanes.

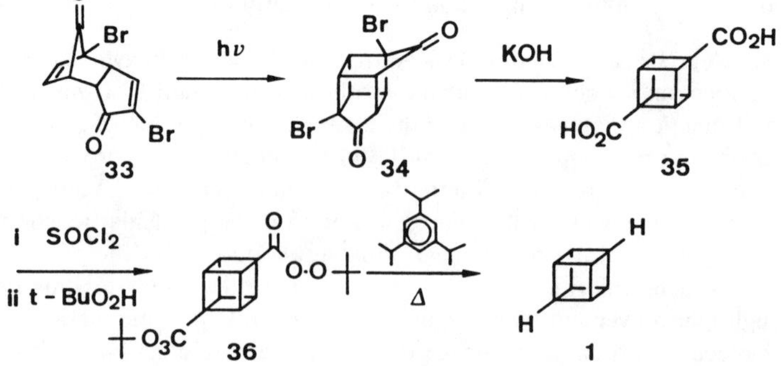

Scheme 8.10 ■ Synthetic route of the parent cubane 1 established first.

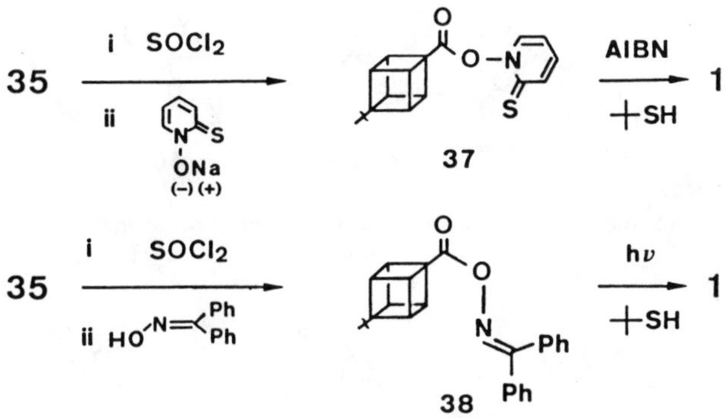

Scheme 8.11 ■ New decarboxylation procedures from **35** to **1**.

pounds in good yields, using radical degradation methodology of 2-mercaptopyridine-*N*-oxycarbonyl compounds **37**, which are easily obtained from the acid chloride and 2-mercaptopyridine-*N*-oxide sodium salt quantitatively (Scheme 8.11). This is widely applicable for preparation of substituted cage molecules as well as the usual alkanes from the corresponding carboxylic acid derivatives.[21,39] Hasebe and co-workers also have reported a convenient procedure for photosensitized radical degradation of the oxime ester **38**, which can be prepared from carboxlic acid or acid chloride and benzophenone oxime.[22] Although this can be regarded as a modified Barton's procedure, both of them will manifest their versatile utilities.

8.3. Reactions of Cubanes with an Electron-Deficient Species

Although there are few differences in C—H and C—C bond lengths in cubane and the usual alkane molecules, the extremely strained structure with an angle of almost 90° in all C—C—C bonds[5] results in giving cubane the name of high energy storage compound (strain energy = 166 kcal mol^{-1}).[23]

Therefore this structural feature easily causes various types of strain-releasing reactions, such as valence isomerization[24] and ring enlargement reactions, under certain conditions. Recently, interesting reactions caused by this strained skeletal property in the cubane derivatives bearing carbonium cation, nitrene, and carbene at the α position have been reported.

8.3A. Ring Enlargement Reactions of Cubylcarbinols under Acidic Conditions—Cubylmethyl Carbonium Cation

It is well known that *tert*-alcohols **39** and **41** bearing cubyl and homocubyl moieties suffer from ready dehydroxylation under acidic conditions to leave the corresponding carbonium cations that drive the ring enlargement reactions of the cubyl skeletons, giving homocubanes and bishomocubanes **40**, **42**, and **43** in high yields, respectively[25] (Scheme 8.12).

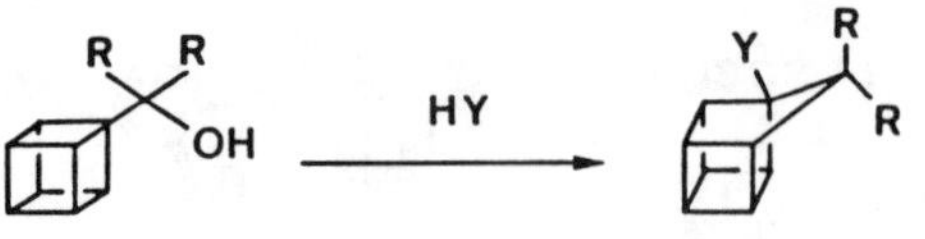

Scheme 8.12 ▪ Monohomocubanes from cubyl carbinols **39**.

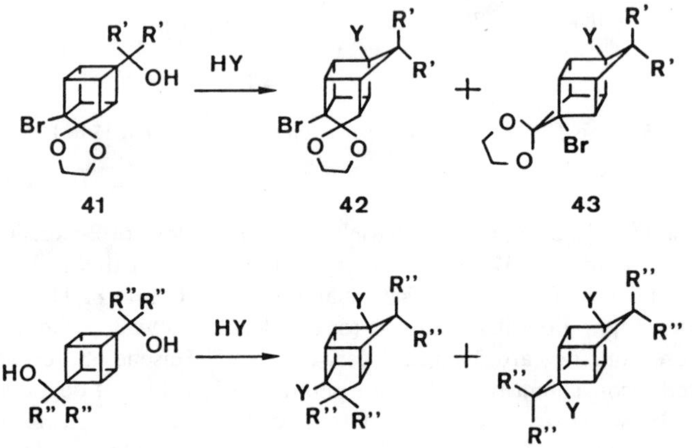

Scheme 8.13 ▪ Bishomocubanes from monohomocubane carbinols **41** and cubane-1,4-bis-carbinols **44**.

Hasegawa and co-workers studied the regioselectivity of the ring enlargement reactions from the cubane **44** with two α-carbinols at 1 and 4 positions to the bishomocubane skeletons **45** and **46** in a mixture of CH_2Cl_2 and MeOH in the presence of *p*-TsOH (Scheme 8.13). They showed the first synthetic example where the 1,4-bishomocubane skeleton is directly formed via ring enlargement reaction of these cubyl carbinols, although it is possibly formed statistically. Results are given in Tables 8.3 and 8.4.[26]

Table 8.3 ▪ Yield and Ratio (*K*) of Bishomocubanes (42/43)

R′	*K*
Me	100/0[a]
Ph	91/ND[b]
p-$MeOC_6H_4$	83/15

[a]Ref. 25.
[b]ND—not detected; ref. 25.

Table 8.4 ▪ Yield and Ratio (K) of Bishomocubanes (45 / 46)

R″	K
Me	80/18
p-ClC$_6$H$_4$	50/12
Ph	42/27
p-MeOC$_6$H$_4$	10/80

Although 1,3-bishomocubane had been concluded to be thermodynamically more stable than the 1,4 isomer by means of empirical methods such as MM2 calculation[27] and, actually, no formation of the 1,4-bishomocubane skeletons from monohomocubanes to bishomocubanes had been reported except for the 1,3 isomers,[25] the ratio of 1,3- to 1,4-bishomocubane was found to depend strongly upon substituents on the phenyl ring. Their results may indicate that more favorable formation of 1,4-bishomocubane can be attained when at least the two structural elements (1) electron-donating groups such as OMe function on the aromatic ring at the para position and (2) two α-cationic reaction sites at 1 and 4 positions are satisfied in the starting material.

During the mechanistic study of its ring enlargement reaction, Kimura and co-workers got the unexpected product 6(H)-azulenone derivative **48** as one of main products, from the same cubane carbinol **47** that was treated with a large excess of p-TsOH in an aqueous medium.[28] The mechanism for the azulenone formation is tentatively proposed as follows (Scheme 8.14). Cubane biscarbonium cation formed via dehydroxylation from the starting biscarbinol **47** under acidic conditions, at the first stage, undergoes ring enlargement to the 1,4-bishomocubane bisalcohol **49**. This

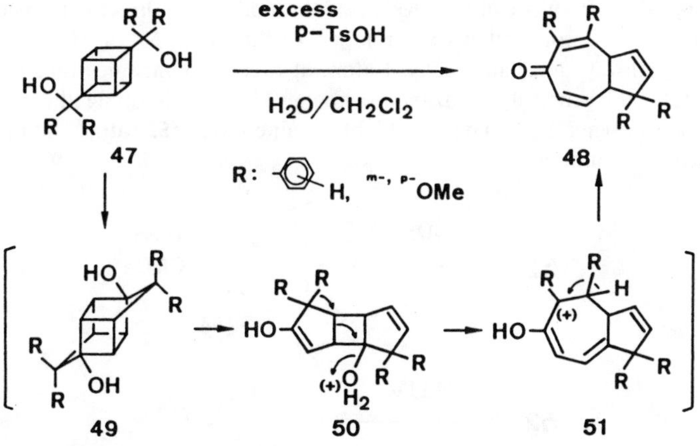

excess P-TsOH

H₂O/CH₂Cl₂

R: H, m-, p-OMe

47 48 49 50 51

Scheme 8.14 ▪ 6(H)-Azulenone derivatives **48** from cubane-1,4-biscarbinols **47** and a tentative reaction mechanism via 1,4-bishomocubane diols **49**.

isolable intermediate **49** then opens one of the four-membered rings, thermally or through the proton-assisted process, to give tricyclic system **50**, followed by dehydration from it to form decisively the carbonium cation **51** leading to the azulenone skeleton via Wagner–Meerwein rearrangement and 1,5-H shift isomerization processes.

8.3B. Twisted Olefins in the Cubane System

8.3B(1). Photolysis and Thermolysis of Cubylazide—Cubyl Nitrene. Direct ultraviolet excitation of saturated azide results in dinitrogen loss to generate the nitrene species that may lead to the formation of imine via rearrangement.[29] When the azide group is a substituent on certain polycyclic systems, the reaction can lead to extraordinary strained, anti-Bredt, bridgehead imines.[30] Although such imines must be expected to be unstable under ordinary conditions, Eaton and Hormann tried to investigate the formation of the imine 9-azahomocubene **53** on the basis of careful analyses of the products obtained from photolysis and thermolysis of cubylazide **52**.[31]

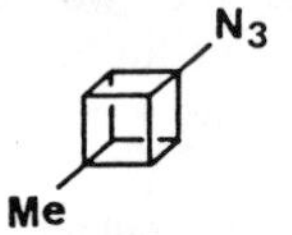

N_3 Me **52** Me **53**

Isolation of the methanol adduct **54** from thermolysis of **52** (100°C, CD_3OD) strongly suggested the formation of the imine **53**. Because the C=N stretching frequency for **53** is estimated to appear at about 1420 cm^{-1},[32] in order to investigate this absorption photolysis of **52** was carried out in a solid argon matrix at 12 K, followed by warming to and holding at 32 K. Absorptions did appear near the predicted value, but these were so weak as to be ignored and the main product from the photolysis was homoprismane nitrile **55** (Scheme 8.15).

Although these results can decisively explain neither the direct conversion of **52** to **55** nor the stepwise conversion via formation of the imine **53**, nor those competitive conversion pathways mechanistically, Eaton, at present, concludes that photolysis of **52** does not lead primarily to azahomocubene **53** by mentioning the more critical point of the idiosyncratic behavior of highly strained azide **52** rather than the difficult geometry of the imine **53**.

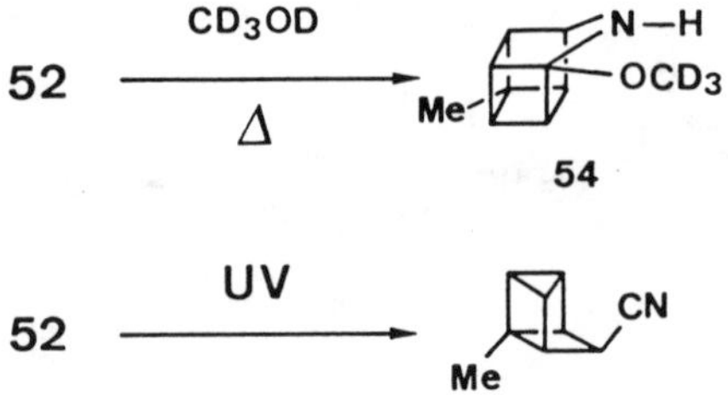

52 —(CD_3OD, Δ)→ Me, N–H, OCD_3 **54**

52 —(UV)→ CN, Me **55**

Scheme 8.15 ▪ Thermal decompositions and photodecompositions of 4-methylcubyl azide **52**.

8.3B(2). Photolysis and Thermolysis of Cubylphenyldiazomethane—Cubylmethyl Carbene. In connection with the study on the azahomocubene **53**, Eaton also tried to investigate the formation of homocubene **57**, which is the most twisted olefin yet known [the relevant dihedral angle (2,1,9,8) in the sigma frame of **57** is approximately 41°, N. L. Allinger, Quantum Chemistry Program Exchange, program MMP2(85)], through the ring enlargement isomerization process of the cubylmethylcarbene generated from thermal and photoinduced decomposition and cubylphenyldiazomethane **56**.[33]

Thermolysis of tosylhydrazone **58** as the practical precursor of the cubylmethylcarbene under irradiation (80–120°C, 0.005 torr, Hg arc, Pyrex filter in EtOH) gave a mixture of the ethanol adducts **59** and its 1,2-phenyl shifted isomer **60** in a ratio of 1.5 : 1, suggesting the addition reaction of ethanol molecule to the formed homocubene **57**. In order to attempt to trap this highly strained olefin **57** as a Diels–Alder adduct, photolysis of its tosylhydrazone salt in monoglyme containing 9,10-diphenylisobenzofuran was carried out and was found to give an adduct of the expected formula; however, it was obvious from its NMR spectra that it had not been formed by a [4 + 2] cycloaddition but only one of the double bonds of the ring had reacted (Scheme 8.16). On the other hand, photolysis of cubylphenyldiazomethane **56** in neat *cis*-2-butene at $-78°C$ gave clearly two spiro-form adducts **62** and **63** ($—N_2$) in a ratio of 2 : 1, whereas in neat *trans*-2-butene, it gave only the single product **64**, as shown in Scheme 8.17.

These results strongly indicate that 9-phenyl-1(9)-homocubene **57** is formed on thermolysis or photolysis of cubylphenyldiazomethane **56**, that it lives long enough even in refluxing ethanol to undergo intermolecular addition of solvent, and that on a similar time scale it rearranges to the carbene 1-phenyl-9-homocubylidene **61**. Dipolar structures like A and C as reasonable descriptions for 9-phenyl-1(9)-homocubene **57** would be drawn. This critical difference of the dipolar property may open a good way to rationalize the observed chemistry; a "simple" Wagner–Meerwein 1,2 shift provides for the skeletal reorganization of **57** into B via A or into D via C (Figure 8.1). In order to elucidate the mechanism for this rearrangement pathway, labeling studied were employed. At the present stage, their results have strongly showed that the mechanism of path 2 is operating rather than a 1,2-phenyl migration mechanism,

N–N(H)–Ts, Ph; EtOH, Δ, hν; OEt, Ph, H; H, OEt, Ph
58 → 59 (1.5 : 1) 60

Scheme 8.16 ▪ Thermolysis and photolysis of the tosylhydrazone **58**.

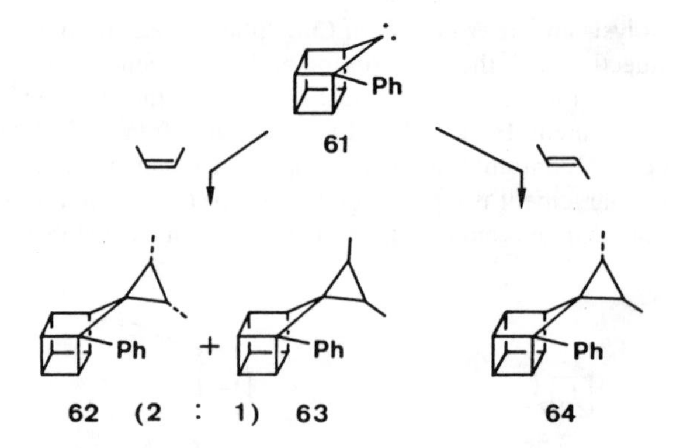

Scheme 8.17 ■ Reactions in neat *cis*- and *trans*-2-butenes.

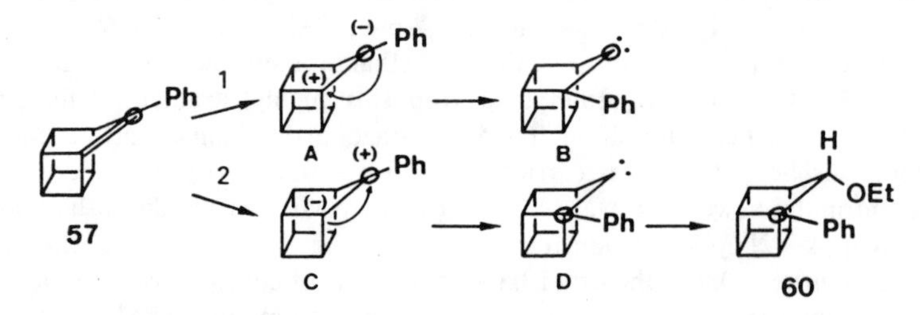

Figure 8.1 ■ Probable rearrangement-pathway mechanisms from **57** to **60**.

path 1.[34] To make these results more confirmed, further examinations of cubyldiazomethanes not only by ordinary techniques but also by matrix isolation spectroscopy have been continued by Eaton.

8.4. Highly Pyramidalized Olefins in the Cubane System—Homocubene and Cubene

Up to what degree can the double bond be deviated from planarity? (Figure 8.2.) Highly pyramidalized olefins are now a special class of compounds.[35] In the specially strained system, where much of the classical π bonding between parallel p orbitals of adjacent sp^2 hybridized carbon is lost, the rehybridization necessary to accommodate the geometric demands of the skeleton in such systems equates to the introduction of substantial s character into the olefin bond. This results in pyramidalization giving its

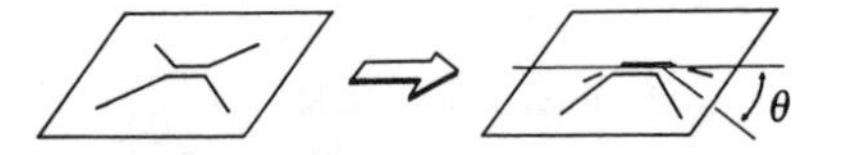

Figure 8.2 ■ Pyramidalized angle (θ) in bent olefins.

Scheme 8.18 ▪ Reaction of bromoiodohomocubane **70** with *n*-BuLi and trapping of homocubene **68** with DPIBF.

angle of approximately 27° in **65**[37(a)], 53° in **66**[37(b)], 48° in **67**[37(c)], and 84° in cubene **69** by means of ab initio calculations at the 6-31 G* TCSCF level by Hrovat and Borden.[36]

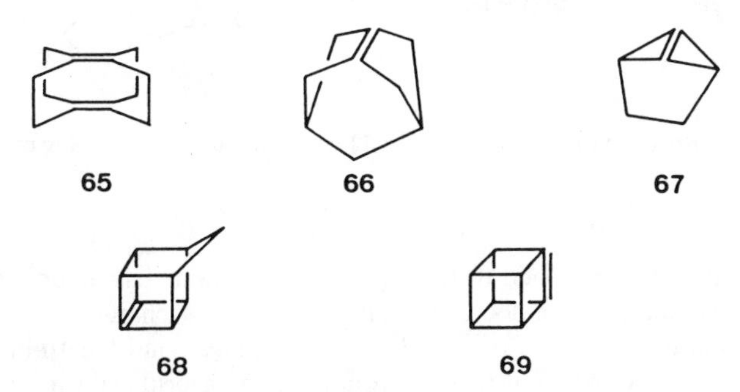

8.4A. Homocubene (4,5-Dehydrohomocubane)

Borden and co-workers have been engaged in the study of the syntheses and properties of a homologous series of pyramidalized alkenes[37] and have more recently been successful in trapping homocub-4(5)-ene **68** by treating bromoiodohomocubane **70** with *n*-butyllithium in THF as the Diels–Alder 1:1 adduct **71** with DPIBF (9,10-diphenylisobenzofuran)[38] (Scheme 8.18). From the fact that attempts to dehalogenate the corresponding homocubane dibromide under the same conditions as just given led only to recovered starting material, at the first elimination step the ready formation of *n*-butyliodide must be very important in this system.

8.4B. Cubene (1,2-Dehydrocubane)

Eaton and Maggini were successful in isolating cubylcubanes **74** along with *tert*-butylcubane **73** by treating diiodocubane **72** with *tert*-butyllithium[39] (Scheme 8.19). They all strongly indicate the formation of cubene **69**, the most pyramidalized olefin. The Diels–Alder adduct **76** of **69** with 11,12-dimethylene-9,10-dihydro-9,10-ethanoanthracene **75**[40] or anthracene also could be obtained, undoubtedly supporting the formation of cubene **69**.

In cubylcubanes **74**, the first synthetic example of the linear dimer of cubane (Figure 8.3), the intercage bond lengths are 1.458 Å in **74-H** and 1.474 Å in **74-*t*-Bu**, respectively, which are both significantly shorter by ca. 0.1 Å than those in ordinary

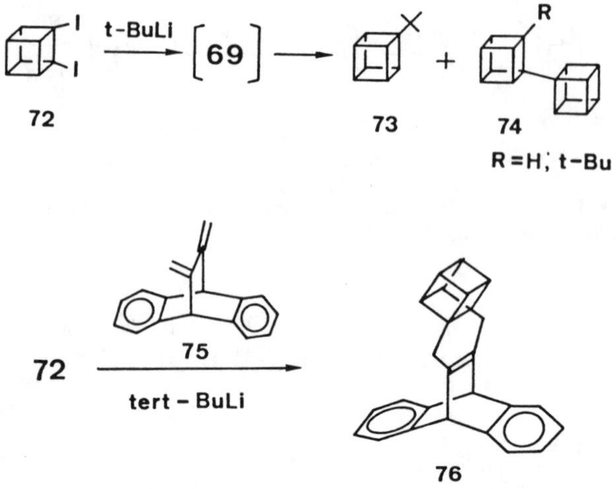

Scheme 8.19 ■ Reaction of 1,2-diiodocubane **72** with *tert*-BuLi and trapping of cubene **69** with exodiene **75**.

alkanes.[41] This can be ascribed to the exceptionally strained skeleton, which leads to increasing p character in endocyclic orbitals, whereas s character increases in the exocyclic orbital in question. Compared to the intercage bond length in 1-adamantyladamantane (1.578 Å), which is formed from pure sp^3 hybridized carbons,[42] the bond length in **74-H** is surprisingly shrunken by up to 0.12 Å, which is the shortest among all recorded to date between carbons fully substituted by other carbon atoms.

8.5. New Type of Cubane Derivative—Combination of Cubane with π-Electronic Ring System

There has been a great deal of effort to synthesize the cubanes and the related cage molecules that are directly attached to the π-electronic ring systems, in connection with the intensive studies on an evaluation of pπ–σ interactions, called vertical delocalization,[43] in those compounds as well as on their reactivities and physiocochemical properties.[44]

8.5A. Phenylcubanes

When we look at how prosperous the chemistry of cubane is, it is surprising that among a large number of cubane derivatives only two synthetic examples of

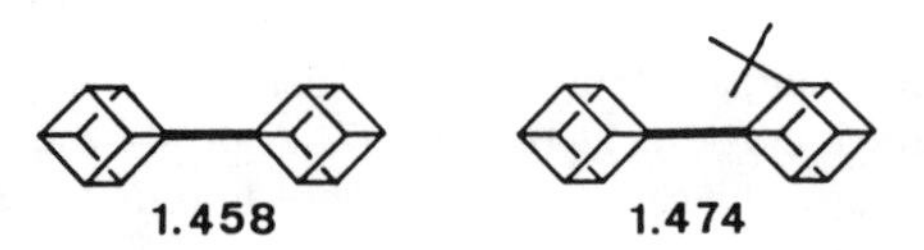

Figure 8.3 ■ Intercage bond lengths (in angstroms) in **74** from X-ray study.

Scheme 8.20 ■ Synthesis of phenylcubane 77.

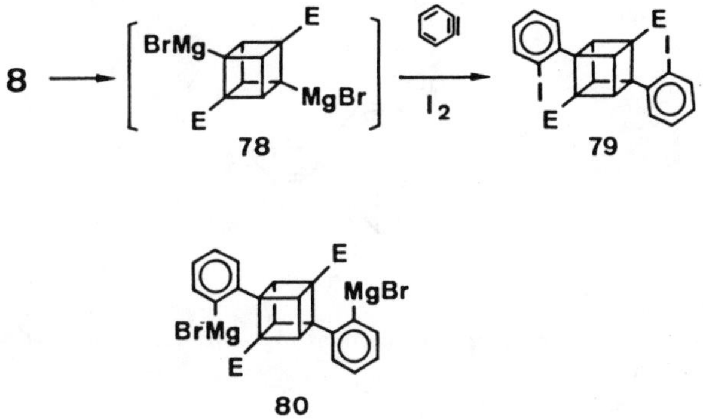

Scheme 8.21 ■ A reasonable mechanism for the formation of 77.

phenylcubanes, where the phenyl ring[45] is incorporated by a reaction of metal-locubane with benzyne[46] or by a free radical arylation of cubanes using cubyl lead acylates[47], have been reported so far. The magnesiated cubane prepared in situ by a metallocubane methodology,[9–13,15] Grignard reagent **78** of cubane, was found to add smoothly to benzyne to give diphenylcubane diamide **77** in a moderate yield (Scheme 8.20). From the fact that when the reaction was quenched with I_2 compound **79** was isolated, the ortho-magnesiated phenyl species **80** is proposed as the crucial intermediate in the benzyne reaction (Scheme 8.21). On the other hand, cubane carboxylic acid **81** upon reaction with lead tetraacetate forms the corresponding isolable lead acylate, which on irradiation in various aromatic solvents such as benzene and holobenzenes gives arylcubanes **82** in high yields (Scheme 8.22). This can be regarded as the first direct C—C bond formation upon cubane via radical species.

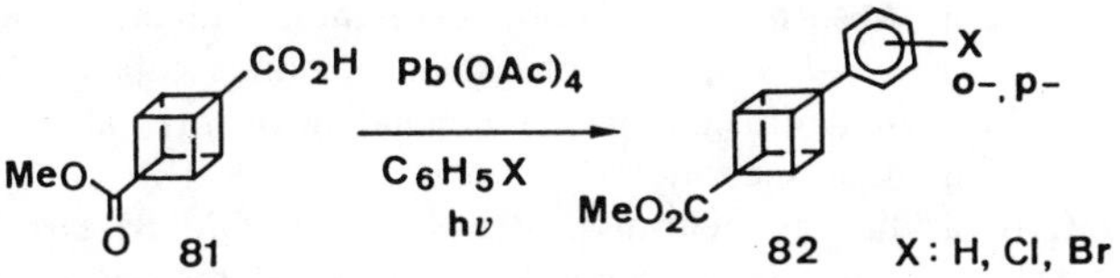

Scheme 8.22 ■ Synthesis of arylcubanes **82** by a combination of lead acylation of cubane carboxylic acid **81** with photolysis in halobenzenes.

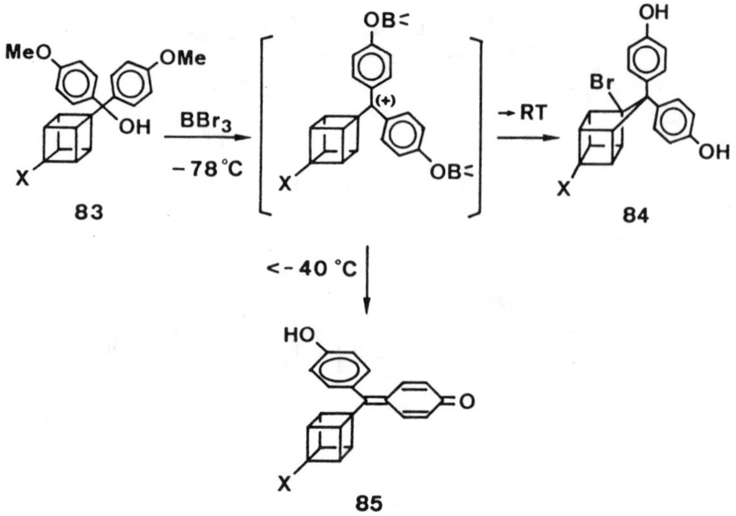

Scheme 8.23 ■ Temperature-dependent reactions of 4-substituted cubyl di(p-methoxyphenyl) carbinols **83** with BBr_3.

8.5B. Quinone Methide Cubanes

8.5B(1). Transformation of the Carboxylic Acid to the Quinone Methide π-Electronic System. Higuchi and co-workers have recently achieved the general conversion method of carboxylic acid group to the quinone methide (QM)[48] π-electronic ring system. This method was successfully applied to the corresponding cubane carboxylic acids in order to form quinone methide cubanes (X-cubane-QM) **85** as a new type of cubane molecule.[49] Because cubane di(p-methoxyphenyl)carbinols **83** prepared from a reaction of the corresponding methyl ester with Grignard reagent (p-$MeOC_6H_4$–MgBr) are treated with the strongly acidic demethylation reagent BBr_3 (Scheme 8.23), this reaction proceeds via formation of cubylmethyl carbonium cation species; thus it always potentially competes with the ring enlargement reaction to give **84** as introduced in Section 8.3A,[4,25,26] it must be completed under controlled conditions at low reaction temperature ($< -40°C$) throughout. This method can be extended to be a general synthetic one for a various type of quinone methide compounds in moderate yields.

8.5B(2). Electronic Absorption of Quinone Methide Cubanes. Similarly to the usual phthaleins,[50] visible–UV spectra of quinone methide cubanes X-cubane-QM thus prepared also exhibit drastic changes depending on pH values owing to the monobasic acid of quinone methide chromophore, as shown in the typical case of H-cubane-QM. In acidic and neutral media, X-cubane-QM **85** largely gives pale yellow solutions (ca. 410 nm), whereas it gives deep reddish purple solutions (540–570 nm) owing to the quinone methide anion species QM(−) in a basic medium (Figure 8.4).

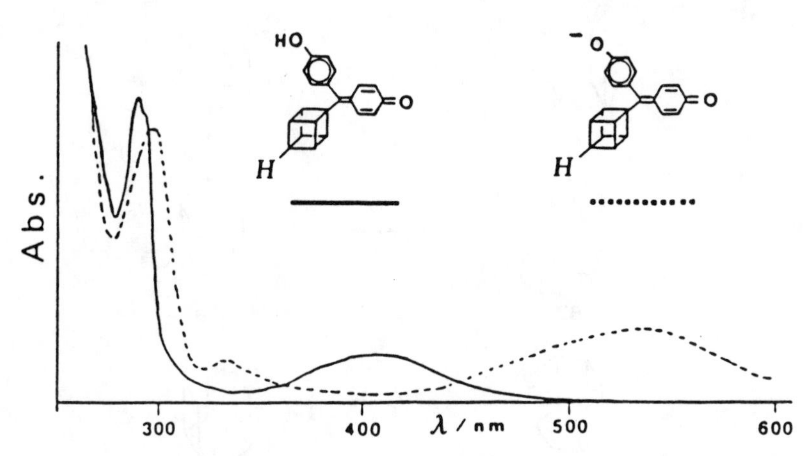

Figure 8.4 ▪ Visible–UV spectra of H-Cubane-QM in H_2O–THF (50% v/v) at 25°C.

Especially in a basic region, the longest-wavelength absorption maxima of quinone methide cubanes observed depend on the substituents X (Table 8.5) and shift to the more bathochromic region with the more electron-withdrawing groups, clearly indicating an effective transmission of the substituent influence on the electronic transition process through the cubane skeleton. These drastic color changes are reversible and their fading rates in a strongly basic medium seem to be fairly slow, unlike phenolphthalein, probably because of a spherical bulkiness of cubane skeleton against intermolecular nucleophilic attacks. It is noteworthy that quinone methide cubane is the first synthetic example of stable "colored cubane" derivatives (cf. **56**[33]).

8.6. Developments of the Related Chemistry of Cubane

8.6A. Photocycloaddition of *syn*-Dienes

Intramolecular [2 + 2] photocycloaddition for the construction of cubane skeleton may be thought to be very useful and due to be employed first. Many attempts, as would be expected, for the synthesis of highly strained prismane molecules such as tetrapris-

Table 8.5 ▪ **Absorption Maxima of X-Cubane-QM(−)**

X	λ (nm)
Br	573
H	543
Me	536
QM	560
QM(−)	546

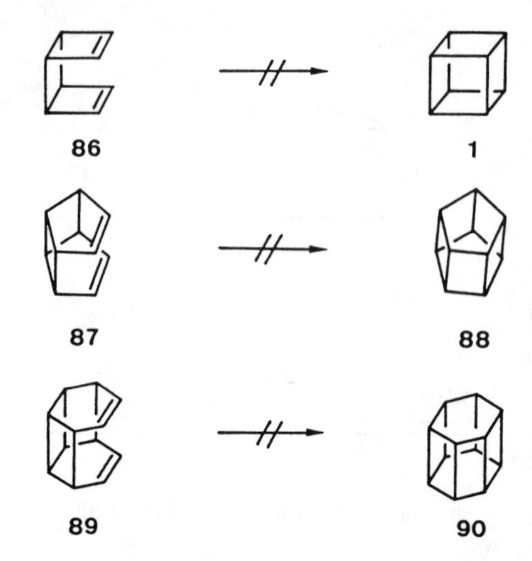

Scheme 8.24 ■ Unsuccessful photocycloadditions of *syn*-dienes.

mane (cubane) **1**,[51] pentaprismane **88**,[52] and hexaprismane **90**[53] by irradiation of the corresponding *syn*-diene compounds **86**, **87**, and **89** have been carried out (Scheme 8.24). Potentially available reactions of those *syn*-dienes, however, except one synthetic example of perfluorooctamethyl cubane,[54] were entirely unsuccessful and resulted in only recovered starting materials or in decomposed products. These results, at present, have been concluded to arise from the following three structural elements:

1. unfavorable geometry and distance between two opposite double bonds for [2 + 2] cycloaddition process;[55]
2. large increment of the molecular strain after the reaction;[55]
3. forbidden process of the photocycloaddition due to T–B interaction.[56]

8.6A(1). Strategy for Intramolecular [2+2] Photocycloaddition in the Cage Molecules. Osawa, however, has predicted by means of molecular orbital calculations that this type of intramolecular photocycloaddition for constructing such strained cage structures can be driven, when these faced double bonds are fixed in an adequate distance and geometry with bridging chains.[57] That is, the number of bonds in a bridging chain between double bonds in question can control the energy relationship between HOMO($\pi+$) and SHOMO($\pi-$) in those *syn*-dienes. A bridging chain with an even number atoms (2, 4, ...) results in the elevation of both $\pi+$ and $\pi-$ energy levels in the double bond with different energy degrees due to the additional through-bond interaction between π-bond and bridging σ-bond orbitals according to the Paddon–Row principle, forming new energy levels of $\pi + - a\sigma +$ and $\pi - - b\sigma -$. This may cause reverse order between these new energy levels HOMO and LUMO in that mixing state and often leads to unsuccessful results of the intramolecular [2 + 2] photocycloaddition in these diene systems. On the other hand, an odd number of atoms (3, 5, ...) in a bridging chain, which consequently lead to the

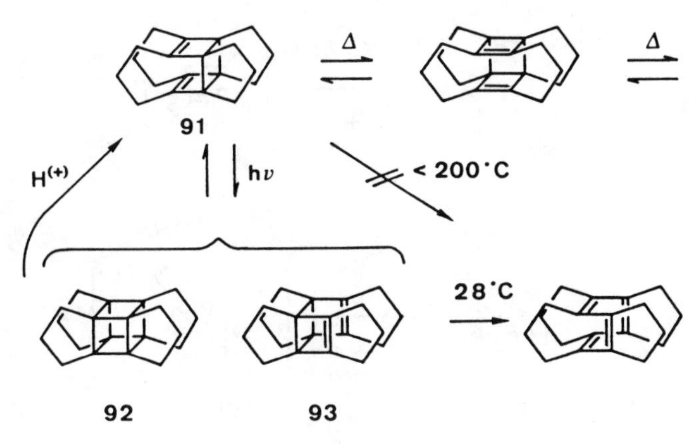

Scheme 8.25 ▪ Reactions of *syn*-tricyclo[4.2.0.0^{2,5}]octa-3,7-diene **97**.

larger energy gap of $\pi+ - a\sigma+$ and $\pi- - b\sigma-$ levels compared with that of $\pi+$ and $\pi-$, increase the possibility of the more effective [2 + 2] photocycloaddition in such diene systems.

8.6A(2). Propellane Type of Cubane. Irradiation (500-W high-pressure Hg lamp, room temperature, pentane) of the *syn*-diene **91** connected with two trimethylene chains, which can be regarded as the first evidence for Osawa's calculations, has been tried by Gleiter and Karcher.[58] In an equilibrated reaction mixture, the propellane-like cubane **92** could be recognized along with the starting diene **91** and its Cope rearrangement isomer **93** in a ratio of 1:10:4 (Scheme 8.25). Although X-ray structure of **91** has not been given yet, the distance between double bonds is estimated to be fairly close within 2.6 Å (3.05 Å in **86**[55]). This [2 + 2] photocycloaddition approach to the cage structures, which is not so theoretically common but technically rather up-to-date, must be very important and may provide a great possibility to reach various types of such compounds (e.g., maybe the last in a series of prismanes, hexaprismane **90**) in the near future.

8.6B. Tetraphosphacubane

In connection with [2 + 2] photocycloaddition of *syn*-diene to cyclobutane ring formation in a cage compound, tetra-*tert*-butyl tetraphosphacubane **97** was directly obtained via an intramolecular [2 + 2] thermal cycloaddition of the tetraphospha *syn*-diene **96**, which was produced in situ through the oligomerization procedure of *tert*-butyl phosphaacetylene **94** (130°C, 65 h without solvent, 8%)[59] (Scheme 8.26). The structure of **97** was determined by ^{13}C NMR spectrum ($\delta = -29.07$, 21.57, 30.64 ppm, C_6H_6) and confirmed by means of X-ray analysis ($P2_1/c$, $\beta = 92.24°$; C—P = 1.881 Å; ∠P—C—P = 94.4°, ∠C—P—C = 85.6°). Although an intermolecular dimerization of the phosphaalkenes has been fully studied,[60] the present dimerization process at the first stage to form diphosphacyclobutadiene **95** can be regarded as the first example of the thermal oligomerization of phosphaalkynes. Strain energy of the phosphacubane **97** has not been studied yet.

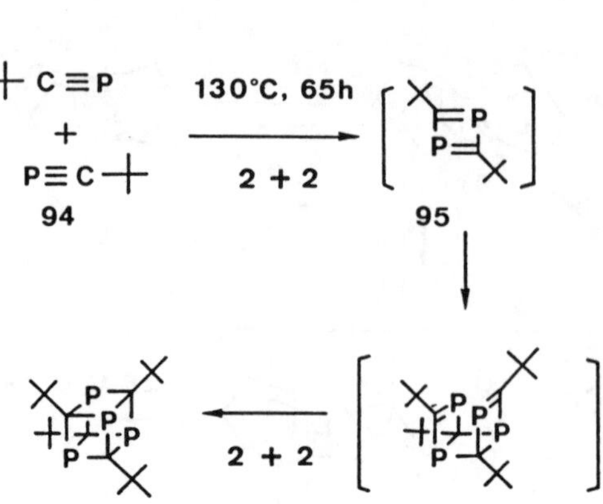

Scheme 8.26 ▪ Preparation of tetra-*tert*-butyl-tetraphosphacubane **97**.

8.7. Future Chemistry of Cubane and Its Perspectives

As described in the introduction of this chapter, during the first 20 years since the conclusive construction of the cubane skeleton by Eaton,[1] there has been reported a large number of works on physiocochemical observations and reactivities to understand its peculiar structure and bond.[2–7] Now, on the basis of the fruits from these studies, the study of cubane chemistry can be recognized to have shifted to the next stage, where it is seeking applicability and challenge to open the door of its new field and era. Of course, to get to these goals, it must be supplemented with a wide variety of results not simply from other fields in organic chemistry but also from inorganic chemistry, biochemistry, physicochemistry, and so on, as well as with the social demands in the modern world.

This section describes, from these applied viewpoints, a few fields of cubane chemistry that have started within the last several years and are about to put forth buds.

8.7A. Synthesis of Polynitrocubanes—High Energy Density Materials

There is a term, detonation pressure (DP), that in short indicates the strength of explosion power and the larger the DP, the more effective the explosive or fuel. According to the "Simple method for calculating detonation properties of C—H—N—O explosives" by Kamlet and Jacobs,[61] the DP value increases in accordance with square of the molecular density (Table 8.6).

Nowadays, due to the crucial situation resulting from the limited amounts of natural resources, intensive research and study on the development of new and useful energy materials to replace natural resources are being carried out all over the world.[62] Especially in the new business including space exploration, creation of cleaner and more effective energy sources is eagerly required, and the chemistry of highly strained three-dimensional molecules such as cubane cannot now have nothing

Table 8.6 ▪ Detonation Pressure (DP) of Polynitrocubanes

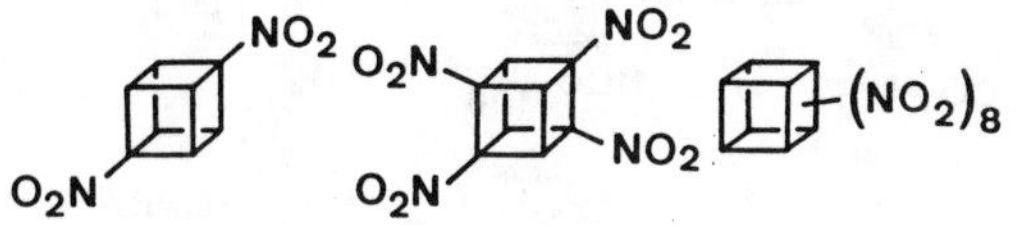

	98	99	100
d (g cm^{-3})	1.66	1.87	2.11
DP (kbar)	205	323	467

to do with this kind of applied research. Thus synthetic chemistry of the compounds that have substantial properties of both enormously highly strained structure and nitro function has started to create a new type of the excellent high-energy-density materials **98** and **101–105**.[63] Yet the molecular density is known to increase with the number of incorporated nitro groups into such strained molecules by calculations.

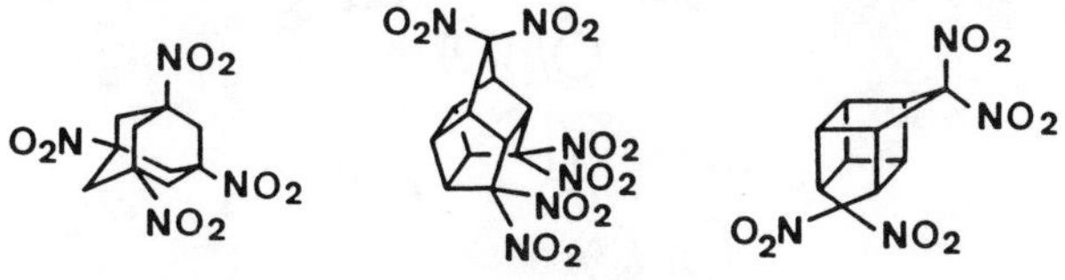

101 102 103

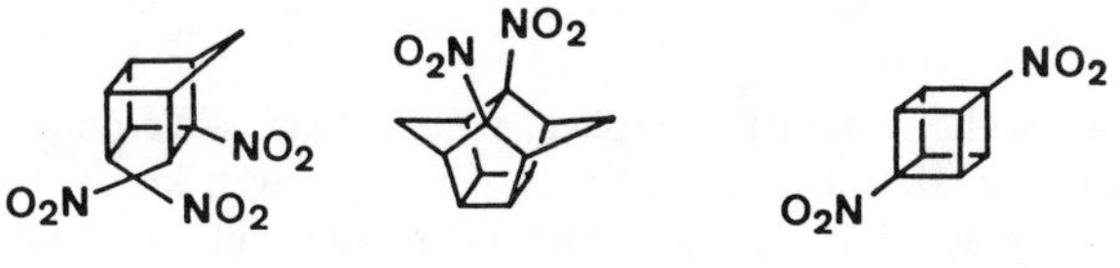

104 105 98

This work has been going on mainly in the United States, and although among cubane derivatives at present only 1,3- and 1,4-dinitrocubanes are known,[64] syntheses of polynitrocubanes with conventional methods and/or the improved method for a direct conversion of isocyanate **106** to the nitro function **98** using dimethyldioxolane **108**[65] have been energetically carried out[66] (Scheme 8.27).

8.7B. Research for New Types of Pharmaceuticals Containing Cubane Skeleton

As introduced in Section 8.3, for a cubane with an electron-deficient species at the α position the ring enlargement reaction readily takes place, to form the corresponding homocubane skeleton.[4,25,26] The amount of strain energy loss in this process comes up to and in some cases exceeds the bond energy of many covalent bonds that play important roles in many organic reactions.[23]

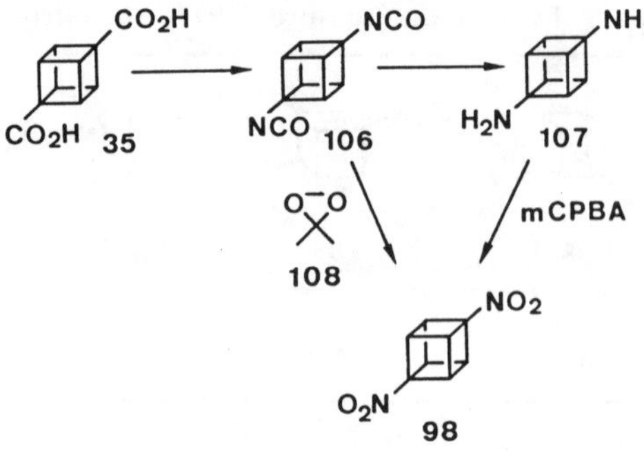

Scheme 8.27 ■ Synthesis of 1,4-dinitrocubane **98**.

On the other hand, cubane derivatives with diarylcarbinol at the α position were observed to form complexes with inorganic and organic molecules such as water, alcohol, chloroform (**111**), and so on.[67]

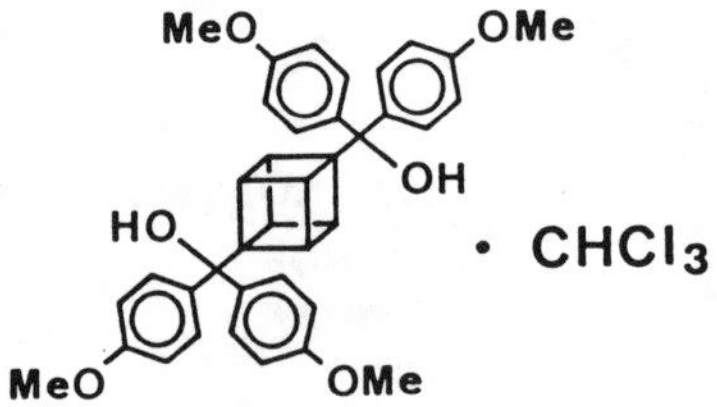

111

This property is another interesting aspect of the cubane skeleton physicochemically comparable to that of diacetylene carbinols, which are well known to be useful and effective host molecules.[68] In addition, the cubane molecule has higher similarity to the corresponding benzene molecule in various properties, such as hydrophobicity, molecular size, s character of the C—H bond,[6] pK_a of its carboxylic acid (5.94 vs. 5.74 for benzoic acid),[69] and so on.

Although we are curious to understand whether or not and, if any, how such cubane derivatives exhibit biological activities, there has been reported no systematic study so far.[70] Such large strain energy release as shown in Figure 8.5 (cubane **1** → homocubane **109** → bishomocubane **110**) can be expected to affect various reactions as well as many redox pathways in a living body, when it transfers directly or indirectly to the critical bonds for biological activity. Elucidation of this biological and mechanistic subject is of importance in that it not only results in an evaluation of the strain effect on biological activities but also may possibly lead to an opportunity to produce a new type of pharmaceuticals containing such strained cubane skeletons.

Ueda and Higuchi[71] have been trying to synthesize a wide variety of cubane derivatives systematically and then to study an effect of the highly strained structures on the biological response and an action mechanism, based on their physicochemical properties and behavior, for example, through the metabolic process and the affinity of those cubane derivatives with the reaction site in local tissues.

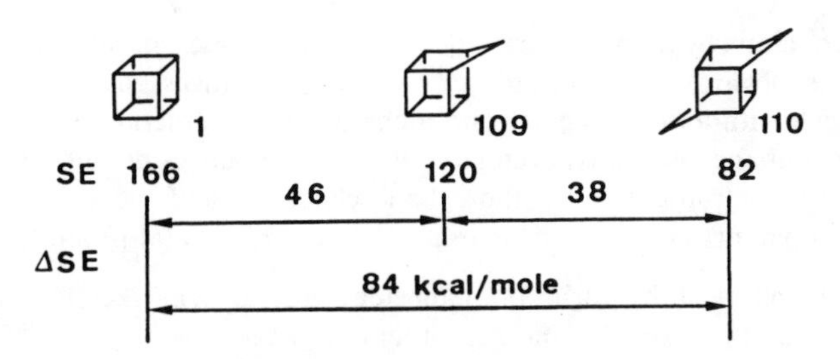

Figure 8.5 ■ Strain energies of related cubanes.

8. Conclusion

In the future chemistry of cubane, what will fascinate our eyes and minds to encourage us to continue to meet the challenge of achieving our dreams? As repeated several times, it apparently can be recognized that the field of cubane chemistry has entered the more developed stage, as if having grown into adolescence from childhood.

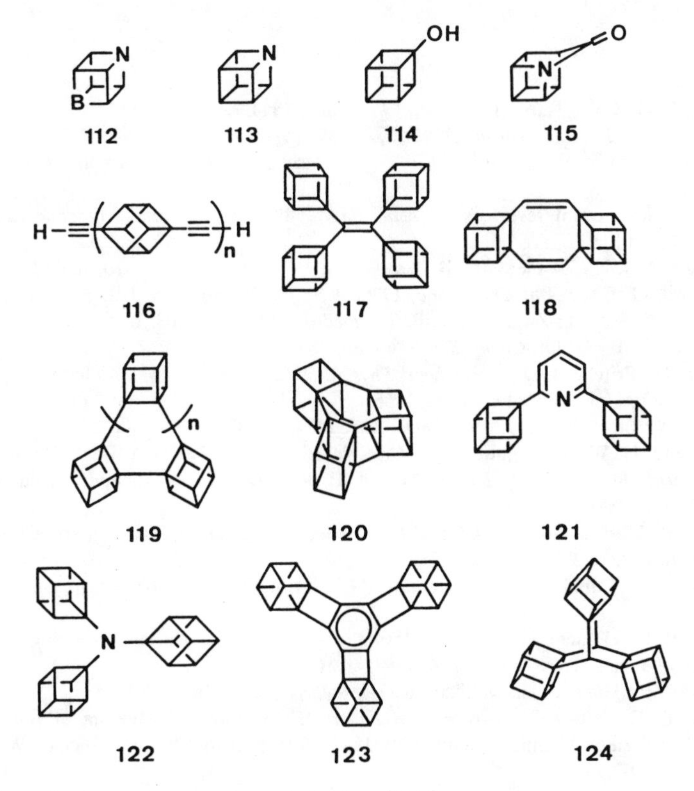

Figure 8.6 ■ Curious unknown cubane derivatives (1989).

Hereafter, along with many new ideas and techniques produced as the world changes, it is obvious that we will see cubane derivatives that have never been imagined, each with a new property and a characteristic structure, steadily adding to the history of cubane chemistry, even if all of these derivatives do not satisfy the social demands. At the turning point of the cubane chemistry and the starting point of its further development, we should bear in mind the fundamental properties of cubane:

1. It is a spherical hydrocarbon molecule with a rigid skeleton, keeping the distance and geometry of the substituents constant.
2. It has potentially high reactivity due to the exceptionally highly strained skeleton, where every C—C—C bond angle is 90°.

On the basis of these two points, one must construct a new cubane derivative and put additional value into this skeleton by calling on one's own skill and ideas. Finally, we present several cubane derivatives (Figure 8.6) that are, of course, unknown as yet but that have been intensely interesting to chemists involved in their synthesis and properties. The scientific significance and value of each target molecule will be left to the reader's evaluation.

Acknowledgment. H. H. is grateful to Prof. Philip E. Eaton for useful discussions and suggestions.

References

1. Eaton, P. E.; Cole, T. W., Jr. *J. Am. Chem. Soc.* **1964**, *86*, 962, 3157.
2. (a) Eaton, P. E. *Tetrahedron* **1979**, *35*, 21989. (b) *Tetrahedron* **1986**, *42*(6) (Tetrahedron Symposia-In-Print No. 26, *Synthesis of Non-Natural Products: Challenge and Reward*; Eaton, P. E., Ed.).
3. For the most recent review, see *Chem. Rev.* **1989**, *89*(5), (*Strained Organic Compounds*; Michl, J., Ed.).
4. Klunder, A. J. H.; Zwanenburg, B. *Tetrahedron* **1972**, *28*, 4131. Edward, J. T.; Farrell, P. G.; Langford, G. E. *J. Am. Chem. Soc.* **1976**, *98*, 3075. Klunder, A. J. H.; Ariaans, G. J. A.; Loop, E. A. R. v. d.; Zwanenburg, B. *Tetrahedron* **1986**, *42*, 1903.
5. Fleischer, E. B. *J. Am. Chem. Soc.* **1964**, *86*, 3889.
6. Muller, N.; Pritcherd, D. E. *J. Chem. Phys.* **1959**, *31*, 768 and 1471. Cole, T. W., Jr. Ph.D. Dissertation, University of Chicago, 1966.
7. Luh, T.-Y.; Stock, L. M. *J. Am. Chem. Soc.* **1974**, *96*, 3712.
8. Gschwend, H. W.; Rodriguez, H. R. *Org. React.* (*N.Y.*) **1979**, *26*, 1. Beak, P.; Snieckus, V. *Acc. Chem. Res.* **1982**, *15*, 306. Wakefield, B. J. *Organolithium Methods*; Academic Press: New York, 1988.
9. Eaton, P. E.; Castaldi, G. *J. Am. Chem. Soc.* **1985**, *107*, 724. Compare *Chemical and Engineering News* **1985** (Feb. 25), 31.
10. Eaton, P. E.; Cunkle, G. T.; Marchioro, G.; Martin, R. M. *J. Am. Chem. Soc.* **1987**, *109*, 948.
11. Eaton, P. E.; Higuchi, H.; Ueda, I. *Abstracts of Papers*, 56th National Meeting of Chemical Society of Japan, Tokyo, April 1–4, 1988; abstract p. 1684.
12. Eaton, P. E.; Higuchi, H.; Millikan, R. *Tetrahedron Lett.* **1987**, *28*, 1055.
13. Eaton, P. E.; Xiong, Y. *Abstracts of Papers*, 198th National Meeting of the American Chemical Society, Miami, September 10–15, 1989; American Chemical Society: Washington, DC, 1989; ORGN 17.

14. Ferris, J. P.; Gerwe, R. D.; Gapski, G. R. *J. Am. Chem. Soc.* **1967**, *89*, 5270. Ferris, J. P.; Gerwe, R. D.; Gapski, G. R. *J. Org. Chem.* **1968**, *33*, 3493. Monkovic, I.; Wong, H.; Bachand, C. *Synthesis* **1985**, 770.
15. Eaton, P. E.; Higuchi, H.; Ueda, I. *Abstracts of Papers*, 56th National Meeting of Chemical Society of Japan, Tokyo; April 1–4, 1988; abstract p. 1280.
16. Kropp, P. J.; Poindexter, G. S.; Pienta, N. J.; Hamilton, D. C. *J. Am. Chem. Soc.* **1976**, *98*, 8135. Kropp, P. J. *Acc. Chem. Res.* **1984**, *17*, 131.
17. Reddy, D. S.; Sollott, G. P.; Eaton, P. E. *J. Org. Chem.* **1989**, *54*, 722. Reddy, D. S. *Abstracts of Papers*, 1989 International Chemical Congress of Pacific Basin Societies, Honolulu, December 17–22, 1989; American Chemical Society: Washington, DC, 1989; ORGN 168.
18. Eaton, P. E.; Cunkle, G. T. *Tetrahedron Lett.* **1986**, *27*, 6055.
19. Moriarty, R. M.; Khosrowshahi, J. S.; Penmasta, R. *Tetrahedron Lett.*, **1989**, *30*, 791. Moriarty, R. M.; Tuladhar, S. M.; Penmasta, R.; Awasthi, A. K. *Abstracts of Papers*, 198th National Meeting of the American Chemical Society, Miami, September 10–15, 1989; American Chemical Society: Washington, DC, 1989; ORGN 18.
20. Barton, D. H. R.; Crich, D.; Motherwell, W. B. *J. Chem. Soc., Chem. Commun.* **1983**, 939.
21. Della, E. W.; Tsanaktsidis, J. *Aust. J. Chem.* **1986**, *39*, 2061.
22. Hasebe, M.; Tsuchiya, T. *Tetrahedron Lett.* **1987**, *28*, 6207, and references cited therein. Higuchi, H. Private result: methyl cubane carboxylate was obtained form cubane-1,4-dicarboxylic acid monomethyl ester **81** with Hasebe's procedure in 55–75% yields.
23. Allinger, N. L.; Tribble, M. T.; Miller, M. A.; Wertz, D. *J. Am. Chem. Soc.* **1971**, *93*, 1637. Engler, E. M.; Andose, J. D.; Schleyer, P. v. R. *J. Am. Chem. Soc.* **1973**, *95*, 8005.
24. Cassar, L.; Eaton, P. E.; Halpern, J. *J. Am. Chem. Soc.* **1970**, *92*, 3515, 6366. Gassman, P. G.; Yamaguchi, R.; Koser, G. F. *J. Org. Chem.* **1978**, *43*, 4392.
25. Luh, T.-Y. Ph.D. Dissertation, University of Chicago, 1974. Mak, T. C. W.; Yip, Y. C.; Luh, T.-Y. *Tetrahedron* **1986**, *42*, 1981. See also ref. 4.
26. Hasegawa, T.; Higuchi, H.; Ueda, I. *Abstracts of Papers*, 58th National Meeting of Chemical Society of Japan, Kyoto; April 1–4, 1989; abstract p. 1336. Hasegawa, T.; Kimura, Y.; Higuchi, H.; Ueda, I. *Abstracts of Papers*, 6th International Symposium on Novel Aromatic Compounds, Osaka, August 20–25, 1989; abstract. p. 103(B-1).
27. Osawa, E.; Aigami, K.; Inamoto, Y. *J. Chem. Soc. 2* **1979**, 181.
28. (a) Kimura, Y.; Higuchi, H.; Ueda, I. *Abstracts of Papers*, 58th National Meeting of Chemical Society of Japan, Kyoto, April 1–4, 1989; abstract p. 1337. (b) Higuchi, H.; Hasegawa, T.; Kimura, Y.; Ueda, I. *Abstracts of Papers*, 1989 International Chemical Congress of Pacific Basin Societies, Honolulu, December 17–22, 1989; American Chemical Society: Washington, DC, 1989; ORGN 177.
29. Wentrup, C. In *Azides and Nitrenes*; Scriven, E. F. V., Ed.; Academic Press: Orlando, FL, 1984; pp. 399–402. Kyba, E. P. In *Azides and Nitrenes*, 26–28. Scrive
30. Szeimies, G. In *Reactive Intermediates*; Abramovitch, R. A., Ed.; Plenum: New York, 1983; pp. 329–359.
31. Eaton, P. E.; Hormann, R. E. *J. Am. Chem. Soc.* **1987**, *109*, 1268.
32. Radziszewski, J. G.: Downing, J. W.; Wentrup, C.; Kaszynski, P.; Jawdosiuk, M.; Kovacic, P.; Michl, J. *J. Am. Chem. Soc.* **1984**, *106*, 7996; **1985**, *107*, 594, 2799, and the footnote 17 cited in ref. 31.
33. Eaton, P. E.; Hoffmann, K.-L. *J. Am. Chem. Soc.* **1987**, *109*, 5285.
34. Eaton, P. E.; White, A. J. *Abstracts of Papers*, 198th National Meeting of the American Chemical Society, Miami, September 10–15, 1989; American Chemical Society: Washington, DC, 1989; ORGN 1968. Appell, R. B.; Eaton, P. E. *Abstracts of Papers*, 1989 International Chemical Congress of Pacific Basin Societies, Honolulu, December 17–22, 1989; American Chemical Society: Washington, DC, 1990; ORGN 152.

35. Szeimies, G. In *Reactive Intermediates*; Abramovitch, R. A., Ed.; Plenum: New York, 1983; Vol. 3, p. 299.
36. Schubert, W.; Yoshimine, M.; Pacansky, J. *Phys. Chem.* **1981**, *85*, 1340. Hrovat, D. A.; Borden, W. T. *J. Am. Chem. Soc.* **1988**, *110*, 4710.
37. (a) Wiberg, K. B.; Adams, R. D.; Okarma, P. J.; Matturo, M. G.; Segmuller, B. *J. Am. Chem. Soc.* **1984**, *106*, 2200. (b) Renzoni, G. E.; Yin, T.-Y.; Borden, W. T. *J. Am. Chem. Soc.* **1986**, *108*, 7121. (c) Renzoni, G. E.; Yin, T.-Y.; Miyake, F.; Borden, W. T. *Tetrahedron* **1986**, *42*, 1581. (d) Yin, T.-Y.; Miyake, F.; Renzoni, G. E.; Borden, W. T.; Radziszewski, J. G.; Michl, J. *J. Am. Chem. Soc.* **1986**, *108*, 3544. (e) Hrovat, D. A.; Miyake, F.; Trammell, G.; Gilbert, K. E.; Mitchell, J.; Clardy, J.; Borden, W. T. *J. Am. Chem. Soc.* **1987**, *109*, 5524.
38. Hrovat, D. A.; Borden, W. T. *J. Am. Chem. Soc.* **1988**, *110*, 7229.
39. Eaton, P. E.; Maggini, M. *J. Am. Chem. Soc.* **1988**, *110*, 7230.
40. Hart, H.; Bashir-Hashemi, A.; Luo, J.; Meador, M. A. *Tetrahedron* **1986**, *42*, 1641, and references cited therein.
41. Gilardi, R.; Maggini, M.; Eaton, P. E. *J. Am. Chem. Soc.* **1988**, *110*, 7232. Compare *Chemical and Engineering News* **1988** (Nov. 14), 45. Ermer, O; Lex, J. *Angew. Chem., Int. Ed. Engl.* **1987**, *25*, 447.
42. Alden, R. A.; Kraut, J.; Traylor, T. G. *J. Am. Chem. Soc.* **1968**, *90*, 74. Kennard, O.; Watson, D.; Allen, F. H.; Motherwell, W. B.; Town, W.; Rodgers, J. *J. Chem. Brit.* **1975**, *11*, 213. Allen, F. H.; Kennard, O.; Watson, D. G.; Brammer, L.; Orpen, A. G.; Taylor, R. *J. Chem. Soc., Perkin Trans. 2* **1987**, S1.
43. Baker, F. W.; Parish, R. C.; Stock, L. M. *J. Am. Chem. Soc.* **1967**, *89*, 5677. Martin, J. C.; Ree, B. R. *J. Am. Chem. Soc.* **1969**, *91*, 5882. Hoffmann, R.; Heilbronner, E.; Gleiter, R. *J. Am. Chem. Soc.* **1970**, *92*, 706. Clinton, N. A.; Brown, R. S.; Traylor, T. G. *J. Am. Chem. Soc.* **1970**, *92*, 5228. Traylor, T. G.; Hanstein, W.; Berwin, H. J.; Clinton, N. A.; Brown, R. S. *J. Am. Chem. Soc.* **1971**, *93*, 5715. Brown, R. S.; Traylor, T. G. *J. Am. Chem. Soc.* **1973**, *95*, 8025. Cole, T. W., Jr.; Mayers, C. J.; Stock, L. M. *J. Am. Chem. Soc.* **1974**, *96*, 4555. There are many other references in the literature.
44. Buchi, G.; Perry, C. W.; Rob, E. W. *J. Org. Chem.* **1962**, *27*, 4106. Tsutsui, M. *Chem. Ind. (London)* **1962**, 780. Cookson, R. C.; Jones, D. W. *Proc. Chem. Soc.* **1963**, 115. Pawley, G. S.; Lipscomb, W. N.; Freedman, H. H. *J. Am. Chem. Soc.* **1964**, *86*, 4725. Throndsen, H. P.; Wheatley, P. J.; Zeiss, H. *Proc. Chem. Soc.* **1964**, 357. Slobodin, Y. M.; Khitrov, A. P. *Zh. Org. Khim.* **1970**, *6*, 1751. Slobodin, Y. M.; Aleksandrov, I. V.; Khitrov, A. P. *Zh. Org. Khim.* **1977**, *13*, 1377. Ikeda, H.; Yamashita, Y.; Kabuto, C.; Miyashi, T. *Tetrahedron Lett.* **1988**, *29*, 5779. There are many other related references, including ref. 43.
45. Freedman, H. H. *J. Am. Chem. Soc.* **1961**, *83*, 2195. Braye, E. H.; Hubel, W.; Caplier, I. *J. Am. Chem. Soc.* **1961**, *83*, 4406. Freedman, H. H.; Petersen, D. R. *J. Am. Chem. Soc.* **1962**, *84*, 2837. Maitlis, P. M.; Stone, F. G. A. *Proc. Chem. Soc.* **1962**, 330. Although it had been reported that the high-melting-point and insoluble product formed via dimerization of the radical intermediate from the thermal decomposition of (4-bromo-1,2,3,4-tetraphenyl-*cis*,cis-1,3-butadienyl)dimethyltinbromide in 85% yield was octaphenylcubane, a three-dimensional X-ray diffraction study of this material yielded the octaphenylcyclooctatetraene structure.
46. Bashir-Hashemi, A. *J. Am. Chem. Soc.* **1988**, *110*, 7234. Bashir-Hashemi, A.; Stec, D., III; Gilbert, E. E. *Abstracts of Papers*, 1989 International Chemical Congress of Pacific Basin Societies, Honolulu, December 17–22, 1989; American Chemical Society: Washington, DC, 1989; ORGN 173.
47. Moriarty, R.; Khosrowshahi, J. S. *Abstracts of Papers*, 198th National Meeting of the American Chemical Society, Miami, September 10–15, 1989; American Chemical Society: Washington, DC, 1989; ORGN 230.

48. Orndorff, W. R.; Purdy, A. C. *J. Am. Chem. Soc.* **1926**, *48*, 2212. Turner, A. B. *Quart. Rev.* **1964**, *18*, 347.
49. Higuchi, H.; Ueda, I. *Abstracts of Papers*, 58th National Meeting of Chemical Society of Japan, Kyoto, April 1–4, 1989; abstract p. 1337. See also ref. 28b.
50. Sager, E. E.; Maryott, A. A.; Schooley, M. R. *J. Am. Chem. Soc.* **1948**, *70*, 732.
51. Criegee, R. *Angew. Chem.* **1962**, *74*, 703, and *Angew. Chem., Int. Ed. Engl.* **1962**, *1*, 519. Iwamura, H.; Morio, K.; Kihara, H. *Chem. Lett.* **1973**, 457.
52. McKennis, J. S.; Brener, L.; Ward, J. S.; Pettit, R. *J. Am. Chem. Soc.* **1971**, *93*, 4957. Paquette, L. A.; Davis, R. F.; James, D. R. *Tetrahedron Lett.* **1974**, *15*, 1615.
53. Eaton, P. E.; Chakraborty, U. R. *J. Am. Chem. Soc.* **1978**, *100*, 3634. Higuchi, H.; Takatsu, K.; Otsubo, T.; Sakata, Y.; Misumi, S. *Tetrahedron Lett.* **1982**, *23*, 671. Mehta, G.; Srikrishna, A.; Nair, M. S.; Cameron, T. S.; Tacreiter, W. *Ind. J. Chem.* **1983**, *22B*, 621. Higuchi, H.; Misumi, S. *KAGAKU* **1983**, *39*, 646. Yang, N. C.; Horner, M. G. *Tetrahedron Lett.* **1986**, *27*, 543. Higuchi, H.; Kobayashi, E.; Sakata, Y.; Misumi, S. *Tetrahedron* **1986**, *42*, 1731.
54. Pelosi, L. F.; Miller, W. T. *J. Am. Chem. Soc.* **1976**, *98*, 4311.
55. Osawa, E.; Aigami, K.; Inamoto, H. *J. Org. Chem.* **1977**, *42*, 2621.
56. Schmidt, W.; Wilkins, B. T. *Tetrahedron* **1972**, *28*, 5649. Gleiter, R.; Heilbronner, E.; Hekman, M.; Martin, H.-D. *Chem. Ber.* **1973**, *106*, 28. Iwamura, H.; Kihara, H.; Morio, K.; Kunii, T. *Bull. Chem. Soc. Jpn.* **1973**, *16*, 3248.
57. Osawa, E. *KAGAKU* **1989**, *44*, 200.
58. Gleiter, R.; Karcher, M. *Angew. Chem.* **1988**, *100*, 851.
59. Wettling, T.; Schneider, J.; Wagner, O.; Kreiter, C. G.; Regitz, M. *Angew. Chem.* **1989**, *101*, 1035.
60. Becker, G.; Massa, W.; Mundt, O.; Schmidt, R. *Z. Anorg. Allg. Chem.* **1982**, *485*, 23. Appel, R.; Casser, C.; Knoch, F. *Chem. Ber.* **1984**, *117*, 2613.
61. Kamlet, M. J.; Jacobs, S. J. *J. Chem. Phys.* **1968**, *48*, 23.
62. There are many reviews, for example: Laird, T. *Chem. Ind.* **1978**, *18*, 186. Grutsch, P. A.; Kutal, C. *J. Am. Chem. Soc.* **1979**, *101*, 4228. King, R. B. *J. Org. Chem.* **1979**, *44*, 385.
63. Sollott, G. P.; Gilbert, E. E. *J. Org. Chem.* **1980**, *45*, 5405. Marchand, A. P.; Suri, S. C. *J. Org. Chem.* **1984**, *49*, 2041. Marchand, A. P.; Reddy, D. S. *J. Org. Chem.* **1984**, *49*, 4078. Paquette, L. A.; Fischer, J. W.; Engel, P. *J. Org. Chem.* **1985**, *50*, 2524. Marchand, A. P.; Dave, P. R.; Rajapaksa, D.; Arney, B. E., Jr.; Flippen-Anderson, J. L.; Gilardi, R.; George, C. *J. Org. Chem.* **1989**, *54*, 1769.
64. Eaton, P. E.; Shanker, B. K. R.; Price, G. D.; Pluth, J. J.; Gilbert, E. E.; Alster, J.; Sandus, O. *J. Org. Chem.* **1984**, *49*, 185.
65. Murray, R. W.; Jeyaraman, R. J. *J. Org. Chem.* **1985**, *50*, 2847. Adam, W.; Chan, Y. Y.; Cremer, D.; Gauss, J.; Scheutzow, D.; Schindler, M. *J. Org. Chem.* **1987**, *52*, 2800.
66. Eaton, P. E.; Wicks, G. E. *J. Org. Chem.* **1988**, *53*, 5353.
67. For example, biscarbinol **111** forms 1 : 1 complex with chloroform. Higuchi, H. Unpublished observation.
68. Toda, F.; Akagi, K. *Tetrahedron Lett.* **1968**, *9*, 3695. Toda, F.; Tanaka, K.; Yagi, M. *Tetrahedron* **1987**, *43*, 1495. Toda, F. *J. Synth. Org. Chem. Jpn.* **1989**, *47*, 1118.
69. Cole, T. W., Jr.; Mayers, C. J.; Stock, L. M. *J. Am. Chem. Soc.* **1974**, *96*, 4555.
70. Only two patents (Loeffler, L. L. U.S. Patents 1,203,528 and 1,909,666, Merck and Co., Inc.; Gregory, W. A. U.S. Patent 710,126, E. I. duPont deNemours and Co.) appear in the footnote of ref. 2a. No report on the physiological or biological activity of cubane derivatives was given.
71. Ueda, I; Higuchi, H. Unpublished results.

9

Recent Advances in Selected Aspects of Bishomocubane Chemistry*

Wendell L. Dilling

Central Michigan University, Michigan, USA

Contents

9.1. Introduction
9.2. Theoretical Calculations
 9.2A. 1,1-Bishomocubanes
 9.2B. 1,3-Bishomocubanes
 9.2C. 1,3′-Bishomocubanes
9.3. Formation Reactions
 9.3A. 1,1-Bishomocubanes
 9.3B. 1,2-Bishomocubanes
 9.3C. 1,3-Bishomocubanes
 9.3D. 1,3′-Bishomocubanes
 9.3E. 1,4-Bishomocubanes
9.4. Reactions
 9.4A. 1,1-Bishomocubanes
 9.4B. 1,3-Bishomocubanes
 9.4C. 1,3′-Bishomocubanes
 9.4D. 1,4-Bishomocubanes
References

9.1. Introduction

Rigid polycyclic aliphatic organic compounds have been a source of fascination for organic chemists for many years. Because of their rigidity, strain, and symmetry, these

---

* Bishomocubane (Pentacyclodecane) Chemistry. XIII. For Part XII see Dilling, W. L. *J. Org. Chem.* **1975**, *40*, 2380–2384.

compounds are often valuable for discovering and testing concepts of bonding, reactivity, structure–activity, and structure–property relationships. Some natural products and biologically active compounds have this polycyclic type of structure. Current research activity indicates that interest in these compounds is likely to remain undiminished in the near future.

This chapter concerns recent developments in one segment of the chemistry of rigid polycyclic aliphatic compounds, namely, that of the bishomocubanes. These bishomocubanes, of which there are five carbon-framework isomers, can be formed conceptually by inserting two methylene groups into one or two bonds of cubane. The trivial nomenclature, 1,*n*-bishomocubane,[1] for these five isomers refers to the shortest path along the edges of a cube (**1**) between the positions of the two methylene bridges. Thus the five possible bishomocubanes are as follows:

pentacyclo[4.4.0.$0^{2,5}$.$0^{3,8}$.$0^{4,7}$]decane, 1,1-bishomocubane (**2**);
pentacyclo[4.4.0.$0^{2,5}$.$0^{3,9}$.$0^{4,7}$]decane, 1,2-bishomocubane (**3**);
pentacyclo[5.3.0.$0^{2,5}$.$0^{3,9}$.$0^{4,8}$]decane, 1,3-bishomocubane (**4**);
pentacyclo[4.4.0.$0^{2,5}$.$0^{3,9}$.$0^{4,8}$]decane, 1,3′-bishomocubane (**5**);
pentacyclo[5.3.0.$0^{2,6}$.$0^{3,9}$.$0^{4,8}$]decane, 1,4-bishomocubane (**6**).

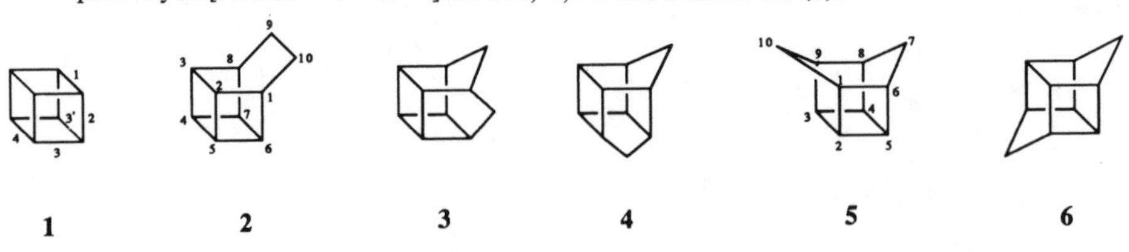

1 **2** **3** **4** **5** **6**

These bishomocubane isomers are sometimes interconvertible[1,2] and hold a special relationship to cyclopentadienes and *endo*-dicyclopentadienes from which many of them have been synthesized.[3] Several polyhalogenated derivatives of 1,4-bishomocubane have become significant commercial materials, primarily as pesticides. The subsequent environmental problems that followed and the attendant laboratory and field studies add another dimension to the chemistry of bishomocubanes. Bishomocubanes have often been critical intermediates in the synthesis of numerous other more highly strained compounds such as cubane itself.[4]

This chapter reviews selected work on bishomocubanes published from 1986 to 1990. Some aspects of this subject have been reviewed recently.[5–9]

9.2. Theoretical Calculations

9.2A. 1,1-Bishomocubanes

Iterative maximum overlap calculations were performed on both basketane (**2**) and basketene (**7**).[10] Calculated bond lengths and bond angles were compared with experimental values for various derivatives of **2** and **7**.

7

The largest discrepancy for **2** appeared to be a C(9)—C(10) bond length that was longer than the calculated value. In **7** both the C(2)—C(3) and C(4)—C(5) bond lengths were longer than the calculated values. For both **2** and **7** the calculated C(2)C(3)C(4)C(5) cyclobutane ring is planar whereas the C(1)C(2)C(5)C(6) cyclobutane ring has a dihedral angle of 152°. The shortest C—H bonds in **2** are those in which the carbon atoms are part of the maximum number of cyclobutane rings: C(4), 1.079 Å; C(2), 1.083 Å; C(1), 1.092 Å; C(9), 1.095 Å. This is presumably related to angle strain in which an increase in p character of the C—C bonds gives the C—H bond more s character. In **7** the calculated C(4)—H bond distance (1.079 Å) is shorter than the vinylic C(9)—H bond (1.081 Å), and both bonds have higher s character than the C—H bond of ethylene. The calculations indicate that the C(2)—C(5) bonds in both **2** and **7** are the weakest of all the bonds in these molecules, which agrees with previous hydrogenolysis results. The total calculated strain energies of **2** and **7** are 125.0 and 123.6 kcal mol^{-1}, respectively.

The homocubyl-1,1-bishomocubyl cation **8** was calculated to be ~8 kcal mol^{-1} less strained than the bis(homocubyl) cation **9** as indicated by the MM2 procedure (229.87 vs. 237.01 or 238.16 kcal mol^{-1} depending on geometry).[11] However, generation of **9** did not lead to its rearrangement to **8**, presumably because of unfavorable overall energetics [ΔH_f(**8**, 2° cation) = 403.87 kcal mol^{-1}, ΔH_f(**9**, 3° cation) = 390.28 or 391.43 kcal mol^{-1}).

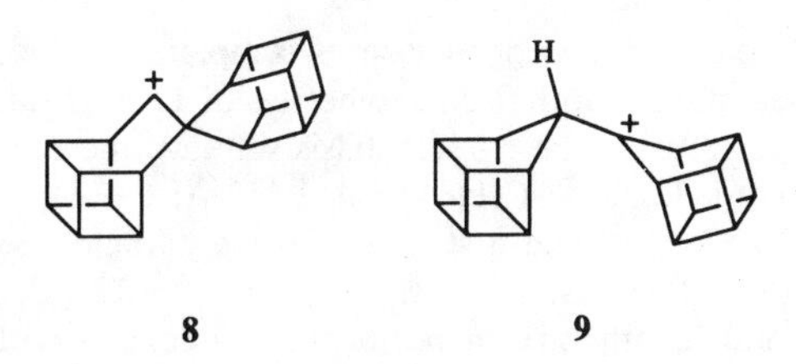

9.2B. 1,3-Bishomocubanes

MM2 calculations indicate a strain energy of 76.63 kcal mol^{-1} for the parent hydrocarbon (**4**) and an increase in strain energy of 41.50 kcal mol^{-1} for the transformation of **4** to homocubane (**10**).[12]

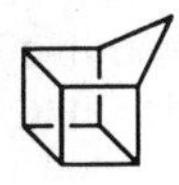

10

9.2C. 1,3′-Bishomocubanes

AM1 calculations have been performed on the 1,3′-bishomocubadiene **11** and several homologs.[13] In this hypothetical molecule both the orbital interaction through space and the orbital interaction via two bonds appear to be strong and to overwhelm the orbital interaction via three bonds to give the natural frontier molecular orbital ordering.

11

Quantum mechanical calculations have been performed at various levels on pentaprismane (**12**), a 1,3′-bishomocubane in which C(7) and C(10) are directly bonded.[14–17] Molecular mechanics calculations predict D_{5h} symmetry with little opportunity for conformational flexibility.[18]

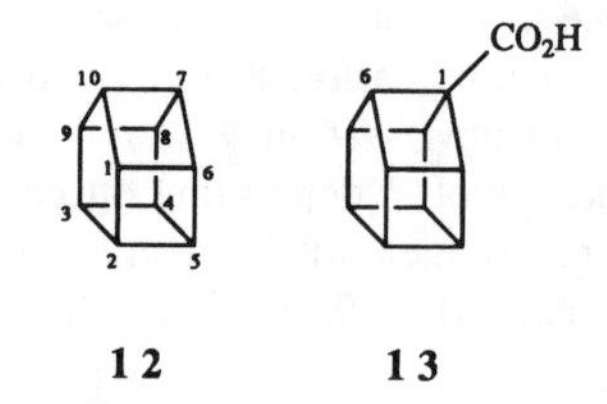

12 **13**

Each face is planar, and each hydrogen atom is completely eclipsed by its immediate neighboring hydrogen atoms. Optimized geometries of **12** constrained to D_{5h} symmetry at the SCF level and using the 6-31G* basis set gave the following bond lengths and angles: C(1)—C(6), 1.558 Å; C(1)—C(2), 1.552 Å; C—H, 1.082 Å; C(6)C(1)H, 123.3°; C(2)C(1)H, 119.4°.[15,17] The first four of these parameters obtained using the STO-3G basis set were 1.559 Å, 1.557 Å, 1.089 Å, and 123.2°, respectively.[17] The calculated C—C bond lengths are in reasonable agreement with the experimental values for the corresponding bonds in pentaprismanecarboxylic acid (**13**), and the calculated C(6)C(1)H bond angle is similar to that of the C(6)C(1)C(O_2H) angle in **13** (124.7°). The heat of formation of **12** was calculated to be 119.6 kcal mol^{-1} by two different 6-31G* (RMP2) homodesmic cycles[17] or 121.2 kcal mol^{-1} using the method of Wiberg.[15] The calculated strain energy of **12** was 140.1 (ref. 17) or 140.2 (ref. 15) kcal mol^{-1}, the lowest of all the [n]prismanes, $n = 3, \ldots, 9$.

Earlier, ab initio calculations were performed to model **12** using the 3-21G basis set on eclipsed ethane with two HCH angles constrained to 108° and the four CCH angles that involve these hydrogens constrained to 90°.[14] The remaining CCH angles, equivalent to C(6)C(1)H in **12** were calculated to be 122.5°. Calculations on **12** itself, using the 3-21G basis set, gave the following parameters: C(1)—C(6), 1.582 Å; C(1)—C(2), 1.561 Å; C—H, 1.077 Å; C(6)C(1)H, 123.24°; $\Delta H_f = 115.4$ kcal mol^{-1} (ref. 15). AM1 calculations gave the following parameters: C(1)—C(6), 1.593 Å; C(1)—C(2), 1.546 Å; C—H, 1.094 Å; C(6)C(1)H, 122.4°; C(2)C(1)H, 119.8°; $\Delta H_f = 117.7$ kcal mol^{-1}; ionization potential = 10.2 eV; zero-point energy = 111.9 kcal mol^{-1}; lowest frequencies, 549, 550, 623 cm^{-1}.[16] The first eight of these parameters calculated by the MNDO method gave the following values: 1.580 Å, 1.559 Å, 1.091 Å, 123.1°, 119.5°, 72.2 kcal mol^{-1}, 10.7 eV, 115.9 kcal mol^{-1}, respectively.[16] MM2 calculations gave the following bond lengths and angles:

C(1)—C(6), 1.566 Å; C(1)—C(2), 1.540 Å; C(6)C(1)H, 116.6°; C(2)C(1)H, 121.7°.[14] This calculation reverses the relative magnitude of the two CCH angles from all the other calculations. The MM2 ΔH_f was 114.7 kcal mol^{-1},[14] and the strain energy was 136.8–143.6 kcal mol^{-1}.[13,14,18]

Attempts at photochemical conversion of hypostrophene (**14**) to pentaprismane (**12**) have been unsuccessful.[18,19] Empirical force field calculations indicate an increase in strain energy of 65.4 kcal mol^{-1} for the hypothetical **14** to **12** conversion.[19] However, the calculated increase in strain energy for conversion of the diene **15** to **12** is only 20.0 kcal mol^{-1}; therefore the photochemical conversion of **15** to **12** may be possible.

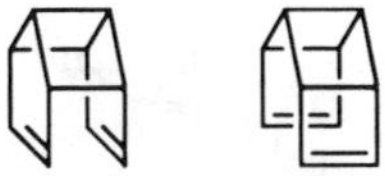

14 **15**

As part of a broader study, the intramolecular hydrogen atom abstraction in the 1,3′-bishomocuboxy radical **16** to give **18** was examined by force field calculations.[20] A force field was developed to model transition states of hydrogen atom abstraction by alkoxy radicals. The MM2 force field was modified to incorporate force constants for atoms involved in the bonding changes. A C(10)—O distance of 2.93 Å and a steric energy of 103.3 kcal mol^{-1} was calculated for **16** whereas the corresponding values for the transition state **17** were 2.41 Å and 107.8 kcal mol^{-1}, respectively. With a correction of 2 kcal mol^{-1}, the calculated activation energy for the conversion of **16** to **17** was 6.5 kcal mol^{-1}, which translated to a rate constant of 6.0×10^6 s^{-1}. This rate was a factor of 5×10^4 slower than that of the related reaction of the less-constrained compound **19**. The formation of a cyclic array in the transition state **17** induces considerable strain.

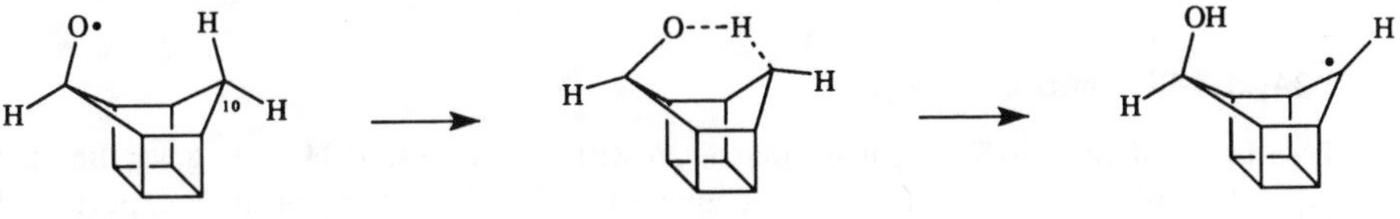

O• H H 10 H → O---H H H → OH H H •

16 **17** **18**

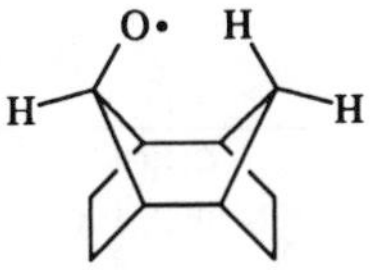

O• H H H

19

Ab initio methods (STO-3G basis set) were used to calculate the geometry of excited triplet states of the 1,3′-bishomocubanones **20** and **21**.[21] Two low-energy structures (**22** and **23**, **24** and **25**) were located for each, syn and anti nonplanar carbonyl forms.

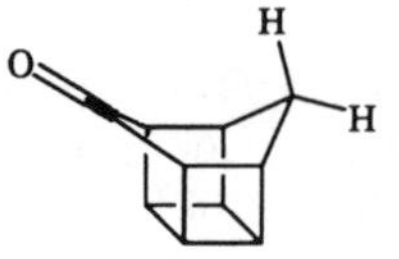

20 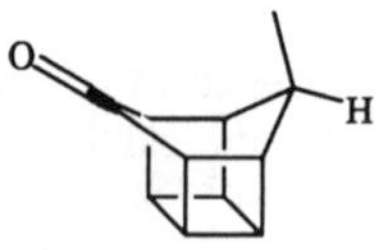 **21** 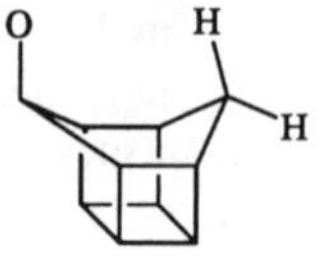 **22**

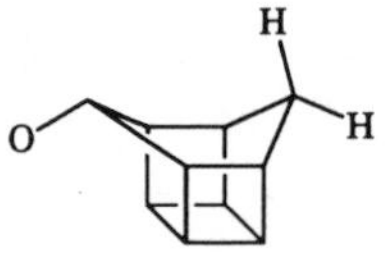

23 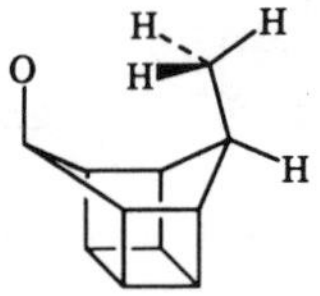 **24** 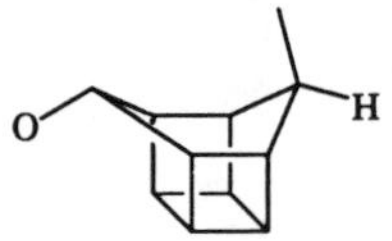 **25**

The calculated C—O out-of-plane angles were as follows: **22**, 37.3°; **23**, 41°; **19**, 31.2°; **25**, 42.9°. The near and far O—H(methyl) distances were 2.08 and 3.64 Å, respectively. MM2 parameters were developed that duplicated these four structures (C—O distance 1.40 Å, C—O out-of-plane force constant, $K_b = 0.13$ mdyne Å deg^{-2}). The reduced carbonyl double bond character allowed for pyramidalization. The MM2 calculated C—O out-of-plane angles for **22–25** were 33.2°, 42.8°, 31.0°, and 42.1°, respectively.

9.3. Formation Reactions

9.3A. 1,1-Bishomocubanes

Treatment of the bis(homocubyl)pinacol **26** with concentrated H_2SO_4 gave the ring expanded spirohomocubyl-1,1-bishomocubanone **27**.[11] This is the product expected of a normal pinacol arrangement.

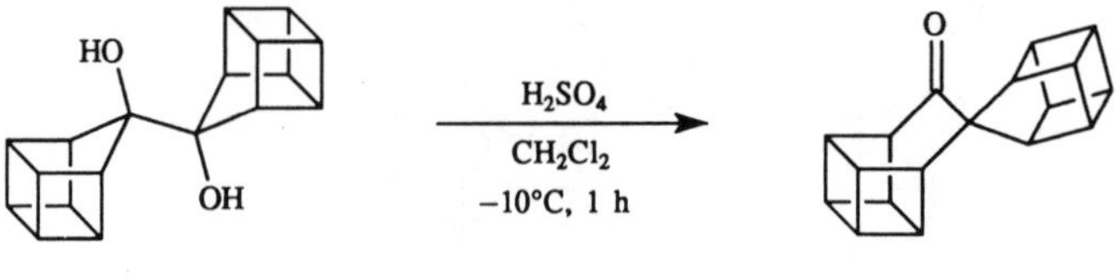

H_2SO_4, CH_2Cl_2, −10°C, 1 h

26 **27** (61%)

The 1,1-bishomocubane **29**, which incorporates the lactone function, was formed on treatment of the secocubane derivative **28** with aqueous base.[22] Likewise, the acid

30 and ester **32** analogs gave the lactone acid **31**. The latter was characterized as the lactone ester **33**.

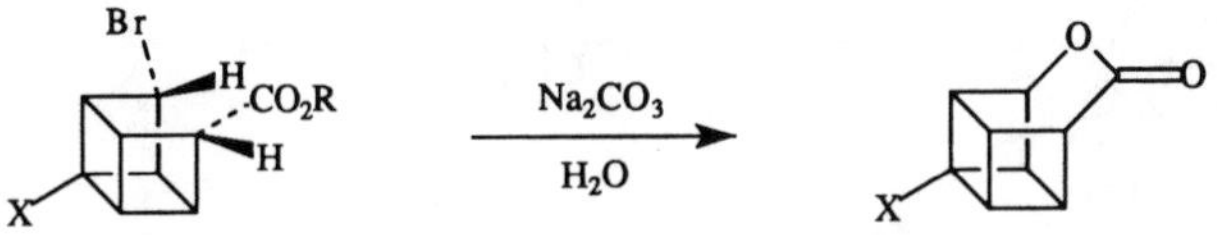

Na_2CO_3, H_2O

28 (R = X = H)
30 (R = H, X = CO_2H)
32 (R = Me, X = CO_2Me)

29 (X = H)
31 (X = CO_2H)
33 (X = CO_2Me)

In a related reaction, the cubanedicarboxylic acid **34** gave the lactone ester **33** on stepwise reactions with HBr, base, acid, and CH_2N_2.

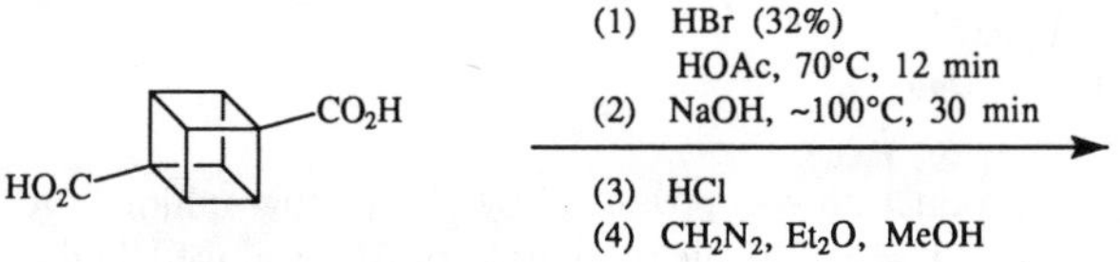

34 → **33** (48%)

(1) HBr (32%) HOAc, 70°C, 12 min
(2) NaOH, ~100°C, 30 min
(3) HCl
(4) CH_2N_2, Et_2O, MeOH

The secocubane **30** and lactone acid **31** are presumably intermediates in this process.

9.3B. 1,2-Bishomocubanes

Very few examples of carbocyclic systems having this skeleton have been reported. Several novel heterocyclic bishomocubyl systems including four compounds that contain the 1,2-bishomocubane skeleton have been reported from the reactions of the phosphaalkyne **35** and the iron sandwich complexes **36** and **48**. Reaction of a 22 molar excess of **35** with **36** gave the ferratetraphospha-1,2-bishomocubane derivatives **37** and **38** in addition to the iron–phosphorus sandwich complexes **39–42** and a reaction residue **43**.[23]

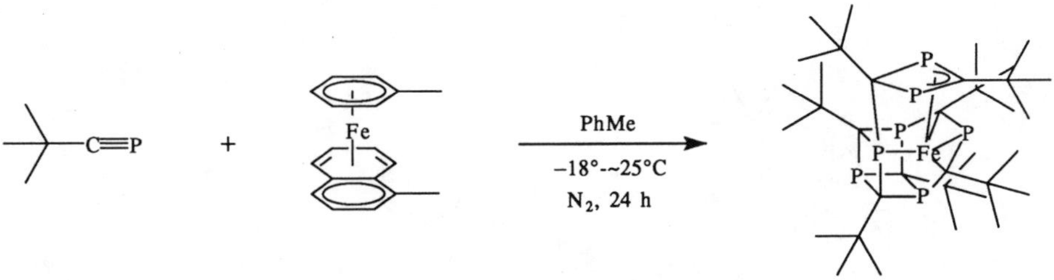

35 + **36** → **37**

PhMe, −18°–25°C, N_2, 24 h

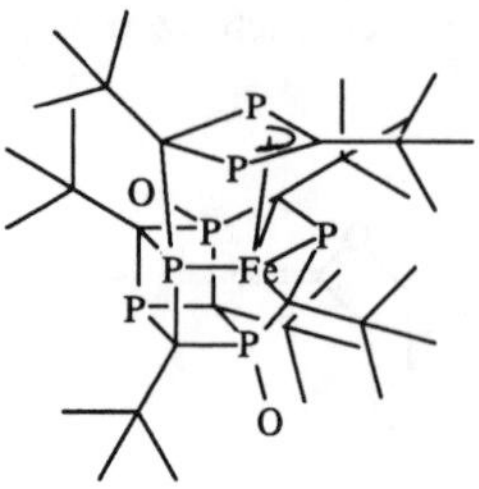

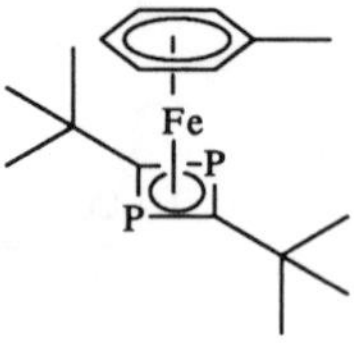

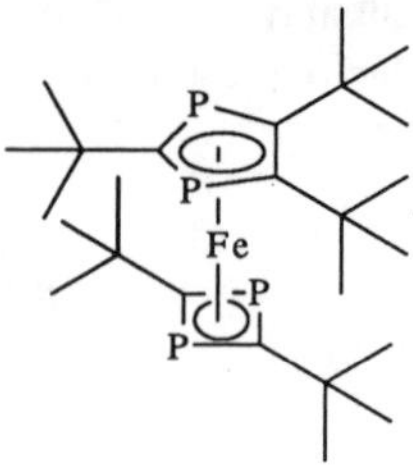

+ 38 (37 + 38, 1%) + 39 + 40

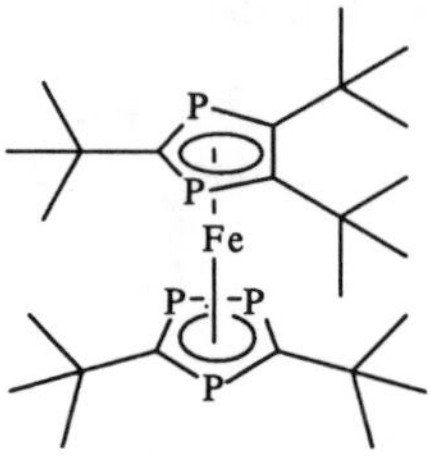

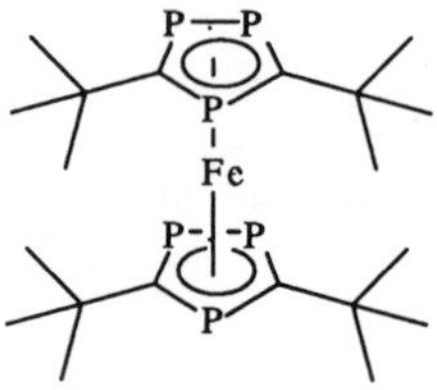

+ 41 + 42 + Residue

43

The partially oxidized product **38** was produced by chromatographing **37** on alumina that contained 5% water. Use of smaller amounts of **35** produced higher yields of **39–42**. Heating the residue **43** at 100°C under vacuum gave low yields of the pentaphospha-1,2-bishomocubane **44** and the hexaphospha-1,3′-bishomocubane **45** in addition to several other products. All of the bishomocubanes **37**, **38**, **44**, and **45** are greater than pentacyclic because of the extra bridging bonds or atoms.

Residue (43) —100°C, 1 Pa→ 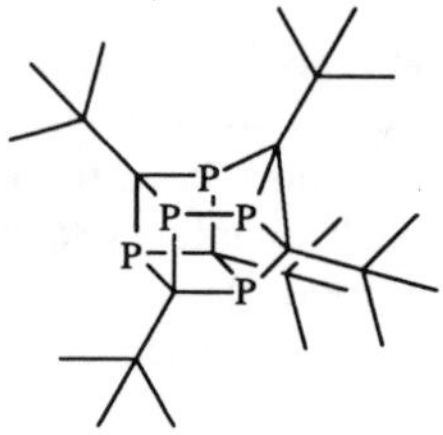 **44** (~1%) + **45** (~1%)

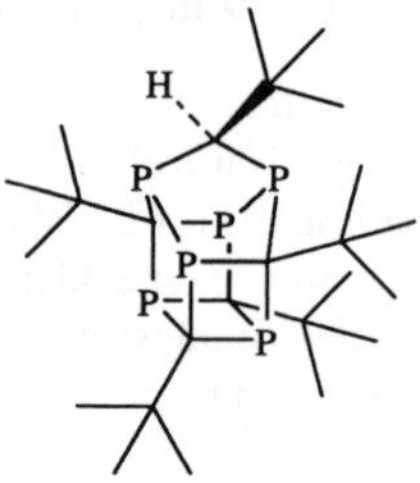

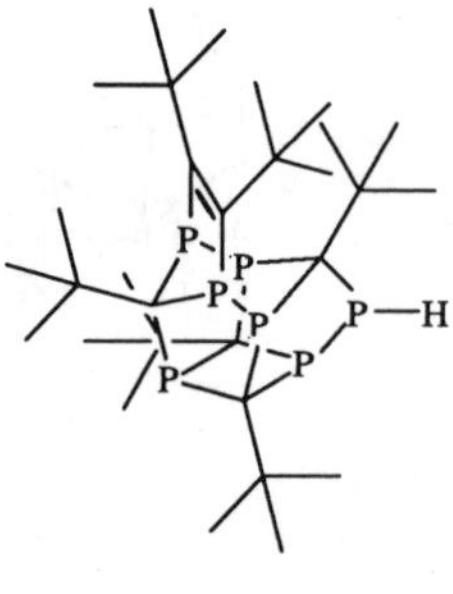

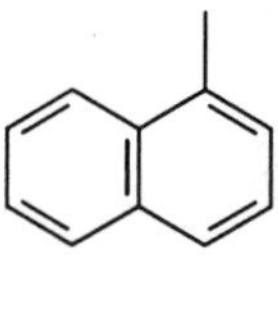

+ **46** (~1%) + **47** + 38 + 39

The closely related reaction of excess phosphaalkyne **35** with the complexes **36** and **48** at 0– ~ 25°C gave a small quantity of the ferrahexaphospha-1, 3′-bishomocubane **49** and several other polycyclic compounds.[24]

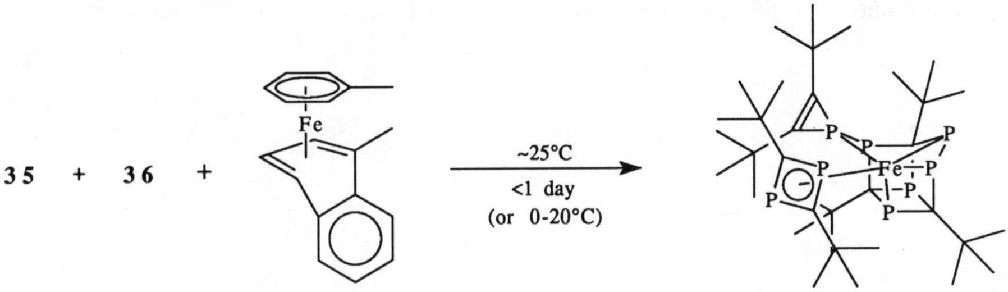

35 + 36 + 48 —(~25°C, <1 day (or 0-20°C))→ 49

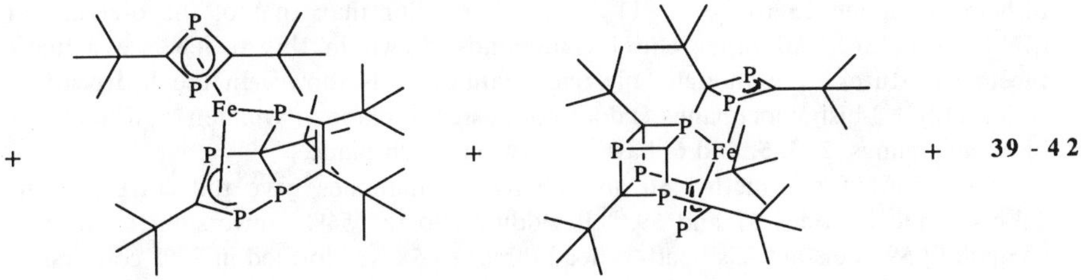

\+ 50 + 51 + 39 - 42

Heating this reaction mixture at 100°C under vacuum gave the previously mentioned bishomocubanes **44**, **45**, and **49**, the hexaphospha-1,2-bishomocubane **52** and several other products.

35 + 36 + 48 —(0-20°C)→ —(100°C, Vacuum)→ 44-46 + 49-51

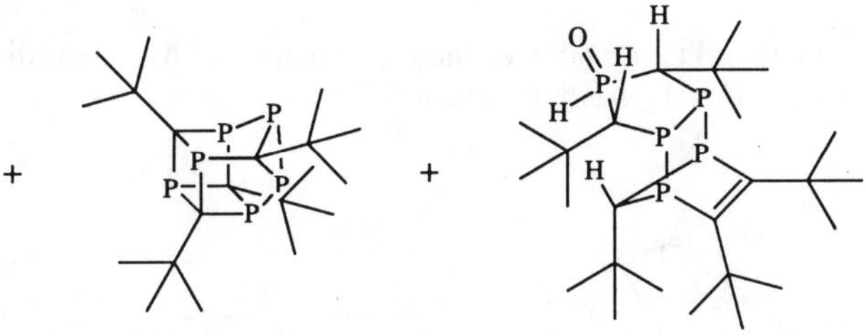

\+ 52 + 53

Both **49** and **52** are also greater than pentacyclic because of the extra bridging bonds.

9.3C. 1,3-Bishomocubanes

The most common method for preparing 1,3-bishomocubanes remains the intramolecular photocycloaddition reaction of *endo*-dicyclopentadiene derivatives. The optically pure dienone (+)-**54** gave the ketone (+)-**55** with no loss of optical purity.[25]

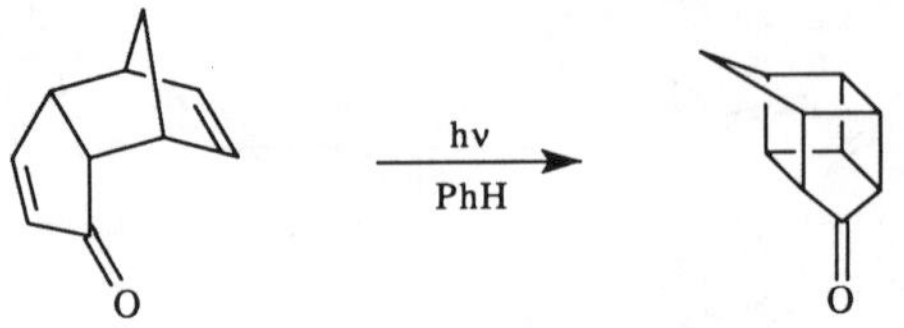

hν, PhH

(+)-**54** (+)-**55** (100%)

The absolute configurations are as shown. The specific rotation of the bishomocubanone **55** ($[\alpha]_D^{23} = +11°$) is much smaller than that of the dienone **54** ($[\alpha]_D^{23} = +141.6°$). All other chiral compounds shown in this review are actually racemic mixtures even though only one enantiomer is shown. In the hydrocarbon series, only 1,3-bishomocubane (**4**) does not possess a plane of symmetry; all the other bishomocubanes, **2**, **3**, **5**, and **6**, have at least one such plane.

Irradiation of the methyl **56** and phenyl **58** dienones gave the corresponding 1,3-bishomocubanones **57** and **59**.[26] In addition to the 54% conversion to the cage compound **59**, a cis-anti-cis head-to-head dimer of **58** was formed in 22% conversion.

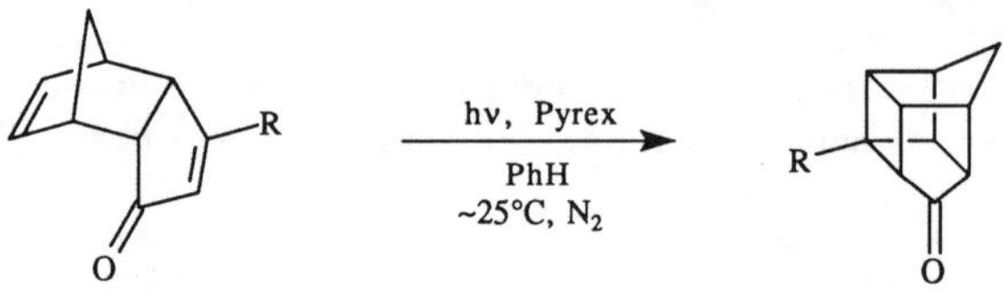

hν, Pyrex; PhH; ~25°C, N_2

56 (R = Me) — 21 h — **57** (R = Me) (65%)
58 (R = Ph) — 6.5 h — **59** (R = Ph) (54%)

The methoxyketone **60** gave the cage compound **61** on irradiation; the reaction in the absence of NH_3 was not reported.[27]

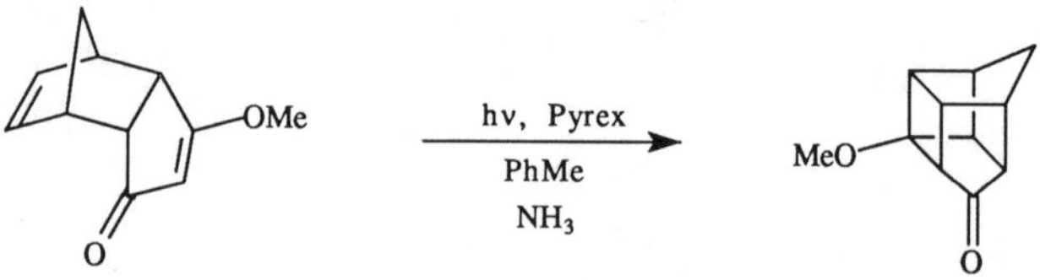

hν, Pyrex; PhMe; NH_3

60 **61** (83%)

Irradiation of the bromomethoxyketone **62**, under conditions where no special precautions were taken to remove traces of acid, gave the ring-opened dione **64** as the major product along with a smaller amount of the expected cage compound **63**.

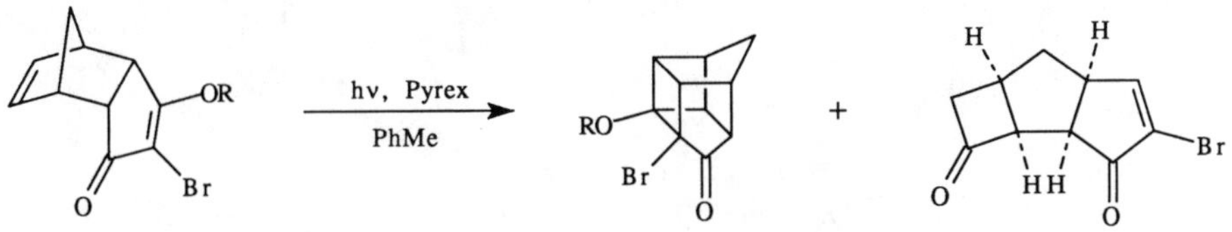
hν, Pyrex
PhMe
OR
Br
O
RO
+
H
HH

62 (R = Me)		**63** (R = Me) (20%)	**64** (60-70%)
65 (R = Ac)	4.5 h	**66** (R = Ac) (61%)	
67 (R = SO_2Me)	~4.5 h	**68** (R = SO_2Me)	

When this reaction was performed in the presence of NH_3, the yield of **63** increased to 73%.

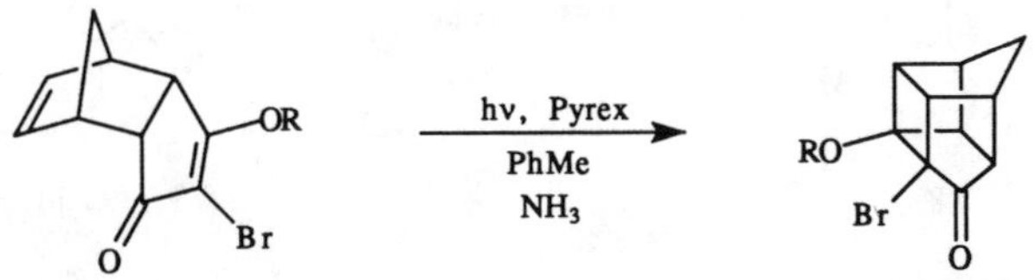
hν, Pyrex
PhMe
NH3
OR
Br
O
RO

62 (R = Me)
69 (R = CH_2OMe)
71 (R = H)

63 (R = Me) (73%)
70 (R = CH_2OMe)

Irradiation of the acetate **65** and methanesulfonate **67** esters in the absence of NH_3 gave the cage compounds **66** and **68**, respectively, as the only reported products. Irradiation of the methoxymethoxybromodieneone **69** in the presence of NH_3 gave the bishomocubane **70**. No bishomocubane was formed on irradiation of the enol **71** in MeOH. The explanation for many of these results follows from the finding that the methoxyketone **63** was converted to the dione **64** by acid. This aspect will be discussed in the section on reactions of 1,3-bishomocubanes.

Irradiation of the variously functionalized dienes **72**, **74**, **76**, and purified **78** gave the 1,3-bishomocubanes **73**,[28] **75**,[29] **77**,[6] and **79**,[27] respectively, mostly in good yields.

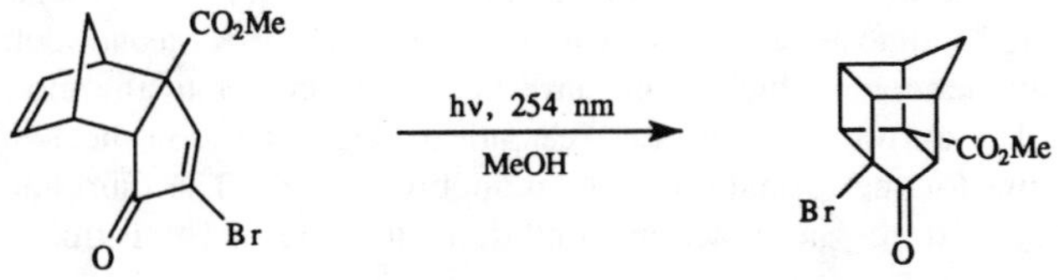
hν, 254 nm
MeOH
CO2Me
Br
O

72

73 (87%)

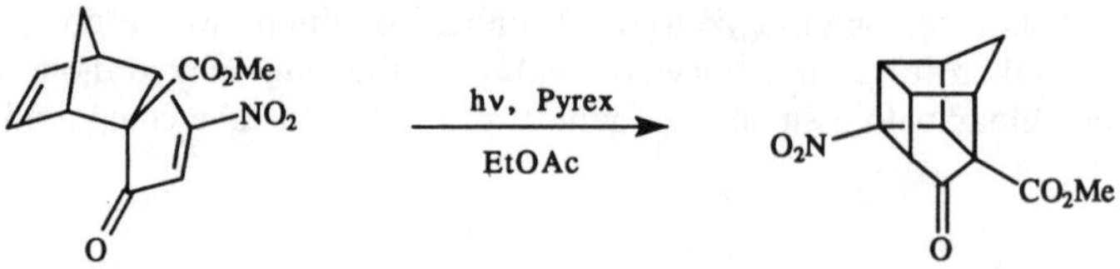

74 → (hν, Pyrex; EtOAc) → 75 (20%)

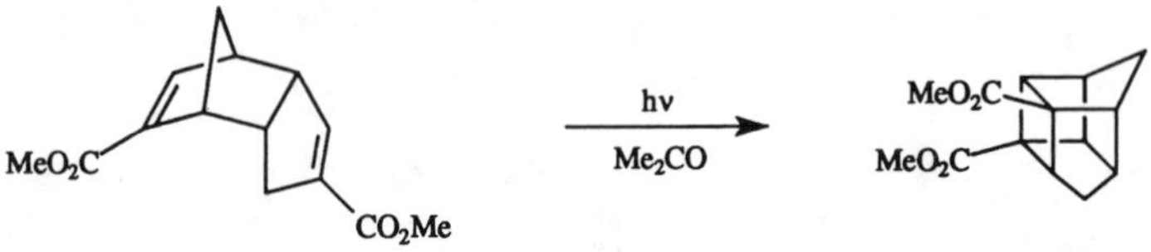

76 → (hν; Me_2CO) → 77 (85%)

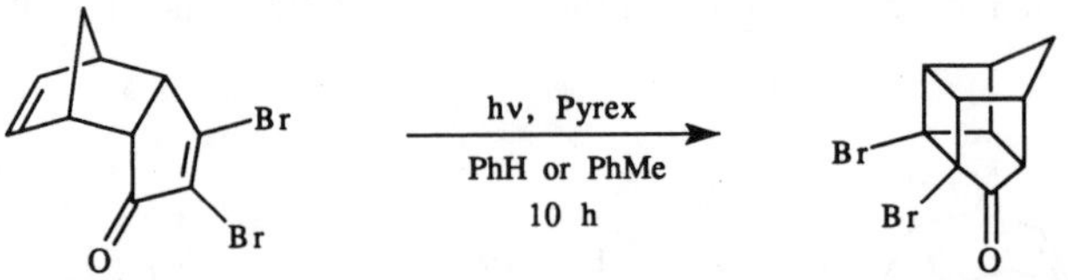

78 → (hν, Pyrex; PhH or PhMe; 10 h) → 79 (95%)

Use of impure dibromide **78**, which contained benzyltriethylammonium bromide (**80**), in PhMe solution gave low yields of both the dibromobishomocubanone **79** and the monobromo analog **81**.[27]

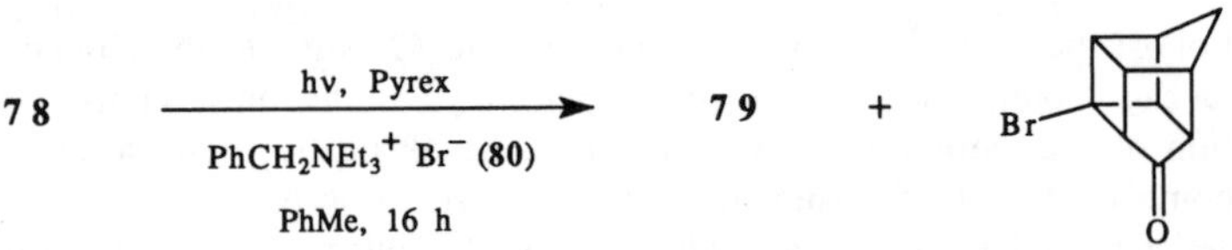

78 → (hν, Pyrex; $PhCH_2NEt_3^+ Br^-$ (80); PhMe, 16 h) → 79 + 81

The same impure material **78** in PhH solution produced a higher yield of the dibromo product **79**, but still also gave the monobromide **81**. Addition of a small amount of the quaternary ammonium salt **80** to the dienone **78** in PhH and irradiation gave the monobromide **81** almost exclusively. The dibromodienone **78** was not debrominated to a monobromodieneone, which could have been a precursor to the monobromo cage compound. The authors proposed a free radical debromination mechanism (Scheme 9.1) to account for the formation of the monobromide **81**. The dibromide **79** was not actually subjected to the reaction conditions to determine if this mechanism is possible.

Irradiation of the dienone complex **83** gave a ~36:64 photostationary state mixture of the isomer **84** and **83**[30]; none of the cage complex **85** was formed.

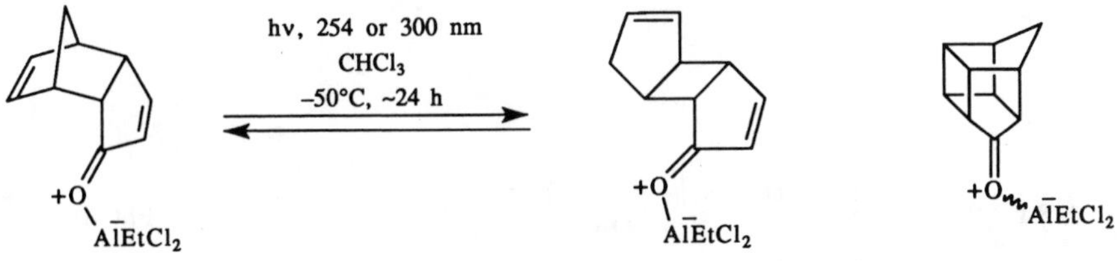

hv, 254 or 300 nm
$CHCl_3$
−50°C, ~24 h

83 **84** **85**

The latter was formed from racemic **55** and $(AlEtCl_2)_2$. No reaction occurred on irradiation of the cage complex **85** under the conditions indicated for **83**. Thus **85** is not an intermediate in the interconversion of **83** and **84**. Irradiation of uncomplexed racemic dienone **54**, presumably in $CHCl_3$ solution, at 254 nm did give racemic **55**, presumably via a triplet state of **54**. The conversion of the complex **83** to **84** may occur via the π,π^* singlet state of **83**.

The novel hexaphospha-1,3-bishomocubane **87** was formed on careful acidification of the triphospha anion **86**.[31]. The suggested mechanism for this reaction involved 2 + 4 dimerization of the triphosphacyclopentadiene **88** to give the hexaphospha-*endo*-dicyclopentadiene **89** and nonphotochemical intramolecular 2 + 2 cycloaddition.

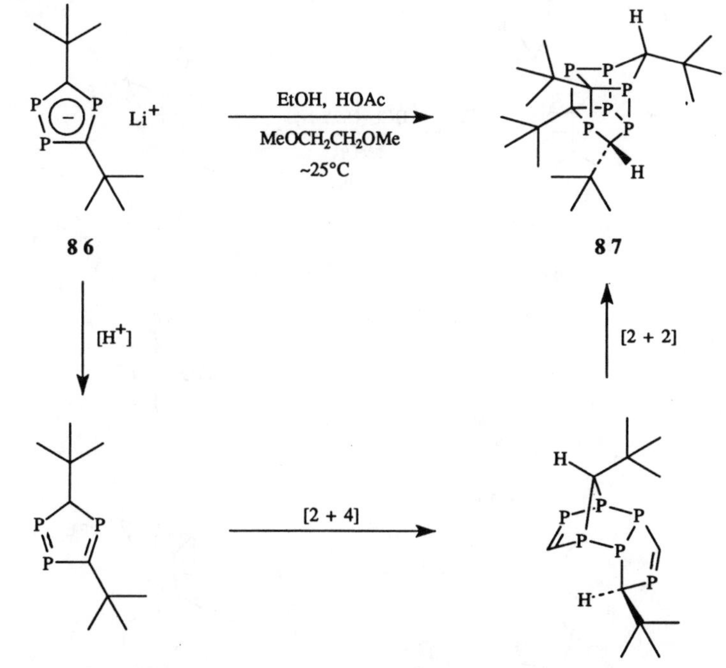

EtOH, HOAc
$MeOCH_2CH_2OMe$
~25°C

86 **87**

$[H^+]$ [2 + 2]

[2 + 4]

88 **89**

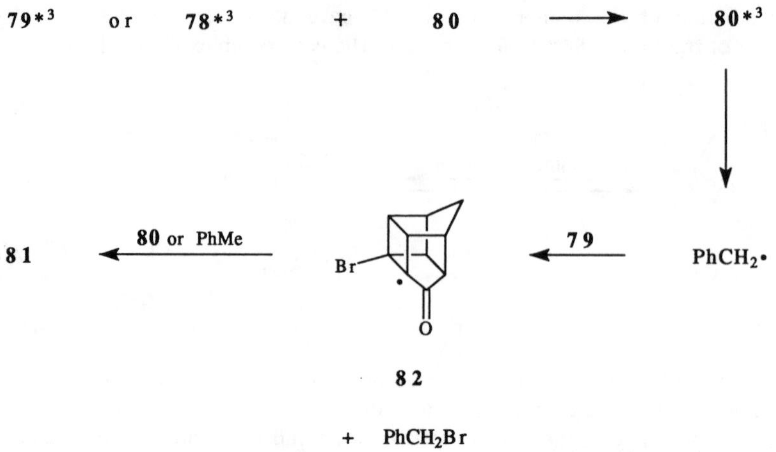

79*3 or 78*3 + 80 → 80*3

80*3 → PhCH2•

PhCH2• + 79 → 82 + PhCH2Br

82 → (80 or PhMe) → 81

Scheme 9.1 ■

Another route to the 1,3-bishomocubyl system involves ring expansion of a homocubyl system. Treatment of the homocubylbenzhydrol **90** with methanolic HCl gave the 1,3-bishomocubyl ether **91**, in high yield, presumably via the cations **92** and **93**.[32]

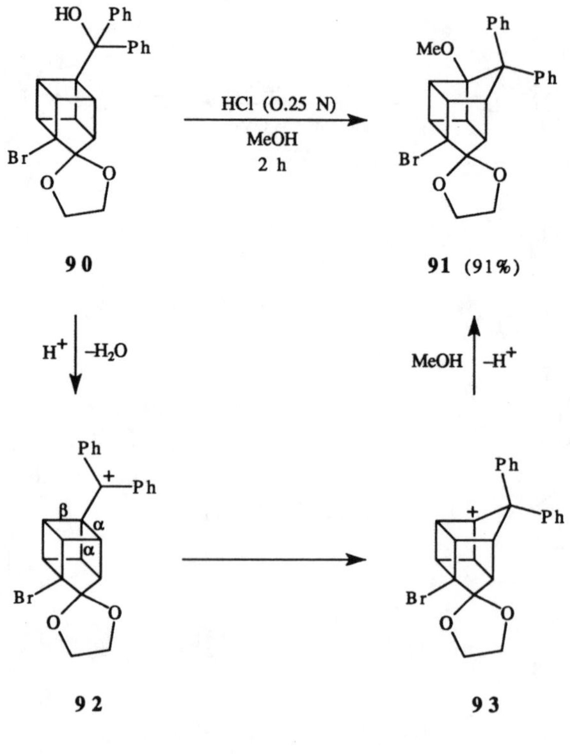

90 → (HCl (0.25 N), MeOH, 2 h) → 91 (91%)

90 → (H^+, $-H_2O$) → 92 → 93 → (MeOH, $-H^+$) → 91

The bond calculated to be more highly strained, α, migrated in preference to the β bond, which could have produced a 1,4-bishomocubyl product. Earlier calculations had indicated that the 1,4-bishomocubyl skeleton is more strained than the 1,3 system. That the ring expansion occurs at all is an indication of the greater strain in the homocubyl skeleton than in the bishomocubyl system.

Ring contractions in trishomocubyl systems can also form 1,3-bishomocubyl products. Several studies of quasi-Favorskii reactions of α-haloketones with strong base have presumably produced bishomocubyl derivatives on the way to homocubyl and cubyl products. A sequential ring-contraction approach with the trishomocubanedione **94** gave the homocubanediacid **95**, presumably via the 1,3-bishomocubanone **96**.[33,34]

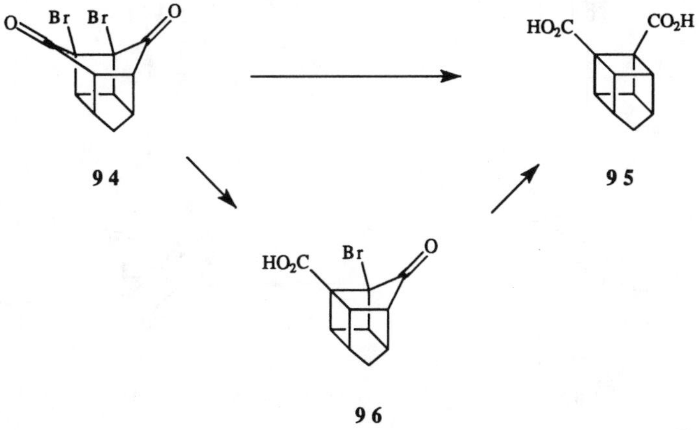

O Br Br O

HO₂C CO₂H

94 **95**

HO₂C Br O

96

Attempted dual ring contractions gave only ring-opened products. Similarly the trishomocubane **97** was converted to the cubane derivative **98**,[35,34] presumably via the initial ring-contraction product **99** and several other intermediates.

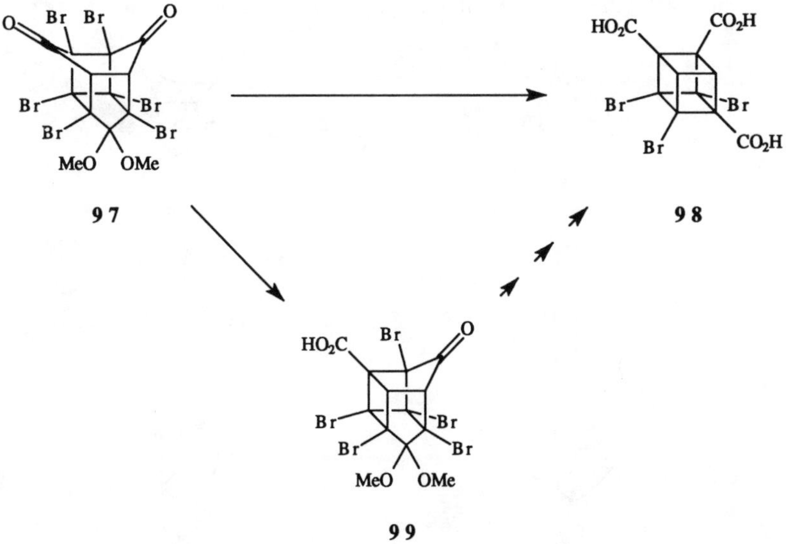

O Br Br O

Br Br Br Br MeO OMe

HO₂C CO₂H Br Br Br CO₂H

97 **98**

HO₂C Br O

Br Br Br Br MeO OMe

99

In the bromination of the diene **14**, rearrangement of the cation **100** to **101** was considered likely to have been a concerted process because no intermediates were intercepted.[36] If the rearrangement was not completely concerted, then cations **102**, a 1,3-bishomocubyl system, and **103** may have had fleeting existence.

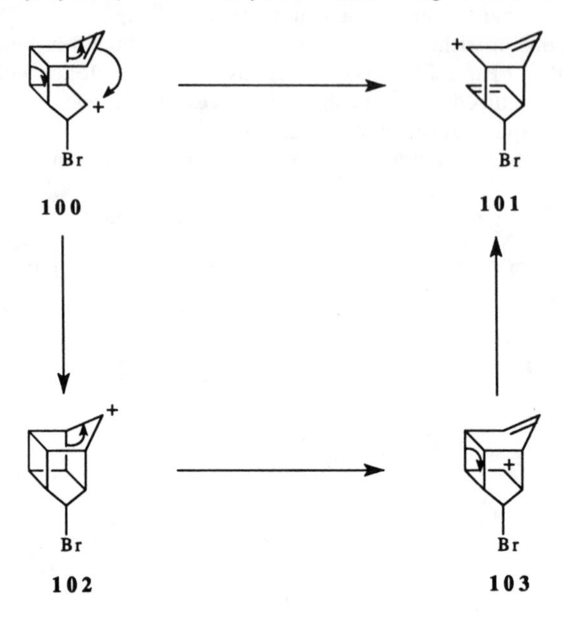

Likewise in the acetolysis of the tosylate **104**, an analogous rearrangement of cation **106** was proposed to lead to *endo*-dicyclopentadienyl acetate **105**.[36] In this case, if the cation **107** is an intermediate, it cannot be exactly the same (e.g., different ion pair) as one of the intermediates formed in the acetolysis of the corresponding 1,3-bishomocubyl tosylates because different products are formed.[1]

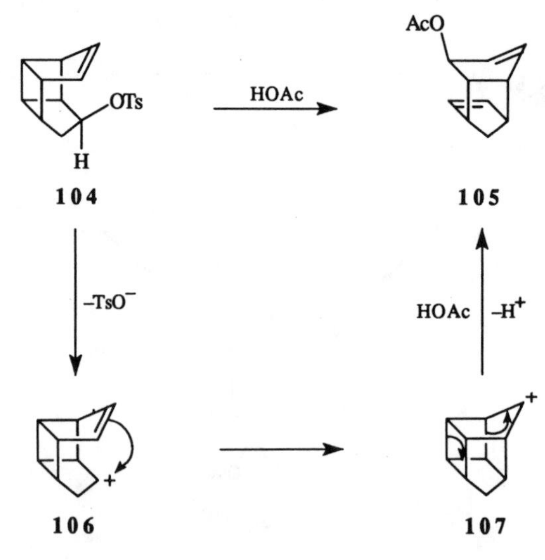

9.3D. 1,3′-Bishomocubanes

Intramolecular photocycloaddition has also been applied to the synthesis of the 1,3′-bishomocubane system. Irradiation of the diene **108** gave the homopentaprismane **109**, a methylene-bridged 1,3′-bishomocubane, in good yield.[18]

MeO OMe
hv, Vycor
Me_2CO
N_2, 45 min
MeO OMe

108 **109** (92%)

The same yield was obtained when xanthone was used as a photosensitizer in benzene solution with a Pyrex filter. As noted previously, when the methylene bridge in **108** is replaced by single bond **14**, the intramolecular photocycloaddition was unsuccessful.

In a ring-contraction reaction similar to the quasi-Favorskii noted previously, the α-tosyloxyketone **110** reacted with strong base to give the bridged 1,3′-bishomocubyl acid **111** in good yield.[37]

OTs
O
KOH (20%)
H_2O
Reflux, 7 h
CO_2H

110 **111** (94%)

Other ring contractions involved elimination of nitrogen from the azo compounds **112** and **115** to give the diesters with the same skeleton **113** and **116**.[38]

RCO_2 N=N Ts
hv, Pyrex
PhH, Ar
RCO_2 Ts
AcO Ts

112 (R = Me) 50 min **113** (R = Me) (52%) **114** (14%)
115 (R = Ph) 1.5 h **116** (R = Ph) (71%)

In addition to the ring-contracted acetate **113**, the ring-opened acetate **114** was also formed from **112**. The diene **114** was not an intermediate in the formation of the cage **113** as shown by the observation that the latter was not formed on irradiation of the former.

Flash pyrolysis of the diazirine **117** gave a mixture of the olefin **118** (~ 105 mg from 200 mg **117**) and the bridged trishomocubane **119**.[39] The suggested mechanism involved elimination of N_2 to give the carbene **120** followed by addition to the double bond to give the dibridged 1,3′-bishomocubane **121**. Addition of two hydrogen atoms (possibly from **118**) to **121**, either directly or via the diradical **122** would account for the formation of **119**. The trishomocubane **119** was also formed on pyrolysis of the dry tosylhydrazone salt **123**.[39]

Flash pyrolysis
310°C
N_2, 100 Pa
117 **118** + **119** (15%)
$-N_2$
RH
120 → **121** → **122** → Polymer

123 200°C / 0.01 Pa → **119** (5%) + Polymer + Olefins

The formation of the heterocyclic 1,3′-bishomocubane derivatives **45** and **49** has been discussed in a previous section on 1,2-bishomocubanes.

9.3E. 1,4-Bishomocubanes

1,4-Bishomocubane (**6**) has been suggested to be a potential energy minimum for the $(CH)_8(CH_2)_2$ valence isomers despite the presence of two four-membered rings.[9] The possibility of isomerization of *endo*-dicyclopentadiene (**124**) to **6** has been speculated on.[9] 1,4-Bishomocubane (**6**) has the highest degree of symmetry (D_{2h} point group, not C_{2h}[40]) of all the bishomocubanes.

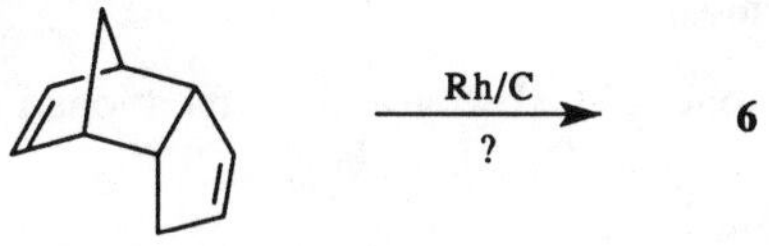

Rh/C
?
6

124

Treatment of the bromohomocubyl acetate **125** with methanolic NaOMe gave the oxa-1,4-bishomocubane **126**.[27] The mechanism proposed for this transformation involved attack by methoxide, elimination of methyl acetate, and endo protonation to give the bromoketone **129**. Attack by methoxide and displacement of bromide gives the ketal **126**.

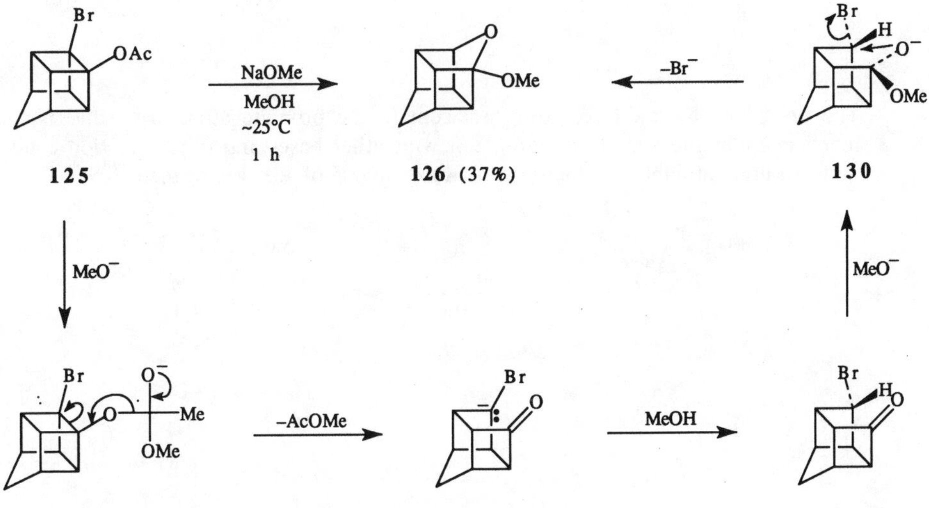

Br OAc
NaOMe
MeOH
~25°C
1 h
O OMe
$-Br^-$
Br H O^- OMe

125 **126** (37%) **130**

MeO^- MeO^-

Br O Me OMe
–AcOMe
Br O
MeOH
Br H O

127 **128** **129**

Addition of the bis(silaphosphine) **131** to the bis(chloroborane) **132** gave the novel tetrabora-tetraphospha-1,4-bishomocubane **133**.[41]

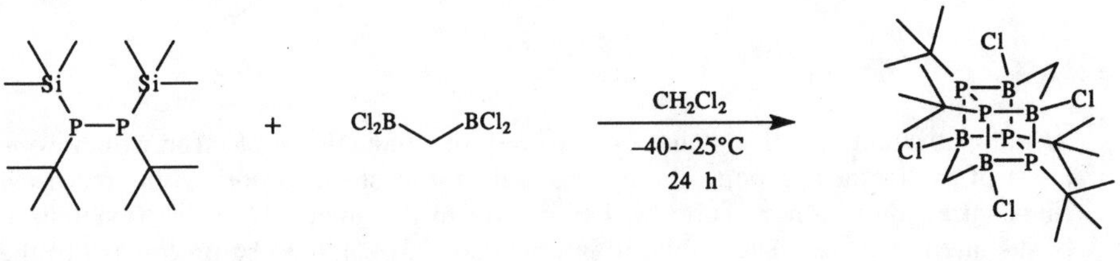

Si Si P—P
+
Cl_2B BCl_2
CH_2Cl_2
–40–~25°C
24 h
Cl P B Cl P B Cl B P B P Cl

131 **132** **133** (42%)

9.4. Reactions

9.4A. 1,1-Bishomocubanes

Esterification of the lactone acid **31** to give **33** is mentioned in an earlier section.

9.4B. 1,3-Bishomocubanes

The quasi-Favorskii reaction of α-halo ketones, mentioned previously, is probably the most extensively studied reaction of this series of compounds because of its usefulness in transforming these compounds to other novel, more highly strained products. The simplest example **134** reacts smoothly to give homocubanecarboxylic acid **135**.[27]

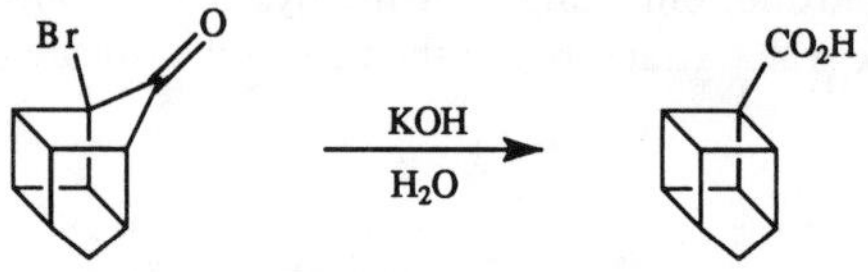

Br O KOH H_2O CO_2H

134 **135**

The presence of a methoxy group adjacent to the bromine atom in **63** diverts the reaction under the conditions shown, or with other bases and other reaction conditions, to unidentifiable products; no methoxy analog of **135** was formed.[27]

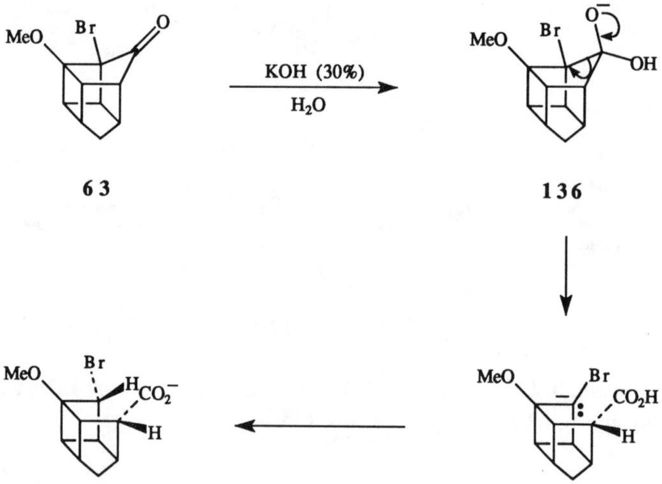

MeO Br O KOH (30%) H_2O MeO Br O^- OH

63 **136**

MeO Br H CO_2^- H MeO Br CO_2H H

138 **137**

A rationalization for this behavior involved the inductively electron-withdrawing effect of the methoxy group. The intermediate **136**, instead of undergoing the usual ring-contraction to a homocubyl system, cleaves to the anion **137**, which is stabilized by the methoxy group. The stabilized intermediate **137** can then be protonated to the product **138** that was suggested to undergo further cage degradation.

Likewise the α-halo-1,3-bishomocubanones **68** and **70** with inductively electron-withdrawing groups adjacent to the bromine atoms gave only decomposition products on treatment with aqueous base.[27] A low yield of the desired homocubanecarboxylic acid **139** was obtained on reaction of **70** in PhMe at reflux with the removal of water.

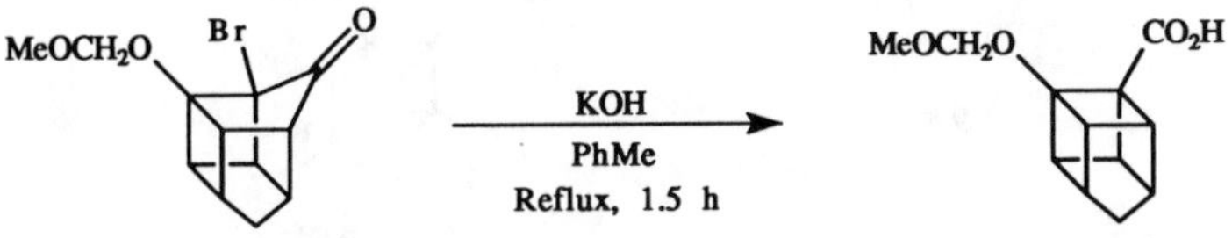

MeOCH$_2$O Br O → KOH, PhMe, Reflux, 1.5 h → MeOCH$_2$O CO_2H

70 **139** (28%)

The formation of the diacid **95** from the presumed intermediate **96** was mentioned previously. The isomeric diacid **140**, the isomeric bromo acids **141** and **145**, and the chloro acid **143** were formed by analogous reactions of the α-bromoketones **73**,[28] **79**,[27] **144**,[28] and **123**,[28] respectively.

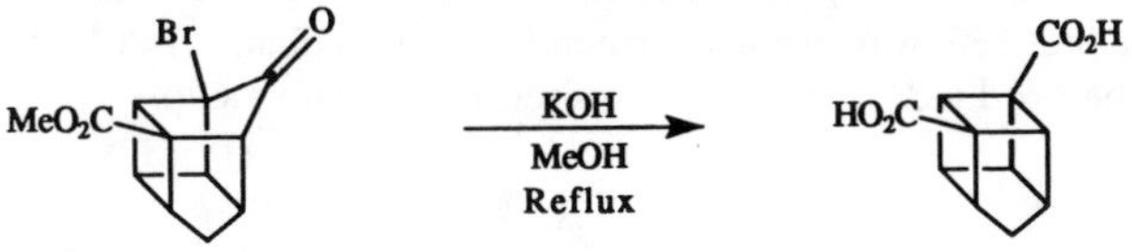

MeO$_2$C Br O → KOH, MeOH, Reflux → HO$_2$C CO_2H

73 **140** (76%)

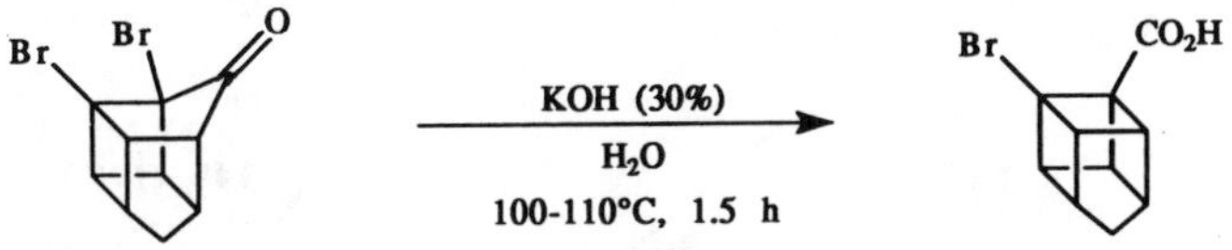

Br Br O → KOH (30%), H_2O, 100-110°C, 1.5 h → Br CO_2H

79 **141** (67%)

X Br O → Strong alkali → X CO_2H

142 (X = Cl)
144 (X = Br)

143 (X = Cl) (66%)
145 (X = Br) (44%)

In the reaction sequence of the pentabromo ketone **99** to the tribromo triacid **98**, noted previously, the tetrabromo diacid **146** is presumably one of the intermediates.[34]

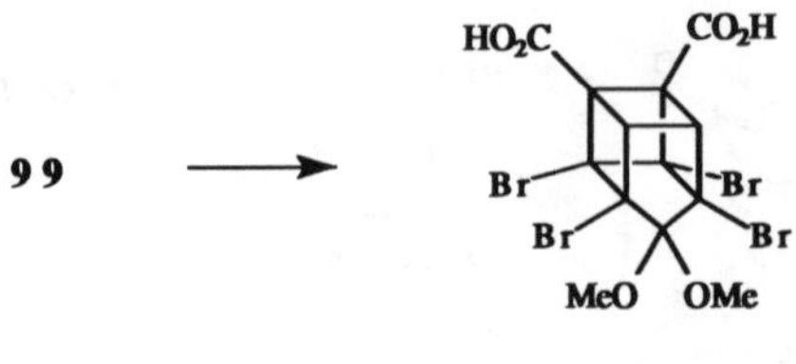
99
HO2C
CO2H
Br
Br
Br
Br
MeO
OMe
146

The perbromodiketone **147** gave the cubane diacid **148** on reaction with aqueous base,[42,43] followed by careful acidification.[44] Interestingly, **148** was the only isomer formed,[43] implying that only the carbonyl bond that is nearest to the second carbonyl of **147** or the carboxyl group of **149** breaks. Occasionally under these conditions the cyclooctatetraene **150** was formed instead of the cubane **148**.[42,44] This product appeared to be produced during the required acidification step.

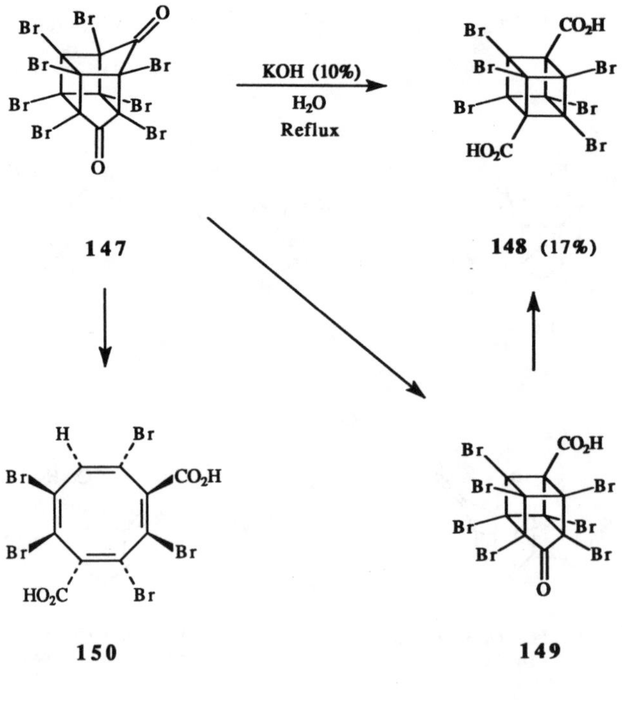
KOH (10%)
H2O
Reflux
147
148 (17%)
150
149
CO2H
HO2C
O
H

Skeletal rearrangements also occurred on treatment of the methyl **57** and phenyl **59** ketones with BF_3–Et_2O.[26] Formation of the dicyclopropyl products **151** and **152** was rationalized by cationic rearrangements of the intermediates **153–155**.

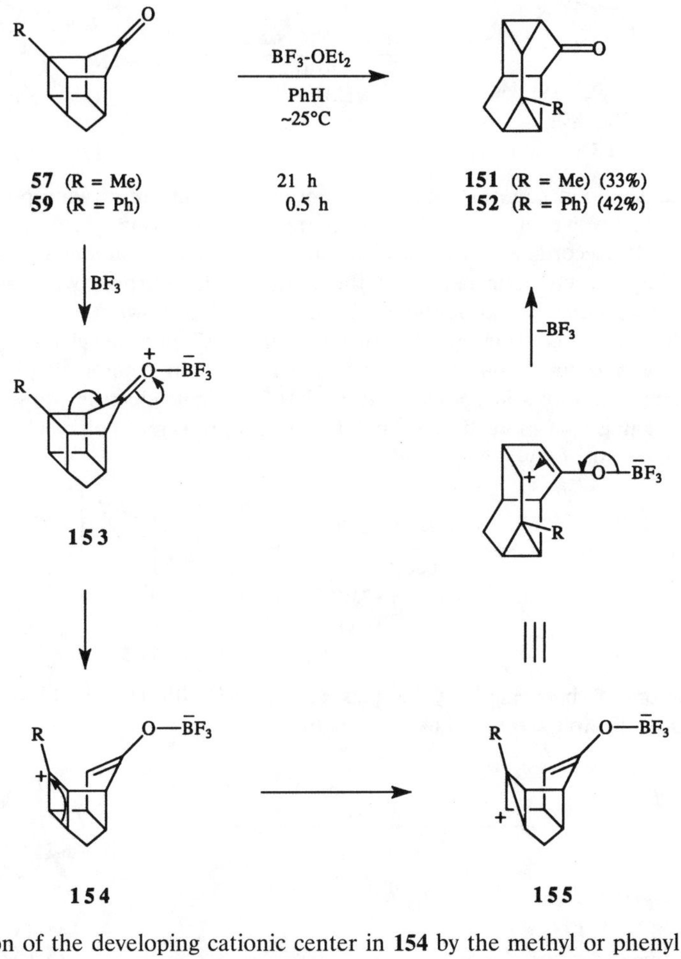

BF_3-OEt_2, PhH, ~25°C

57 (R = Me) — 21 h — **151** (R = Me) (33%)
59 (R = Ph) — 0.5 h — **152** (R = Ph) (42%)

BF_3

153

$-BF_3$

154

155

Stabilization of the developing cationic center in **154** by the methyl or phenyl groups in the probable rate-determining step was necessary, as indicated by the absence of a similar rearrangement of the unsubstituted racemic ketone **55** when subjected to the same reaction conditions for 55 h. MM2 calculations on **4** (*vide supra*) and the unsubstituted hydrocarbon having the skeleton of **151** and **152** indicated a decrease in strain energy of 6.57 kcal mol^{-1} for this process. Thus, the driving force for the conversion of **57** and **59** to **151** and **152** may be explained by thermodynamic control. Rearrangement of the phenyl derivative **59** to **152** was also accomplished by $AgClO_4$.

59 $\xrightarrow{AgClO_4}$ (PhH, dark, ~25°C, 5 h) **152** (26%)

Although no skeletal rearrangement was apparent, the transformation of the dinitro-1,3-bishomocubyl ester **156** to the dinitro ester ether **157** was proposed to occur via a novel skeletal rearrangement.[29]

O_2N CO_2Me H NO_2 **156** — $K_3Fe(CN)_6$, $NaNO_2$; NaOH; MeOH, H_2O → MeO_2C O_2N O_2N H OMe **157** (57%)

A mechanism to account for the apparent changes in substituent positions was suggested to involve proton and electron transfers, and ring opening and closure (Scheme 9.2). According to this mechanism, if the single enantiomer of **156** shown (disregarding the epimeric nature of the nitromethylene group) were used in this reaction, the product would be the enantiomer of **157** as drawn.

Another proposed ring-opening reaction of a 1,3-bishomocubane, the parent hydrocarbon **4**, occurred on dissolution in superacid.[45] No stable ionic product could be detected. The formation of cation-radical **165** by single electron transfer followed by ring opening to a more flexible, kinetically less-protected, polycycle that decomposed or polymerized was suggested.

4 — SbF_5 or FSO_3H/SbF_5; SO_2ClF; −130°C or −78°C → **165**

Treatment of both methoxy ketones **61** and **63** with HCl in PhMe gave the ring-opened diketones **166** and **64** after hydrolysis.[27]

61 (X = H), **63** (X = Br) — HCl (g); PhMe → **166** (X = H) (45%), **64** (X = Br)

61/**63** → **167** → **168** → **169** (X = H), **170** (X = Br) — H_2O → **166**/**64**

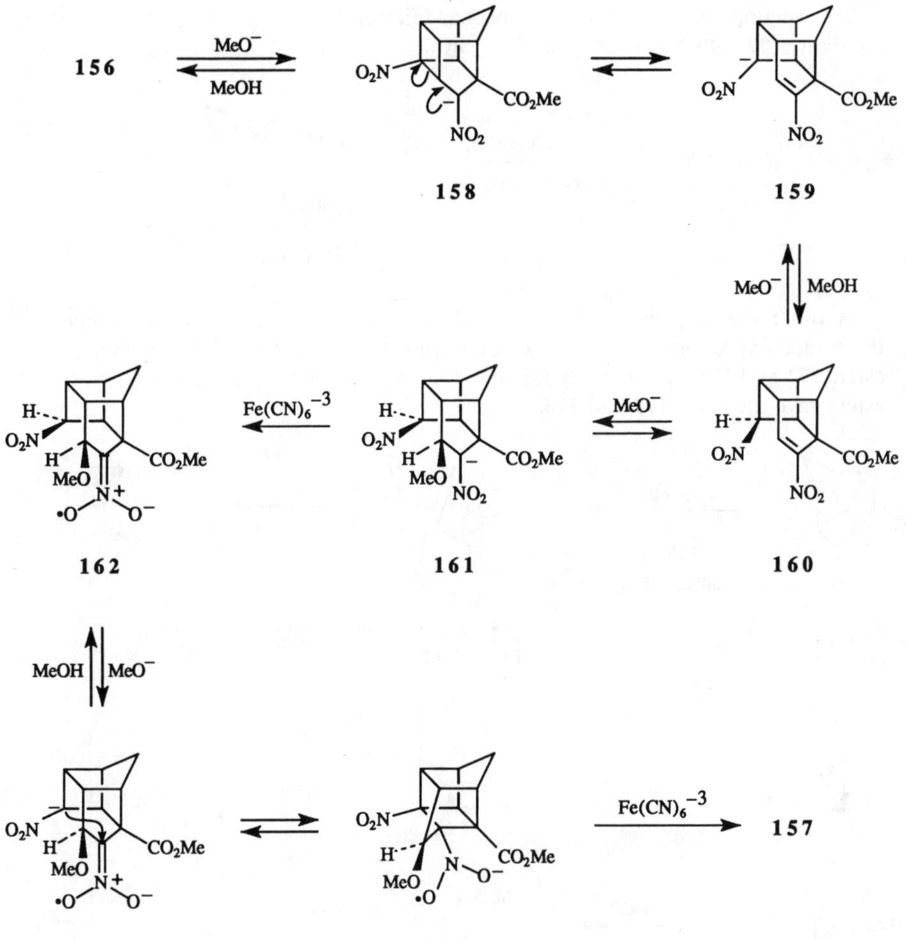

Scheme 9.2 ■

None of the related bromo ketones **66**, **68**, or **70** gave a derivative of **64** on treatment with acid. Thus the electron-releasing, by resonance, effect of the methoxy groups was required for this reaction to occur. The initial ring opening of **167** to **168** was accompanied by release of strain and was considered to be the rate-determining step whereas the proton loss and further ring opening of **168** to **169** or **170** was thought to be very fast. The enol ether **170** was detected on treatment of the bromo ketone **63** with HCl in $CDCl_3$. Thus the overall process of converting the dienones **60** and **62** to **169** ad **170**, respectively, via the 1,3-bishomocubanones **61** and **63**, respectively, can be formulated as two-step metathesis reactions.

The ring-opened *seco*-1,3-bishomocubanediester **171** was formed on reaction of the diester **77** with Na–K alloy and Me_3SiCl.[6]

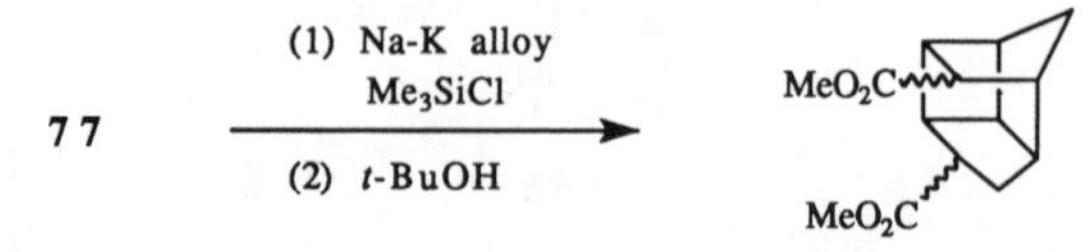

(1) Na-K alloy Me_3SiCl (2) *t*-BuOH

171 (20%)

Another rearrangement of the 1,3-bishomocubyl system occurred on reaction of the β-methoxy ketones **61** and **63** with aqueous $AgNO_3$ and KOH.[27] The cyclopropyl esters **172** and **175** were initially formed along with silver metal. Saponification of the esters gave the acids **173** and **176**.

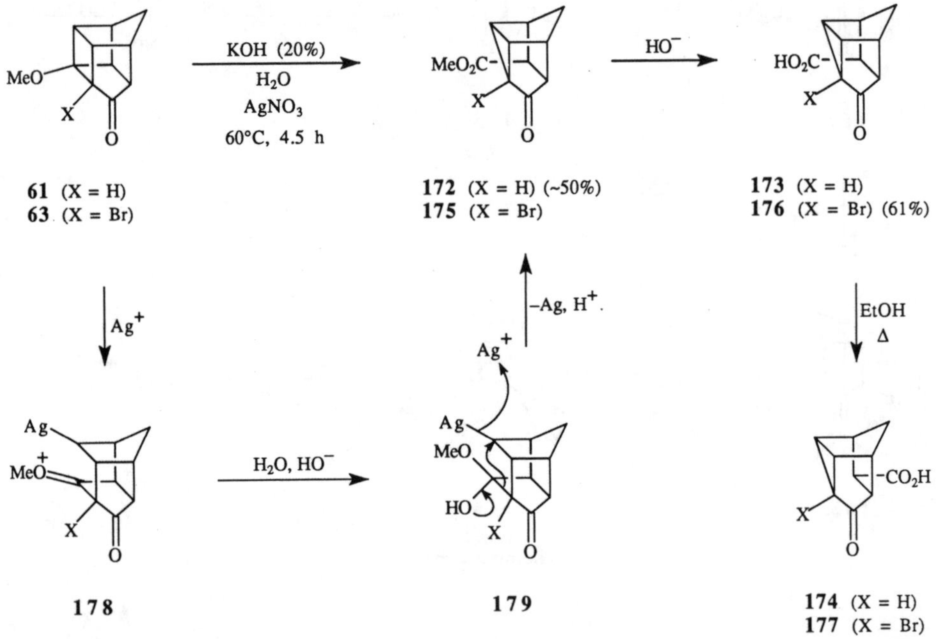

KOH (20%) H_2O $AgNO_3$ 60°C, 4.5 h

61 (X = H)
63 (X = Br)

172 (X = H) (~50%)
175 (X = Br)

173 (X = H)
176 (X = Br) (61%)

178 **179**

174 (X = H)
177 (X = Br)

The bromo ester **176** underwent slow epimerization to **177** on heating in EtOH. The methoxy ketone **61** was also reported to give the epimerized acid **174** on reaction with aqueous $AgNO_3$ and KOH. The proposed mechanism for these rearrangements involved formation of the organosilver compound **178** by cleaving the most strained bond and by bond migration in **179**. The alkoxy group was required to stabilize the cation **178** so that its lifetime was sufficient to allow adding a hydroxyl group rather than cleaving to the dienone **62**. No cage opening occurred on similar treatment of the ketone **55**. The methoxymethoxy ketone **70** also gave the epimerized acid **177** under these conditions.[27]

Extensive rearrangement occurred on reduction of the saturated hydrocarbon **4** with sodium borohydride and triflic acid to adamantane (**180**).[46] The diene **124** also underwent the same reaction (96% yield).

4 → ($NaBH_4$, CF_3SO_3H, $CF_2ClCFCl_2$ / Dry Ar or N_2, −78–∼25°C, 3-5 h, ∼25°C, 18 h) → **180** (97%)

Photoinduced electron transfer from 0.02 M 1,-3-bishomocubane (**4**) or bisketal **183** to 0.02 M quinone **181** gave CIDNP spectra that indicated the radical cations **182** or **184** were formed.[47]

4 + **181** → (hν, Pyrex or 355 nm / $(CD_3)_2CO$ or CD_3CN, Ar) → **182**

183 + **181** → (hν, Pyrex or 355 nm / $(CD_3)_2CO$ or CD_3CN, Ar) → **184**

Irradiation was provided by either a Pyrex-glass-filtered high-pressure mercury arc lamp or a 355-nm laser (13-ns pulse width, 30–40 mJ $pulse^{-1}$). The methine hydrogen atoms of **182** appeared in emission, and the methylene hydrogen atoms showed no polarization. Therefore, no appreciable spin density resided at the carbon atoms adjacent to the methylene carbon atoms. In the CIDNP experiment with the bisketal **183**, the 2.85-ppm signal appeared in emission, whereas the 2.45-ppm signal showed enhanced absorption. Some delocalization of spin and charge into C-2 and C-8 could not be ruled out. However, clearly, only one bond was weakened or broken. Significant strain energy should be released by weakening or breaking the C(3)—C(4) bond. The suggested mechanism for the processes occurring in the CIDNP experiment is shown in Scheme 9.3.

Another process involving electron transfer occurred on electron impact in mass spectrometry. The polynitro compounds **185** and **186** were subjected to collisionally induced dissociation studies in which helium, the collision gas, was present at a pressure such that the main beam intensity was reduced to one-half of its initial

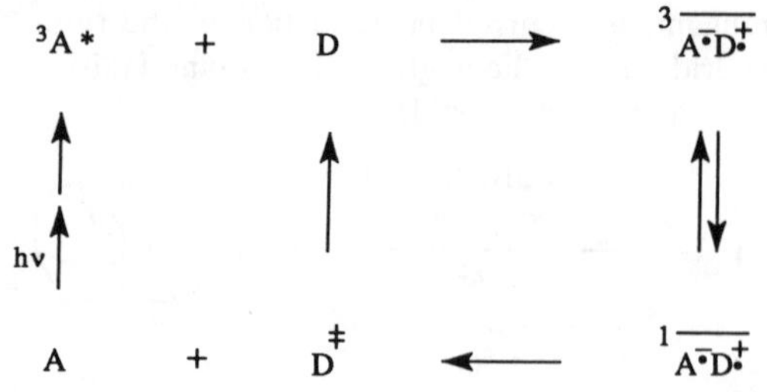

$$^{3}A^{*} + D \longrightarrow {}^{3}\overline{A^{\bullet-}D^{\bullet+}}$$

$$A + D \xrightarrow{h\nu} {}^{3}A^{*} \qquad D \rightarrow D \qquad {}^{3}\overline{A^{\bullet-}D^{\bullet+}} \rightleftarrows {}^{1}\overline{A^{\bullet-}D^{\bullet+}}$$

$$A + D^{\ddagger} \longleftarrow {}^{1}\overline{A^{\bullet-}D^{\bullet+}}$$

A = **181**

D = **4** or **183**

Scheme 9.3 ■

value.[48] Schemes 9.4 and 9.5 show the major dissociation pathways for the first two steps from the presumed initially formed molecular ions. Only those processes that produced daughter ions with relative abundances of at least 50% are shown. The first step in each case is a presumed pathway; the molecular ion was not detected for either compound. Further steps beyond those shown were also reported.[48] For the trinitro compound **185**, the three nitro groups apparently were lost before fragmentation of the hydrocarbon structure occurred. As expected the C—N bonds broke before the C—C bonds did.

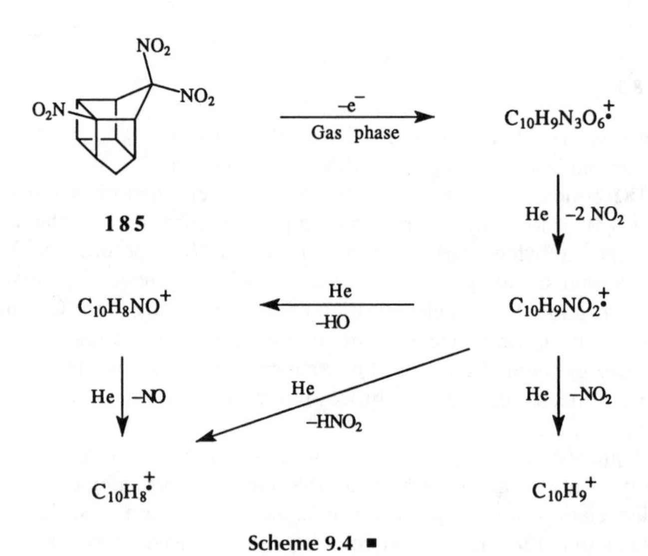

Scheme 9.4 ■

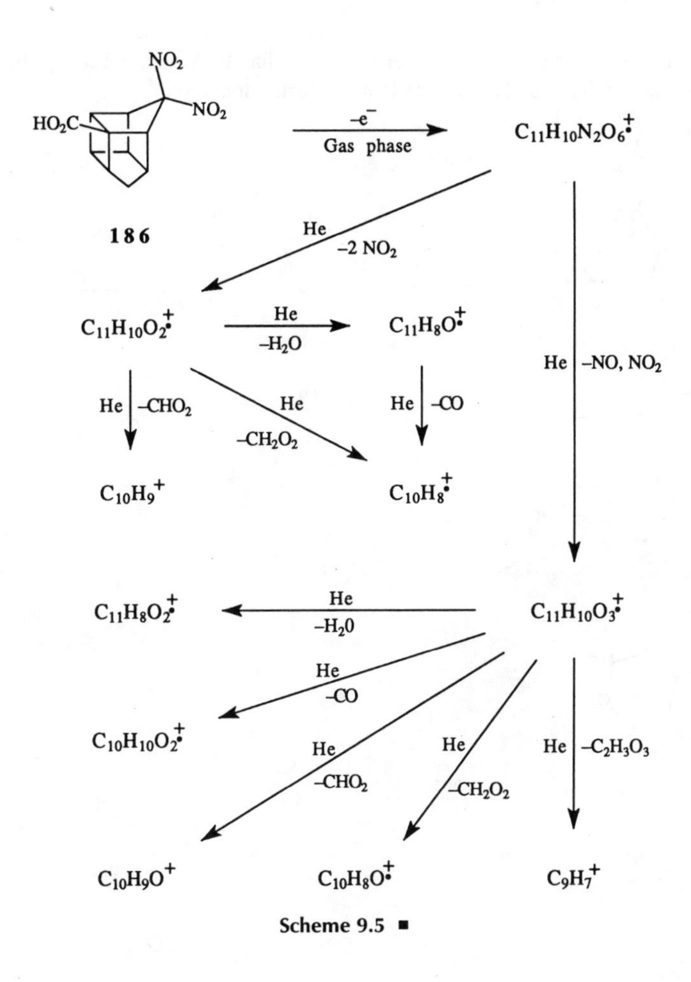

Scheme 9.5 ■

Flash vacuum thermolysis of the methoxy ketones **61** and **63** gave the enol ethers **169** and **170**, presumably via the capto-datively stabilized intermediates **187**.[27]

61, 63 $\xrightarrow[\text{0.35 torr}]{400°\text{C}}$ **169, 170**

MeO X O

187 (X = H, Br)

Benzyne (**189**), in addition to other products, has been generated by the pyrolysis of the lactone derivatives **188** and **191** of 1,3-bishomocubane.[49]

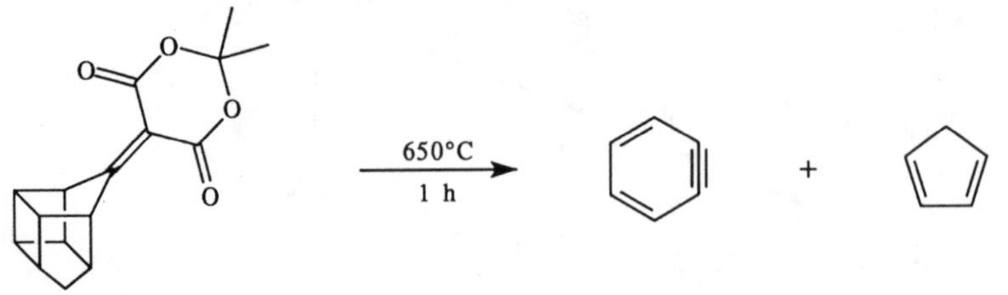

650°C
1 h

188 **189** **190**

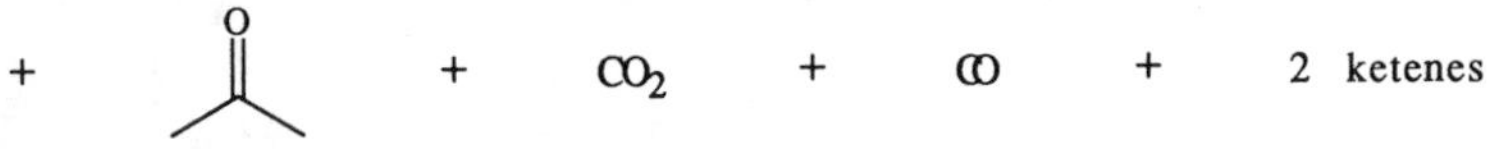

+ + CO_2 + CO + 2 ketenes

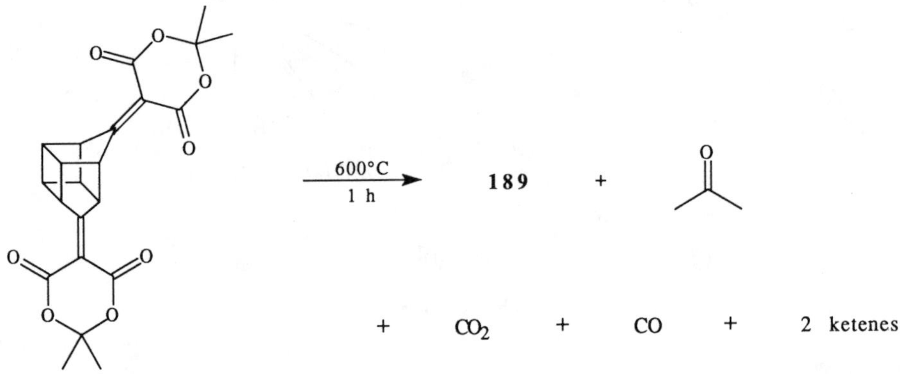

600°C
1 h

189 +

+ CO_2 + CO + 2 ketenes

191

The products were collected in Ar at ~ 9.5 K. The pyrolysis of the monomethylene derivative **188** was also performed at 500, 550, 600, and 700° C in addition to 650° C. The amount of CO increased as the temperature increased, and the relative amounts of the two unidentified ketenes changed with a change in temperature. Collecting the products at nearly ambient temperature from the pyrolysis of the bismethylene derivative **191** gave biphenylene (**192**).

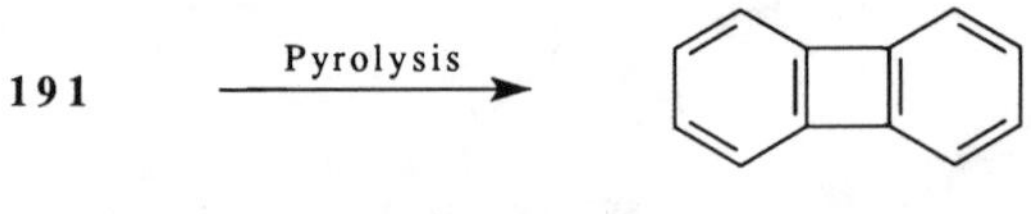

191 Pyrolysis

192

The proposed mechanisms for these reactions involved the ketene **193** and the carbene **194**.

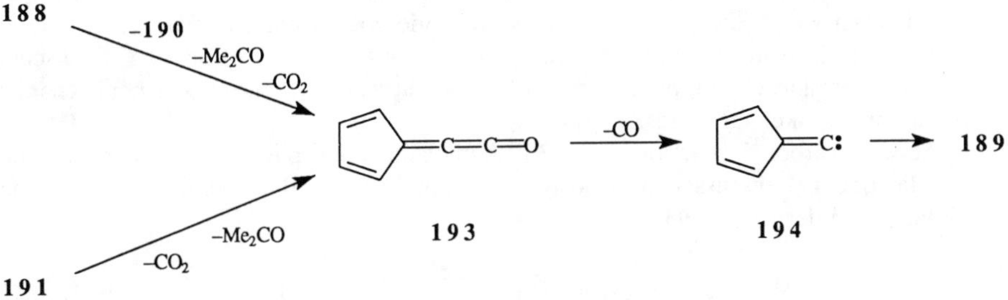

188 —(−190, $-Me_2CO$, $-CO_2$)→ 193 (C=C=O) —(−CO)→ 194 (=C:) → 189

191 —($-Me_2CO$, $-CO_2$)→ 193

Decomposition temperatures were determined for several 1,3-bishomocubane derivatives by differential scanning calorimetry (DSC): **4**, 168 ± 2° C; **185**, 291° C; **186**, 268° C; **195**, 279° C; **196**, 267° C; **197**, 262° C; **198**, 282° C.[50]

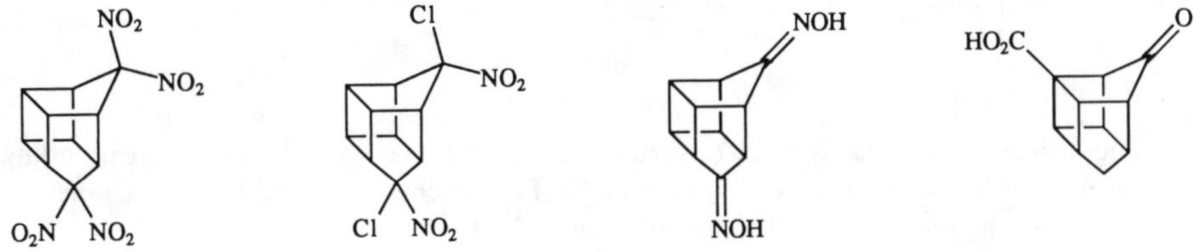

195 (NO_2, NO_2, O_2N, NO_2) 196 (Cl, NO_2, Cl, NO_2) 197 (NOH, NOH) 198 (HO_2C, O)

The authors rationalized the relative order of decomposition temperatures and determined an activation energy for decomposition of the dinitro acid **186** based only on DSC data. These rationalizations must be regarded as highly speculative until more definitive data are available. Crystal transition temperatures and enthalpies of fusion have been determined for several of these compounds. Low values for ΔS_f for **4** (10.6 J mol^{-1} K^{-1}), **185** (6.3 J mol^{-1} K^{-1}), and **195** (5.0 J mol^{-1} K^{-1}) compared to that for the acid **186** (27 J mol^{-1} K^{-1}) was attributed to a high degree of motional freedom in the solid state.

The reasons for the higher reactivity of the bishomocubanones **199** compared to the homologs **200** and **201** with CH_2N_2 have been discussed.[51]

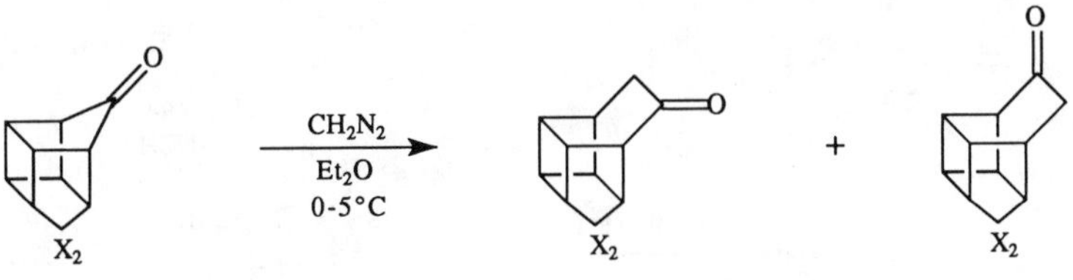

199 (X_2, O) —(CH_2N_2, Et_2O, 0-5°C)→ 200 (X_2, O) + 201 (X_2, O)

The original explanation was the large strain that would be introduced in the seven-membered-ring products that contain the bicyclo[2.2.0]hexane moiety. This explanation implies reversibility in the addition of CH_2N_2 to **200** and **201**. Perhaps the addition to **199** is considerably more facile than to **200** and **201**, that is, kinetic rather than thermodynamic control.

In this section on reactions of 1,3-bishomocubanes, we have thus far covered only reactions in which the pentacyclic skeleton undergoes a change, that is, a skeletal bond is involved in the transformation. In the remainder of this section we discuss reactions of side chains or groups on the pentacyclic nucleus in which no change occurs on the skeletal bonds.

Several carbonyl reactions of 1,3-bishomocubanones have been reported. The tetralactone **191** discussed previously was prepared from the diketone **202** and the glutaric acid derivative **203**.[49]

202 + **203** → (piperidine·HOAc) **191**

The oximes **204**, **206**, and **207** were prepared by reactions of the corresponding ketones **75**,[29] **205**,[52] and **147**[5,53] with $HONH_2$. Racemic ketone **55** did not react on irradiation at 254 nm in $CHCl_3$ solution at $-50°$ C.[30]

75 → ($NH_2OH·HCl$, NaOAc) **204** (80%)

205 → ($NH_2OH·HCl$, NaOAc, MeOH, ~25°C, 24 h) **206** (94%)

147 → ($NH_2OH·HCl$, pyridine, PhMe) **207**

Much interest has been shown recently in the preparation of polynitro-bishomocubanes and related compounds as high-energy materials. The 1,3-bishomocubyl oximes **204** and **206** were oxidized to the corresponding nitro compounds **156**,[29] and the mixture of isomers **208** and **209**.[52]

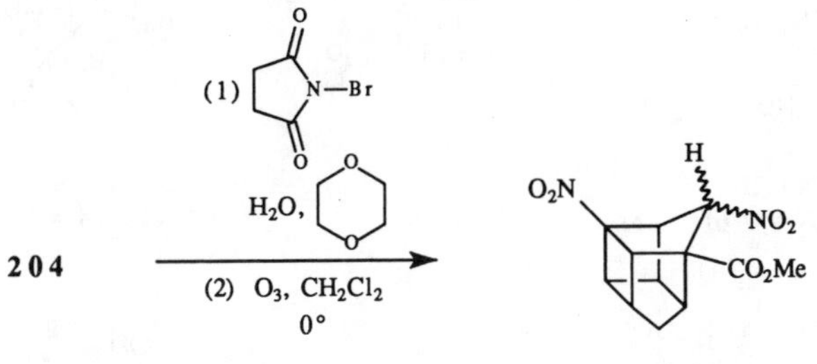

204 —(1) N-bromosuccinimide, H_2O, dioxane; (2) O_3, CH_2Cl_2, 0°→ **156** (16%)

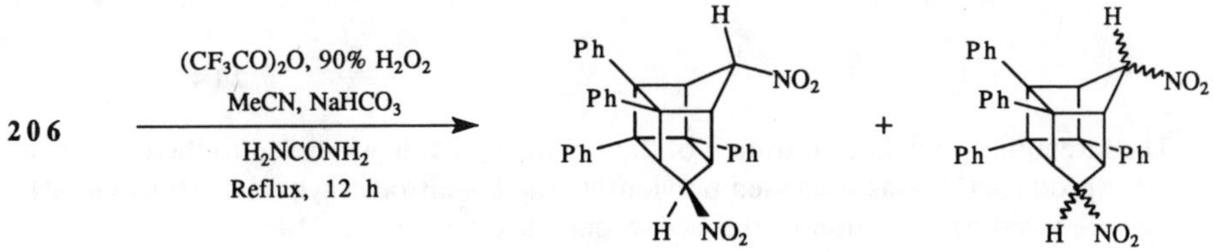

206 —$(CF_3CO)_2O$, 90% H_2O_2, MeCN, $NaHCO_3$, H_2NCONH_2, Reflux, 12 h→ **208** + **209** (77%)

Bromination and oxidation of the polybromo-bis-oxime **207** gave the perbromodinitro derivative **210**.[5,53]

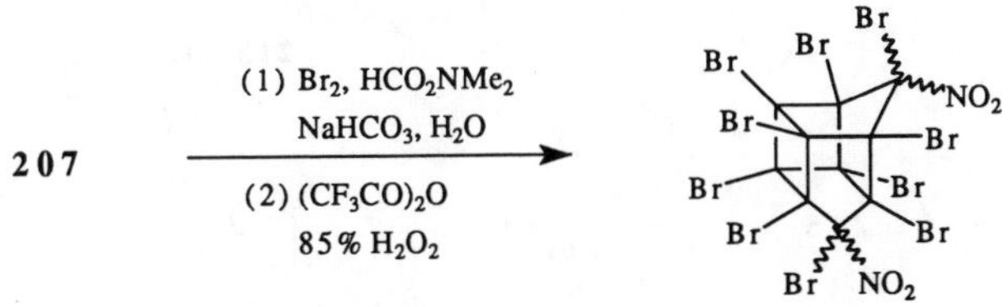

207 —(1) Br_2, HCO_2NMe_2, $NaHCO_3$, H_2O; (2) $(CF_3CO)_2O$, 85% H_2O_2→ **210**

Nitration of the nitromethylene derivatives **208**, **211**, and **213** gave the *gem*-dinitro products **211**,[52] **212**,[52] and **214**,[5,53] respectively.

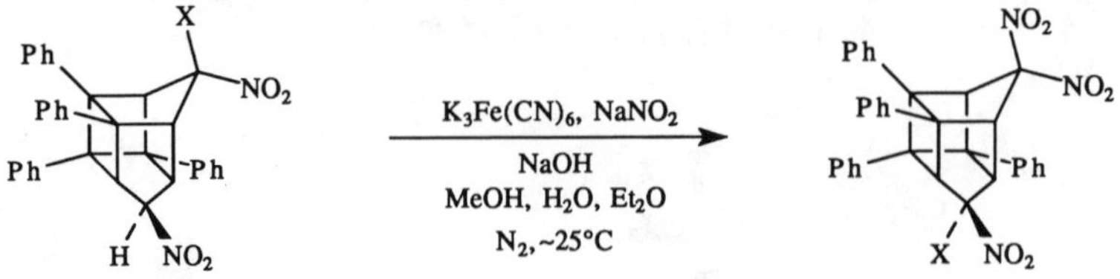

$K_3Fe(CN)_6$, $NaNO_2$
NaOH
MeOH, H_2O, Et_2O
N_2, ~25°C

208 (X = H) — 6 h — **211** (X = H) (65%)
211 (X = NO_2) — 24 h — **212** (X = NO_2) (50%)

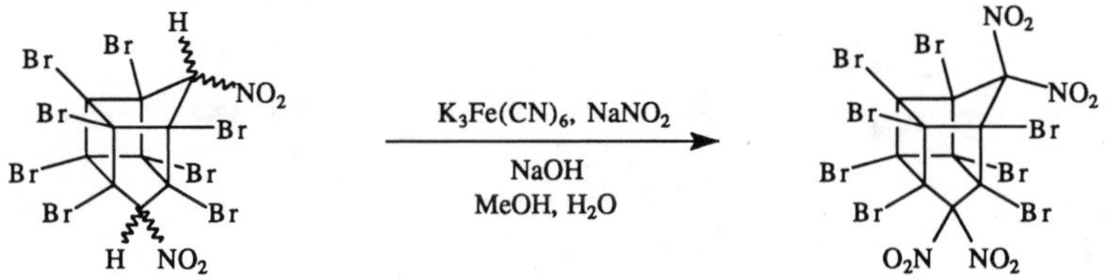

$K_3Fe(CN)_6$, $NaNO_2$
NaOH
MeOH, H_2O

213 → **214**

The attempted analogous nitration of the dinitroester **156**, which led to the rearrangement product **157**, was discussed previously. The bis-nitromethylene intermediate **213** was prepared by reduction of the *gem*-bromonitro compound **210**.[5,53]

210 →($NaBH_4$, EtOH) **213**

Reduction of the phenyl ketone **59** with $NaBH_4$ gave a single alcohol,[26] presumably **215** if steric approach control were in operation. Acetylation of this alcohol gave a single acetate,[26] presumably **216**.

59 →($NaBH_4$, MeOH, 0°C, 30 min) **215** (92%)

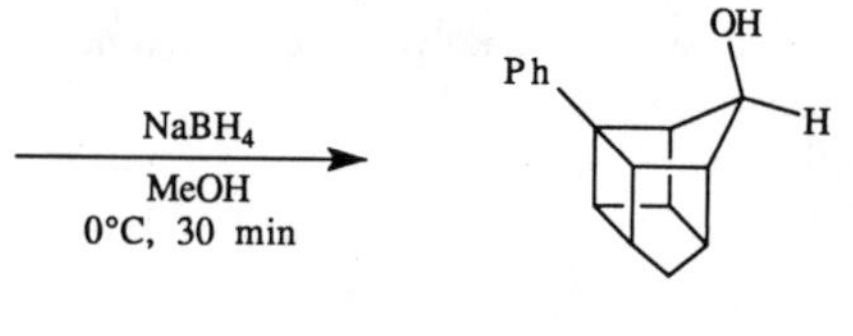

215 →(Ac_2O, Reflux, 1 h) **216** (78%)

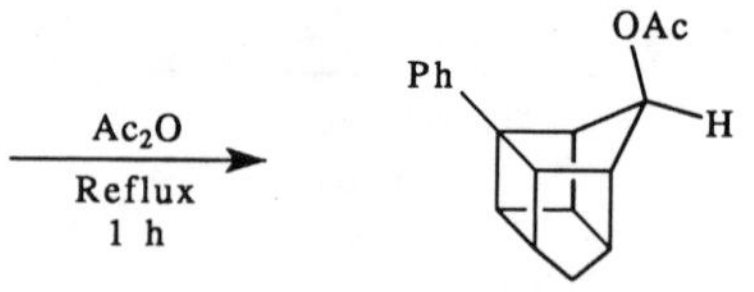

The acid chloride **218** was prepared from the acid **217** by reaction with oxalyl chloride.[28] The acid chloride **218** was then converted to the chlorobromoketone **142** and the dibromoketone **144**.[28]

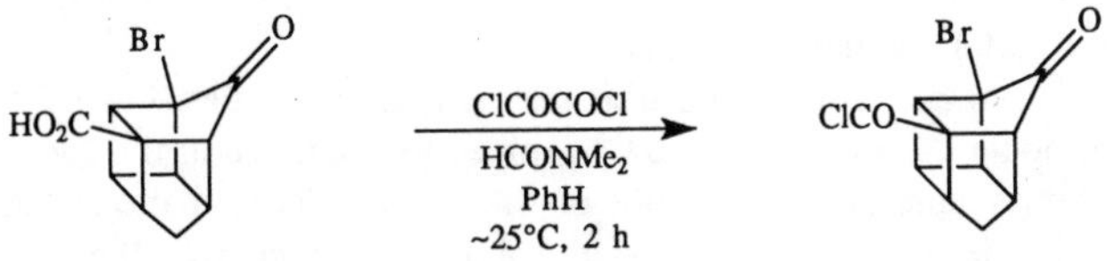

$ClCOCOCl$, $HCONMe_2$, PhH, ~25°C, 2 h

217 → **218**

218 → **142** (47%) with (219), (220), CCl_4, Reflux

218 → **144** (79%) with 219, 220, $BrCCl_3$, Reflux

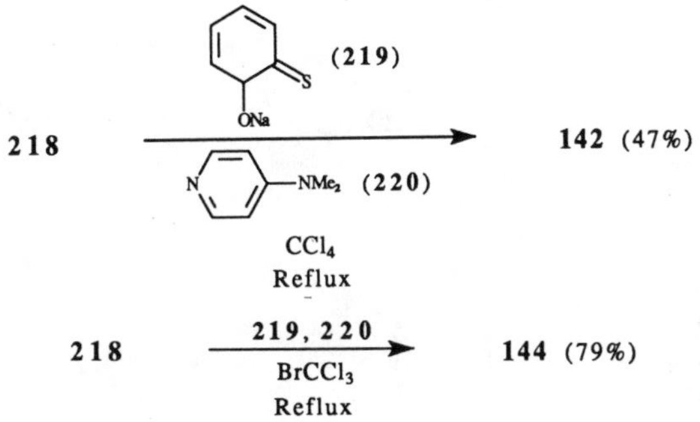

In marked contrast to the corresponding keto acetate, where ring opening occurred immediately, the epimeric alcohols **221** and **223** gave the unrearranged diols **222** and **224**.[7]

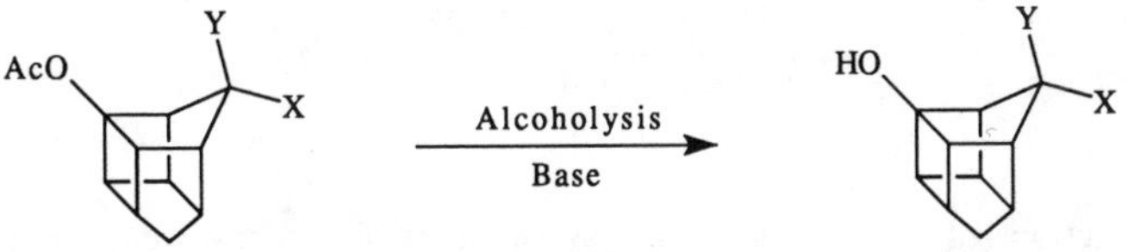

Alcoholysis, Base

221 (X = OH, Y = H) → **222** (X = OH, Y = H)
223 (X = H, Y = OH) → **224** (X = H, Y = OH)

9.4C. 1,3′-Bishomocubanes

A reaction that brings about a change in the carbon skeleton in this series of compounds is the base-induced stepwise quasi-Favoriskii ring contraction of the α-haloketone **225**,[34] which presumably provides **226**.

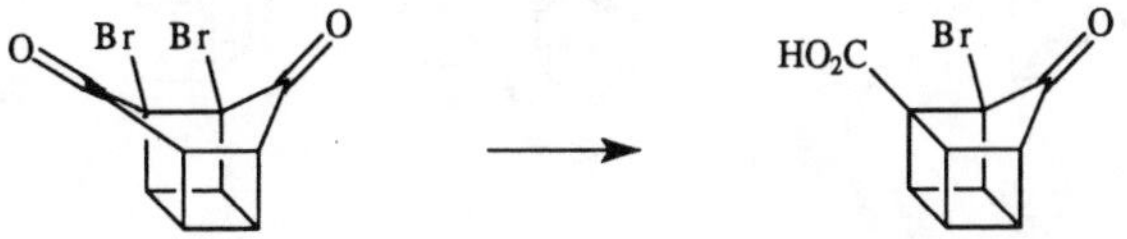

225 → **226**

Another is the ring opening of pentaprismane **12** to give hypostrophene **14** by the action of Rh(I) complexes.[18] The rate of the latter reaction was about one-half that of the corresponding cubane reaction. The proposed ring opening of the dibridged 1,3′-bishomocubane **121** to give the reduction product **119** and polymer is another example of a skeletal change reaction.[39]

The formation of pentaprismanecarboxylic acid **13** by a quasi-Favorskii-type reaction of the sulfonate esters **227** and **229**[18] represents a change in the total skeleton of the homopentaprismane system but not of the 1,3′-bishomocubane system, which is contained within the larger system. A significant by-product from the reaction of the methanesulfonate ester **227** is the Haller–Bauer cleavage product **228**. Other reaction conditions using solid NaOH in an organic solvent gave only the cleavage product **228**.

O, OSO_2R —KOH (20%), H_2O→ **13** + **228**

227 (R = Me) 80°C, 0.5 h (≥30%) (≥10%)
229 (R = 4-MeC_6H_4) 110°C, 5 h (≥60%) (trace)

227 —NaOH (s), PhMe, Reflux→ **228**

227 —NaOH (s), THF, ~25°C→ **228**

Other reactions of 1,3′-bishomocubyl systems that do not involve skeletal changes have also been described. Several hydrolysis reactions were reported in conjunction with the synthesis of pentaprismane (**12**).[18] An attempt to prepare the *tert*-butyl perester, using wet *t*-BuO_2H with the acid chloride **230** gave a substantial quantity of the anhydride **231**.

COCl —*t*-BuO_2H, H_2O, pyridine→ [CO–]$_2$O

230 **231**

Saponification of the lactone **228** gave the hydroxy carboxylate **232**, which was not isolated but was subjected to the oxidation described later. The ketal **109** gave the ketone **233** on hydrolysis.

228 $\xrightarrow[H_2O]{KOH}$ 232 (≥81%)

109 $\xrightarrow[H_2O,\ Et_2O,\ \sim25°C,\ \sim10\text{-}15\ h]{H_2SO_4\ (30\%)}$ 233 (95%)

Esterification of the acyl chloride **230** gave the perester **234**.[18]

230 $\xrightarrow[\text{pyridine},\ Et_2O,\ \sim0°C,\ \sim25°C,\ \sim10\text{-}15\ h]{t\text{-}BuO_2H\ (dry)}$ 234 (≥98%)

The lactol **235**, a tautomer of the keto acid **237**, gave the methyl ester **236** on reaction with CH_2N_2.[18]

235 $\xrightarrow[Et_2O,\ CH_2Cl_2,\ 0\text{-}5°C]{CH_2N_2}$ 236 (96%)

235 ⇌ 237 $\xrightarrow{CH_2N_2}$ 236

237

Enough acid **237** was present in equilibrium with the lactol **235** to allow reaction with CH_2N_2 to proceed. The *p*-toluenesulfonate ester **229** was prepared from the ketol **238** by a standard procedure, as was the methanesulfonate **227**.[18] Treatment of the acid **13** with oxalyl chloride gave the acyl chloride **230**.[18]

238 → **229** (Me–C$_6$H$_4$–SO_2Cl, pyridine; ~100°C, 0.5 h; 0°C, ~10-15 h)

13 → **230** (ClCOCOCl; Reflux, 1 h)

Oxidation of the diol **239** with chlorine and dimethyl sulfide was the source of the ketol **238**.[18] Most other oxidants gave the lactol **235** from the diol **239**. The oxidation of the hydroxy carboxylate **232** mentioned previously gave the ketol **235**.[18]

239 → **238** (77-85%) ((1) Cl_2, Me_2S, CH_2Cl_2, -25°C, 3 h; (2) Et_3N, CH_2Cl_2, -25°C, 10 min)

232 → **235** (≥81%) ($NaIO_4$, RuO_2, H_2O, ~25°C, ~10-15 h)

The lactone **228** resulted from the peracid oxidation of the ketone **233**.[18]

233 → **228** (90%) (m-$ClC_6H_4CO_3H$, CH_2Cl_2, ~25°C, ~10-15 h)

Reduction of the keto ester **236** with sodium in liquid ammonia gave the diol **239**.[18] Reduction of the ketol **238** also gave the diol **239**.

236 → **239** (83%) (Na, NH_3 (l), Et_2O, Reflux, 0.5+ h; ~25°C, ~10-15 h)

The final step in the synthesis of pentaprismane (**12**) was pyrolysis of the perester **234** in triisopropylnitrobenzene.[18] The ^{13}C—H coupling constant of 148 Hz for **12** indicates ~30% s character in the C—H bonds. Substantial rehybridization of the C_{10} frame of **12** away from tetrahedral geometry causes an increase in the p character of the C—C bonds and consequently an increase in the s character of the C—H bonds.

NO_2

234 → **12** (42% from **13**)

150°C, 7 h

N_2

9.4D. 1,4-Bishomocubanes

In the discussion of the results for the homoaromatic anion formed by deprotonation of bicyclo[3.2.1]octa-2,6-diene, the authors suggest that the bridge anion **240** of 1,4-bishomocubane **6** could reorganize slightly to the anion **241**.[54] This anion is an excellent candidate for homoaromaticity and might also result from deprotonation of *endo*-dicyclopentadiene (**124**). Simple models show that the C(3)—C(9) and C(5)—C(8) distances in anion **242** are nearly the same as the corresponding distances in the allylic anion from the bicyclooctadiene. Further, the p orbitals in the allylic and olefinic moieties of **242** are oriented almost directly toward one another. Results in this area are eagerly awaited.

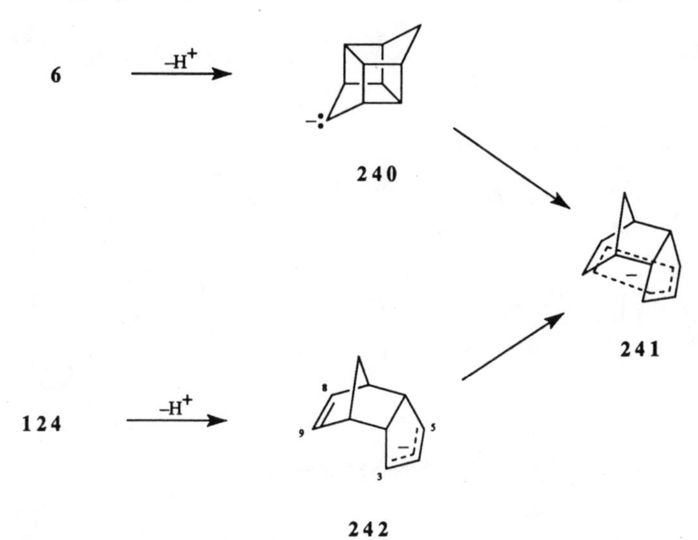

Collisionally activated dissociation experiments, with NH_3 as the collision gas, have been performed on perchloro-1,4-bishomocubane **243**, also known as the pesticide mirex.[55] At an NH_3 pressure of 350 mtorr and a source temperature of 170° C, mirex **243** produced the cations and cation-radicals **244–250**; proposed routes are shown in Scheme 9.6[55,56] Under these conditions, $H_3NH_4^+$ is believed to be the reagent ion. The route from **243** to **248** is questionable; the ion peak for **248** had very

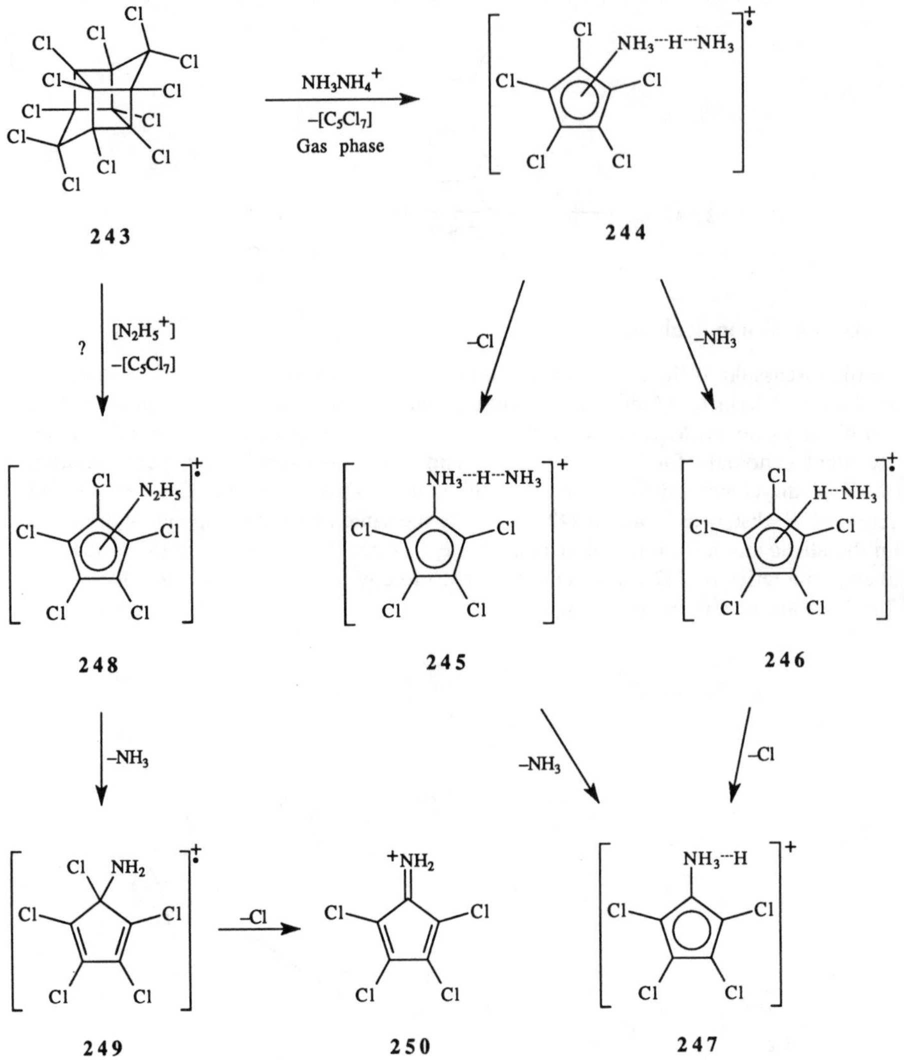

Scheme 9.6 ■

low relative abundance. No parent ion for **248** was detected in MS/MS experiments. The structure of cation-radical **248** was speculated to be NH_3 π-bonded to the cation-radical **249**. Hexachlorocyclopentadiene gave a spectrum identical to that of mirex **243** under these conditions.

Research results continue to be reported on the environmental behavior of pesticidal polychloro derivatives of 1,4-bishomocubane. Mirex **243** (60–100 ng l^{-1})

was reductively dechlorinated to the monohydro **251** and dihydro **252** derivatives on ultraviolet or visible irradiation (250–700 nm) in aqueous solution.[57,58]

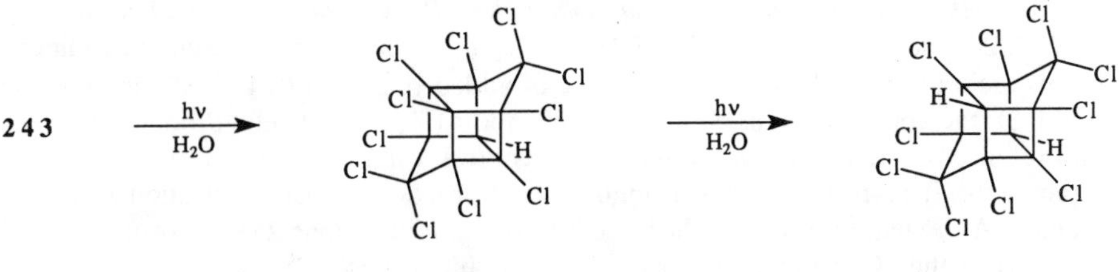

243 $\xrightarrow[H_2O]{h\nu}$ 251 $\xrightarrow[H_2O]{h\nu}$ 252

Reaction occurred both on irradiation with monochromatic laboratory light sources and with sunlight. The kinetics of these photochemical reactions were determined in water containing various natural materials. The presence of humic acids in purified water or water from Lake Ontario caused a more rapid conversion of **243** to **251** than purified water alone. The mechanism by which this acceleration occurred is not known. Three weeks of solar radiation produced a ratio of **251** to **243** of ~0.7 in Lake Ontario water and ~0.2 in purified water alone. The rate of destruction of **251** was either faster or slower than its formation rate, depending on the other materials present and the light intensity. The rate of the **243**-to-**251** transformation with visible light (400–700 nm) was faster in the presence of humic acids than it was in either purified or Lake Ontario water. Photoproducts of mirex **243** have been found in biota and water from Lake Ontario, presumably because of the action of sunlight on mirex from anthropogenic sources. An X-ray crystal structure determination of the dihydro compound **252** has been reported.[59]

Kepone **253**, another pesticide, was also reductively dechlorinated to isomercially undefined products **254** and **255** and was converted to other products by another environmental process, biodegradation.[60]

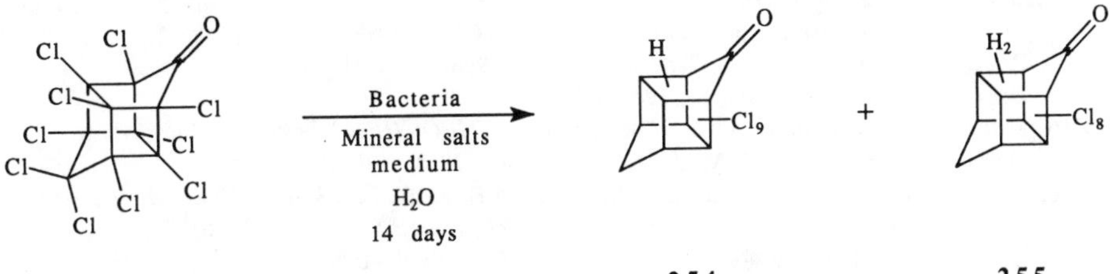

253 $\xrightarrow[\text{Mineral salts medium, } H_2O, \text{ 14 days}]{\text{Bacteria}}$ 254 + 255 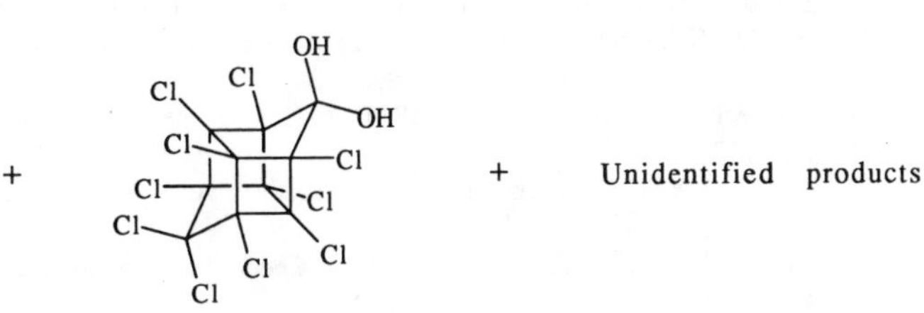 + 256 + Unidentified products

Three types of bacteria, grown from sediment collected from the James River at Jordan Point near Hopewell, VA, in media that contained **253**, effected this transformation. *Pseudomonas putida*, *P. maltophilia*, and *P. vesicularis* produced conversions of 33, 26, and 16%, respectively, of **253** (50 μg ml^{-1}) in water that contained mineral salts medium in 14 days. In the presence of added yeast extract (0.01%) and glucose (0.1%), the conversions of **253** were 38, 33, and 41%, respectively. Products **254–256** were also detected in an uninoculated control even though 110% of **253** was recovered. This result implies a nonbiological process or misidentification of HPLC peaks. A discussion of the synthesis and destruction of kepone **253** has been included as part of the "Consumer Chemistry in an Organic Course."[61]

References

1. Dilling, W. L.; Reineke, C. R.; Plepys, R. A. *J. Org. Chem.* **1969**, *34*, 2605–2615.
2. Dauben, W. G.; Whalen, D. L. *J. Am. Chem. Soc.* **1966**, *88*, 4739–4740.
3. Dilling, W. L. *Chem. Rev.* **1966**, *66*, 373–393.
4. Eaton, P. E.; Cole, T. W., Jr. *J. Am. Chem. Soc.* **1964**, *86*, 3157–3158.
5. Marchand, A. P. *Tetrahedron* **1988**, *44*, 2377–2395.
6. Marchand, A. P. *Chem. Rev.* **1989**, *89*, 1011–1033.
7. Klunder, A. J. H.; Zwanenburg, B. *Chem. Rev.* **1989**, *89*, 1035–1050.
8. Anand, N.; Bindra, J. S.; Ranganathan, S. *Art in Organic Synthesis*, 2nd ed.; Wiley: New York, 1988; pp. 134–135, 255.
9. Balaban, A. T. *Rev. Roumaine Chim.* **1989**, *34*, 1745–1752.
10. Kovacevic, K.; Maksic, Z. B.; Vuckovic, D. L.; Vujisic, L. *J. Mol. Struct.* (*THEOCHEM*) **1987**, *151*, 233–243.
11. Marchand, A. P.; Vidyasagar, V.; Watson, W. H.; Nagl, A.; Kashyap, R. P. *J. Org. Chem.* **1991**, *56*, 282–286.
12. Osawa, E.; Barbiric, D. A.; Lee, O. S.; Kitano, Y.; Padma, S.; Mehta, G. *J. Chem. Soc., Perkin Trans. 2*, **1989**, 1161–1165.
13. Osawa, E.; Rudzinski, J. M.; Xun, Y.-M. *Struct. Chem.* **1990**, *1*, 333–344.
14. Reddy, V. P.; Jemmis, E. D. *Tetrahedron Lett.* **1986**, *27*, 3771–3774.
15. Dailey, W. P. *Tetrahedron Lett.* **1987**, *28*, 5787–5790.
16. Miller, M. A.; Schulman, J. M. *J. Mol. Struct.* (*THEOCHEM*) **1988**, *163*, 133–141.
17. Disch, R. L.; Schulman, J. M. *J. Am. Chem. Soc.* **1988**, *110*, 2102–2105.
18. Eaton, P. E.; Or, Y. S.; Branca, S. J.; Shankar, B. K. R. *Tetrahedron* **1986**, *42*, 1621–1631.
19. Jemmis, E. D.; Leela, G. *Proc. Indian Acad. Sci., Chem. Sci.* **1987**, *99*, 281–282.
20. Dorigo, A. E., Houk, K. N. *J. Org. Chem.* **1988**, *53*, 1650–1664.
21. Sauers, R. R.; Krogh-Jespersen, K. *Tetrahedron Lett.* **1989**, *30*, 527–530.
22. Eaton, P. E.; Millikan, R.; Engel, P. *J. Org. Chem.* **1990**, *55*, 2823–2826.
23. Hu, D.; Schaufele, H.; Pritzkow, H.; Zenneck, U. *Angew. Chem.* **1989**, *101*, 929; *Angew. Chem., Int. Ed. Engl.* **1989**, *28*, 900–902.
24. Zenneck, U. *Angew. Chem.* **1990**, *102*, 171; *Angew. Chem., Int. Ed. Engl.* **1990**, *29*, 126–137.
25. Klunder, A. J. H.; Huizinga, W. B.; Hulshof, A. J. M.; Zwanenburg, B. *Tetrahedron Lett.* **1986**, *27*, 2543–2546.
26. Ogino, T.; Awano, K.; Fukazawa, Y. *J. Chem. Soc., Perkin Trans. 2* **1990**, 1735–1738; **1991**, 437.
27. Klunder, A. J. H.; Ariaans, G. J. A.; Loop, E. A. R. M. v. d.; Zwanenburg, B. *Tetrahedron* **1986**, *42*, 1903–1915.
28. Schafer, J.; Szeimies, G. *Tetrahedron Lett.* **1990**, *31*, 2263–2264.

29. Marchand, A. P.; Jin, P.; Flippen-Anderson, J. L.; Gilardi, R.; George, C. *Chem. Commun.* **1987**, 1108–1109.
30. Childs, R. F.; Duffey, B. M.; Mahendran, M. *Can. J. Chem.* **1986**, *64*, 1220–1223.
31. Bartsch, R.; Hitchcock, P. B.; Nixon, J. F. *Chem. Commun.* **1989**, 1046–1048.
32. Mak, T. C. W.; Yip, Y. C.; Luh, T.-Y. *Tetrahedron* **1986**, *42*, 1981–1988.
33. Breyer, R. A.; Griffin, G. W.; Reichel, L. W.; Wu, J. C.; Stevens, E. D. *Abstracts of Papers*, 194th National Meeting of the American Chemical Society; American Chemical Society: Washington, DC, 1987; ORGN 91.
34. Griffin, G. W.; Breyer, R. A.; Reichel, L. W.; Stevens, E. D. *Abstracts of Papers*, 195th National Meeting of the American Chemical Society; American Chemical Society: Washington, DC, 1988; ORGN 34.
35. Griffin, G. W.; Reichel, L. W.; El-Hajj, T.; Breyer, R. A.; Stevens, E. D. *Abstracts of Papers*, 192nd National Meeting of the American Chemical Society; American Chemical Society: Washington, DC, 1986; ORGN 102.
36. Osawa, E.; Kanematsu, K. *Mol. Struct. Energet.* **1986**, *3*, 329–369.
37. Marchand, A. P.; Deshpande, M. N. *J. Org. Chem.* **1989**, *54*, 3226–3229.
38. Marchand, A. P.; Reddy, G. M.; Watson, W. H.; Kashyap, R. P.; Nagl, A. *J. Org. Chem.* **1991**, *56*, 277–282.
39. Majerski, Z.; Veljkovic, J.; Kaselj, M. *J. Org. Chem.* **1988**, *53*, 2662–2664.
40. Hargittai, I.; Hargittai, M. *Symmetry through the Eyes of a Chemist*; VCH: Weinheim, 1986; p. 87.
41. Driess, M.; Pritzkow, H.; Siebert, W. *Angew. Chem.* **1988**, *100*, 410–411; *Angew. Chem., Int. Ed. Engl.* **1988**, *27*, 399–400.
42. Griffin, G. W.; Chaudhuri, A. L.; Reichel, L. W.; Breyer, R. A.; Condon, M. M.; Elhajj, T.; Lankin, D. C.; Stevens, E. D.; Li, Y. J. *Abstracts of Papers*, 194th National Meeting of the American Chemical Society; American Chemical Society: Washington, DC, 1987; ORGN 11.
43. Griffin, G. W.; Marchand, A. P. *Chem. Rev.* **1989**, *89*, 997–1010.
44. Griffin, G. W.; Breyer, R. A.; Chaudhuri, A.; Condon, M. M.; Elhajj, T.; Lankin, D. C.; Reichel, L. W.; Stevens, E. D. *Abstracts of Papers*, 195th National Meeting of the American Chemical Society; American Chemical Society: Washington, DC, 1988; ORGN 362.
45. Prakash, G. K. S.; Krishnamurthy, V. V.; Herges, R.; Bau, R.; Yuan, H.; Olah, G. A.; Fessner, W.-D.; Prinzbach, H. *J. Am. Chem. Soc.* **1988**, *110*, 7764–7772.
46. Olah, G. A.; Wu, A.; Farooq, O.; Prakash, G. K. S. *J. Org. Chem.* **1989**, *54*, 1450–1451.
47. Roth, H. D.; Schilling, M. L. M.; Abelt, C. J. *Tetrahedron* **1986**, *42*, 6157–6166.
48. Yinon, J.; Bulusu, S. *J. Energet. Mater.* **1986**, *4*, 115–131.
49. Brown, R. F. C.; Browne, N. R.; Coulston, K. J.; Danen, L. B.; Eastwood, F. W.; Irvine, M. J.; Pullin, A. D. E. *Tetrahedron Lett.* **1986**, *27*, 1075–1078.
50. Weinstein, D. I.; Alster, J.; Marchand, A. P. *Thermochim. Acta* **1986**, *99*, 133–137.
51. Krow, G. R. *Tetrahedron* **1987**, *43*, 3–38.
52. Marchand, A. P.; Annapurna, G. S.; Vidyasagar, V.; Flippen-Anderson, J. L.; Gilardi, R.; George, C.; Ammon, H. L. *J. Org. Chem.* **1987**, *52*, 4781–4783.
53. Griffin, G. W.; Elhajj, T.; Chaudhuri, A.; Stevens, E. D.; Wu, J. C. *Abstracts of Papers*, 194th National Meeting of the American Chemical Society; American Chemical Society: Washington, DC, 1987; ORGN 12.
54. Lee, R. E.; Squires, R. R. *J. Am. Chem. Soc.* **1986**, *108*, 5078–5086.
55. Cairns, T.; Siegmund, E. G. *Rapid Commun. Mass Spectrom.* **1989**, *3*, 340–341.
56. Westmore, J. B.; Alauddin, M. M. *Mass Spectrom. Rev.* **1986**, *5*, 381–465.
57. Mudambi, A. R.; Hassett, J P. *Am. Chem. Soc., Div. Environ. Chem.* **1987**, *27*(1), 201–203.
58. Mudambi, A. R.; Hassett, J. P. *Chemosphere* **1988**, *17*, 1133–1146.
59. Cordes, A. W.; Eubanks, J. R. I. *Acta Crystallogr., Sect. C* **1987**, *43*, 1848–1849.
60. George, S. E.; Claxton, L. D. *Xenobiotica* **1988**, *18*, 407–416.
61. Miller, J. A. *J. Chem. Educ.* **1988**, *65*, 210–211.

10

A New Approach to Adamantane Rearrangements

Camille Ganter

Swiss Federal Institute of Technology, Zürich, Switzerland

Contents

10.1. Introduction
10.2. Aluminum Halide Catalyzed Rearrangements
  10.2A. 1,2-*endo*- and 1,2-*exo*-Trimethylene-8,9,10-trinorbornane
  10.2B. 1,7-Trimethylene-8,9,10-trinorbornane
  10.2C. *endo*-THDCP and *exo*-THDCP
  10.2D. *syn*- and *anti*-Tricyclo [4.2.1.1^{2,5}]decane
  10.2E. Summary
10.3. Rearrangements of Regioselectively Generated Carbocations under Ionic Hydrogenation Conditions
  10.3A. Carbocations at C-2 and C-6 of 1,2-*endo*- and 1,2-*exo*-Trimethylene-8,9,10-trinorbornane
    10.3A(1). Rearrangements Involving a Degenerate Rearrangement
      C(2),C(3)-Olefin (Type A)
      C(2),C(3′)-Olefin (Type C)
      6-*endo*-Alcohol (Type E)
    10.3A(2). Rearrangements not Involving a Degenerate Rearrangement
      6-*exo*-Alcohol (Type F)
      C(5),C(6)-Olefin (Type G)
  10.3B. Carbocations at C-3 and C-5 of 1,2-*endo*-, 1,2-*exo*-, and 1,7-Trimethylene-8,9,10-trinorbornane
  10.3C. Carbocation at C-1′ of 1,7-Trimethylene-8,9,10-trinorbornane
  10.3D. Carbocations at C-2 and C-6 of 1,7-Trimethylene-8,9,10-trinorbornane

and Carbocations at C-2, C-5, and C-1′ of *endo*-THDCP as well as *exo*-THDCP
10.3E. Carbocations at C-3 and C-9 of *syn*- and *anti*-Tricyclo[4.2.1.1^{2,5}]decane
10.4. Product Analyses
10.5. Concluding Remarks
Acknowledgments
References

10.1. Introduction

In 1957 Schleyer[1] reported on "A simple preparation of adamantane," the result of an accidental discovery while studying the Lewis acid ($AlBr_3$ and $AlCl_3$, respectively) catalyzed isomerization of 2-*endo*,3-*endo*-trimethylene-8,9,10-trinorbornane [**7**; *endo*-tetrahydrodicyclopentadiene (*endo*-THDCP)] to 2-*exo*,3-*exo*-trimethylene-8,9,10-trinorbornane [**11**; *exo*-tetrahydrodicyclopentadiene (*exo*-THDCP)] (Scheme 10.1). This, at the time most unexpected and amazingly simple, access to adamantane (**19**) was a milestone in polycyclic hydrocarbon chemistry and opened the area of a new class of compounds, the *cage hydrocarbons*.[2] However, talking about adamantane (**19**), one cannot pass over in silence its isolation[4] from Czechoslovakian petroleum in Hodonín (Moravia) in 1933 and its first synthesis, just 50 years ago, by Prelog and Seiwerth[5] in 1941.

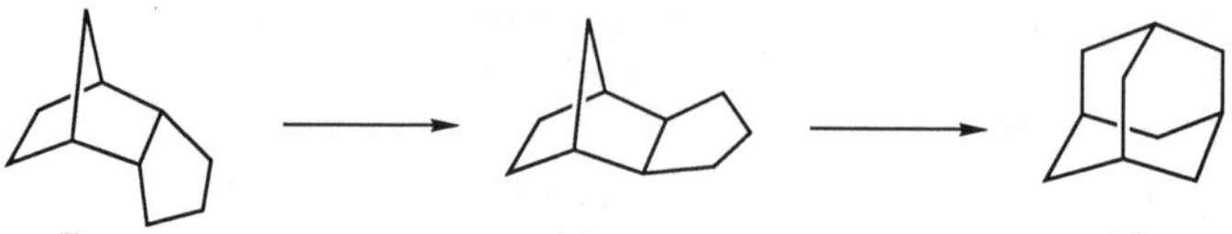

7 **11** **19**

Scheme 10.1 ■

The preparation of adamantane (**19**) by rearrangement via carbocations from an isomeric and readily available precursor (*endo*-THDCP **7**) was the starting point for the many studies on the mechanism of the so-called adamantane rearrangement.[6] A most significant contribution was made by Whitlock and Siefken,[9] a graph intercorrelating 16 isomeric $C_{10}H_{16}$ compounds. Subsequent additions by Schleyer and co-workers[10] led to the enlarged interconversion map of 19 isomers ("adamantaneland") including calculated heats of formation and arrows standing for the most likely pathways by which any of the isomers are predicted to rearrange.

For higher clarity we slightly modified Schleyer's version: The chosen arrangement of the 19 isomers (Scheme 10.2) avoids any crossing of arrows, and the numbering of the isomers follows their calculated heats of formation—**1** representing the isomer highest in energy, adamantane (**19**) the thermodynamically most stable.

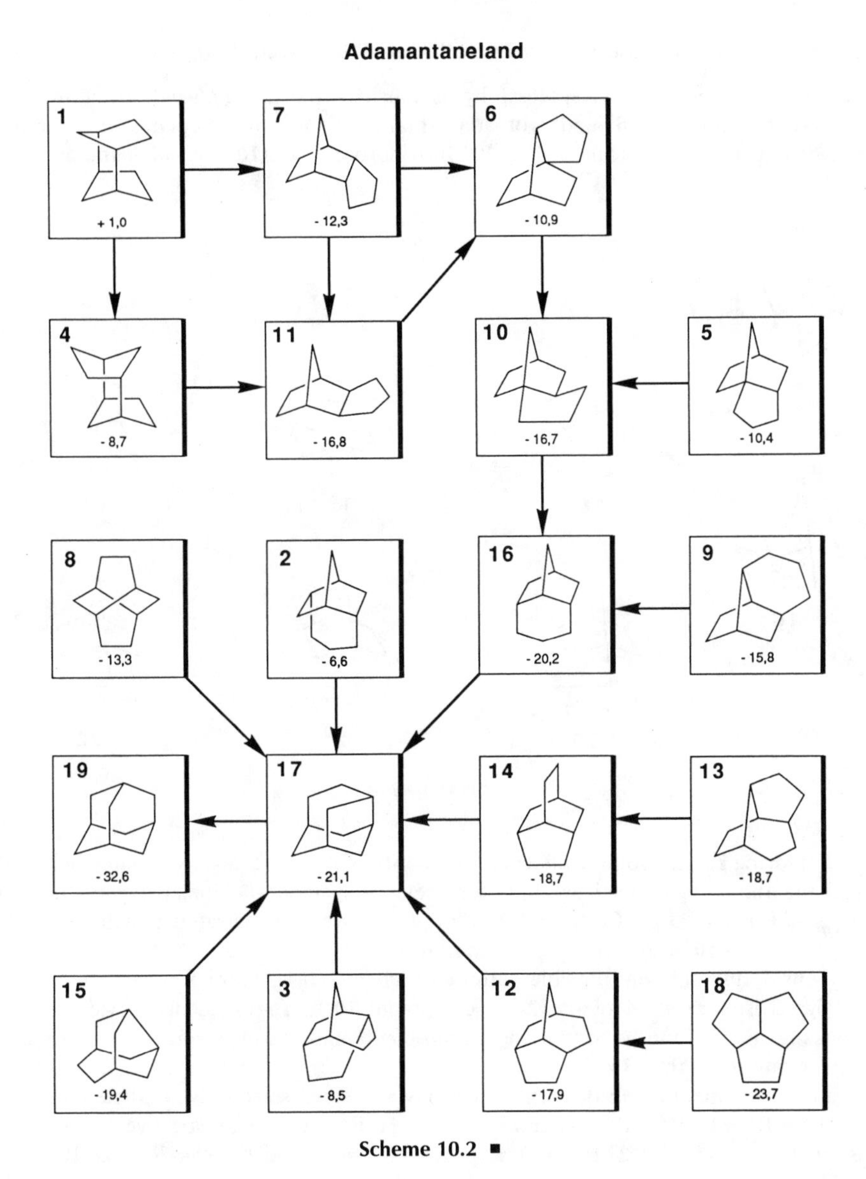

Scheme 10.2 ■

10.2. Aluminum Halide Catalyzed Rearrangements[11]

In view of our studies on rearrangements of regioselectively generated carbocations (Section 10.3), a few representative examples of aluminum halide catalyzed rearrangements shall be mentioned briefly. The reactants concerned are **1**, **4–7**, **10**, and **11**, the compounds of the upper two rows in Scheme 10.2.

10.2A. 1,2-*endo*- (5) and 1,2-*exo*-Trimethylene-8,9,10-trinorbornane (10)

Schleyer and co-workers[10] successfully established that 2-*endo*,6-*endo*-trimethylene-8,9,10-trinorbornane (**16**) and protoadamantane (**17**) are intermediates in the rearrangement of 1,2-*exo*-trimethylene-8,9,10-trinorbornane (**10**) to adamantane (**19**) (Scheme 10.3).

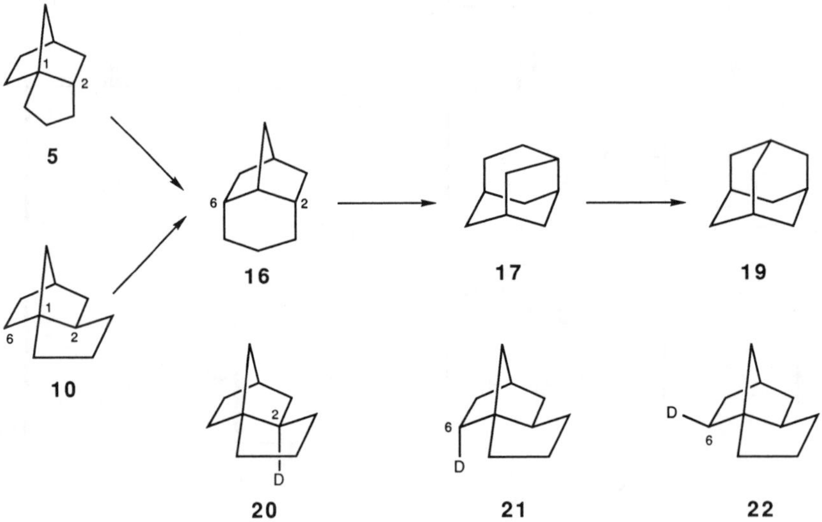

Scheme 10.3 ■

Exploring kinetic isotopic effects in competition experiments between unlabeled **10** and the three regioselectively and stereoselectively mono-D-labeled compounds **20** [D_{endo}-C(2)], **21** [D_{endo}-C(6)], and **22** [D_{exo}-C(6)], we demonstrated that hydride-ion abstraction in **10** occurs at C-6 from the exo side.[12]

Considering different possible pathways (Scheme 10.4), we reexamined the $AlBr_3$-catalyzed isomerization of the 2-*exo* compound **10** to **16**–especially in view of the rearrangement of the corresponding 2-*endo*-compound **5**, which isomerizes at a much higher rate to **16** than **10**.[13]

From treatment of various regioselectively and stereoselectively D-labeled analogs **23** ($R^1 = R^4 = D$, $R^2 = R^3 = H$ and $R^1 = R^4 = H$, $R^2 = R^3 = D$, respectively), **25** ($R^1 = R^3 = D$, $R^2 = R^4 = H$), **27** ($R^1 = D$, $R^2 = R^3 = R^4 = H$), and **28** ($R^1 = R^2 = D$, $R^3 = R^4 = H$) of 1,2-*endo*-, 1,2-*exo*-, and 2-*endo*,2-*endo*-trimethylene-8,9,10-trinorbornane (**5**, **10**, and **16**, respectively), the evidence was gained that a degenerate rearrangement **c1** ⇄ **c2** (Scheme 10.4) is involved in the isomerization of both **5** and **10** to **16**.[14]

Sorensen and Whitworth[8] were able to characterize fully the kinetically quite stable carbocations **c5** and **c6** ($R^2 = R^3 = R^4 = H$), respectively, by various NMR measurements and determined a barrier of 12.5 kcal mol^{-1} for the equilibrium between **c5** and **c6**, these two being at least 2.5 kcal mol^{-1} more stable than **c1** and **c2**, respectively.

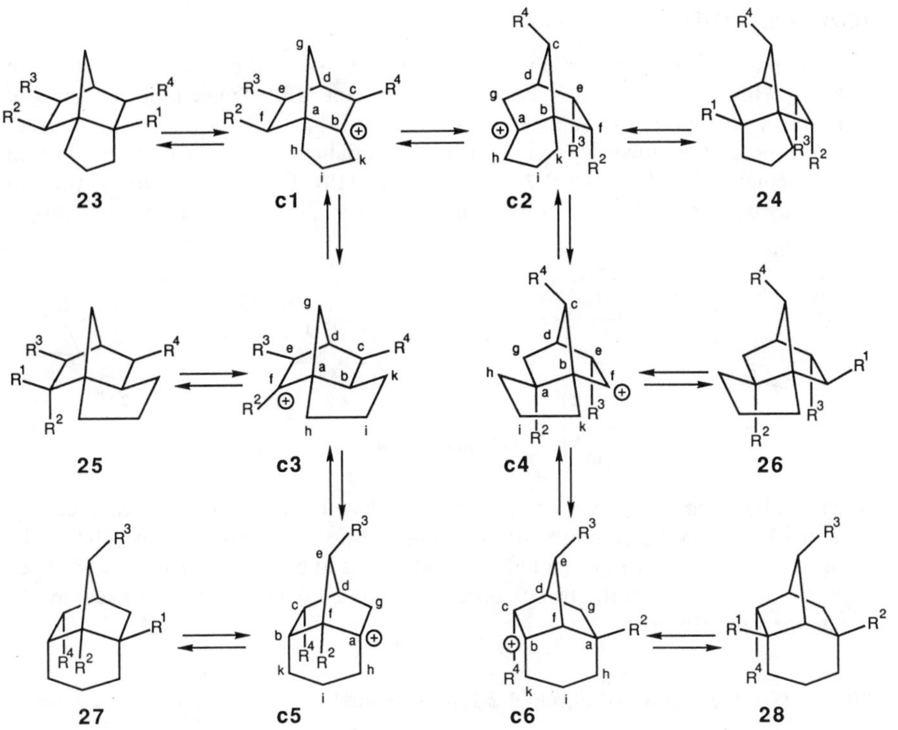

Scheme 10.4 ■

10.2B. 1,7-Trimethylene-8,9,10-trinorbornane (6)

Schleyer et al.[15] described **6** as an "energetic bottleneck" and observed that under $AlBr_3$-catalysis it behaves as a "continental divide," rearranging both "backward" to *exo*-THDCP **11** as well as "forward" to adamantane (**19**). In small amounts (3–5%) a new intermediate, tricyclo[5.3.0.0^{4,8}]decane (**13**), was observed (Scheme 10.5). Tobe et al.[16] confirmed **13** as an intermediate. In their experiments with $AlCl_3$ in CH_2Cl_2, **13** could be accumulated up to 33% within 3 min at 15°. Our experiments[17] with $AlBr_3$ in CS_2 gave a qualitatively similar result. After 4 min at 0°, the composition was 22% **6**, 41% **11**, 11% **13**, and 26% **19**.

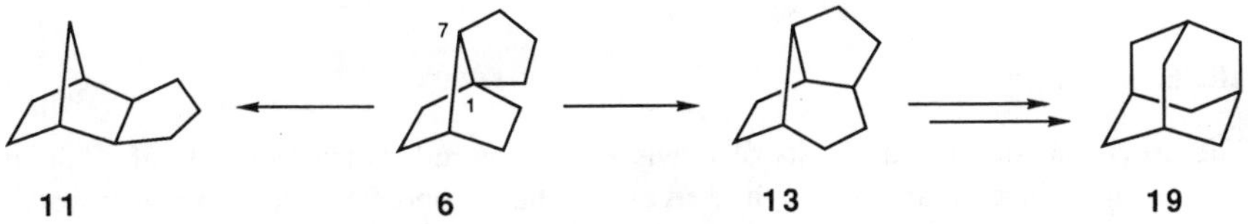

Scheme 10.5 ■

10.2C. *endo*-THDCP 7 and *exo*-THDCP 11

As mentioned in the introduction to this chapter, Lewis acid catalyzed isomerization of *endo*-THDCP **7** to *exo*-THDCP **11** and adamantane (**19**)[1] signaled the beginning of mechanistic studies on adamantane rearrangement.[6]

Sorensen and Whitworth[8] described the potential-energy diagram for the overall rearrangement of *endo*-THDCP **7** to adamantane (**19**). The starting point is the two tertiary carbocations **c7** and **c10** with *endo*- and *exo*-THDCP skeleton, respectively,

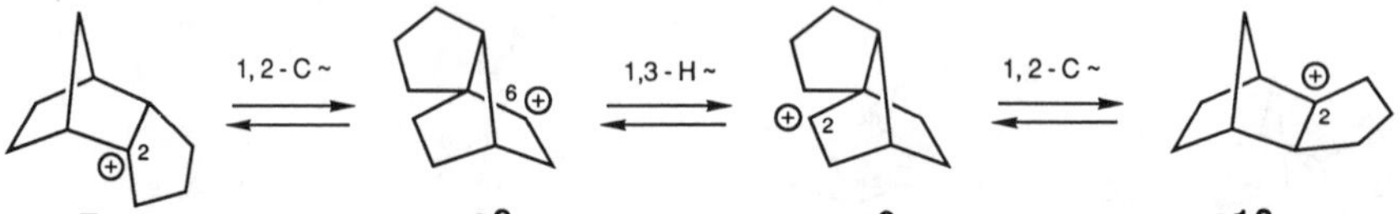

c7 ⇌(1,2-C~) c8 ⇌(1,3-H~) c9 ⇌(1,2-C~) c10

Scheme 10.6 ▪

which are interconnected by the two 1,7-trimethylene-8,9,10-trinorbornyl carbocations **c8** and **c9** via two 1,2-carbon shifts and one 1,3-hydride shift (Scheme 10.6). The energy difference between **c7** and **c10** was estimated to be < 2.5 kcal mol^{-1} (*endo*-**c7** being the more stable of the two carbocations) and a barrier of $\Delta G^{\ddagger} < 10$ kcal mol^{-1} for the process **c10** → **c7**.

10.2D. *syn*- (1) and *anti*-Tricyclo[4.2.1.1^{2,5}]decane (4)

Paquette's results[18] that the *syn*-hydrocarbon **1**, indeed the most highly strained of all the adamantane isomers, should not be able to rearrange to a thermodynamically more stable isomer, seemed very unlikely. Therefore, we reinvestigated the behavior of the syn compound **1** on treatment with $AlBr_3$ in CS_2 at various temperatures and for different reaction times. Included in the studies was the corresponding anti isomer **4**[19] (Scheme 10.7). On the basis of the fact that the *syn*-hydrocarbon **1** rearranges already below 0° and exclusively to the anti isomer **4** as well as the observation that the latter isomerizes only above 0° and at a much higher rate to *exo*-THDCP **11**, the following conclusion must be drawn concerning the mechanistic pathway: Carbocation **c12** with an anti isomer skeleton obtained from the syn compound **1** by hydride abstraction (→ **c11**) and subsequent rearrangement obviously differs from **c13**, formed directly from the anti isomer **4**, that is, both 1,2-carbon migrations, **c11** → **c12** as well as **c13** → **c14**, have to proceed relatively fast in comparison to a possible 1,3-hydride shift **c12** → **c13** and/or a hydride addition **c12** → **4**. Hence, **c12** is the most stable of the three carbocations **c11**, **c12**, and **c13**.

10.2E. Summary

The previously described results of Lewis acid catalyzed rearrangements of adamantane isomers in combination with studies by Olah[20] and Sorensen and Whitworth[8] under stable ion conditions in superacids (although only very few carbocations could

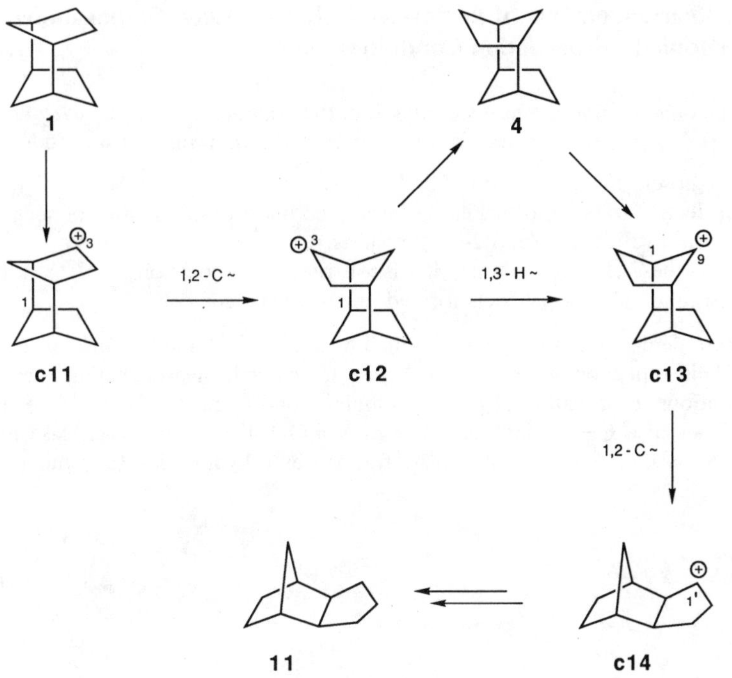

Scheme 10.7 ■

be observed so far) as well as estimated transition-state barriers for several isomerization steps and hence construction of potential-energy diagrams, undoubtedly allow much insight into the complex pathways of the adamantane rearrangement, especially for the interconversion of *endo*-THDCP **7** to adamantane (**19**).

On the other hand, one has to be aware of the specific properties of Lewis acid catalyzed rearrangements:

1. Hydride abstraction may occur regioselectively or even stereoselectively. However, one cannot dictate the regioselective generation of a specific carbocationic center.
2. All 19 members of adamantaneland are a priori candidates for hydride abstraction, that is, not only the initial reactant but also subsequently formed hydrocarbons. In most cases this renders it difficult or even impossible to isolate intermediate isomers.
3. Rearrangements proceed under thermodynamic control. Thus, only the more-stable isomers can be detected.

These three facts limit studies on adamantane rearrangement considerably. Hence, a supplementary approach will be presented in Section 10.3.

10.3. Rearrangements of Regioselectively Generated Carbocations[11] under Ionic Hydrogenation Conditions

To overcome the limitations encountered in the aluminum halide catalyzed rearrangement studies (see Section 10.2E), we assembled the following list of desiderata:

1. regioselective generation of carbocations;
2. selective trapping of primarily formed carbocations as well as of such resulting from hydride and/or 1,2-carbon shifts;
3. inertness of hydrocarbons, that is, of trapped carbocations;
4. isolation also of products formed under kinetic control.

These demands can largely be fulfilled by applying the method of ionic hydrogenation. This reduction method, in contrast to catalytic hydrogenation, proceeds via carbocations. Protonation of a C=Y double bond (e.g., Y = O, NR^1R^2, CR^1R^2) or heterolysis of a C—X single bond (e.g., X = OH, halogen) in each case leads to a carbocation that subsequently can be trapped by a hydride ion (Scheme 10.8). Thus

$$R^1R^2C{=}Y \xrightarrow{H^+} R^1R^2C^{\oplus}{-}YH \xrightarrow{H^-} R^1R^2CH{-}YH$$

$$R^1R^2R^3C{-}X \xrightarrow[\text{or } BF_3]{H^+} R^1R^2C^{\oplus}{-}R^3 \xrightarrow{H^-} R^1R^2R^3C{-}H$$

Scheme 10.8 ■

the required reagents are a pair of a suitable proton donor and a hydride-ion donor. Most successful is the use of trifluoroacetic acid and triethyl silane (Et_3SiH). In cases in which treating the substrate with a protic acid is not opportune or leads to undesired results, boron trifluoride (BF_3 as gas or etherate) may be the reagent of choice to generate the required carbocation. Numerous examples give evidence for the wide applicability of the ionic hydrogenation reaction (for some representative reviews see refs. 21–25).

For our own studies we mainly use unlabeled as well as D-labeled alcohols as reactants, gaseous BF_3 in combination with Et_3SiH as reagents, and methylene dichloride (CH_2Cl_2) as solvent. Fry and co-workers[26] found that under such conditions even simple secondary aliphatic alcohols can be reduced to saturated hydrocarbons. This enables the regioselective generation of many so far inaccessible carbocations and, last but not least, opens new possibilities for gaining more insight into the still fascinating rearrangement of *endo*-THDCP **7** to adamantane (**19**). The fact that secondary carbocations tend to undergo 1,2-carbon migrations and/or hydride shifts to more-stable carbocations offers an additional bonus for such studies. Several examples are discussed in the following sections.

10.3A. Carbocations at C-2 and C-6 of 1,2-*endo*- (5) and 1,2-*exo*-Trimethylene-8,9,10-trinorbornane (10)[27]

The behavior of the regioselectively generated carbocation centers at C-2 and C-6 in 1,2-*endo*- (**5**) and 1,2-*exo*-trimethylene-8,9,10-trinorbornane (**10**) was investigated in order to evaluate whether, under the conditions of ionic hydrogenation, a degenerate

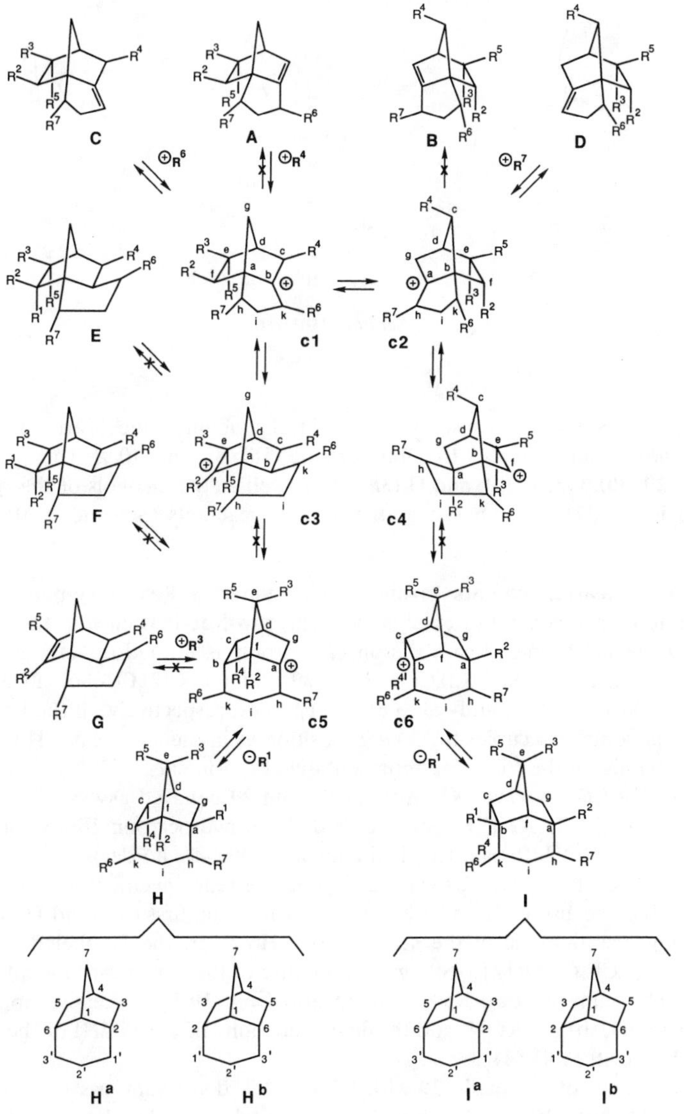

Scheme 10.9 ■

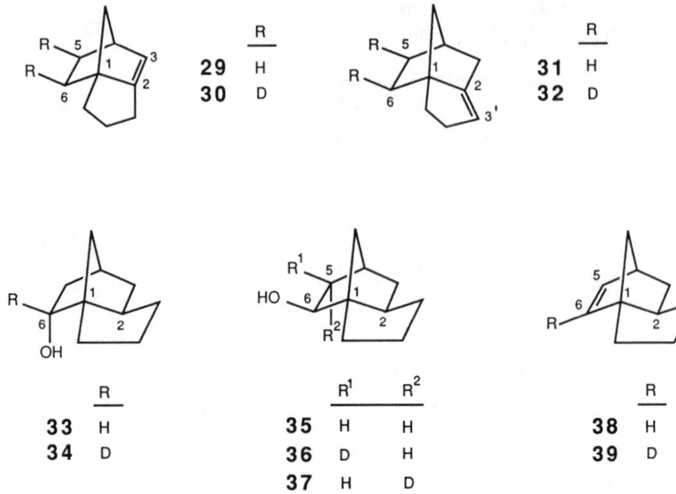

	R
29	H
30	D

	R
31	H
32	D

	R
33	H
34	D

	R^1	R^2
35	H	H
36	D	H
37	H	D

	R
38	H
39	D

Scheme 10.10 ■

rearrangement **c1** ⇄ **c2** is also operative in the adamantane rearrangement to 2-*endo*,6-*endo*-trimethylene-8,9,10-trinorbornane (**16**) (Scheme 10.9). The alkenes of the types **A** (**29**, **30**), **C** (**31**, **32**), and **G** (**38**, **39**) as well as the alcohols of the types **E** (**33**, **34**) and **F** (**35–37**) were chosen as the various reactants (Scheme 10.10 and Table 10.1).

10.3A(1). Rearrangements Involving a Degenerate Rearrangement c1 ⇄ c2. A degenerate rearrangement **c1** ⇄ **c2** is inevitably involved inasmuch as the 1,2-trimethylene-8,9,10-trinorborn-2-yl carbocation **c1** not only is formed directly as manifested by the conversion of the C(2),C(3)-olefin **29** and the C(2),C(3′)-olefin **31** but also indirectly (via 1,3-hydride shift **c3** → **c1** and **c4** → **c2**, respectively), if the leaving group at C-6 to be ionized occupies the 5-*endo* position as in the alcohol **33**. This conclusion is based mainly on the following representative experiments.

C(2),C(3)-Olefin 29 (Type A). Although olefin **29** behaved as expected, that is, on the one hand (CF_3COOH/Et_3SiH) it yielded the hydrocarbon **16** and on the other hand [(a) CF_3COOH, (b) aq. KOH] the alcohols **40**/**40′**, no decision could be made whether the reaction proceeds via the degenerate rearrangement **c1** ⇄ **c2** (**H** + **I**) or not (**H** only), because of **H** and **I** being identical in the first case and $\mathbf{H}^{a}$ (**40**) and $\mathbf{I}^{b}$ (**40′**) being enantiomeric in the second case. However, the D-labeled olefin **30** on reaction with CF_3COOH/Et_3SiH gave a mixture of the two constitutional isomers **41** and **42**. The same mixture was obtained also from **30** by successive application of (a) CF_3COOH, (b) aq. KOH, and (c) final reduction (BF_3/Et_3SiH) of the intermediate alcohol mixture **43**/**44**.

On treatment of the olefin **29** with CF_3COOD, deuterium was not only incorporated at C-3 but also at C-1′ and C-3′, as established by the partial (ca. 15%) formation of **45**. Analogously, D-incorporation was observed when the olefin **29** was

Table 10.1 ▪ Conversions of the Alcohols and Alkenes 29–39 as well as of Related Compounds

	Reactant		Reaction	Products[b,c]								
Section	Compounds	Type	Conditions[a]	Compound	Type	R^1	R^2	R^3	R^4	R^5	R^6	R^7
10.3A(1)	**29**	**A**	*A*	**16**	$\mathbf{H^a \equiv I^b}$	H	H	H	H	H	H	H
	29	**A**	*B*	**40/40′**	$\mathbf{H^a/I^b}$	OH	H	H	H	H	H	H
	30	**A**	*A*	**41/42**	$\mathbf{H^a/I^a}$	H	D	D	H	H	H	H
	30	**A**	*B*	**43/44**	$\mathbf{H^a/I^b}$	OH	D	D	H	H	H	H
	43/44	**A**	*C*	**41/42**	$\mathbf{H^a/I^a}$	H	D	D	H	H	H	H
	29	**A**	*D*	**45**	$\mathbf{H^b \equiv I^b}$	H	H	H	D	H	D	D
	29	**A**	*E*	**46/47**	$\mathbf{H^a/I^b}$	OH	H	H	D	H	D	D
	29	**A**	*F*	**48/49**	**C/D**	—	H	H	D	H	—	H
	31	**C**	*A*	**16**	$\mathbf{H^a \equiv I^b}$	H	H	H	H	H	H	H
	31	**C**	*B*	**40/40′**	$\mathbf{H^a/I^b}$	OH	H	H	H	H	H	H
	32	**C**	*B*	**43/44**	$\mathbf{H^a/I^b}$	OH	D	D	H	H	H	H
	31	**C**	*D*	**48**	$\mathbf{H^a \equiv I^b}$	H	H	H	H	H	D	D
	31	**C**	*E*	**49/49′**	$\mathbf{H^a/I^b}$	OH	H	H	H	H	D	D
	32	**C**	*G*	**32/50**	**C/D**	—	D	D	H	H	—	H
	33	**E**	*C*	**16**	$\mathbf{H^a \equiv I^b}$	H	H	H	H	H	H	H
	34	**E**	*C*	**51/52**[d]	$\mathbf{H^a/I^a}$	H	D	H	H	H	H	H
	33	**E**	*H*	**53/53′**	$\mathbf{H^a/I^b}$	Cl	H	H	H	H	H	H
	34	**E**	*H*	**54/55**	$\mathbf{H^a/I^b}$	Cl	D	H	H	H	H	H
	54/55	**E**	*I*	**56/57**	$\mathbf{H^a/I^b}$	OH	D	H	H	H	H	H
10.3A(2)	**35**	**F**	*C*	**16**	$\mathbf{H^a \equiv I^b}$	H	H	H	H	H	H	H
	35	**F**	*H*	**53**	$\mathbf{H^a}$	Cl	H	H	H	H	H	H
	36	**F**	*J*	**58**	$\mathbf{H^a}$	Cl	H	D	H	H	H	H
	37	**F**	*H*	**59**	$\mathbf{H^a}$	Cl	H	H	H	D	H	H
	38	**G**	*A*	**16**	$\mathbf{H^a \equiv I^b}$	H	H	H	H	H	H	H
	38	**G**	*K*	**52**	$\mathbf{H^a}$	D	H	H	H	H	H	H
	38	**G**	*D*	**60**	$\mathbf{H^a}$	H	H	D	H	H	H	H
	39	**G**	*A*	**51**	$\mathbf{H^a}$	H	D	H	H	H	H	H
	38	**G**	*L*	**42**[e]	$\mathbf{H^a}$	D	H	D	H	H	H	H
	38	**G**	*E*	**61**	$\mathbf{H^a}$	OH	H	D	H	H	H	H

[a] *A*, CF_3CO_2H/Et_3SiH; *B* (1) CF_3CO_2H, (2) aq. KOH; *C*, BF_3/Et_3SiH; *D*, CF_3CO_2D/Et_3SiH; *E* (1) CF_3CO_2D, (2) aq. KOH; *F*, ≤ 0.5 mol-equiv. CF_3CO_2D; *G*, ≤ 0.5 mol-equiv. CF_3CO_2H; *H*, $SOCl_2$, room temperature; *I*, aq. K_2CO_3; *J*, $SOCl_2$, $-20°$ C; *K*, CF_3CO_2H/Et_3SiD; *L*, CF_3CO_2D/Et_3SiD.

[b] For a given product of the types **H** and **I**, its enantiomer is specified by a dash (e.g., **40** and **40′**, respectively).

[c] The different numbering in $\mathbf{H^a}$ and $\mathbf{H^b}$, respectively, $\mathbf{I^a}$ and $\mathbf{I^b}$, respectively, of the same C atoms follows from the correct IUPAC nomenclature. In order to ease the comprehension, the substituents R^1 to R^7 are not drawn in $\mathbf{H^a/H^b}$ and $\mathbf{I^a/I^b}$; their positions are the same as in *formulae* **H** and **I**, respectively, and are independent of the different numbering.

[d] It must be noted that for **52**, type $\mathbf{I^a}$ with $R^2 = D$ is identical to type $\mathbf{H^a}$ with $R^1 = D$.

[e] It must be noted that for **42**, type $\mathbf{H^a}$ with $R^1 = R^3 = D$ is identical to type $\mathbf{I^a}$ with $R^2 = R^3 = D$.

reacted with CF_3COOD followed by base hydrolysis, which among others yielded ca. 15% of the alcohols **46** and **47**. Finally, **29** was treated with only 0.5 mol-equiv of CF_3COOD. The reisolated olefin (ca. 50%) consisted, in addition to unlabeled **29**, of ca. 40% of a 1:1 mixture of **48** [type **C**, R^4 = D-C(3)] and **49** [type **D**, R^4 = D-C(7)]. This proves the equilibrium **C** ⇄ **c1** ⇄ **c2** ⇄ **D**.

It must be mentioned that neither starting from **29** nor from **30** were D-incorporation or scrambling in the recovered reactants detected as in the case of the $AlBr_3$-treatment[14] of **5** and **10**.

C(2),C(3')-Olefin 31 (Type C). That a degenerate rearrangement **c1** ⇄ **c2** occurs also in the reaction of unlabeled olefin **31** with CF_3COOH/Et_3SiH and CF_3COOH followed by base hydrolysis, respectively, yielding **16** and the C(2)-alcohols **40/40'**, respectively, is confirmed by the 1:1 mixture of the constitutionally isomeric dideuterated alcohols **43** (type $\mathbf{H^a}$) and **44** (type $\mathbf{I^b}$), which was obtained from the D-labeled olefin **32** [(a) CF_3COOH; (b) aq. KOH].

Conclusive direct proof for the equilibrium **C** ⇄ **c1** ⇄ **c2** ⇄ **D** was gained from experiments as follows. Unlabeled olefin **31** on treatment with CF_3COOD gave 2-*endo*,6-*endo*-trimethylene-8,9,10-trinorbornanes with D-incorporation at C-1' as well as at C-3': partial (15%) formation of the dideuterated product **48** (CF_3COOD/Et_3SiH) and the alcohols **49/49'** [(a) CF_3COOD; (b) aq. KOH] were observed. Finally, when exposing the 5-*exo*,6-*exo*-dideuterated olefin **32** to only 0.5 mol-equiv. of CF_3COOH, the reisolated reactant consisted of a 1:1 mixture of **32** (type **C**, $R^2 = R^3$ = D) and the 5-*endo*,6-*endo*-dideuterated analog **50** (type **D**, $R^2 = R^3$ = D).

6-endo-Alcohol 33 (Type E). The degenerate rearrangement **c1** ⇄ **c2** is also involved in the conversion (BF_3/Et_3SiH) of the 6-*endo*-alcohol **33** to the hydrocarbon **16**. Under the same conditions the corresponding D_{exo}-C(6)-labeled alcohol **34** led to a ca. 3:1 mixture of **51** [D-C(1)] and **52** [D-C(2)]. This pathway is operative too, treating **33** and **34**, respectively, with thionyl chloride ($SOCl_2$): The former yielded **53/53'**, the latter a 10:1 mixture of the chlorides **54** [D-C(1)] and **55** [D-C(2)], which for further identification was hydrolyzed to the corresponding mixture of alcohols **56** and **57**.

10.3A(2). Rearrangements not Involving a Degenerate Rearrangement. In sharp contrast to the equilibrium **c5** ⇄ **c3** ⇄ **c1** ⇄ **c2** ⇄ **c4** ⇄ **c6** observed in the $AlBr_3$-catalyzed rearrangements (see Section 10.2D), it was found that under ionic hydrogenation conditions no 1,2-hydride shift **c5** → **c3** and, consequently, no degenerate rearrangement **c1** ⇄ **c2** is operative starting from reactants that lead directly to a 2,6-trimethylene-8,9,10-trinorborn-2-yl carbocation **c5**, which is instantaneously trapped by a hydride ion to yield only the product of type $\mathbf{H^a}$. This is the case with both the ionization of the 6-*exo*-alcohol **35**, having the leaving group in a stereoelectronically favored configuration to undergo simultaneous C(1),C(2) bond migration to **c5**, as well as the protonation of the olefin **38**.

6-exo-Alcohol 35 (Type F). The 6-*exo*-alcohol **35** rearranged smoothly (BF_3/Et_3SiH) to the hydrocarbon **16** and its reaction with $SOCl_2$ gave the expected 2-chloro-compound **53**. The following results are conclusive: treatment of the D_{exo}-C(5)-labeled alcohol **36** with $SOCl_2$ led only to the monodeuterated chloride **58** of the general type $\mathbf{H^a}$. No product of type $\mathbf{I^b}$ (R^1 = Cl, R^3 = D), that is, the enantiomer of the C(7)-diastereoisomer **59**, could be detected. Equivalent to this result was the conversion of the D_{endo}-C(5)-labeled alcohol **37** with $SOCl_2$, which gave rise to the monodeuterated chloride **59** (type $\mathbf{H^a}$) as the sole product. Again no compound of type $\mathbf{I^b}$ ($R^1 = Cl$, $R^5 = D$), that is, the enantiomer of the C(7)-diastereoisomer **58**, was observable.

C(5),C(6)-Olefin 38 (Type G). The C(5),C(6)-olefin **38** rearranged quantitatively to 2-*endo*,6-*endo*-trimethylene-8,9,10-trinorbornane (**16**) on treatment with CF_3COOH/Et_3SiH. Analogously, reactions either with Et_3SiD or CF_3COOD gave the expected monodeuterated products **52** and **60**, respectively. Further information was gained as follows. The D-labeled olefin **39** was transformed to the D-C(1) compound **51** of type $\mathbf{H^a}$ as the sole product (CF_3COOH/Et_3SiH), and none of its constitutional isomer of type $\mathbf{I^a}$ ($R^2 = D$) that corresponds to **52** ($\mathbf{H^a}$, $R^1 = D$) was observed. Again, only a type $\mathbf{H^a}$ compound and none of type $\mathbf{I^b}$, that is, the C(7) diastereoisomer of its enantiomer, was obtained by converting the undeuterated olefin **38** (CF_3COOD/Et_3SiD) to the dideuterated product **42** and to the monodeuterated alcohol **61** [(a) CF_3COOD; (b) aq. KOH], respectively.

10.3B. Carbocations at C-3 and C-5 of 1,2-*endo*- (5), 1,2-*exo*- (10), and 1,7-Trimethylene-8,9,10-trinorbornane (6)[28]

The six carbocations **c15–c20**[29] at C-3 and C-5 of 1,2-*endo*- (**5**), 1,2-*exo*- (**10**), and 1,7-trimethylene-8,9,10-trinorbornane (**6**) form a self-contained group, closely interrelated by three 1,2-carbon and three 1,3-hydride shifts (Scheme 10.11). Each of the six carbocations was generated regioselectively in order to learn from trapping experiments about their kinetic stabilities and rearrangement properties. As substrates for our studies we used **62–73**, the six *exo*- and six *endo*-alcohols at C-3 and C-5 of **5**, **6**, and **10**, as well as some D-labeled analogs **74–76** thereof (Scheme 10.12).

First the unlabeled alcohols **62–73** were treated with BF_3/Et_3SiH at various temperatures. The following features are significant: (a) In accordance with the well-known reactivity of norbornane derivatives, heterolysis of the *exo*-C,O bond occurs at a much higher rate than of the corresponding endo bond. (b) At room temperature all the alcohols **62–73** yield 2-*endo*,2-*endo*-trimethylene-8,9,10-trinorbornane (**16**) as the main ($\geq 65\%$) or even the sole product. Its formation requires the carbocations **c1** and **c2** (see Section 10.3A) as intermediates. They may originate either directly by an exo 1,2-hydride shift from **c17** and/or an endo 1,2-hydride shift from **c20** or indirectly by an endo 1,3-hydride shift from **c4**, the result of an exo 1,2-hydride shift from **c19** (Scheme 10.13). (c) At $-78°$, the *endo*-alcohols remaining completely unchanged, the *exo*-alcohols are fully converted to 1,2-*exo*-trimethylene-8,9,10-trinorbornane (**10**) as the main ($\geq 85\%$) or even the sole product.

The previously described results do not allow us to distinguish whether both carbocations **c19** and **c20** are trapped or only one of them, nor whether the carbocations **c15–c18** rearrange clockwise and counterclockwise or only in one specific direction.

To answer the first question, the six *exo*-alcohols **62**, **64**, **66**, **68**, **70**, and **72** were reacted with BF_3 at $-78°$ in the presence of Et_3SiD. In all experiments the monodeuterated 1,2-*exo*-trimethylene-8,9,10-trinorbornane was a mixture of **77** [D_{exo}-C(5)] and **78** [D_{exo}-C(3)], that is, that in each case both carbocations **c19** as well as **c20** were trapped. However, depending on the substrate, the ratio of **77**/**78** varies considerably: on the one hand 1:20 (starting from **62** and **72**) and on the other hand 3:1 (starting from **64**, **66**, **68**, and **70**). The results from treating **70** and **72**, the direct

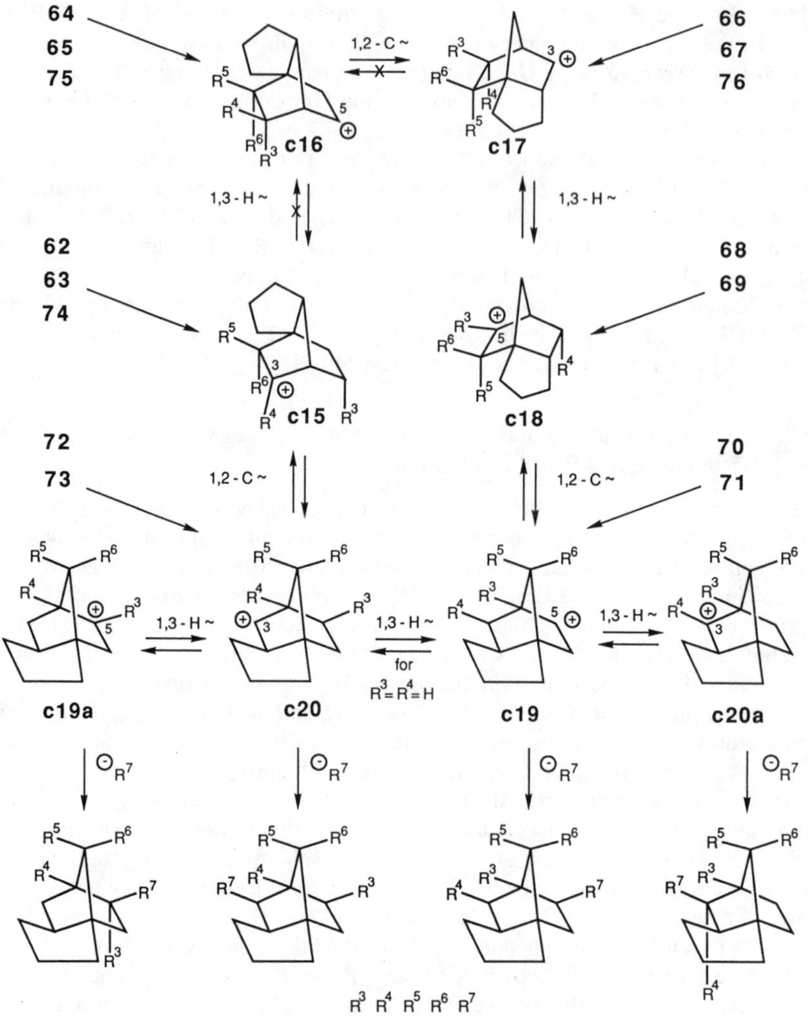

c19a	c20	R^3	R^4	R^5	R^6	R^7	c19	c20a
	10	H	H	H	H	H	10	
	78	H	H	H	H	D	77	
79	79	H	D	D	H	H	80	81
83	84	D	H	H	D	H	82	82
87	87	H	D	H	H	H	85	86

Scheme 10.11 ■

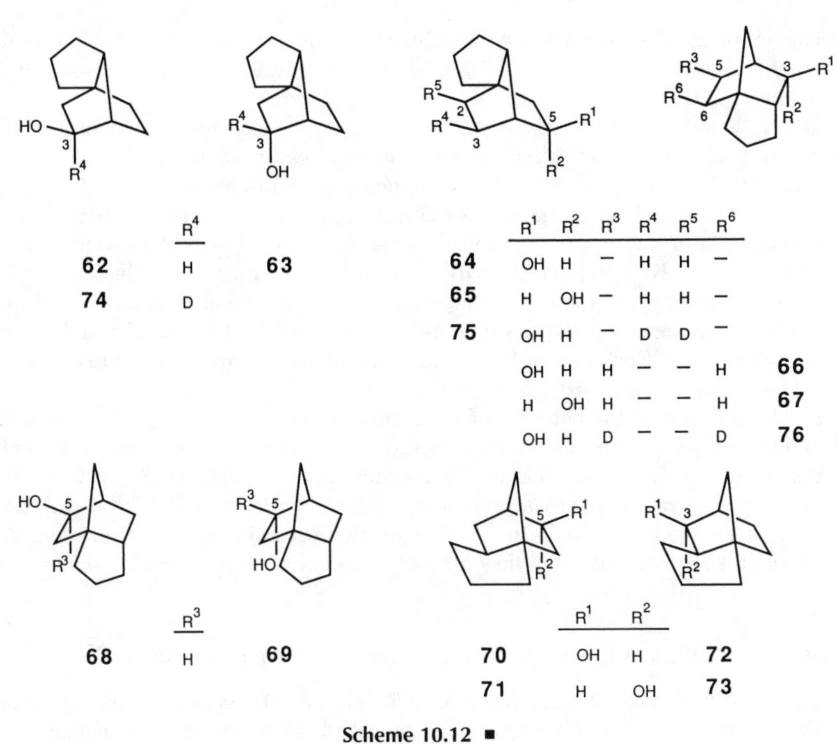

Scheme 10.12 ■

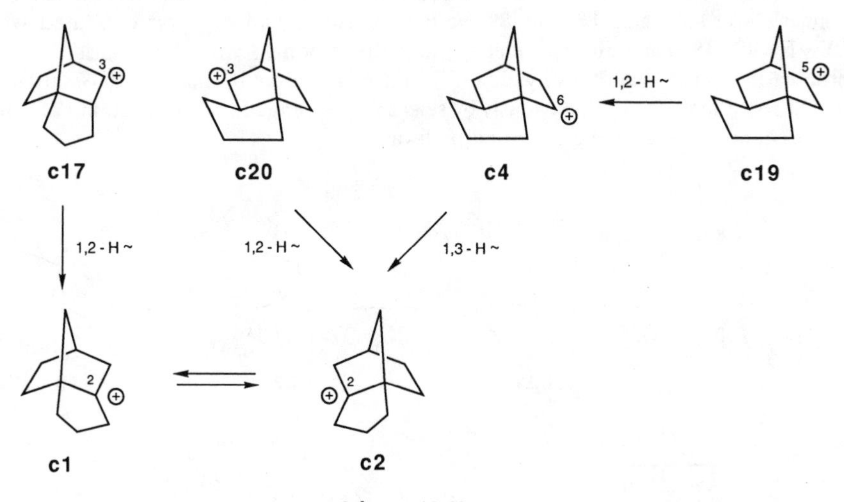

Scheme 10.13 ■

precursors of the carbocations **c19** and **c20**, respectively, demonstrate that the latter are in equilibrium; however, the 1,3-hydride shift **c19** → **c20** is very much favored over the reverse, **c20** → **c19**.

The 1:20 ratio of **77**/**78** using **62** as the reactant leads to the conclusion that carbocation **c15**, if not exclusively, at least predominantly, rearranges counterclockwise to **c20**. To prove this hypothesis unambiguously and to answer at the same time the question which of the carbocations **c15**–**c18** rearrange clockwise and/or counterclockwise, each of the regioselectively and stereoselectively D-labeled alcohols **74**–**76** was treated with BF_3/Et_3SiH at −78°. Alcohol **74** (→ **c15**) led exclusively to **79**, which is the result of the direct rearrangement **c15** → **c20**. None of **80** and/or **81**, the products of a clockwise pathway **c15** → **c16** → **c17** → **c18** → **c19** → **c20a**, could be detected. However, a clockwise pathway was followed starting from **76** (→ **c17**), that is only **82** and none of the mixture **83**/**84** was formed.

Finally, we studied the behavior of the carbocation **c16** by reacting the alcohol **75**. Most interestingly, both pathways are operative: 95% rearranged clockwise (**c16** → **c17** → **c18** → **c19** → **c20a**) to **85**/**86** and 5% counterclockwise (**c16** → **c15** → **c20** → **c19a**) to **87**. Hence carbocation **c16** acts as a watershed. Sorensen and Whitworth[8] estimated that for **c16** ⇄ **c17** "both the barriers for the forward and reverse reactions are less than 10 kcal/mol, although they are not expected to be much less."

10.3C. Carbocation at C-1′ of 1,7-Trimethylene-8,9,10-trinorbornane (6)[17]

As described in Section 10.2B, tricyclo[5.3.0.$0^{4,8}$]decane (**13**) could be isolated as one of the intermediates in the Lewis acid catalyzed adamantane rearrangement of 1,7-trimethylene-8,9,10-trinorbornane (**6**).

To study the behavior of the most probably involved carbocation **c21**, the two epimeric C(1′)-alcohols **88** and **89** were synthesized and both were treated with BF_3/Et_3SiH (Scheme 10.14). Indeed, the hydrocarbon **13** was obtained as the sole product in almost quantitative yield (75% conversion after 10 min and 100% after 30 min at room temperature), obviously by selective trapping of the carbocation **c22**. This result supports strongly the proposed pathway.

c21 c22 13

	R^1	R^2
88	OH	H
89	H	OH

Scheme 10.14 ■

10.3D. Carbocations at C-2 and C-6 of 1,7-Trimethylene-8,9,10-trinorbornane (6) and Carbocations at C-2, C-5, and C-1′ of *endo*-THDCP 7 as well as *exo*-THDCP 11

The $AlBr_3$-catalyzed rearrangement of 1,7-trimethylene-8,9,10-trinorbornane (**6**) (see Section 10.2B) after 1 h at 0° leads to a final ca. 1:1 mixture of *exo*-THDCP **11** and adamantane (**19**).[17] For the formation of the former, several different pathways can be taken into account (Scheme 10.15), for example, hydride abstraction at C-2 (→ **c9**) and subsequent 1,2-carbon migration to the C(2)-carbocation **c10** of the *exo*-THDCP **11**; hydride abstraction at C-6 (→ **c8**) followed either by a 1,3-hydride shift to **c9** (and consequently to **c10**) or by a 1,2-carbon migration to carbocation **c7** of the *endo*-THDCP **7**. The hydrocarbon **7** is known to undergo immediate rearrangement to *exo*-THDCP **11** (see Sections 10.1 and 10.2C).

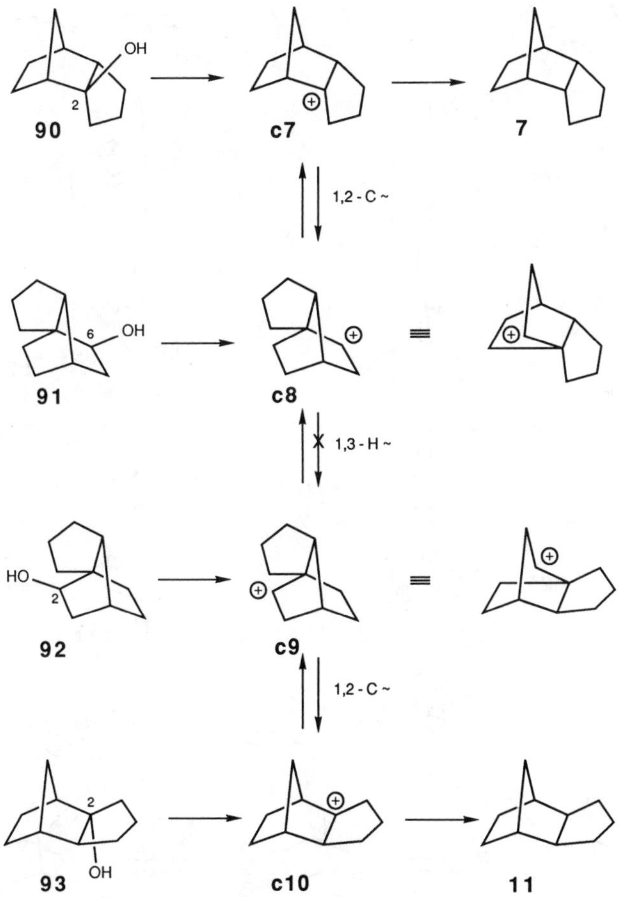

Scheme 10.15 ■

The method of ionic hydrogenation is best suited to be applied to this complex problem. Initial information was gained from the behavior of the four carbocations **c7–c10**. For their regioselective generation, the corresponding alcohols **90–93** were treated with BF_3/Et_3SiH at various temperatures. Whereas the two alcohols **90** and **91** yielded exclusively *endo*-THDCP **7**, a mixture of *endo*-THDCP **7** and *exo*-THDCP **11** was obtained from each of the two alcohols **92** (**7**/**11** = 3:1) and **93** (**7**/**11** = 1:2).

These results clearly show that the 1,3-hydride shift **c8** → **c9** does not occur and consequently the carbocations **c7** and **c8** cannot rearrange to **c9** and **c10**. However, nothing stands in the way of the reverse interconversion from **c10** to the more-stable

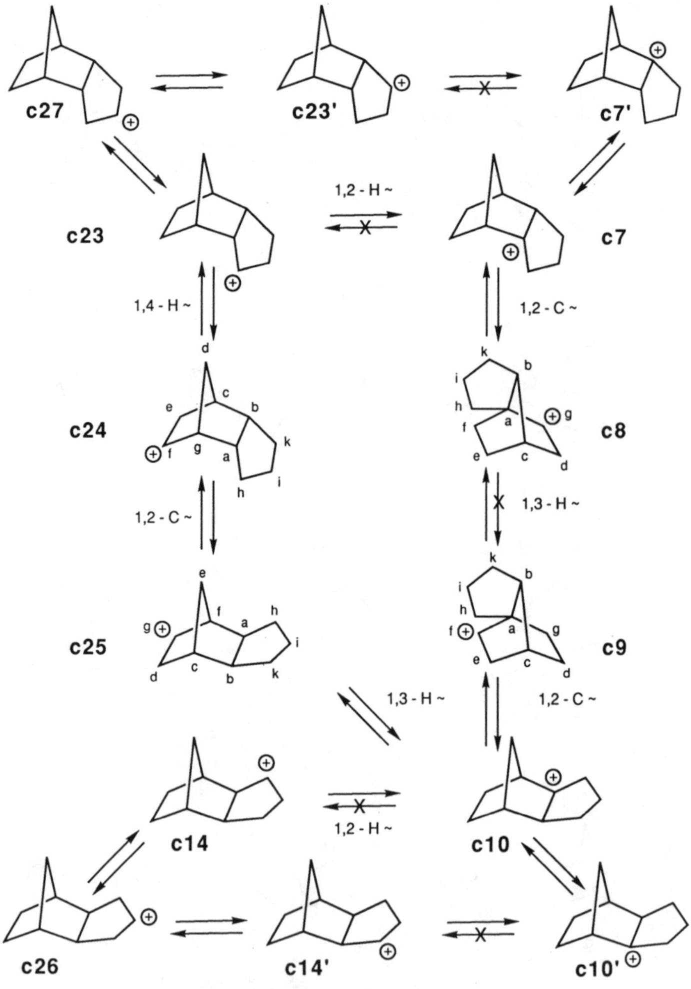

Scheme 10.16 ■

90

93

	R^1	R^2	
94	OH	H	96
95	H	OH	97
104	D	OH	

	R^1	R^2	
98	OH	H	100
99	H	OH	101
	OH	D	102
	D	OH	103

Scheme 10.17 ▪

(by < 2.5 kcal mol^{-1}) carbocation **c7**[8] via **c9** and **c8**, whereby going from **c10** to **c7**, a barrier $\Delta G^{\ddagger} < 10$ kcal mol^{-1} was reported.[8] Neither **c8** nor **c9** can be trapped nor do they undergo a 1,2-hydride shift to the C(5)- and C(3)-carbocation intermediates **c16** and **c15**, respectively, (see Section 10.3B) in the overall cationic rearrangement of *endo*-THDCP **7** to adamantane **(19)**. The combined barrier from **c7** to **c16** was estimated to be ca. 21–22 kcal mol^{-1}.[8]

Further remarkable information was gained from the four carbocations **c23** (C-1′, *endo*-THDCP), **c24** (C-5, *endo*-THDCP), **c25** (C-5, *exo*-THDCP), and **c14** (C-1′, *exo*-THDCP), which are interconnected under themselves and with the previously discussed carbocations **c7–c10** by 1,2-carbon and 1,3-hydride shifts, with the exception of one 1,4-hydride shift (**c23** ⇄ **c24**)[30] (Scheme 10.16). For sterical reasons a 1,3-hydride shift **c24** ⇄ **c7**, analogous to **c25** ⇄ **c10**, is excluded. For the regioselective generation of the carbocations **c23–c25** and **c14**, the corresponding unlabeled alcohols **(94–101)** as well as some D-labled analogues **(102–104)** served as substrates. (Scheme 10.17).

All eight alcohols **94–101** on treatment with BF_3/Et_3SiH gave a mixture of *endo*- and *exo*-THDCP **7/11**, although the ratio differs remarkably depending on the substrate: ca. 3:1 form **94** and **95** (additionally, ca. 10% olefin **105**); ca. 1:4 from **96–99**; ca. 1:1 from **100** and **101** (additionally, ca. 10% olefin **117**). For the first six alcohols **94–99**, the *exo*-alcohols again are much more reactive than the *endo*-alco-

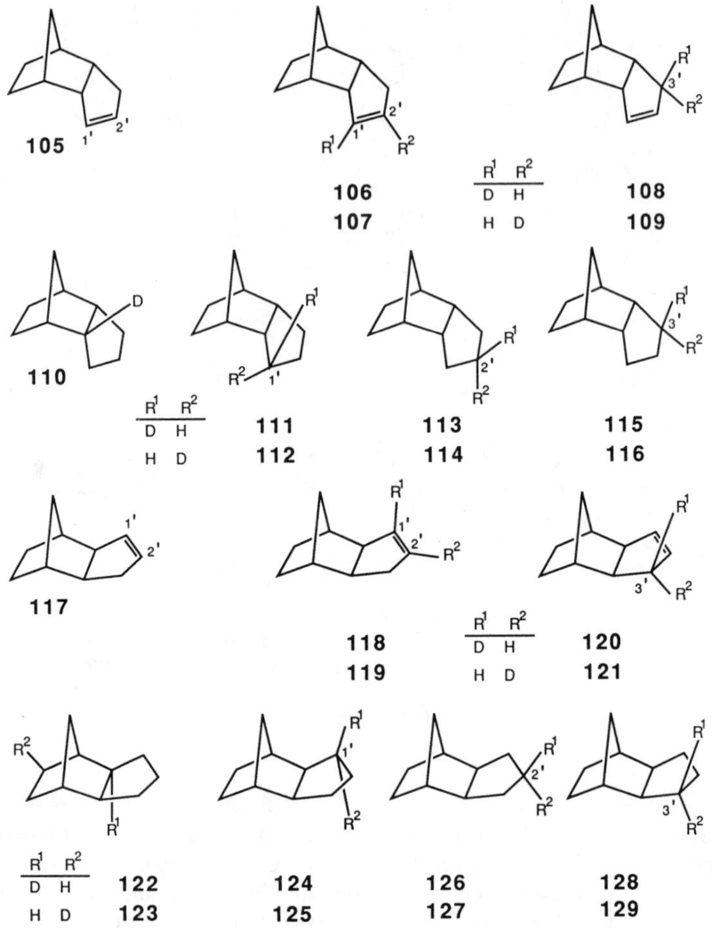

Scheme 10.18 ■

hols, whereas the two C(1′)-alcohols **100** and **101** are of almost equal reactivity. In contrast to the alcohols **94–101** yielding **7**/**11** mixtures, alcohol **90** (→ **c7**), as previously mentioned, leads exclusively to **7**. Hence the 1,2-hydride shift **c7** → **c23** does not take place.

A priori *endo*-THDCP **7** can result from trapping any of the three carbocations **c7**, **c23**, and **c24**. A further question is which carbon migrations and/or hydride shifts operate in both or only in one direction. Some clarifying experiments are discussed later.

Treatment of the unlabeled alcohols **90** and **93–101** with BF_3/Et_3SiD revealed the following features (Scheme 10.18): (a) Whenever *endo*-THDCP is formed, **110** [D_{exo}-C(2)] is found as the sole endo compound, that is, of the three carbocations **c7**,

c23, and **c24**, only the first (**c7**) is trapped. (b) Inasmuch as *exo*-THDCP is a product, both carbocations **c25** [→ **123**, D_{exo}-C(5)] and **c10** [→ **122**, D_{endo}-C(2)] are trapped, that is, there is an equilibrium between **c25** and **c10**. However, it can be suppressed at low temperature, for example, reaction at −75° leads exclusively to **122**. (c) The C(1′)-alcohols **100** and **101** and indeed only these direct precursors of carbocation **c14** give rise to a most noteworthy result: In addition to **110**, **122**, and **123**, to the extent of ca. 10–15% a mixture of the *exo*-THDCPs **124–129** is obtained with D-incorporation not only at all three methylene carbon atoms C-1′, C-2′, and C-3′ but also from the endo side as well as the exo side. In a complementary experiment the D-C(1′)-labeled alcohols **102** and **103** were treated with BF_3/Et_3SiH. Again a mixture of **124–129** (with almost statistical D-distribution) was formed and in addition a mixture of all four possible D-labeled olefins **118–121**. To rationalize this outcome one has to expand the scheme by the carbocations **c10′**, **c14′**, and **c26**. By analogy also for the endo isomers the carbocations **c7**, **c23′**, and **c27** are added (Scheme 10.16). Note that each pair **c10/c10′**, **c14/c14′**, **c7/c7′** and **c23/c23′**, respectively, represent enantiomers and hence **c10** ⇄ **c10′** and **c7** ⇄ **c7′** are degenerate rearrangements. For the latter, Olah, Schleyer, and co-workers determined an energy barrier $\Delta G^{\ddagger} = 7.2 \pm 0.5$ kcal mol^{-1}.[31]

As the formation of **124–129** occurs only by reaction of precursors of carbocation **c14**, the 1,2-hydride shift **c10** → **c14** and **c10′** → **c14′**, respectively, is excluded. As already mentioned, the analogous 1,2-hydride shift **c7** → **c23** of the endo isomer is also not taking place. It should be added that starting from the C(1′)-alcohols **94** and **95** of *endo*-THDCP **7**, none of the carbocations **c23**, **c23′**, and **c27** can be trapped.

That nonetheless **c23**, **c23′**, and **c27** clearly are involved is unambiguously demonstrated by the reaction of the C(1′)-D-labeled alcohol **104** in the presence of Et_3SiH. In analogy to the three products, *endo*-THDCP **7**, *endo*-olefin **105**, and *exo*-THDCP **11**, obtained from the corresponding unlabeled alcohol **95**, again three fractions were isolated. The *endo*-THDCP consisted of a mixture of all possible D-labeled isomers **111–116**, the olefin of a mixture of all D-labeled isomers **106–109** and *exo*-THDCP of a mixture of all D-labeled isomers **124–129**. This result not only proves the participation of **c23**, **c23′**, and **c27** but at the same time demonstrates that the 1,2-hydride shift **c23** → **c27** must occur faster than both the 1,2-hydride shift **c23** → **c7** (as only **c7** can be trapped) and the 1,4-hydride shift **c23** → **c24** or at least in competition to them due to a very similar energy barrier.

10.3E. Carbocations at C-3 and C-9 of *syn*- (1) and *anti*-Tricyclo[4.2.1.1^{2,5}]decane (4)[28,32]

Each of the four possible secondary carbocations at C-3 and C-9, **c11** and **c28** of *syn*-tricyclo[4.2.1.1^{2,5}]decane (**1**) as well as **c12** and **c13** of *anti*-tricyclo[4.2.1.1^{2,5}]decane (**4**), were generated regioselectively by treatment of the corresponding *endo*- and *exo*-alcohols **130–137** with BF_3/Et_3SiH (Scheme 10.19). The following behavior was noticed and the following conclusions could be drawn: (a) Of a given pair of diastereoisomers, especially in the case of the C(3)-alcohols, the *exo*-alcohol is remarkably more reactive than the sterically more hindered *endo*-alcohol (e.g., **130** > **131**, **134** > **135**). However, this difference in reactivity has no influence on the

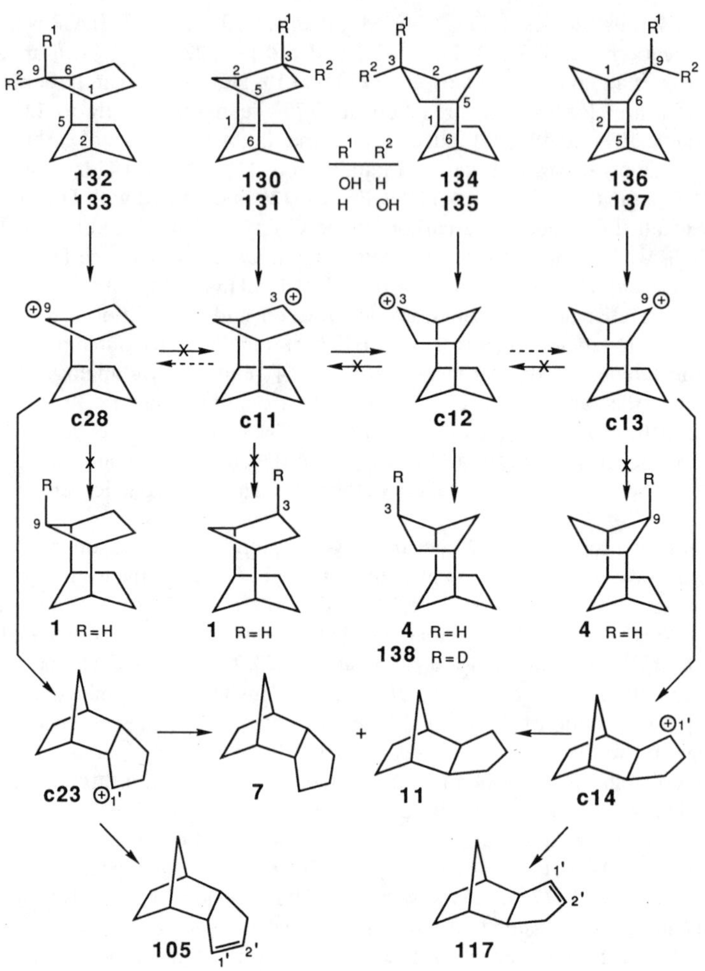

Scheme 10.19 ■

ratio of the products formed. (b) The C(9)-carbocations **c28** and **c13** of the methylene bridges are generated more rapidly than **c11** and **c12**, the C(3)-carbocations of the ethylene bridges. (c) All C(3)-alcohols [**130** and **131** (syn), **134** and **135** (anti)] yield almost exclusively the *anti*-hydrocarbon **4**. For **130** and **131**, this result can best be interpreted, that the *syn*-carbocation **c11** rearranges to the *anti*-carbocation **c12**. If at all, both species undergo a 1,3-hydride shift to **c28** and **c13**, respectively, only to very minor extents (dotted lines in Scheme 10.19). This result is in good agreement with the $AlBr_3$-catalyzed isomerization of **1** to **4** (see Section 10.2D). Reaction of the four C(3)-alcohols **130** and **131** (syn) as well as **134** and **135** (anti) in the presence of Et_3SiD leads to **138** as practically the sole product. This not only proves unequivocally that in each case carbocation **c12** is trapped but, furthermore, that the hydride enters from the exo side. (d) A completely different reaction course is followed starting from the C(9)-alcohols **132** and **133** (syn) as well as from **136** and **137** (anti). No trace of the *anti*-hydrocarbon **4** could be detected, that is, none of the 1,3-hydride shifts **c28** → **c11** and **c13** → **c12** is operative. In each case a mixture of *endo*-THDCP **7** and *exo*-THDCP **11** is obtained and, in addition, depending on the reactant (*syn*- or *anti*-skeleton), the C(1′),C(2′)-olefin **105** or **117**, respectively, is a by-product. These results are consistent with the intermediacy of the carbocations **c23** and **c14** (their behavior is discussed in Section 10.3D), the consequence of a 1,3-carbon migration **c28** → **c23** and **c13** → **c14**, respectively.

10.4. Product Analyses

Capillary and preparative gas–liquid chromatography as well as column chromatography were applied to identify and isolate the compounds. The latter were analyzed spectroscopically (^{1}H, ^{2}H, and ^{13}C NMR and mass spectroscopy).

Of special interest are the D-labeled compounds. The following characteristic features are observed in their ^{13}C NMR spectra: (a) D-labeled C atoms, $t(^{1}J(\mathrm{C,D})$ ca. 20 Hz), ca. 0.4 ppm shifted to higher field; (b) C atoms α to D-labeled C atoms, $t(^{2}J(\mathrm{C,D}) < 1$ Hz), ca. 0.1 ppm shifted to higher field; (c) C atoms β to D-labeled C atoms, $t(^{3}J(\mathrm{C,D}) < 1$ Hz), ca. 0.02 ppm shifted to higher field.

^{2}H NMR spectroscopy allows accurate integration of D-atoms as well as determination of their positions and configurations. D-Labeled adamantane isomers are among the most predestined compounds for application of this method. Even signals with a chemical shift difference of 0.02 ppm (ca. 1 Hz at 45 MHz) are fully resolved. For some cases the signals were assigned by comparing data from ^{1}H and ^{2}H NMR spectra of regioselectively and stereoselectively D-labeled compounds obtained by independent syntheses.

10.5. Concluding Remarks

As demonstrated by the examples discussed in Section 10.3, ionic hydrogenation, applied to the problem of the complex adamantane rearrangement, is a most efficient method for studying the behavior of possibly involved carbocations. It allows as well regioselective generation as the selective trapping of primarily formed and/or rearranged carbocations. Particularly in combination with, for example, D-labeled reactants and/or Et_3SiD as trapping reagent most enlightening information can be gained. Furthermore, it has the advantage that, once trapped, hydrocarbons remain unchanged under the reaction conditions and products formed under kinetic control can be isolated, too.

Acknowledgments. My deep appreciation and warmest thanks go to all of my enthusiastic co-workers, full of scientific curiosity, who for a shorter or longer period of time joined the exciting voyage of discovery, exploring pathways through the jungle of adamantaneland. This review mentions contributions of the following: M. Brossi, K. I.

Ghatak, T. Gögh, O. Holzschuh, A. M. Klester, and M. Valentíny. Also much appreciated is the help of M. Kríz in preparing the schemes.

This work was partially supported by the Swiss National Science Foundation and by Ciba-Geigy AG, Basel.

References

1. (a) Schleyer, P. v. R. *J. Am. Chem. Soc.* **1957**, *79*, 3292. (b) Schleyer, P. v. R.; Donaldson, M. M. *J. Am. Chem. Soc.* **1960**, *82*, 4645.
2. For a detailed overview, see ref. 3.
3. Olah, G. A., Ed. *Cage Hydrocarbons*; Wiley: New York, 1990.
4. Landa, S. *Chem. Listy* **1933**, *27*, 415, 433. Landa, S.; Máchácek, V. *Collect. Czech. Chem. Commun.* **1933**, 5, 1.
5. Prelog, V.; Seiwerth, R. *Chem. Ber.* **1941**, *74*, 1644, 1769.
6. For the most recent extensive reviews, see refs. 7 and 8.
7. Schleyer, P. v. R. My thirty years in hydrocarbon cages: From adamantane to dodecahedrane; in *Cage Hydrocarbons*; Olah, G. A., Ed.; Wiley: New York, 1990.
8. Sorensen, T. S.; Whitworth, S. M. The superacid route to 1-adamantyl cation; in *Cage Hydrocarbons*; Olah, G. A., Ed.; Wiley: New York, 1990.
9. Whitlock, H. W., Jr.; Siefken, M. W. *J. Am. Chem. Soc.* **1968**, *90*, 4929.
10. Engler, E. M.; Farcasiu, M.; Sevin, A.; Cense, J. M.; Schleyer, P. v. R. *J. Am. Chem. Soc.* **1973**, *95*, 5769.
11. All carbocations are characterized by the prefix **c**, for example, **c1**, **c15**, etc., and are drawn as "classical" carbocations, a simplification without any prejudice. For more detailed comments on this aspect, see ref. 8, p. 72. For the purpose of comprehensive discussions, the latter are based on one enantiomeric form of chiral reactants only, although racemates were used in all experiments.
12. Klester, A. M.; Ganter, C. *Helv. Chim. Acta* **1985**, *68*, 104.
13. Klester, A. M.; Jäggi, F. J.; Ganter, C. *Helv. Chim. Acta* **1980**, *63*, 1294.
14. Klester, A. M.; Ganter, C. *Helv. Chim. Acta* **1983**, *66*, 1200.
15. Schleyer, P. v. R.; Grubmüller, P.; Maier, W. F.; Vostrovsky, O.; Skattebøl, L.; Holm, K. H. *Tetrahedron Lett.* **1980**, 921.
16. Tobe, Y.; Terashima, K.; Sakai, Y.; Odaira, Y. *J. Am. Chem. Soc.* **1981**, *103*, 2307.
17. Ghatak, K. L.; Ganter, C. *Helv. Chim. Acta* **1988**, *71*, 124.
18. Paquette, L. A.; Doecke, C. W.; Klein, G. *J. Am. Chem. Soc.* **1979**, *101*, 7599.
19. Brossi, M.; Ganter, C. *Helv. Chim. Acta* **1987**, *70*, 1963.
20. Olah, G. A. Carbocations and electrophilic reactions of cage hydrocarbons, in *Cage Hydrocarbons*; Olah, G. A., Ed.; Wiley: New York, 1990.
21. Kursanov, M.; Parnes, Z. N.; Loim, N. M. *Synthesis* **1974**, 633.
22. Colvin, E. W. In *Silicon in Organic Synthesis*; Butterworths: London, 1981; p. 329.
23. Weber, W. P. Silicon reagents for organic synthesis; in *Reactivity and Structure Concepts in Organic Chemistry*; Vol. 14, editors: Hafner, K.; Lehn, J.-M.; Rees, Ch. W.; Schleyer, P. v. R.; Trost, B. M.; Zahradnik, R., Springer-Verlag: Berlin, 1983; p. 272.
24. Kursanov, D. N.; Parnes, Z. N.; Kalinkin, M. I.; Loim, N. M. *Ionic Hydrogenation and Related Reactions*; Harwood Academic Publishers: CH-Chur, 1985.
25. Pawlenko, S. In *Organosilicon Chemistry*; Walter de Gruyter: New York, 1986; p. 114.
26. Adlington, M. G.; Orfanopoulos, M.; Fry, J. L. *Tetrahedron Lett.* **1978**, 2955.
27. Klester, A. M.; Ganter, C. *Helv. Chim. Acta* **1985**, *68*, 734.
28. Brossi, M.; Gögh, T.; Holzschuh, O.; Klester, A. M.; Valentíny, M.; Ganter, C. Unpublished results.

29. To distinguish **c19** from **c19a** and **c20** from **c20a** it becomes necessary that $R^3 \neq R^4$. However, for $R^3 = R^4 = H$, this is irrelevant because **c19** ≡ **c19a** and **c20** ≡ **c20a**.

30. For reports on 1,4-hydride shifts, see, for example, the following: (a) Gwynn, D. E.; Skillern, L. *J. Chem. Soc., Chem. Commun.* **1968**, 490. (b) Rome, D. W.; Johnson, B. L. *Tetrahedron Lett.* **1968**, 6053. (c) Wilder, P., Jr.; Cash, D. J.; Wheland, R. C.; Wright, G. W. *J. Am. Chem. Soc.* **1971**, *93*, 791. (d) Shephard, J. M.; Singh, D.; Wilder, P., Jr. *Tetrahedron Lett.* **1974**, 2743. (e) Field, M. J.; Hillier, I. H.; Smith, S.; Vincent, M. A.; Mason, S. C.; Whittelton, S. N.; Watt, C. I. F.; Guest, M. F. *J. Chem. Soc., Chem. Commun.*, **1987**, 84.
31. Olah, G. A.; Prakash, G. K. S.; Shih, J. G.; Krishnamurthy, V. V.; Mateescu, G. D.; Liang, G.; Sipos, G.; Buss, V.; Gund, T. M.; Schleyer, P. v. R. *J. Am. Chem. Soc.* **1985**, *107*, 2764.
32. Brossi, M.; Ganter, C. *Helv. Chim. Acta* **1988**, *71*, 848.

11

Reflex and Anti Reflex Effects: Discovery and Developments

Josette Fournier

Université d'Angers, Angers, France

Bernard Waegell

Université d'Aix-Marseille III, Marseille, France

Contents

11.1. Introduction
11.2. Experimental Data
  11.2A. Reflex Effect
    11.2A(1). 3,3,5,5-Tetramethylcyclohexanone
    11.2A(2). Derivatives of 3,3,5-Trimethylcyclohexanone and of 3,3-Dimethylcyclohexanone
    11.2A(3). 2,2,6,6-Tetramethyltetrahydropyran-4-one Derivatives
  11.2B. Anti Reflex Effect
    11.2B(1). Bicyclo[3.2.1]octane Derivatives
    11.2B(2). Bicyclo[3.2.1]oct-6-en-3-one and Derivatives
    11.2B(3). Bicyclo[3.2.1]octan-6-one and Derivatives
    11.2B(4). Derivatives of 6,6-Dimethylbicyclo[3.1.1]heptan-3-one
    11.2B(5). Mechanistic Applications of the Anti Reflex Effect to Favorskii Rearrangement
11.3. Theoretical Approach to the Reflex and Anti Reflex Effects
  11.3A. Molecular Mechanical Calculations
  11.3B. Semiempirical CNDO/2 Calculations

11.4. Conclusion
11.4A. Geometrical Aspects
11.4B. Energetical Aspects
Acknowledgments
References

11.1. Introduction

In the early 1960s, it was already recognized that the perfect chair conformation represents an ideal state, because electron diffraction studies had revealed that cyclohexane itself was flatter than supposed,[1] the internal C—C—C angle being 111.55° instead of 109.40°. Furthermore, it was thought that ring substitution would cause additional deviations from the chair conformation.

In 1958 Sandris and Ourisson[2] observed discrepancies in the conformational behavior of substituted and unsubstituted α,α'-dibromocyclohexanones. In order to explain the differences in the conformational equilibria of these ketones, they postulated the occurrence of a new conformational transmission effect, which they called the *reflex effect*.

In the chemical equilibrium between *cis*- (equatorial–equatorial) and *trans*- (equatorial–axial) 2,6-dibromocyclohexanone, the latter is the more abundant. This is the result of a compromise between steric and electrostatic effects: The cis isomer, where the two bulky bromines are equatorial, is favored from the steric point of view but not from the electrostatic one, as the three dipoles (two C—Br and one C═O) are practically parallel. This situation is reversed in the trans isomer because the dipoles (at least one C—Br and one C═O) are now almost orthogonal.[3] While studying the same kind of equilibrium in 2,6-dibromo-3,3,5,5-tetramethylcyclohexanone, Sandris and Ourisson were quite surprised to find that the cis diequatorial isomer was prevailing in the cis–trans equilibrium. They proposed the following mechanical explanation, which they named reflex effect. As the 1,3-diaxial interaction between the bulky 3,5-methyl group cannot be avoided by conformational equilibrium, they repel each other and move apart from one another. This displacement is mechanically transmitted by the two C(2)—C(3) and C(5)—C(6) bonds of the cyclohexane ring so that the axial bonds (bromines, for instance) on C(2) and C(6) are pinched together. Consequently, the corresponding axial substituents must be destabilized, especially if they are bulky like bromines. To summarize, it was possible to say that a steric constraint on one side of a cyclohexane ring will result (reflect) in a steric constraint on the other side of the ring.[4] (See Figure 11.1.)

Such structural features, or 1,3-diaxial interactions, occur in numerous natural products, particularly tetracyclic triterpenes and steroids.[5] This observation provided a good reason to carry on a systematic study of the reflex effect.[6]

On the other hand and quite rapidly, the possibility of the occurrence of a reverse reflex effect—that is, the *anti reflex effect*—was taken into consideration. If spreading

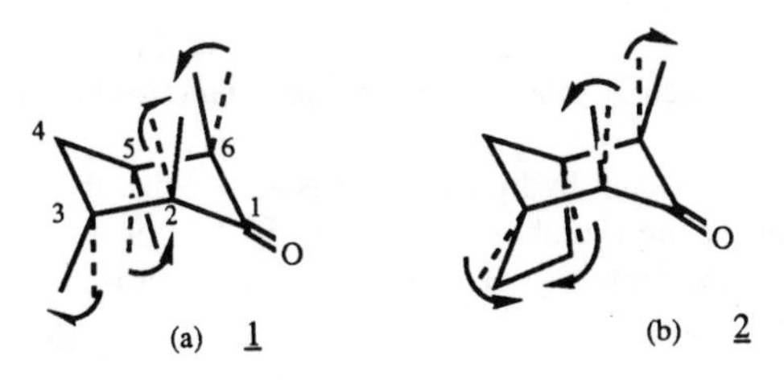

Figure 11.1 ■ Operation of (a) the reflex effect and (b) its inverse, that is, the anti reflex effect.

apart of axial substituents on one side of the cyclohexane ring destabilizes the axial substituents on the other side by bringing them together, logically the pinching of axial substituents on one side of a cyclohexane ring should stabilize the axial substituents on the other side. This is exactly what can be done by the introduction of a 1,3-ethano bridge in the axial position of cyclohexane, which results in the formation of the bicyclo[3.2.1]octane.[7] Indeed, as expected the 2,6-diaxial dibromobicyclo-[3.2.1]octan-3-one is now the stable isomer.

The reflex effect and anti reflex effect have been studied on various model systems by means of structural analysis both from the spectroscopic and chemical points of view,[8] by molecular mechanical calculations,[9] and by semiempirical quantum mechanical calculations.[10] These studies have used and confirmed (and even allowed the discovery of) various rules that can be applied to the conformational and stereochemical analysis in the cyclohexane derivatives series; for instance, the following:

1. There is an approximately linear relationship between the carbonyl frequency in infrared and the C—(C=O)—C angle in cyclohexanone.[11]
2. An axial bromine next to the carbonyl group induces a $\nu_{C=O}$ shift close to zero.[12]
3. An equatorial bromine next to the carbonyl group induces a $\nu_{C=O}$ shift that ranges from 15 to 22 cm^{-1}.[13] $\Delta\nu_{C=O} = 7 + 13\cos\phi$, where ϕ = dihedral angle of O=C—C—Br.[13]
4. An axial bromine next to the carbonyl induces a $\Delta\lambda$ shift of around +28 nm and a $\Delta\varepsilon$ ranging from +40 to +120 of the $n \rightarrow \pi^*$ band[14] of the carbonyl.
5. An equatorial bromine next to the carbonyl induces a $\Delta\lambda$ shift of around −5 nm and a $\Delta\varepsilon$ ranging from +5 to +20 for the same UV band, $\Delta\lambda = -16 + 44\sin^2\phi$.[15]
6. An axial proton next to a carbonyl in a cyclohexanone gives an NMR signal at a lower field than the corresponding equatorial proton.[10b]
7. An axial proton geminated with an equatorial hydroxyl group on a chair cyclohexane gives an NMR signal that appears at higher field than the corresponding equatorial proton geminated with an axial hydroxyl group.
8. The ^3J coupling between protons obey to Karplus-type rules.[16]
9. The ^4J coupling requires a coplanar M-or W-shaped arrangement.[17,34]
10. The value of the $^2J_{^{13}C-H}$ coupling varies in a linear way as a function of the s character of the C—H bonding orbital.[18]

11. The bromination of cyclohexanones proceeds via the acid-catalyzed enolization of the ketone, which is followed by the axial electrophilic addition of bromine to the enol.[19]
12. The acetylation of axial hydroxyl group is very difficult if not impossible.
13. The reduction of the cyclohexanone carbonyl by borohydride proceeds stereoselectively, with a preferential equatorial approach on the less-hindered side.

11.2. Experimental Data

In order to confirm the reflex and anti reflex effects, it was necessary initially to examine the behavior of suitable molecules, both from the physical and chemical points of view. The following substrates have been studied by means of their bromination and of Favorskii rearrangement on the resulting bromoketones, the reduction of the carbonyl group, the acetylation of the corresponding alcohols, and the acetolysis of the related tosylates: 3,3,5,5-tetramethylcyclohexanone; 2,2,6,6-tetramethyltetrahydropyran-4-one; bicyclo[3.2.1]octan-3-one; and 6,6-dimethylbicyclo[3.1.1]-heptan-3-one.

11.2A. Reflex Effect

11.2A(1). 3,3,5,5-Tetramethylcyclohexanone. The axial and equatorial conformer ratio of the α-brominated ketone was measured by UV and IR spectroscopy as well as by dipole moment measurements in various solvents of increasing polarity. These results were compared to those reported in the literature for α-bromocyclohexanone (Figure 11.2). In nonpolar trichloromethane, the ratio of axial conformer is 44% when measured by IR for the tetramethylated bromoketone derivative and must be compared to the 53% of axial conformer for the nonmethylated cyclohexanone bromo derivative; in polar acetonitrile the same ratios are 32 and 43%,[4] respectively. For the constrained tetramethylated bromoketone derivative, the axial bromo conformer is destabilized especially in polar solvents (isolating the dipoles), as expected from the reflex effect theory.

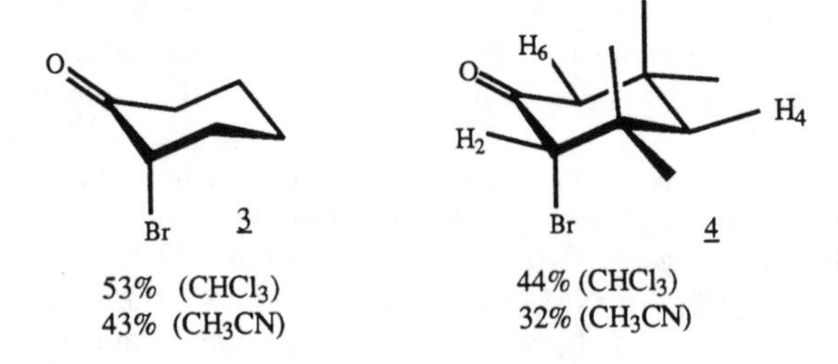

Figure 11.2 ■ Compared proportions of axial bromine conformer in the conformational equilibrium of **3** and **4** in solvents of low ($CHCl_3$) and high (CH_3CN) polarity (estimated by IR spectroscopy).

Later on, it was possible to confirm[20] the "chair symmetry" and the conformational equilibrium of 2-bromo-3,3,5,5-tetramethylcyclohexanone by the observation of two long-range couplings ^4J between the H_2, H_4, and H_6 equatorial protons for the conformer with an axial bromine, which is more abundant in CCl_4. Simultaneously, this observation allowed confirmation that, in α-brominated cyclohexanone, the NMR signal of the geminated equatorial proton appears at higher field than the axial proton, contrary to what is observed in cyclohexanes.[21] This observation, which could be extended to 39 α-brominated ketones,[22] was assigned to the carbonyl anisotropy by calculating the shielding constants for axial and equatorial protons in nonbrominated cyclohexanones.[23] Accordingly, the chemical shift of a proton geminated with a substituent next to a cyclohexanone carbonyl allows assignment of the conformation (axial or equatorial) of the substituent and the position of the corresponding conformational equilibrium.[24] Further information on the deformation of tetramethylated cyclohexanone derivatives can be gained from the analysis of the solvent effects on the methyl group that is in the β position relative to the carbonyl group.[25] The ^{1}H NMR spectrum of *trans*-2,6-dibromo-3,3,5,5-tetramethylcyclohexanone exhibits a single signal for the protons geminated with the bromines. This observation is in agreement with a 50% conformational equilibrium between the two identical conformers. Conversely, according to the ^{1}H NMR spectrum of the corresponding cis isomer, there is a single diequatorial conformer present in tetrachloromethane, with no conformational equilibrium. These observations are in perfect agreement with the reflex effect theory.

In the corresponding α-fluorinated derivative, the equatorial conformer is the only one present in polar solvents or in concentrated solutions (10% in DMSO, 50% in CCl_4); the axial conformer is the most abundant in tetrachloromethane or in diluted solutions (1%). As in brominated derivatives, similar variations are observed for the $\Delta\nu_{C=O}$ in the infrared ($+25$ cm^{-1} for an equatorial conformation, $+15$ cm^{-1} for an axial conformation). The W-shaped long-range $^4J_{HH}$ coupling that was observed in the ^{1}H NMR spectrum of the brominated ketone disappears and is replaced by a U-shaped $^4J_{HH}$ coupling (1.6 Hz) between the 4 and 6 axial protons. This observation emphasizes the sensitivity of the $^4J_{HH}$ coupling to weak deformations of the cyclohexane ring due to 1,3-*syn*-diaxial interactions, when a bulky bromine is replaced by a fluorine with a much weaker van der Waals radius. Mixed dihalogenated derivatives have been studied. The *cis*-2-fluoro-6-bromo derivative is exclusively diequatorial whereas in the corresponding trans isomer and in the geminated -2,2-fluorobromo derivative, the bromine is axial. The quite important deformations resulting therefrom are at the origin of quite important modifications in the magnitude of the coupling constants. Thus $^4J_{F_{eq}-H_{4eq}}$ varies from 5.8 Hz in the equatorial α-fluorinated ketone to 3.6 Hz in the α,α-gemfluorobrominated ketone.

Using ^{13}C NMR and the increments related to the ^{13}C chemical shift changes resulting from the replacement of a hydrogen by a halogen, it was possible to study the conformational equilibrium of 2-bromo-3,3,5,5-tetramethylcyclohexanone in $CDCl_3$ and C_6D_{12}.[26] Similar results are obtained by considering either the β (43% of axial bromine conformer in $CDCl_3$, 57% in C_6D_{12}) or γ effect (respectively, 49 and 66%). Using the same approach it was shown that the *cis*-2,6-fluorobromo-3,3,5,5-tetramethylcyclohexanone is essentially diequatorial (91%).

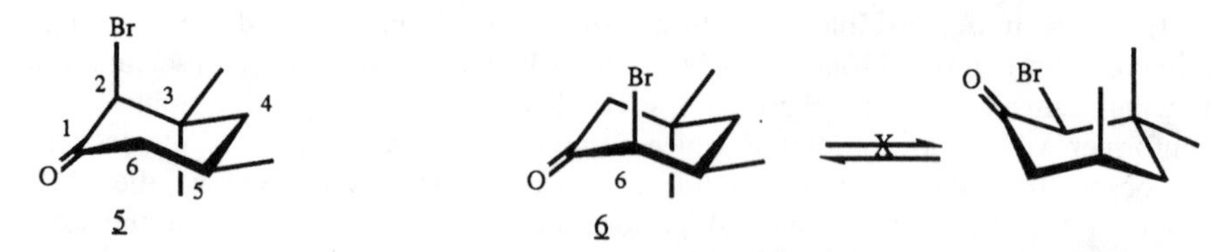

Figure 11.3 ■ 2- and 6-bromo derivatives of 3,5,5-trimethylcyclohexanone feature an exclusively axial bromine.

11.2A(2). Derivatives of 3,3,5-Trimethylcyclohexanone and of 3,3-Dimethylcyclohexanone. It could be shown[27] that the 2- and 6-monobrominated derivatives of the trimethylated cyclohexanone are exclusively axial (Figure 11.3). This observation clearly shows that is it the 1,3-diaxial interaction between the methyl groups that is decisive in the reflex effect.

The 2,6-dibromo-3,3,5-trimethylcyclohexanone obtained by bromine addition to the corresponding ketone was shown—both chemically (reduction into the corresponding bromohydrin) and spectroscopically (UV and IR)—to be exclusively the trans thermodynamically more stable epimer, as in the case of the nonmethylated ketone. Consequently the 1,3-diaxial interaction between a methyl group and one hydrogen is not sufficient to produce a significant reflex effect. However, the analysis of the bromination of the 3,3-dimethylcyclohexanone,[28] as well as the conformational equilibria of the corresponding monobrominated derivatives obtained either in 2 or 6 position, provides some more information (Figure 11.4).

The fact that the dimethylated axial bromo conformer are destabilized relative to the nonmethylated cyclohexanone whether the gem dimethyl group is in the 3 or in the 5 position shows that the reflex effect is due to 1,3-diaxial methyl interactions, not to interactions between the 2-axial bromine and the 3-equatorial methyl group.

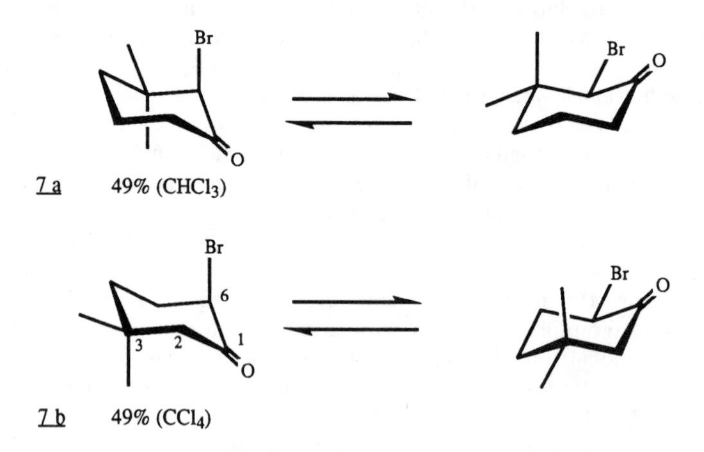

Figure 11.4 ■ Conformational equilibrium of α-brominated 3,3-dimethylcyclohexanone.

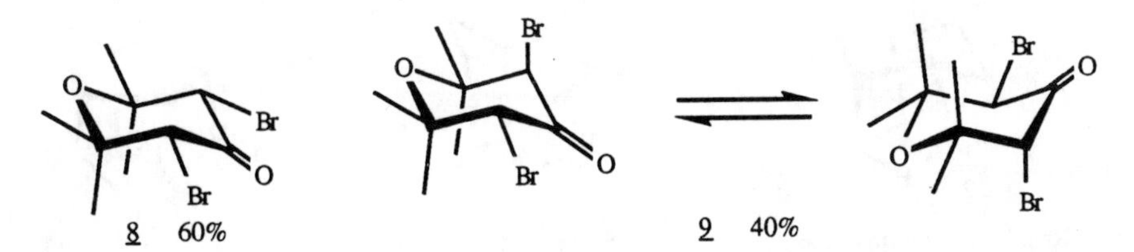

Figure 11.5 ▪ Epimerization equilibrium between cis (**8**) and trans (**9**) 3,5-dibromo-2,2,6,6-tetramethyltetrahydropyranone. The trans isomer undergoes conformational equilibrium.

11.2A(3). 2,2,6,6-Tetramethyltetrahydropyran-4-one Derivatives. As the ^{1}H NMR spectrum of this molecule is much simpler than that of 3,3,5,5-tetramethylcyclohexanone, this was in principle an ideal model to test the reflex effect.[29]

However, because the C—O bond length is shorter than the C—C bond and also introduces additional dipoles, the real situation is more complex. In the α-monobrominated derivative, the bromine is axial and the overall conformation is probably twisted.[4,22,29] The carbon–bromine has a position intermediate between the axial and the equatorial positions. Accordingly, there is no bromine γ effect on the carbon chemical shifts. The structure of the cis α,α'-dibrominated derivative was shown to be symmetrical, with two equatorial bromines; this assignment is confirmed by the fact that this molecule easily undergoes Favorskii rearrangement. The trans α,α'-dibrominated isomer, which undergoes a rapid conformational equilibrium, can be epimerized into the cis (60%, as estimated by IR spectroscopy) isomer in the presence of hydrobromic acid at 40°C. This can be assigned to the reflex effect (Figure 11.5).

When reduced with sodium borohydride in ethanol, the 3,3,5-tribromo-2,2,6,6-tetramethyltetrahydropyran-4-one gives rise to two bromohydrins: one with an axial hydroxyl group, the other with an equatorial hydroxyl group. This can be rationalized on the basis of a skewed conformation of the ketone, which provides a relief for the steric and dipolar interactions. This conformational analysis could be further confirmed by ^{1}H NMR and IR spectroscopy using the hydroxyl group as a conformational probe, particularly by showing that intramolecular hydrogen bonding occurs between the hydroxyl group and the bromine when they are sufficiently close to each other. Furthermore, the equatorial hydroxyl group can be acetylated when equatorial. Finally, it could be shown that during the reduction of various bromoketones, aimed toward the formation of the corresponding bromohydrins, the hydride attack occurs exclusively on the opposite side of the halogen when the latter is axial and only preferentially when it is equatorial. These mechanistic considerations are in agreement with theory.[30]

The replacement of a hydrogen by a halogen induces variations of ^{13}C chemical shifts relative to reference cyclohexanones. Using these changes it was possible[26] to establish the position of the conformational equilibrium of 4 (Figure 11.2), in $CDCl_3$ and C_6D_{12}. Results are self-consistent whether one uses γ effects (49 and 66% of axial bromine conformer, respectively, in $CDCl_3$ and C_6D_{12}) or β effects (43 and 57% of axial bromine conformer, respectively, in $CDCl_3$ and C_6D_{12}). Using the same

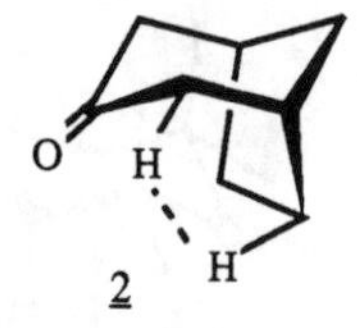

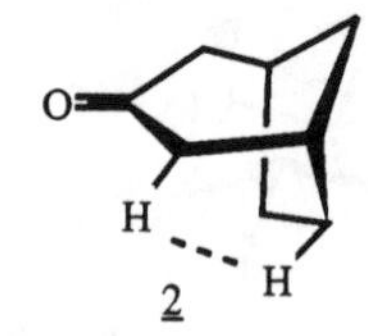

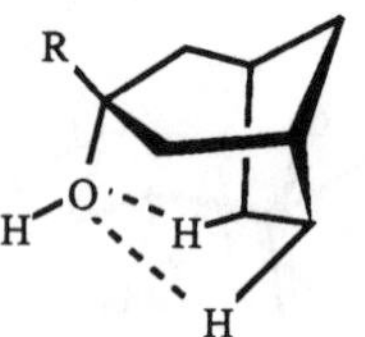

Figure 11.6 ■ Possible nonbonded interactions between endo hydrogens in bicyclo[3.2.1]octan-3-one and corresponding alcohols.

measurements, *cis-α′,α*-fluorobromo-3,3,5,5-tetramethylcyclohexanone was estimated to be 91% diequatorial.

11.2B. Anti Reflex Effect

Pushing apart two 1,3-diaxial 3,5-substituents (methyl group) on 3,3,5,5-tetramethylcyclohexanone, brings 2,6-axial substituents together on the other side of the cyclohexane ring; as a consequence they are destabilized. Conversely, it was expected that bringing together such 1,3-diaxial 3,5-substituents—by linking them together—would stabilize the axial position for bulky substituents on the other side of the ring. This was called the anti reflex effect. Bicyclo[3.2.1]octan-3-one, and more generally bridged bicyclo[3.2.1] or -[3.1.1] derivatives appeared as suitable models to test these ideas.

11.2B(1). Bicyclo[3.2.1]octane Derivatives. The first experimental proofs of the occurrence of the anti reflex effect were brought in 1964[7,31] after it had been shown that bicyclo[3.2.1] octan-3-one could be prepared univocally in high yields by carbene addition to norbornene.[32] This bridged bicyclic ketone is a satisfactory model in the sense that the pinching due to the ethano bridge compares well to the spreading apart due to inescapable steric interaction occurring between the two axial methyl groups of 3,3,5,5-cyclohexanone, visualized by their van der Waals spheres. However, the bicyclic ketone is more rigid than its flexible nonbridged counterpart and is therefore less prone to undergo dynamic conformational equilibria (Figure 11.6).

The fact that the bridgehead substituents are hydrogens, removes the possibility of steric interactions with axial bromines, as was the case with the equatorial methyl groups in **1**. The bridged model nevertheless features some drawbacks as nonnegligible interactions are likely to occur between the endo substituents of the cyclohexane ring and the ethano bridge endo hydrogens.

Bromination of **1** with pyridinium perbromide in THF occurs extremely rapidly, so that to yield easily 75% of *cis-α,α′*-diaxial dibromide **11** (Figure 11.7). The bromination is so easy that it is difficult to stop the reaction at the axial monobromide **10** stage, which therefore can only be obtained with 30% yield. The extreme ease with which this bromination occurs, is in agreement with Corey's axial bromination theory.[19] The ring flattening near the carbonyl favors enolization so that further bromination can occur, without any great structural modifications. In the presence of traces of hydrobromic acid at 40°C, dibromo derivative **11** epimerizes into the trans

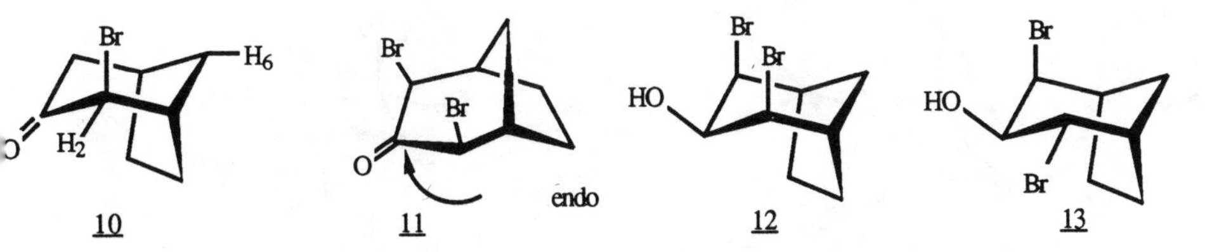

Figure 11.7 ■ Axial monobromo (**10**) and dibromo (**11**) derivatives of bicyclo[3.2.1]octan-3-one and bromohydrines **12** and **13** obtained by sodium borohydride reduction of **11** and its trans epimer (not represented).

axial–equatorial dibromide with 80% yield. In both **10** and **11**, the spectroscopic data are in agreement with the axial position of the bromine[33] on the cyclohexanone ring with essentially a flattened chair conformation. For the dibromo derivatives **11**, the UV absorption is fairly high (342 nm in $CHCl_3$ as compared to 280 nm for the corresponding nonbrominated ketone **2**) and the dipole moment is slightly too weak (4.20 D measured) as compared to the expected calculated value of 4.98 D. Nevertheless the conformations of **10** and **11** are in agreement with the observation of W-shaped long-range coupling $^4J_{H_2-H_6}$.[34] The assignment of an equatorial position of the hydroxyl in bromohydrins **12** and **13** (obtained by potassium borohydride reduction of **11** in the presence of boric acid) results from the observation of an easy acetylation, a strong OH----Br bond with the nearby axial bromine, and the appropriate vicinal coupling constants of the proton geminated with the hydroxyl group.

Obviously the hydride attack on the carbonyl of **11** and its trans epimer must have occurred from the endo side, which is the less hindered, because of the cyclohexanone ring flattening resulting from the anti reflex effect.

The two α-fluorinated ketones **14** and **15** have been prepared by opening of 3-cyano-2,3-epoxybicyclo[3.2.1]octane with triethylammonium fluoride. The axial α-fluorinated ketone **14**, which exhibits a narrow IR band at 1730 cm^{-1}, is the kinetic product. In presence of acid, it isomerizes into **15**, where the IR $\nu_{C=O}$ has now shifted to 1740 cm^{-1}.

The chemical shifts and coupling constants of the equatorial H_2 proton of **14** and axial proton H_2 of **15** (reported in Figure 11.8) have been measured and have later been used to study conformational equilibria of flexible fluorinated cyclohexanones.

As reasonably expected, the bromination of the α-fluorinated bicyclo[3.2.1]octan-3-one **15** with an equatorial fluorine yields the trans α-fluorinated α'-brominated derivative with an axial bromine. On the other hand, the bromination of **14** yields the diaxial derivative **16**, when carried out in buffered medium. But an epimerization occurs in nonbuffered acetic acid so as to form the trans epimer with an equatorial fluorine. If we compare (Figure 11.9) the $\nu_{C=O}$ shifts of both **16** and **17** relative to the nonhalogenated ketones **2** and **1**, one would expect a much weaker value than the 20 cm^{-1} observed for **16** considering the axial position of the two halogens.[11b,13]

This observation and the ^{1}H NMR analysis can be interpreted by assigning to the carbon–halogen bonds in **16** a position that is intermediate between axial and equatorial and that results from the cyclohexanone ring flattening already discussed.

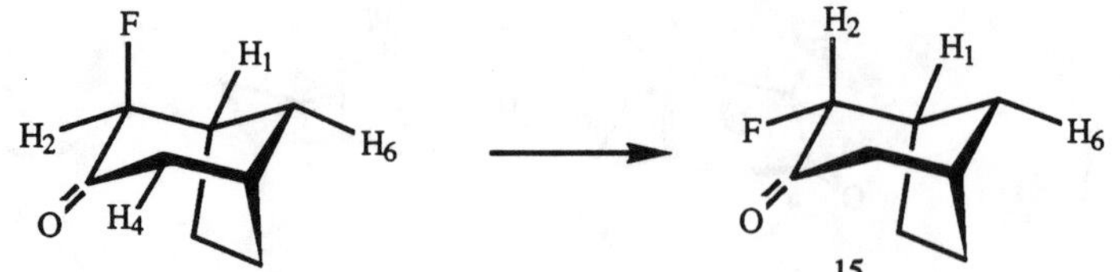

F H1 H2 H6 O H4 14 → H2 H1 F H6 O 15

δ_{H_2} = 4.25 ppm

$^{2}J_{H_2F}$ = 51.0 Hz

$^{3}J_{H_2H_1}$ = 4.8 Hz

$^{4}J_{H_2H_1}$ = $^{4}J_{H_2H_4}$ = 1Hz

δ_{H_2} = 4.69 ppm

J_{H_2F} = 48.6 Hz

Figure 11.8 ■ Use of fluorine as a conformational probe to study the anti reflex effect.

Figure 11.9 ■ Ring flattening in the bicyclo[3.2.1]octan-3-one system as it appears from $\Delta\nu_{C=O}$ in infrared.

The introduction of a bridgehead methyl group suppresses the symmetry of the nonsubstituted bicyclo[3.2.1]octan-3-one and removes the degeneracy in the ^{1}H NMR spectra.[22] The overall chemical and spectroscopic behavior of the corresponding monobromo and dibromo derivatives and of the related bromohydrins completely confirms the anti reflex effect. The α,α'-axial dibrominated derivative **18** forms rapidly and is more easily obtained than the corresponding monobrominated derivatives in 2 and 4 positions, which could be obtained univocally by the bromination of the corresponding independently prepared enol acetates (Figure 11.10). In acetic acid and

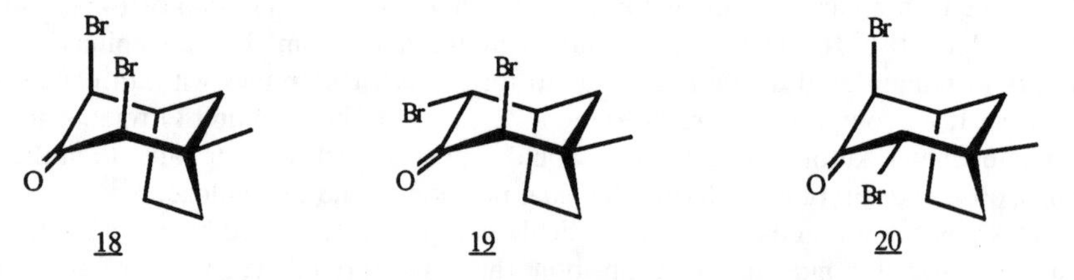

Figure 11.10 ■ α,α'-Dibromo derivatives of 1-methylbicyclo[3.2.1]octan-3-one.

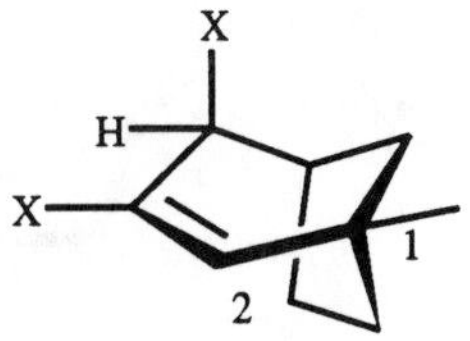

X = Br 21 / 22 = 80 / 20
X = Cl 21 / 22 = 47 / 53

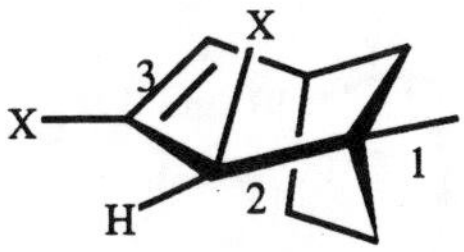

Figure 11.11 ■ Stabilization of the bulky halogen on the exo side of the bridged bicyclic system.

in the presence of hydrobromic acid the α,α'-dibromo derivative **18** epimerizes in a 60 : 40 mixture of the trans derivatives **19** and **20**.

Dibromocarbene and dichlorocarbene addition to 1-methylnorbornene produces, after thermal electrocyclic cyclopropane ring opening, the two compounds **21** and **22**—the ratio of which depends on the halogen nature[35] (Figure 11.11).

In both derivatives the allylic bulky halogen is axial and trans to the ethano bridge, as expected from the anti reflex effect. The lithium aluminum hydride reduction of dibromo derivative **21** gives rise to an 84 : 16 mixture of vinylic bromides **23** and **24** (Figure 11.12).

When lithium aluminum deuteride is used instead of the hydride it can be seen that the deuterium is exo in the major compound **23** and endo in the minor one **24**. Consequently, the major product results from an S_N2'-type attack of the hydride anti to the ethano bridge, as predicted by the anti reflex effect.[36] The minor derivative **24** is the product of a straightforward S_N2 substitution where the hydride approach occurs syn to the ethano bridge.

Variable-temperature recordings of the circular dichroism of 1,8,8-trimethylbicyclo[3.2.1]octane-3-one **25** provided evidence for a conformational equilibrium between a chair (80%) and a boat conformation (20%)[37] (Figure 11.13). According to the vicinal coupling constants, the chair is flattened near the carbonyl. The energy

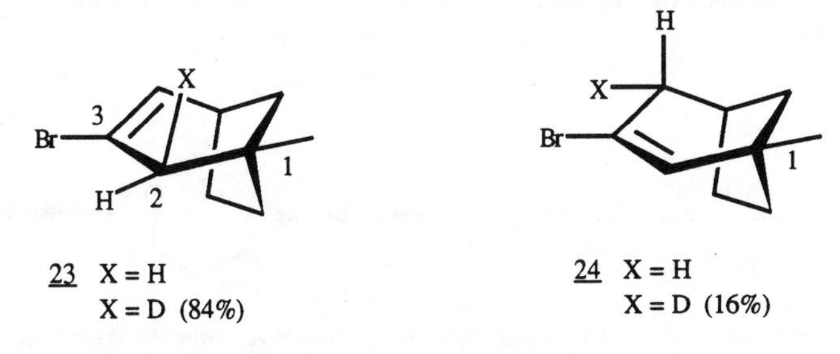

Figure 11.12 ■ Preferential exo attack of the hydride (or deuteride) on **21**.

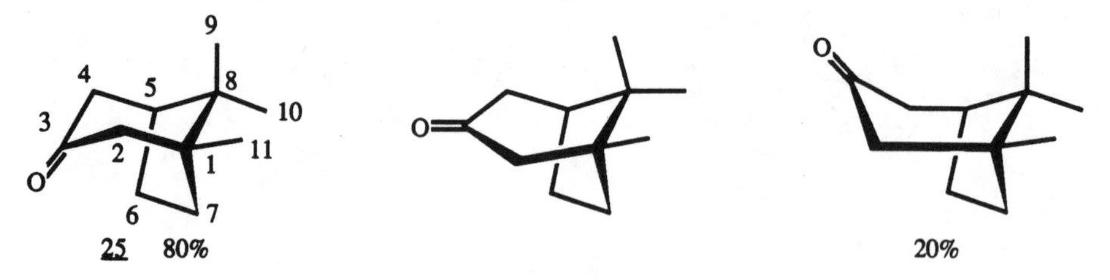

Figure 11.13 ■ Conformational equilibria in 1,8,8-trimethylbicyclo[3.2.1]octan-3-one as estimated by variable-temperature circular dichroism measurements.

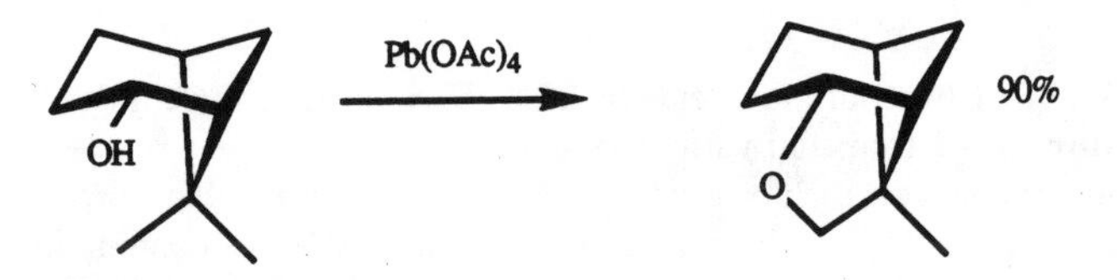

Figure 11.14 ■ Transannular lead tetraacetate cyclization.

difference of 0.74 kcal mol^{-1} between the chair and boat conformations is much lower than the 5.5 kcal mol^{-1} generally admitted for cyclohexanone.

Provided the appropriate reactions were used, it was expected that the anti reflex effect should also be operating in reactivity. Lead tetraacetate oxidation of alcohols was a good candidate as it was known that it was possible to observe intramolecular through-space transannular reactions[38] depending on the distance between the reactive sites (Figure 11.14).

A cyclic transition state with an intermediate alkoxy radical is generally considered as the mechanism most likely to be involved.[39] The important point to consider here is that the distance between the oxygen and the δ-carbon (Figure 11.15), must be in the range of 240 to 270 pm in order to be able to observe this transannular reaction. This is the reason why lead tetraacetate oxidation of alcohols derived from bicyclo[3.2.1]octan-3-one were studied. The reactivity of the carbonyl toward various nucleophiles was also of interest. For instance, the addition of methyl magnesium

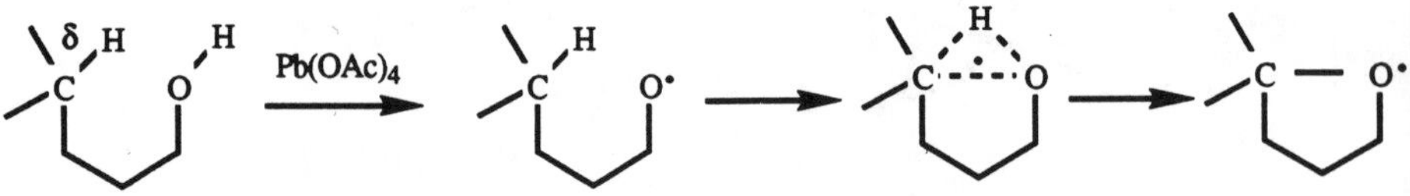

Figure 11.15 ■ Importance of the distance between the oxygen and the δ-carbon in the lead tetraacetate cyclization mechanism.

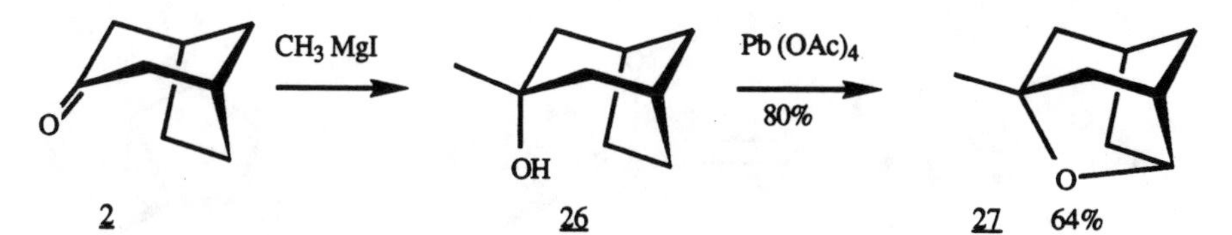

Figure 11.16 ■ Exo Grignard addition to bicyclo[3.2.1]octan-3-one.

iodide on **2** occurs exclusively on the exo side, anti to the ethano bridge so as to yield the tertiary alcohol **26** (Figure 11.16).

When the latter is submitted to lead tetraacetate oxidation, ether **27**, which results from a transannular cyclization, is the major compound (64%) of the reaction mixture formed with 80% yield. As far as the carbonyl reduction of **2** is concerned, the best stereoselectivity (97–98%) in *endo*-alcohol **28** is observed with catalytic hydrogenation in the presence of PtO_2 (Figure 11.17).

With sodium borohydride in ethanol, approximately equal amounts of **28** and **29** are obtained, whereas with an excess of $LiAlH_4$ (4 : 1 relative to the ketone) in THF, it is the *exo*-alcohol **29** that becomes the major product of the reaction. Formation of **29**, which results from an axial attack of the hydride, can again be rationalized by considering the cyclohexanone ring flattening.[41] These alcohols have been characterized by ^1H NMR spectroscopy.[42]

Quite interestingly it is only the *endo*-alcohol **28** that cyclizes when submitted to lead tetraacetate oxidation. The cyclization giving rise to ether **30** occurs with a somewhat lower yield than the one observed for the formation of **27**. This can be rationalized not only by the fact that the acetylation (giving **31**) and the alcohol oxidation (giving **32**) are now more easy because **28** is a secondary alcohol, but also on the basis of a conformational effect of the methyl group in **26**.[40] Because the *exo*-alcohol **29** does not cyclize into **33** under lead tetraacetate oxidation conditions, a conformational equilibrium can reasonably be excluded (Figure 11.18).

As can be seen in Figure 11.19, neoisocedranol **33** features the same bicyclo[3.2.1]octan skeleton in a somewhat more complex structure. Furthermore, the

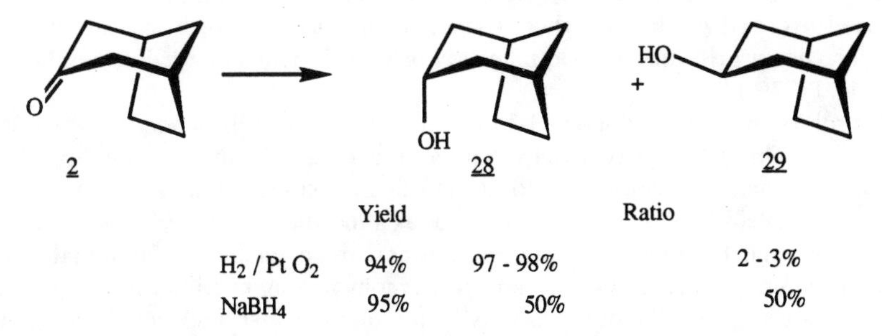

Figure 11.17 ■ Comparison of stereoselectivity of the reduction of the bicyclo[3.2.1]octane-3-one with H_2–PtO_2 and $NaBH_4$.

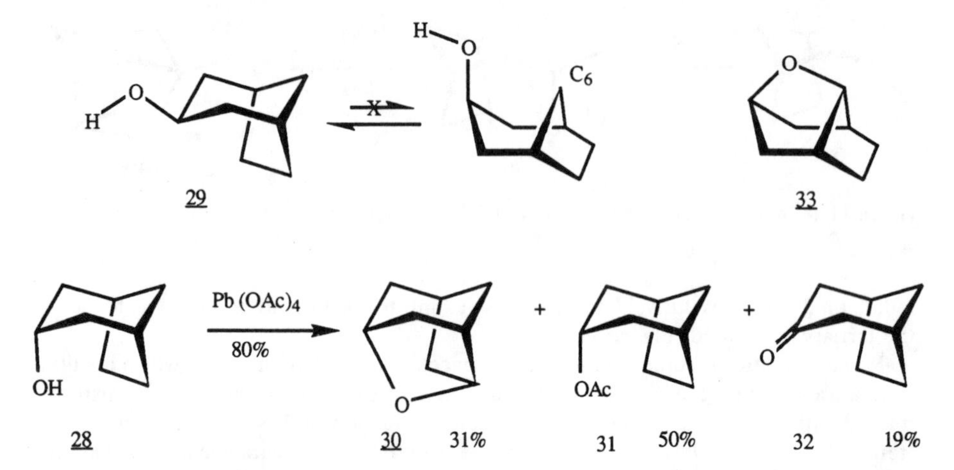

Figure 11.18 ■ Lead tetraacetate oxidation of bicyclo[3.2.1]octan-3-ols (exo and endo).

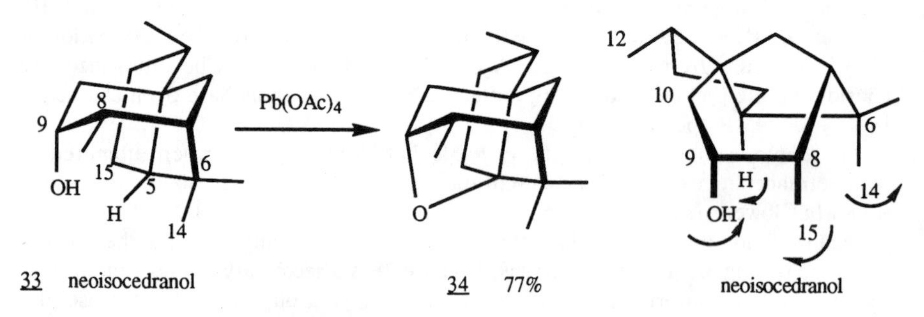

Figure 11.19 ■ Lead tetraacetate oxidation of neoisocedranol, which features both the bicyclo[3.2.1]octane skeleton and the reflex effect between the 14- and 15-methyl groups.

two 14- and 15-methyl groups on carbons 6 and 8 exert on each other a 1,3-diaxial interaction as in 3,3,5,5-tetramethylcyclohexanone, thus bringing by the reflex effect the axial hydroxyl group on carbon 9 close to the *endo*-hydrogen on carbon 5. It is therefore no surprise that lead tetraacetate oxidation yields the cyclic ether **34** with a high yield (77%).[40]

Cedrol **35** and 2-methylbicyclo[3.2.1]octan-2-ol do not cyclize under these oxidation conditions. This absence of reactivity is as much (and perhaps more) due to the reaction mechanism involved as to structural and conformational features of the substrates. It had been observed[43] as early as 1964 that heterocyclizations could be induced by using bromine in the presence of mercuric oxide in a reaction that involves a hypobromite, an alkoxy radical, and a bromohydrin as key intermediates (Figure 11.20). It is the cyclization of the latter via an intramolecular nucleophilic substitution that gives rise to the final cyclized oxo-7-dimethyl-1,4-bicyclo[3.2.1]octane. Indeed both alcohols **26** and **28** can be cyclized with high yields using this experimental

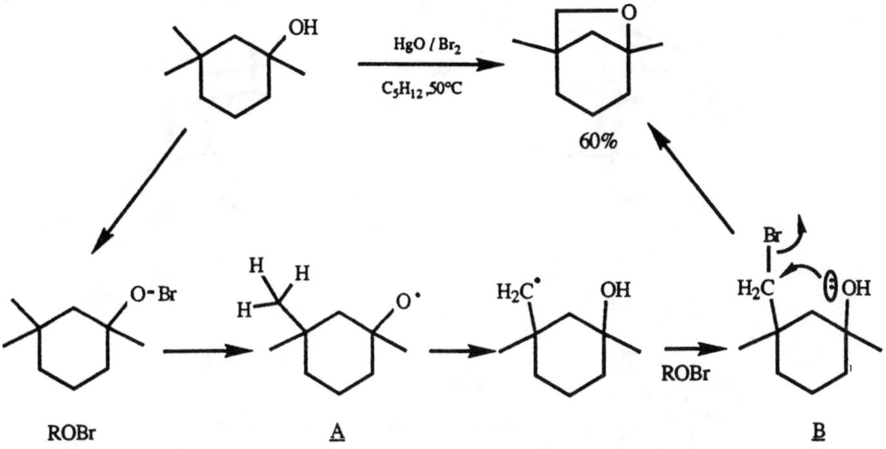

Figure 11.20 ■ Oxidation mechanism of 1,3,3-trimethylcyclohexanol in the presence of mercuric oxide–bromine reagent. An intermediate alkoxy racidal **A** as well as bromohydrin is involved in the key steps.

technique.[40b] With mercuric oxide and bromine, cedrol **35** can also be cyclized easily into **36** (Figure 11.21) as the intramolecular nucleophilic substitution is now easy on the 14-methyl group that is brominated in one of the intermediate reaction steps (see Figure 11.20).

This would not be the case if the *endo*-hydrogen on carbon 5 of neoisocedranol **33** would be substituted by bromine; inversion of this carbon during the intramolecular substitution is now impossible, so that cyclization of neoisocedranol with these reagents, indeed, does not occur.

Solvolysis, or more precisely acetolysis of tosylates is another reaction that was likely to provide interesting information on the influence of structural changes due to the reflex or anti reflex effect on reactivity. The acetolysis rate of the *exo* and *endo*-bicyclo[3.2.1]octan-3-ol tosylates **37** and **40** and of the corresponding equatorial tosylate **43** in the 3,3,5,5-tetramethylcyclohexanone series. The activation enthalpies have been measured and are reported in Figure 11.22 as well as the product distribution and the nature of the products.

The results are quite surprising and not easy to rationalize. For instance, it is remarkable that the *exo*-tosylate **37** gives exclusively the *endo*-acetate **38** whereas the *exo*- and *endo*-acetates **41** and **38** are formed from the *endo*-tosylate **40**. This might be due to the occurrence of more or less intimate ion pairs, depending on the stereochemistry of the initial tosylate. In any event, the exclusive formation of **38** from **37** implies a flattening of the cyclohexane ring, which will facilitate the *endo* attack on the intermediate carbonium ion. As far as the solvolysis of **43** is concerned, the small amount of acetate **45** formed might be due to either steric hindrance of the methyl group, which renders the approach of the carbonium ion difficult, or to the ease of the elimination process.

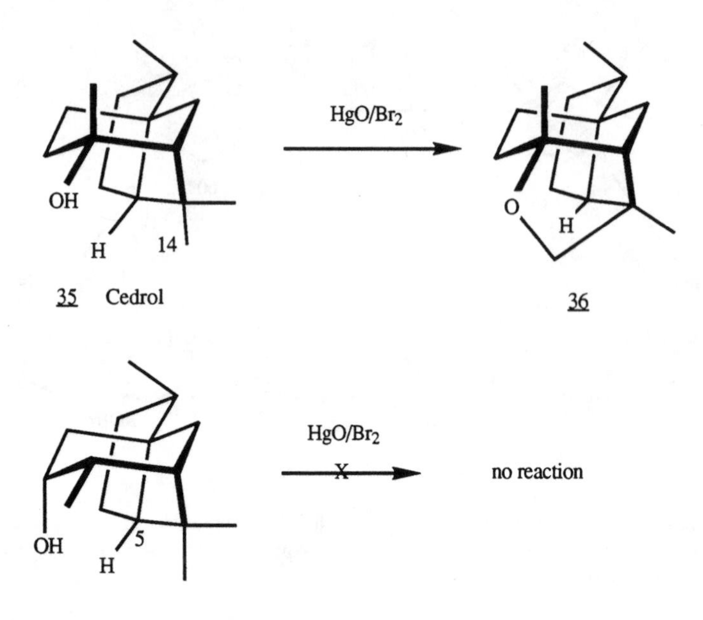

Figure 11.21 ■ Difference of reactivity of cedrol and neoisocedranol toward oxidation with mercuric oxide–bromine reagent.

These results have been reexamined at least for **37** and **40**[48,49] using buffered solvolysis media (acetic and formic acid, 98 and 50% aqueous ethanol). From high kinetic isotopic effects and from rate-constant values, it was concluded that these solvolyses proceed essentially via an S_N1 mechanism. Under these experimental conditions, the olefin **39** now becomes the major compound during the acetolysis (Figure 11.23), whereas the substitution essentially proceeds with inversion of configuration. This was interpreted by considering that the *endo*-tosylate **40** reacts with the cyclohexane boat conformation under these experimental conditions. This ring must be fairly flat in the intermediate species, so that it easily accommodates the double bond of the major compound formed, that is, olefin **39**.

11.2B(2). Bicyclo[3.2.1]oct-6-en-3-one Derivatives. As shown by Figure 11.24, sodium borohydride reduction of unsaturated **44** occurs preferentially from the exo side, giving the *endo*-alcohol **46** as the major product of the reaction.[46] Reduction of the corresponding saturated ketone **2** yields almost equal quantities of the *endo*-alcohol **28** and the *exo*-alcohol **29**. The etheno bridge imposes a more important pinching than the ethano bridge. According to the anti reflex effect, the exo face of **44** should therefore be more accessible to reagents like sodium borohydride, thus favoring the formation of *endo*-alcohol **46** as a major product of the reaction. According to the Halford relation, the carbonyl angle of **44** ($\nu_{C=O} = 1700\ cm^{-1}$) is larger than the one of **2** ($\nu_{C=O} = 1711\ cm^{-1}$ for **2**). Acetolysis of the corresponding

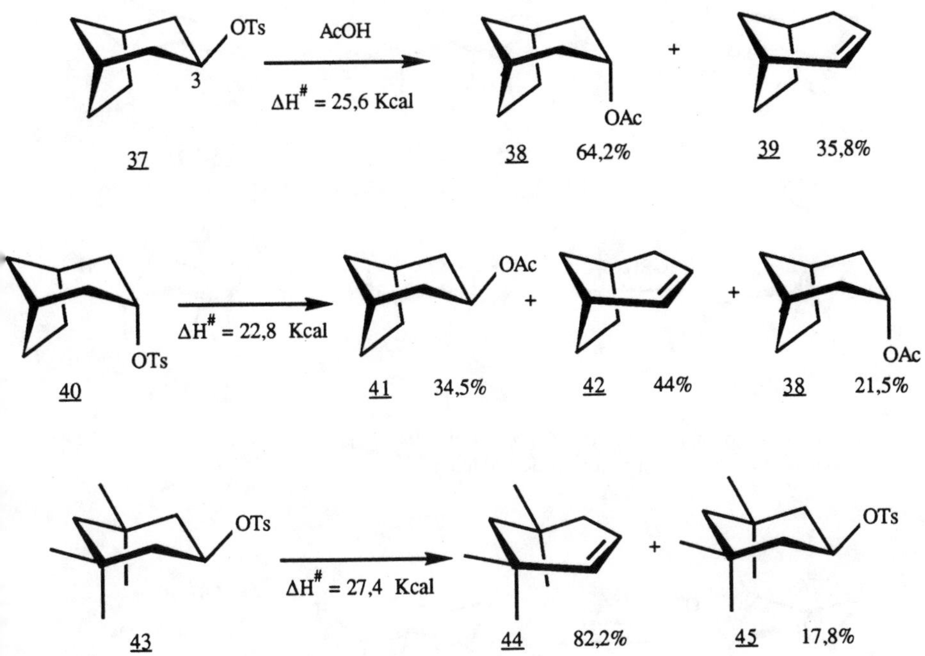

OTs AcOH 3 ΔH# = 25,6 Kcal + OAc

37 38 64,2% 39 35,8%

OAc + + ΔH# = 22,8 Kcal OTs OAc

40 41 34,5% 42 44% 38 21,5%

OTs ΔH# = 27,4 Kcal + OTs

43 44 82,2% 45 17,8%

Figure 11.22 ■ Influence of the anti reflex and reflex effects on tosylate acetolysis.

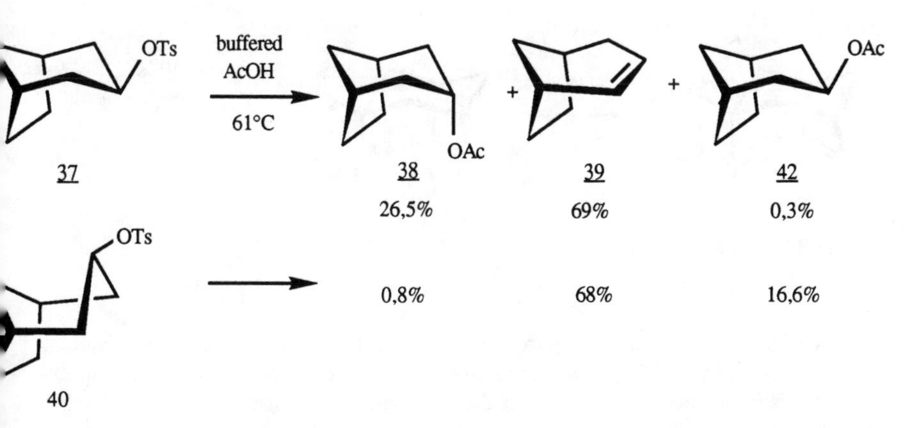

Figure 11.23 ■ Acetolysis of the tosylates derived from *exo*- and *endo*-bicyclo[3.2.1]octan-3-ol in buffered acetic acid.[48]

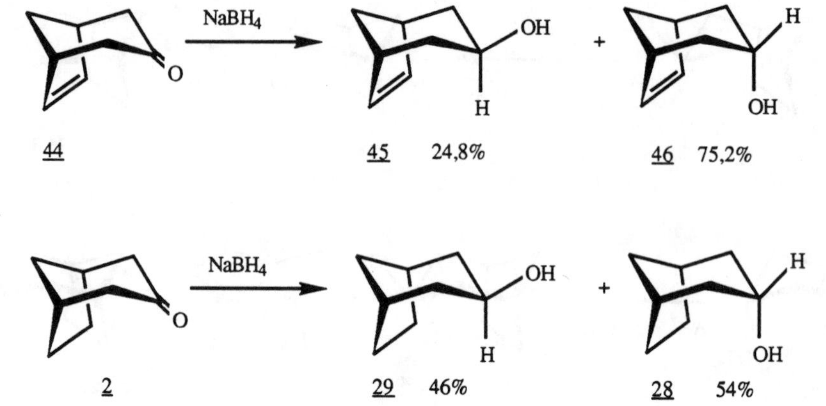

Figure 11.24 ▪ Comparison of the behavior of bicyclo[3.2.1]octan-3-one **2** and bicyclo[3.2.1]oct-6-en-3-one **44** toward sodium borohydride reduction.

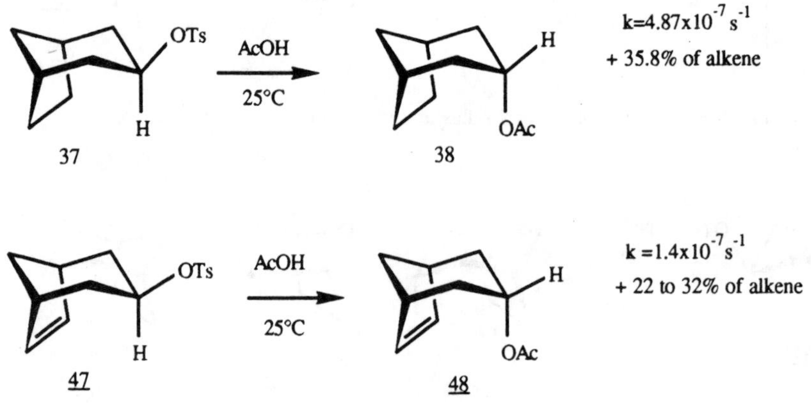

Figure 11.25 ▪ Compared rates of acetolysis of **37** and **47**.

tosylates **37** and **47** has been compared,[42,46] (Figure 11.25). The slower rate observed for **47** as compared to **37** clearly shows that there is no participation of the double bond. This is in perfect agreement with the ring flattening that renders this participation unlikely.

By reacting cyclopentadiene with 2,4-dibromopentan-3-one in the presence of a Zn–Cu couple, it was possible[47] to obtain the two *cis*-2,6-dimethylbicyclo[3.2.1]oct-6-en-3-ones **49** and **50** (Figure 11.26) by a cycloaddition reaction. Here again the comparable stability of these two stereoisomers is in agreement with the anti reflex effect.

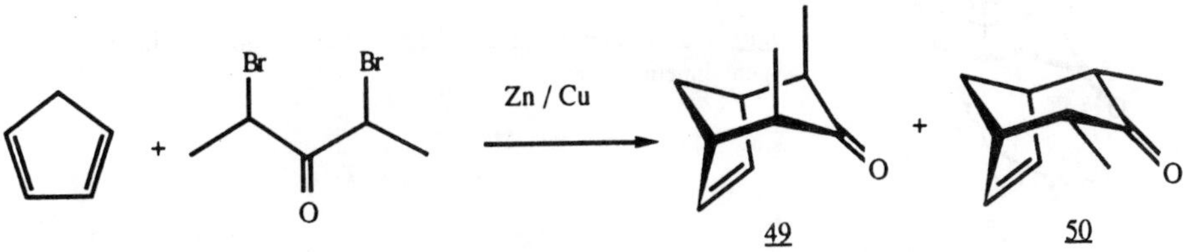

Br Br Zn / Cu 49 50

Figure 11.26 ■ Obtention of *cis-exo* and *-endo*-2,6-dimethylbicyclo[3.2.1]oct-6-en-3-ones.

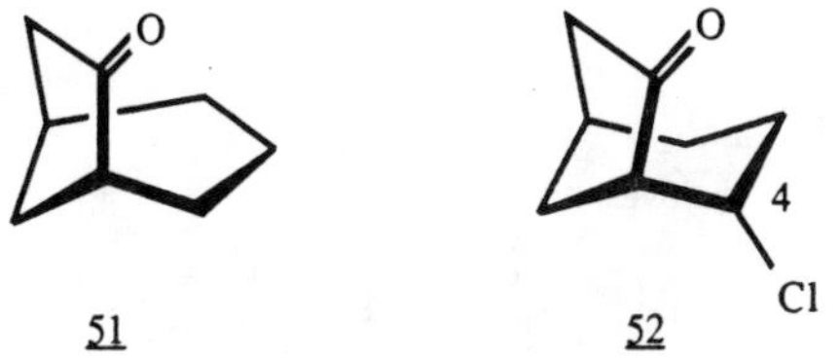

51 52 4 Cl

Figure 11.27 ■ Flattening of the cyclohexane ring in bicyclo[3.2.1]octan-6-one.

11.2B(3). Bicyclo[3.2.1]octan-6-one and Derivatives. The $^{13}C—^{13}C$ coupling constants of **51** and **52** (Figure 11.27) have been measured. The weak values of the vicinal coupling constants (relative to carbon 3), which are known to be dependent of the dihedral angle values,[50] reflect the flattening of the cyclohexane ring [the dihedral angle $C_3—C_4—C_5—C_6$ is close to 90°]. A single isomer with an axial chlorine **52** could be isolated and does not undergo any conformational equilibrium.

Comparison of the chemical shifts of carbons 2 and 8 in **51** and **52** shows a typical γ shielding effect.

11.2B(4). Derivatives of 6,6-Dimethylbicyclo[3.1.1]heptan-3-one. The apopinane derivatives that feature a bicyclo[3.1.1] skeleton, provide a good support to study the anti reflex effect. As compared to the bicyclo[3.2.1]octan system, the pinching is increased and steric hindrance is introduced by the gem dimethyl group, which introduces a reference to locate the cis or trans substituents on the propano bridge. We will consider arbitrarily the chair or boat conformation of the cyclohexane ring that contains this *gem*-dimethyl group (Figure 11.28).

In fact, the analysis of bicyclo[3.2.1]heptane by electronic diffraction[51] as well as the ^{1}H NMR analysis of a complete series of pinane derivatives clearly shows that these molecules should be considered rather as a cyclobutane with a propano bridge than as a cyclohexane with a methano bridge. Considering the projections shown in Figure 11.29, the *i* angle has a value around 37°,[51] as compared to the value of 50° for cyclohexane.

In apopinane **55** (or norpinane), analysis of the rearrangement products leads to the conclusion that this six-membered ring containing the *gem*-dimethyl group is in a chair conformation[52] (Figure 11.30).

Nopinone **56** was assumed to be a rigid molecule not undergoing any conformational equilibrium, with the six-membered ring containing the *gem*-dimethyl group

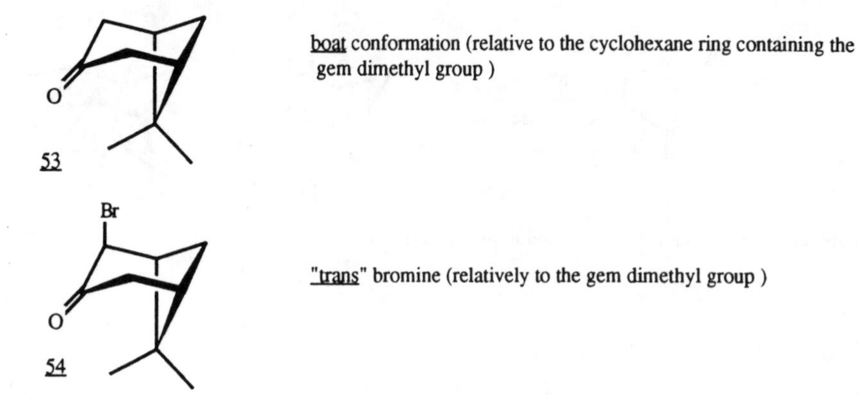

O
53
boat conformation (relative to the cyclohexane ring containing the gem dimethyl group)
Br
O
54
"trans" bromine (relatively to the gem dimethyl group)

Figure 11.28 ■ Nomenclature used in apopinane derivatives.

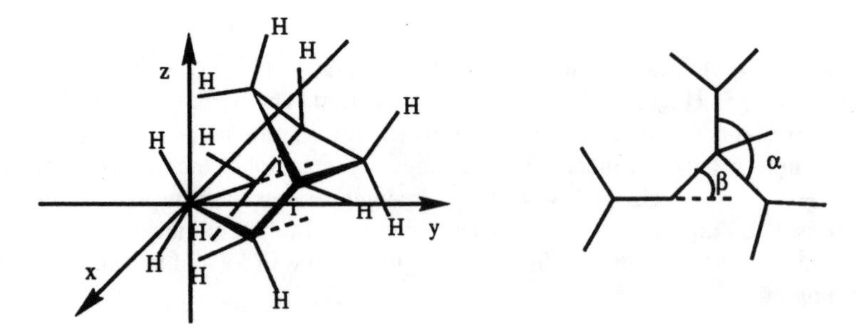

Figure 11.29 ■ Flattening of the projection in the *x–y* plane as appreciated by the *i* angle value.

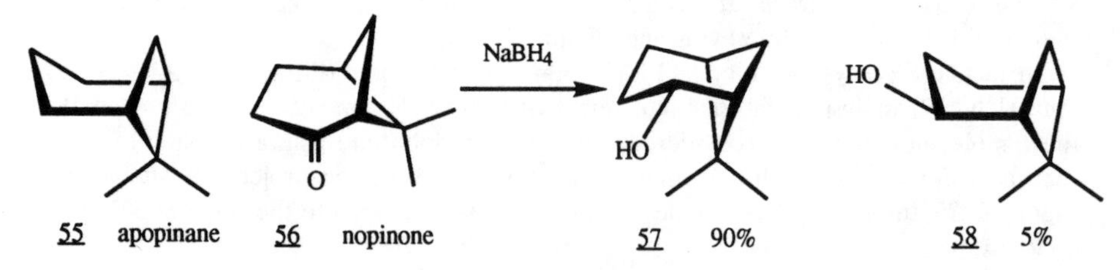

Figure 11.30 ■ Chair conformation of apopinane and sodium borohydride reduction products of nopinone.

favoured axial attack of the nopinone carbonyl resulting in the formation of cis nopinol 57

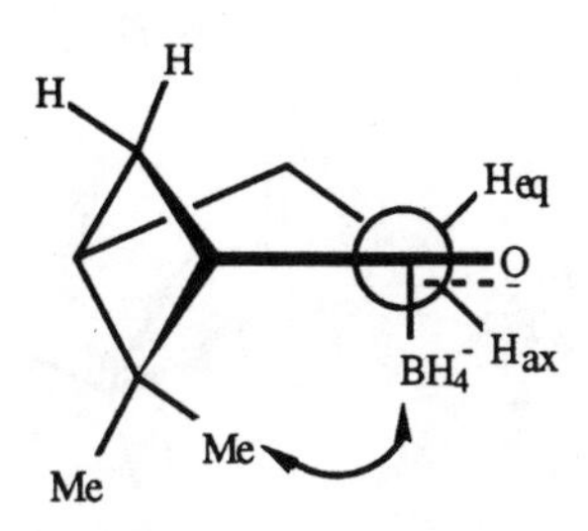

disfavoured equatorial attack of the nopinone carbonyl yielding the trans nopinol 58

Figure 11.31 ■ Steric interaction occuring during the sodium borohydride reduction of nopinone.

being in a chair conformation.[53] Analysis of the ^{1}H NMR spectrum[8,54] showed that this conformation was considerably flattened. Reduction of nopinane **56** using the Meerwein–Ponndorf–Verley reaction gave 36% *cis*-nopinol **57** and 20% *trans*-nopinol **58**; these ratios are, respectively, 90 and 5% when sodium borohydrure is used as reducing agent. The configuration and the conformations of these nopinols and their corresponding acetates have been determined by ^{1}H NMR and are shown in Figure 11.30. The sodium borohydride reduction proceeds by a stereoselective approach on the less-hindered side of the carbonyl. As shown in Figure 11.31, the results observed favor a flattened boat conformation for the *gem*-dimethylated six-membered ring.[55,9c]

The 4-methyl group of verbanone is *trans* to the 6,6-*gem*-dimethyl group, its reduction with $AlNaH_2(MeO)(EtO)$ gives the alcohol cis with the *gem*-dimethyl group. This observation supports a boat conformation for **59**, so that the hydride approaching the carbonyl does not interfere with the syn axial or pseudoaxial 4-methyl group.[56] The anti reflex effect, which is precisely at the origin of this decrease of steric interactions, provides a good explanation for these experimental observations (Figure 11.32). A partial ^{1}H NMR analysis of the *gem*-dimethylated six-membered ring of isoverbanone **60** leads to the same conclusions about the flattening of the boat conformation.[57]

There is a general agreement[8,53,55] on a unique boat conformation (of the *gem*-dimethylated six-membered ring) for isonopinone **53** (Figure 11.28). When the latter is reduced with sodium borohydride, it is *cis*-isonopinol **61** that is the major compound (97%). This observation is consistent with the carbonyl hydride approach on the isonopinone flattened boat conformation, as predicted by the anti reflex effect

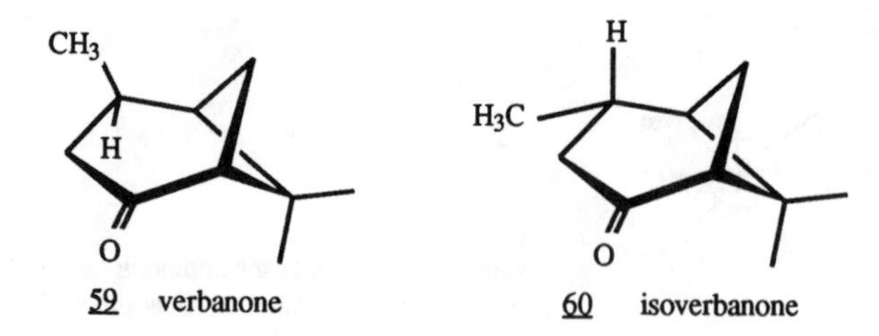

Figure 11.32 ▪ Flattened boat conformation of verbanone and isoverbanone.

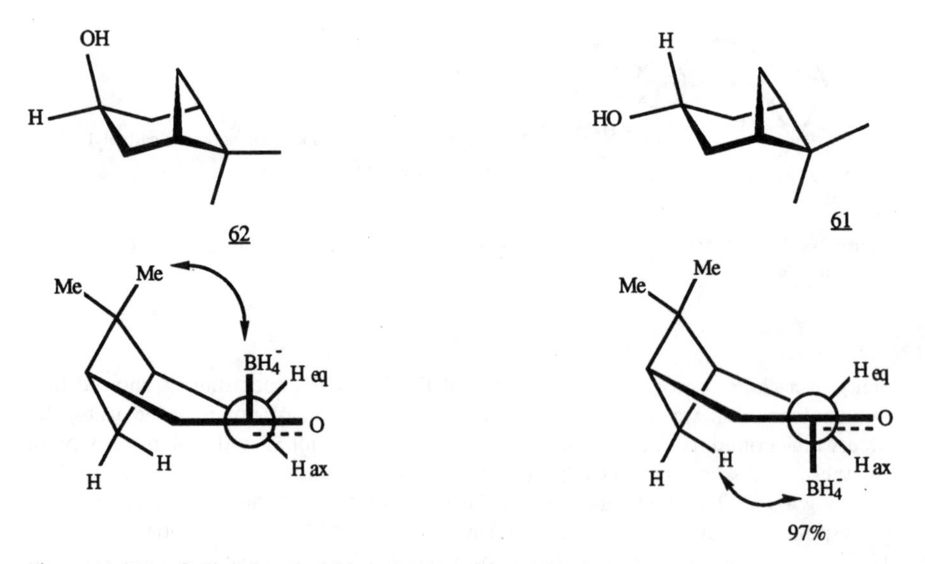

Figure 11.33 ▪ Sodium borohydride reduction of isonopinone.

(Figure 11.33). According to their ^{1}H NMR spectra, nopinols **61** and **62** as well as the corresponding acetates are exclusively in chair conformation.

Bromination with pyridinium perbromide in anhydrous THF or bromine in acetic acid[58] yields the α,α'-dibromoketone **63**, which is the only reaction product (26%) even when there is a lack of brominating agent. This *cis*-dibromoketone where the two bromines are trans to the *gem*-dimethyl group is simultaneously the kinetic and thermodynamic product of the reaction. Only trace amounts of a *trans*-α,α'-derivative **64** can be obtained in the presence of hydrobromic acid (Figure 11.34).

The ^{1}H NMR data as well as the IR and UV spectra clearly show that the two bromines are axial in **63**. However, the fairly high difference $\Delta\nu_{C=O}$ ($+18$ cm^{-1}) between the nonbrominated isopinone **53** and dibromo derivative **63** favors a conformation with a flattened cyclohexane ring, where the two bromines are pseudoaxial. Furthermore, the facile formation of **63** shows that the intermediate enol formation,

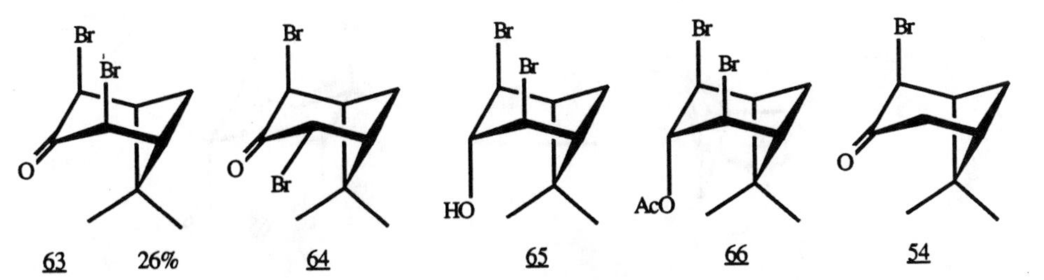

Figure 11.34 ■ Stabilization of pseudoaxial bromines, resulting from the anti reflex effect in isonopinone derivatives.

which is the bromination rate-determining step, does not require much energy and is therefore, geometrically speaking, close to the initial isonopinone **53**, which must therefore also feature a flattened cyclohexane ring. Bromination of the isonopinone enolacetate in acetic acid yields a single monobrominated derivative **54**. The structure shown with a pseudoaxial bromine, results from the analysis of the spectral data (UV, IR, ^{1}H NMR). The axial bulky bromines trans to the *gem*-dimethyl group of **53** and **54** are stabilized by the *anti* reflex effect, which acts in such a way as to spread apart the two bromines and to flatten the cyclohexane ring in **63**. The pseudoaxial position of these two bromines is further confirmed by the reduction of **63** with sodium borohydride, which can approach from the trans side, to yield the bromohydrin **65** where the hydroxyl group is cis relative to the *gem*-dimethylated methano bridge as shown by the analysis of its ^{1}H NMR spectrum. Accordingly the **65** bromohydrin hydroxyl group, which is also pseudoaxial, can be acetylated into **66** without any problem.

The ^{13}C NMR spectra of halogenated derivatives of bicyclo[3.2.1]octan-3-one and 6,6-dimethylbicyclo[3.2.1]heptan-3-one have been analyzed and compared.[59] The α and γ gauche values are reported in Figure 11.35. Only the γ gauche effects vary continuously with the corresponding pinching, which increases when going from

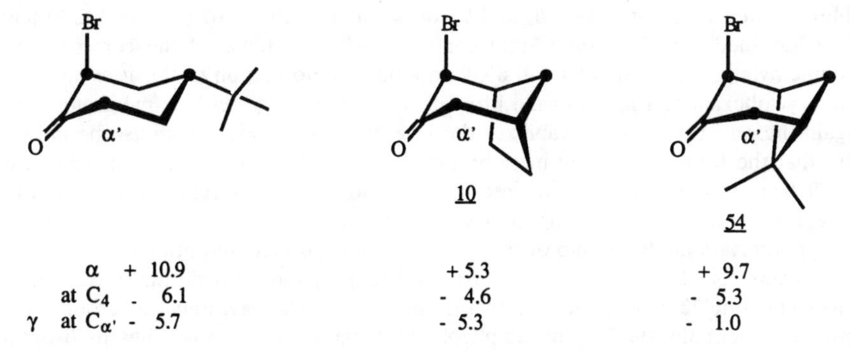

α	+ 10.9	+ 5.3	+ 9.7
at C_4	- 6.1	- 4.6	- 5.3
γ at $C_{\alpha'}$	- 5.7	- 5.3	- 1.0

Figure 11.35 ■ Comparison of α and γ gauche effects; $\Delta J_{^{13}C}$ values are in parts per million for the axial bromides as compared to the nonbrominated ketones.

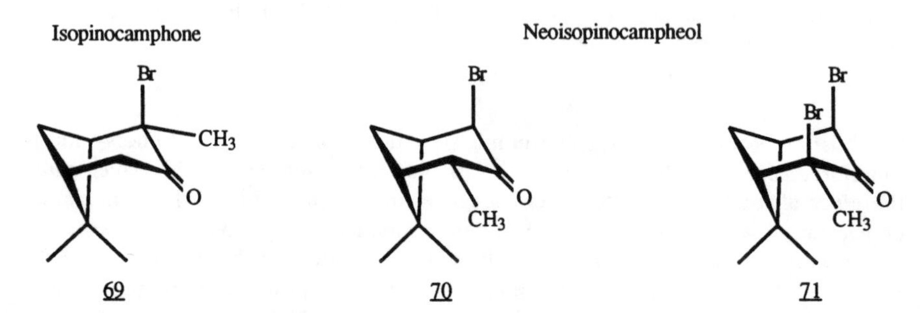

Figure 11.36 ■ Conformations and configurations of isopinocamphone and related derivatives.

2-bromo-4-tertiobutylcyclohexane to **10** and **54** although it is not possible to assign this effect exclusively to the anti reflex effect.

The conformation of the *gem*-dimethylated six-membered ring of isopinocamphone **67** was thought to be boat,[52,60] but the ^{1}H NMR analysis favored a much more flattened conformation,[61] which allows the 2-methyl group to escape the 1,3-diaxial interaction with the axial methyl of the methano bridge (Figure 11.36). It is therefore no surprise that the reduction gives essentially (96%) neoisopinocampheol **68** whether the reaction is carried out with $AlLi(MeO)_3H$ (ref. 53b) or $LiAlH_4$ (ref. 61). These results compare well with the obtention of isonopinol **61** as a major product of the isonopinone **53** (Figure 11.33) sodium borohydride reduction. Monobrominated and dibrominated derivatives **69**, **70**, and **71** of isopinocamphone **67** (Figure 11.36) have been assigned[62] conformations that compare well with those of the corresponding nonmethylated derivatives **54** and **63**, with a boat conformation of the *gem*-dimethylated cyclohexanone ring and axial bromine(s) trans to the *gem*-dimethyl group. Here again the anti reflex effect stabilizes the α,α'-axial bromines. It can also be pointed out that the formation of the monobrominated derivative **69** is favored as compared to **70**, because the corresponding intermediate enol forms preferentially, thus allowing escape from the previously mentioned interaction between the 2-methyl group and the appropriate methyl group of the *gem*-dimethylated methano bridge.

Because of the flattened structure of the propano bridge, the reduction of pinocamphone **72** (where the 2-methyl is now trans to the *gem*-dimethyl group) again gives preferentially (90%) pinocampheol **73**[63] (Figure 11.37), which has its hydroxyl group cis to the *gem*-dimethyl group as in **61** and **68**. Although a chair conformation had been assigned to **72**,[61] on the basis of deshielding of the *endo*-methyl on carbon 6

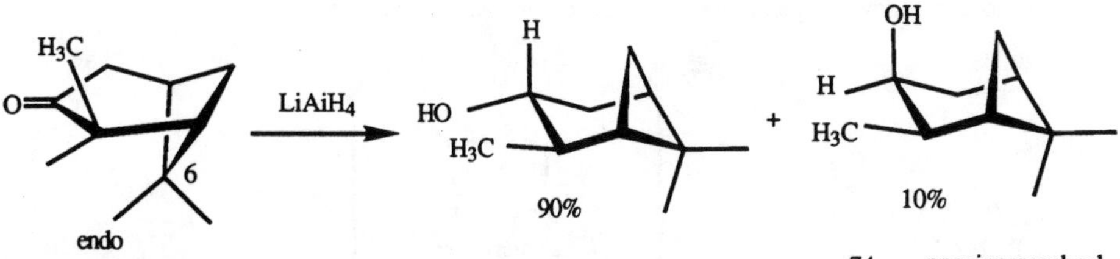

72 pinocamphone 73 pinocampheol 74 neopinocampheol

Figure 11.37 ■ Lithium aluminum hydride reduction of pinocamphone.

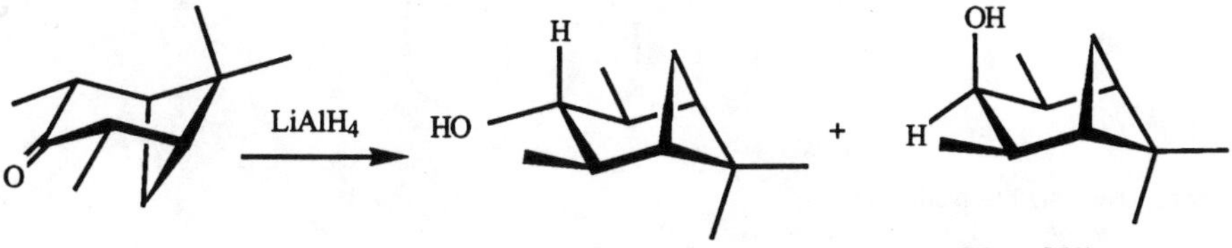

75 76 65% 77 35%

Figure 11.38 ■ Lithium aluminum hydride reduction of 2,4,6,6-tetramethylbicyclo[3.1.1]heptan-3-one anti stereoisomer.

($\delta = 0.91$) as compared to isopinocamphone **67** ($\delta = 0.88$), the overall results are in agreement with a flattened boat conformation. In such a conformation the 2-methyl group (because of its pseudoaxial character) has little interaction with the hydride approaching the carbonyl.

The formation of **76** as a major reduction product (65%) of the 2,4,6,6-tetramethyl-bicyclo[3.1.1]heptan-3-one **75** can be rationalized on the same basis, by considering a flattened boat form (Figure 11.38).

11.2B(5). Mechanistic Applications of the Anti Reflex Effect to Favorskii Rearrangement. Various mechanisms[8] have been proposed to rationalize Favorskii rearrangement, which allows the ring contraction of α-halogenated cyclic ketones. One of the most spectacular applications of this contraction is cubane synthesis,[64] where it was observed that an α-bromocyclopentanone could ring-contract efficiently to a cyclobutane carboxylic acid. The two major mechanisms that are generally taken into consideration, that is, the cyclopropanic mechanism on the one hand and the semibenzylic mechanism on the other, are shown in Figure 11.39.

The occurrence of either mechanism[8] is favored by the solvent nature, the substrate strain, and the stereochemistry of the halogen. In the cyclohexane series, the formation of an intermediate zwitterion is favored by an axial halogen. Substitution products are then formed with strained systems, or in protic media, whereas ring contraction occurs in aprotic media. An equatorial halogen favors the formation of an intermediate cyclopropanone, whatever the nature of the solvent. If the ring strain is too high to allow the cyclopropanone formation or if the base is too weak, the

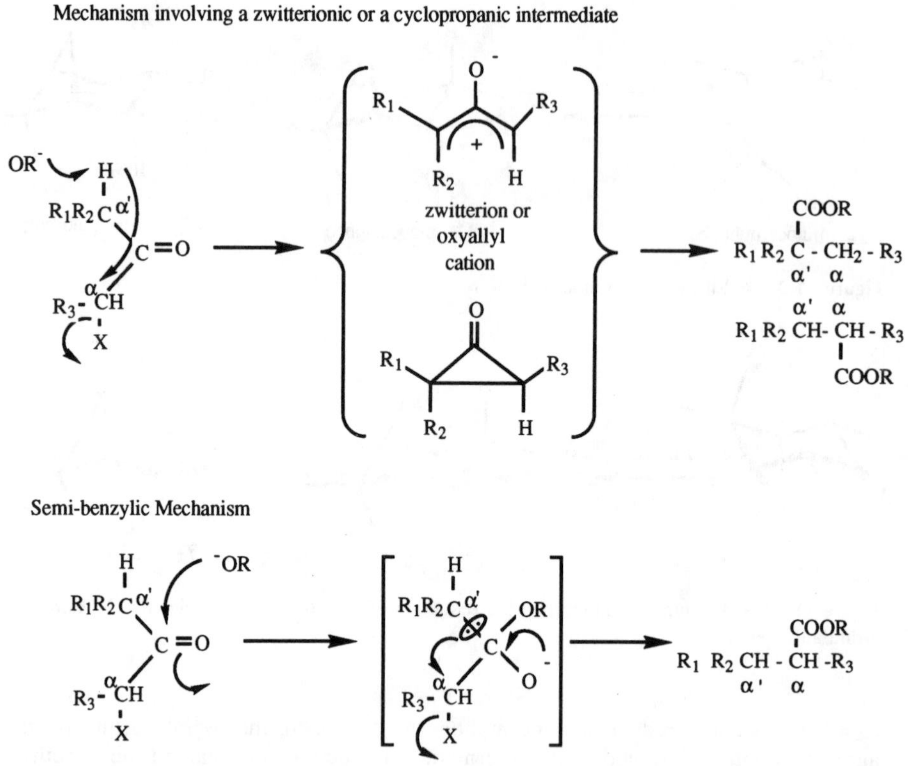

Mechanism involving a zwitterionic or a cyclopropanic intermediate

zwitterion or oxyallyl cation

Semi-benzylic Mechanism

Figure 11.39 ■ Currently admitted mechanisms for Favorskii rearrangement.

semibenzylic mechanism is favored by an equatorial halogen (Figure 11.39). When monobrominated isonopinone **54** is treated under nitrogen atmosphere by sodium methylate in methanol or in dimethoxyethane at 0 or 20°C, a major proportion of substituted products **78** and **79** (80–90%) is obtained (Figure 11.40).

Epimerization on the α-carbon, as well as conformational equilibria, is not allowed by the anti reflex effect, so that practically only substitution products can be formed. When α-brominated bicyclo[3.2.1]octan-3-one **10** is treated under nitrogen with sodium methoxide in methanol at 20°C, one obtains 14% ring-contracted ester **80** after 1 h, when the base concentration is high (2.2 mol l^{-1}). However, with a lower base concentration (0.089 mol l^{-1}), 100% of the substitution products **81** and **82** (Figure 11.41) is formed.

It is known that the axial 2-bromobicyclo[3.2.1]octan-3-one **10** does not epimerize. Furthermore, because of the conformational rigidity of α-bromoketones **10** and **54** and the stabilization of the axial bromide by the anti reflex effect, it is no surprise that these derivatives essentially undergo substitution reaction under Favorskii rearrangement conditions.[8] Furthermore, as the ring strain decreases when going from the

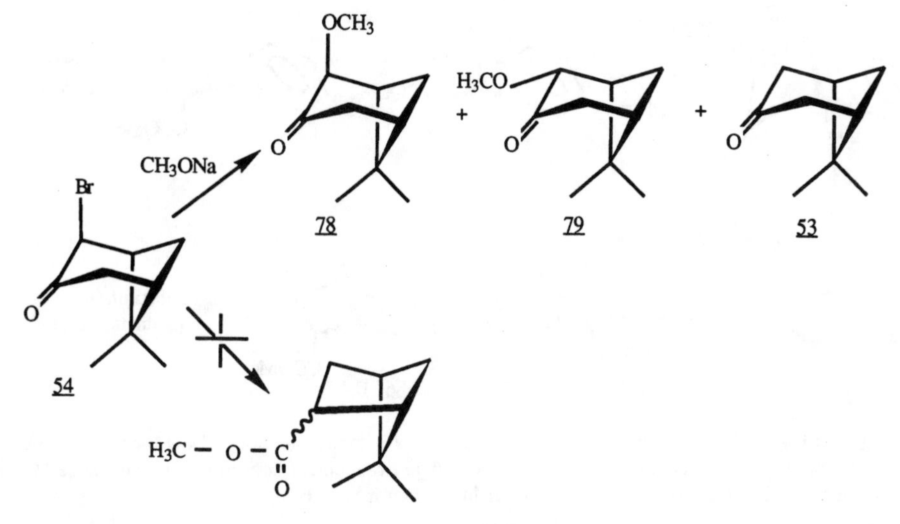

Figure 11.40 ■ Formation of ring-contracted derivatives of α-bromoisonopinone under Favorskii rearrangement conditions. Ring contraction is not observed.

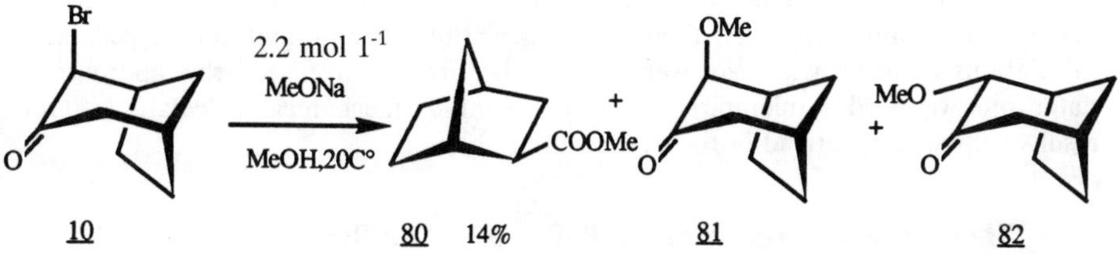

Figure 11.41 ■ Formation of ring-contracted derivatives of α-bromobicyclo[3.2.1]octan-3-one under Favorskii rearrangement conditions. Substitution products can become the exclusive reaction products when low base concentrations are used.

bicyclo[3.2.1]heptan system to the bicyclo[3.2.1]octan system, it was expected that the latter should undergo some ring contraction, as experimentally observed.

During a comparative study of Favorskii ring contraction on 2-bromo[6.7]benzobicyclo[3.2.1]oct-6-en-3-one **83** and on **10** (Figure 11.42), it was shown[65] that the ring contraction is much more important on **83** than it is on **10** under similar conditions. These differences in reactivity might be due to electronic effects of the aromatic ring in **83**. However, some steric effects might be involved, as the aromatic ring could act so as to maintain the two bridgehead carbons of **83** farther apart than in **2**, so that the anti reflex effect decreases in the former product, thus facilitating a more important ring contraction.

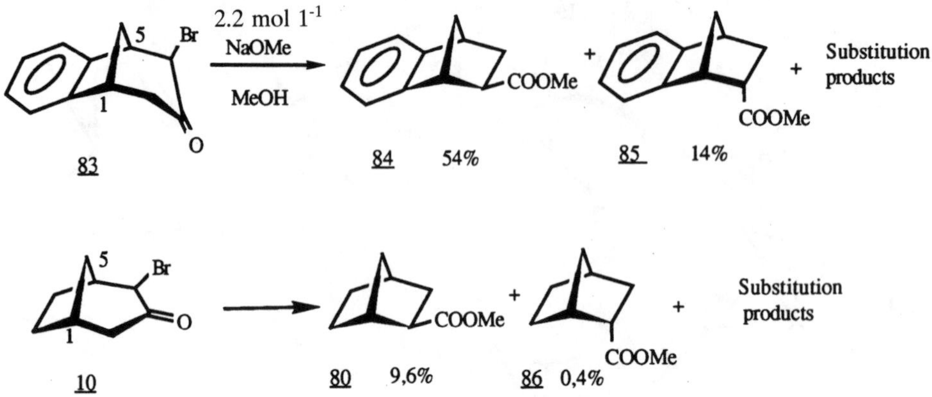

2.2 mol l^{-1} NaOMe / MeOH

83 → 84 54% + 85 14% + Substitution products

10 → 80 9,6% + 86 0,4% + Substitution products

Figure 11.42 ▪ Comparison of reactivity of 2-bromo[6.7]benzobicyclo[3.2.1]oct-6-en-3-one and of 2-bromocyclo[3.2.1]octan-3-one under Favorskii rearrangement conditions. The distance between bridgehead carbons is probably larger in **83** than it is in **10**.

At this stage it can be said that the theory of reflex and anti reflex effects is well supported by experimental observations, both from spectroscopic and reactivity points of view. While these investigations were going on in our laboratories and in others, we confirmed and often predicted the outcome of some experiments using conformational analysis and particularly theoretical calculations, that is, molecular mechanical calculations to begin with. We were among the first to have used this approach.[9a] Later on, we used semiempirical CNDO quantum mechanical calculations. Our results will be summarized in the next section.

11.3. Theoretical Approach to the Reflex and Anti Reflex Effects

11.3A. Molecular Mechanical Calculations

We reported our first results concerning molecular mechanical calculations on 3,3,5,5-tetramethylcyclohexane and the reflex effect in 1961.[9a] Using a Bull γ computer we could show that the 1,3-diaxial interaction between the two 3- and 5-methyl groups could be absorbed or compensated by a general deformation of the valence angles.

The bond lengths were considered as constants. The energy was minimized using the valence angle deformation energy only:

$$E_d = \sum k_i(\theta_i - \theta_0)^2 \qquad \theta_0 = 109°28'$$

where

$$k_i\ (\text{kcal mol}^{-1}\ \text{deg}^{-2}) = \begin{cases} 17.5 & \text{for } \theta_i\ (\text{C—C—C}) \\ 12.0 & \text{for } \theta_i\ (\text{H—C—C}) \\ 6.9 & \text{for } \theta_i\ (\text{H—C—H}) \end{cases}$$

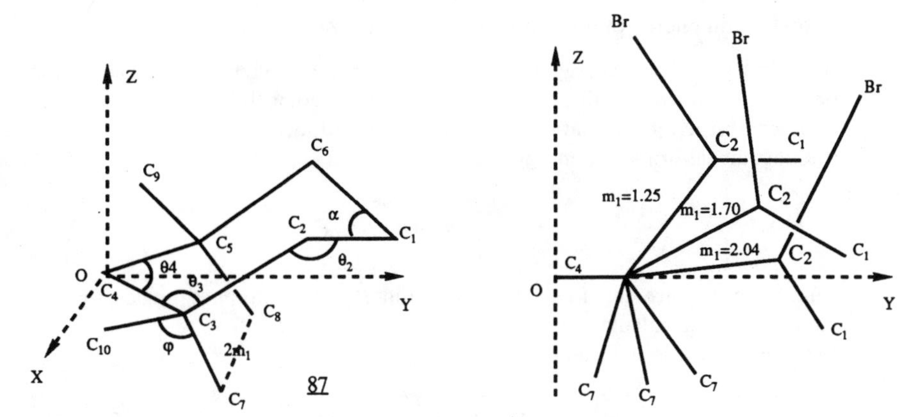

Figure 11.43 ■ Flattening of the cyclohexane ring of 1,1,3,3-tetramethylcyclohexane by the reflex effect as a function of different values of m_1, corresponding to the van der Waals radius ($m_1 = 2.04$ Å), or to the Braude radius ($m_1 = 1.7$ Å) of the methyl group.[9b]

Furthermore, discrete values m_1 (1.25, 1.70, and 2.04 Å) of the distance between the two axial 3- and 5-methyl carbons were imposed on the system, for example, the θ_4 value was included in the 114–117° range (Figure 11.43). Within this interval we found 650 solutions that could be accepted from the geometrical point of view.

The θ_4 angle is opened relative to the tetrahedral angle, the α angle is pinched. The cyclohexane ring is a deformed chair, as the energy increase is not sufficient to switch to a boat form.

The torsional angle energy was calculated on the molecular shape that had been found by valence angle energy minimization. Similar calculations were made to calculate the ideal conformation of 3,3,5,5-tetramethylcyclohexanone **87** and its brominated derivatives.[9,66]

Finally, it could be seen that the six-membered ring is progressively flattened as the distance d increases. This is precisely what had been observed by Allinger and Da Rooje[67] and by Lehn, Levisalles, and Ourisson[68] on 4,4′-dimethylated steroids. Furthermore, in spite of this very simplified approach, the agreement between the geometry thus found and the one obtained in the solid state by X-ray analysis[69] was satisfactory.

Later on, when more-sophisticated computers became available to us we developed a more-refined approach[9c,66] in order to reach three goals:

1. to find satisfactory functions that would give good results on various molecules of known conformations;
2. to calculate the conformations of various derivatives in the apopinane series;
3. to write a flexible program that could be adapted to other molecules such as bicyclo[3.2.1]octane derivatives.

The Westheimer approach was used.[68] The electrostatic and resonance effects generally could be neglected on strained molecules, so that the difference between the steric effects could account for the enthalpy difference between two conformers.[70]

The total strain energy of a given molocule is a sum of factors:

1. the bond stretching energy W_1, which need not be taken into consideration because the bond length is fairly constant when going from one molecule to another (in the approximation of independent and localized bonds);
2. the angular deformation energy W_b,

$$W_b = \sum_{1}^{i} \tfrac{1}{2} k_{bi}(\theta_i - \theta_{0i})^2$$

   where the k_{bi} are the force constants and θ_{0i} the reference angles for the molecules to be calculated;
3. the dihedral angle deformation energy W_d,

$$W_d = \sum_{1}^{i} \sum_{1}^{j} \tfrac{1}{2} V_{ij}(1 + \cos 3D_{ij})$$

   where V_{ij} is the rotational barrier around the ith bond that is included in the jth dihedral angle;
4. the nonbonded interaction energy W_a.

An additional dipolar interaction energy term has been added to these four energy terms for the molecules that feature a permanent dipole such as C=O or C—Br. Three techniques have been used for energy minimization.[71] A set of functions for the nonbonded interactions W_a was found and gave satisfactory results for cyclohexane, methylcyclohexane, bicyclo[2.2.2]octane, and cyclohexane. This means that the results obtained by calculations were in good agreement with those obtained experimentally, especially from the geometrical and energetical points of view.[9c, 72, 73] The method has been analyzed critically from the theoretical point of view[74] and a survey of applications has been provided.[75]

These theoretical studies provided new insight into the theory of the reflex and anti reflex effects.

For instance, it could be observed in 1,3-dimethylcyclohexane **88** that the steric effect of the equatorial methyl group was resulting in a pinching of the C(6)—C(5)—C(4) angle and an aperture of the H—C—H angle on carbons 4 and 6 (Figure 11.44).

The pinching of the C(4)—C(5)—C(6) angle and the aperture of the C(1)—C(2)—C(3) angle provides the basis of the reflex effect. The axial carbon methyl

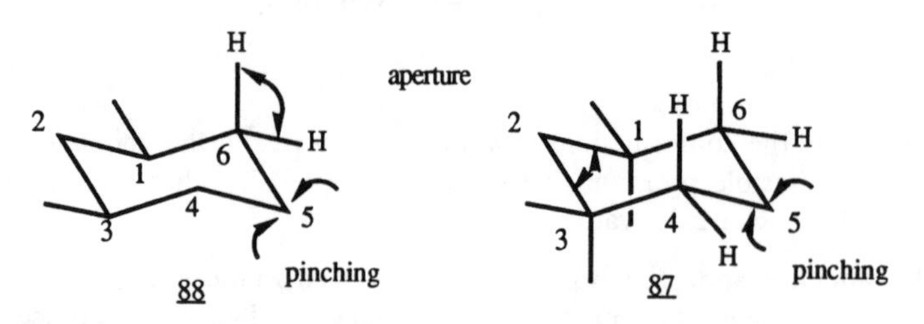

Figure 11.44 ▪ Influence of the methyl groups on valence angles in the reflex effect.

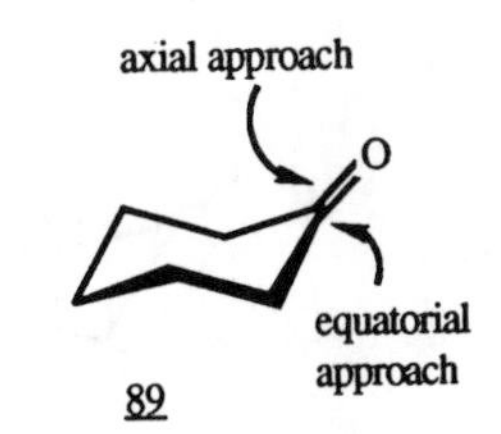

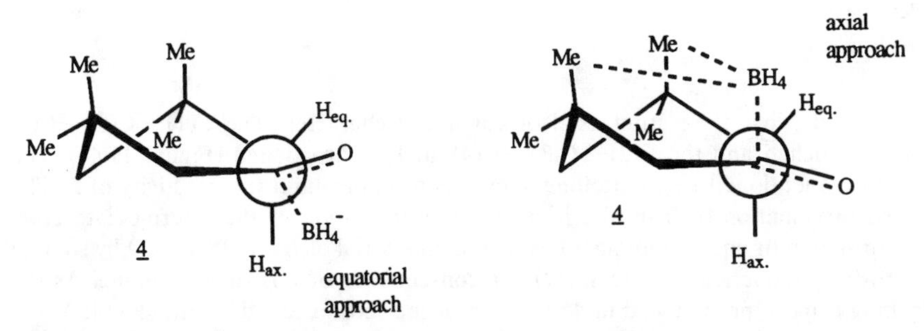

Figure 11.45 ■ Carbonyl approach of sodium borohydride during the reduction of 3,3,5,5-tetramethylcyclohexanone **4**.

bonds move away from each other whereas the axial carbon hydrogen bond on carbons 4 and 6 are brought nearer. Meanwhile one observes an increase of the equatorial character of the equatorial carbon–hydrogen bond on the same carbons (Figure 11.44).

The calculated energy difference between the unsubstituted and substituted hydrocarbons **87**, where carbon 5 is a tetrahedron and the corresponding ketone (with a carbonyl in 5) yields values of 0.52 and 8.11 kcal, respectively, for the nonsubstituted and substituted derivatives. This provides a nice basis to explain the difference in the sodium borohydride reduction rates[76] of cyclohexanone (2.35×10^{-2} l mol^{-1} s^{-1} and 3,3,5,5-tetramethylcyclohexanone (0.0034×10^{-2} l mol^{-1} s^{-1}) carried out under similar experimental conditions. The origin of the rate difference lies in the difficulty of forming an intermediate complex[77] where carbon 1 is a tetrahedron[78] (Figure 11.45).

The calculations confirm that the conformational equilibrium of α-bromocyclohexanone (**4**) must be shifted toward the equatorial conformer in the tetramethylated derivative.

Similarly, it can be explained why the *cis*-α,α'-dibrominated diequatorial isomer is more stable when the cyclohexanone is tetramethylated in 3 and 5 positions. However, the importance of the electrostatic effects in these changes is also stressed by these calculations.

For bicyclo[3.2.1]octan-3-one **2** (Figure 11.46), the calculations show the flattening of the cyclohexane ring on the C-1, C-2, C-3, C-4, and C-5 carbons so that the 2- and

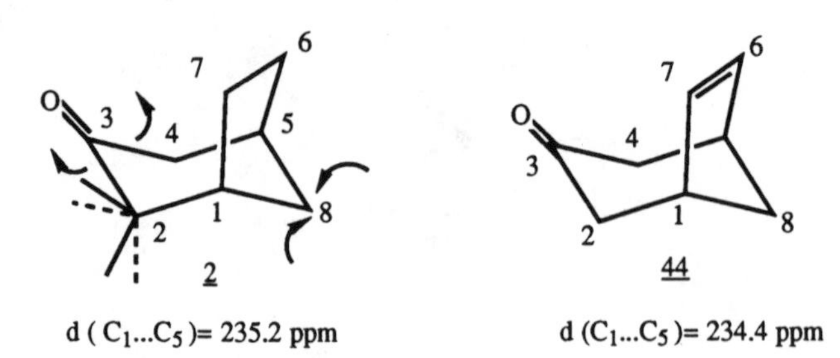

Figure 11.46 ■ Influence of an ethano or an etheno bridge on the valence angles in the anti reflex effect.

4-substituents lose their strict axial or equatorial character. The C(1)—C(8)—C(5) angle is pinched and the C(2)—C(3)—C(4) angle is broadened (Figure 11.46). The previously mentioned ring flattening allows us to understand the reactivity of **2**. The fast *cis*-bromination of **2** results from the easy formation of the intermediate enol. The formation of equal amounts of axial and equatorial alcohols **28** and **29** by sodium borohydride reduction (Figure 11.17) is a consequence of this ring flattening. As the pinching is more pronounced in **44** because of the shorter length of the double bond, the previously mentioned effects are more apparent. Although the C(1)—C(5) distances are not much different in **2** and **44**, the C(1)—C(7)—C(6) and C(7)—C(6)—C(5) angles are larger in **44** than in **2**.

The same calculations show that the preferential position of the bromine in α-bromo and α,α'-dibromo derivatives of bicyclo[3.2.1]octan-3-one, **10** and **11** (Figure 11.7), is axial, with a chair conformation for the cyclohexane ring (90% for **11**). The preferential stability of the trans epimer (80%), which forms by the epimerization of the *cis*-diaxial derivative **11**,[7a] is not shown by the calculations as it probably results from electrostatic effects rather than from steric effects. MM2 calculations[79] carried out on 1,8,8-trimethylbicyclo[3.2.1]octan-3-one[37] show that this ketone has a flattened chair conformation where both α-equatorial and α-axial hydrogens become quasi-axial.

Apopinane **55** (Figure 11.47) appears as a rigid structure with an unique chair conformation for the *gem*-dimethylated six-membered ring. The same ring becomes a flattened boat in isonopinone **53** (Figures 11.28 and 11.47) and in the corresponding bromo derivatives **54** and **63** (Figure 11.34).

Because in such a flattened boat the carbon–bromine bonds are neither equatorial nor axial, it is difficult to distinguish them by using Jones' rule[12] in the interpretation of the $\nu_{C=O}$ frequency in infrared. However, the position of the latter can be calculated by the Halford formula[11a] on the basis of the carbonyl valence angle and of the related dihedral angles calculated by molecular mechanics. Similarly, by using the same data, the UV carbonyl wavelength can be found with the Julg formula[15] and the NMR coupling constants can be estimated. It also was possible to show that the *gem*-dimethylated cyclohexane ring is in a flattened conformation (without any conformational equilibrium) in nopinone **56**, isonopinone **53**, verbanone **59**, isoverbanone **60**, and pinocamphone **72**. Examination of model transition states for the hydride

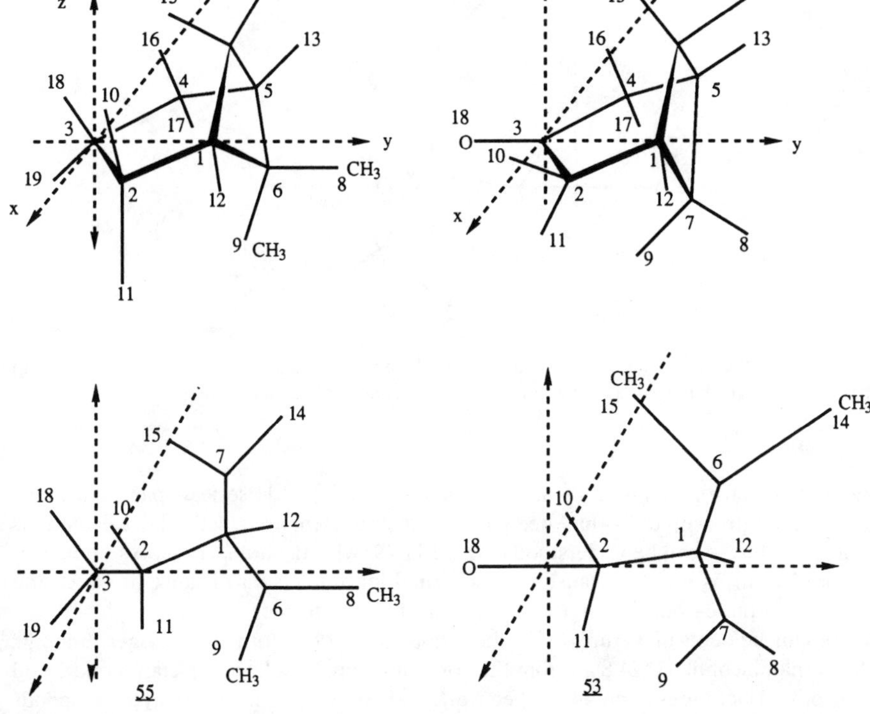

Figure 11.47 ■ Conformation of 6,6-dimethylbicyclo[3.1.1]heptan or apopinane **55** and of 6,6-dimethylbicyclo[3.1.1]heptan-3-one or isonopinone **53**, obtained by molecular mechanical calculations. All bond termini (except those identified) are hydrogens.

reduction of these ketones allows explanation of the formation of an alcohol with an hydroxyl group cis to the *gem*-dimethyl group (**57**, **61**, **73**, and **68**). The controversy reported in the literature about the preferential obtention of **68** from isopinocamphone **67** might be due to the occurrence of a conformational equilibrium in the latter. For the monobromo and dibromo derivatives **54** and **63** of isonopinone, the carbon–bromine bonds are stabilized in a pseudoaxial position trans to the *gem*-dimethyl group (Figure 11.48).

The corresponding calculated energies were compared to those of the related derivatives of cyclohexanone and of tetramethyl-3,3,5,5-cyclohexanone **1**; they are in good agreement with the reflex and anti reflex theory.

The ^{1}H coupling constants measured on various derivatives where the reflex and anti reflex effects operate correspond to dihedral angle values in good agreement with those found by calculation (Table 11.1). Comparison between the conformations of differently strained alcohols going from 3,3,5,5-tetramethylcyclohexanol **89** to isonopinol **61** very nicely illustrates how the reflex and anti reflex effects operate.[80] The solvolysis of the corresponding tosylates have been analyzed in a similar way. The

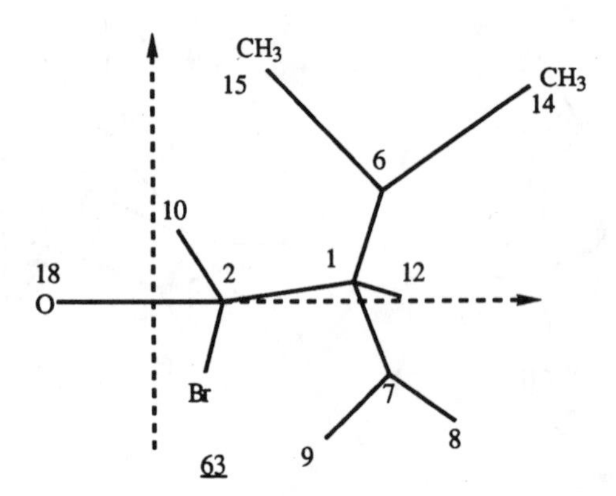

Figure 11.48 ■ *Cis*-Diaxial 2,4-dibromo-6,6-dimethylbicyclo[3.2.1]heptan-3-one **63**; projection in the z–y plane of symmetry as obtained by molecular mechanical calculations.

chemical behavior of different alcohols in the bicyclo[3.2.1] series—particularly toward lead tetraacetate[40]—has been analyzed and compared with the calculations results.[81] Thus, it can be understood (Figure 11.49) why alcohol **26** cyclizes better than alcohol **28**; this is simply because the concerned intramolecular distance (between the oxygen and the carbon in δ position) is shorter in the former.

As can be seen in Figure 11.24 the *exo*-alcohol **29** is formed in larger amounts (46%) than alcohol **45** (24.8%) from the corresponding bicyclo[3.2.1]octan-3-one **2** and bicyclo[3.2.1]oct-6-en-3-one **44**, respectively. This difference in reactivity corresponds to an energy difference between the alcohols and the ketones of 2.9 and 3.0 kcal, respectively. This is because the carbonyl carbon has a structure closer to sp^3 hybridization in the transition state occurring during the reduction of **2** than is the case in the reduction of **44** (Figure 11.50).

As shown in Figure 11.25 acetolysis of tosylate **37** is faster than that of the corresponding unsaturated tosylate **47**. As this solvolysis involves a carbenium ion intermediate that must have a structure similar to that of the corresponding ketones **2** and **44**, the calculated distances shown in Figure 11.51 (respectively, 174 and 180 pm) allow us to understand these rate differences. Furthermore, there is neither anchimeric assistance nor double bond participation during the solvolysis of **47**, because of the cyclohexane ring flattening, contrary to what had been previously published.[46]

11.3B. Semiempirical CNDO/2 Calculations

These semiempirical calculations have been applied to optimized geometries calculated by the Westheimer method.[9c] Interorbital angles have been calculated by the Dishler formula.[82] According to this approach, the C(2)–C(3)–C(4) interorbital angle and valence angle are equal for cyclohexane (Figure 11.52) and become different in bicyclo[3.2.1]octane and bicyclo[3.1.1] heptane. For a given carbon the constraint was associated to the difference Δ between the interorbital angle of the strained compound and that of cyclohexane, taken as a reference.[10a]

Table 11.1 ▪ Coupling Constants and Calculated Geometric Parameters for Some Constrained Cyclohexanols

	J_{aa} (Hz)	J_{ae} (Hz)	Dihedral Angles ϕ_{aa} from J Calc.	Dihedral Angles ϕ_{ae} from J Calc.	β	Angle θ at C_3	d (Å) Calc.	d (Å) Dreiding Models	Refs.
 90	11.2	4.0	170° 164°	52° 55°	47°	112°	2.58	2.64	42, 36
 91	11.1	4.3	167° 163°	49° 53°	46°	112°	2.52	2.52	90
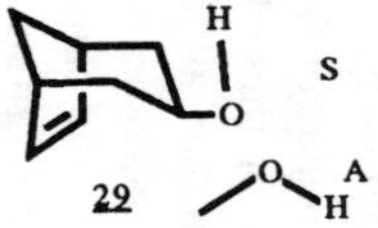 29 (S)	10.0	5.5	158° 148°	44° 40°	40°	114°	2.37	2.40	7, 35, 42
29 (A)					41°	113°			
 45	9.5	6.5	155° 146°	39° 33°	28°	114°	2.36	2.32	46
61 (S)	9.2	7.6	153° 144°	32° 30°	32°	111°	2.03	2.08	80
61 (A)			119°	9°	15°	113°			

Note: All measurements were made in CCl_4 except **45**, which was taken in $CDCl_3$. For **90** and **91** the calculated values were obtained from the corresponding hydrocarbons. Dreiding models were constructed as follows:
90 Using six cyclohexane [sp^3] carbons and putting the *syn*-axial methyl–methyl distance at 8.5 cm (= 3.4 Å).
91 Regular chair.
29 Model composed of five cyclopentane and three cyclohexane [sp^3] carbons.
45 Three cyclopentane and three cyclohexane [sp^3] carbons and two cyclopentene [sp^2] carbons.
61 Four cyclobutane and three cyclopentane [sp^3] carbons.

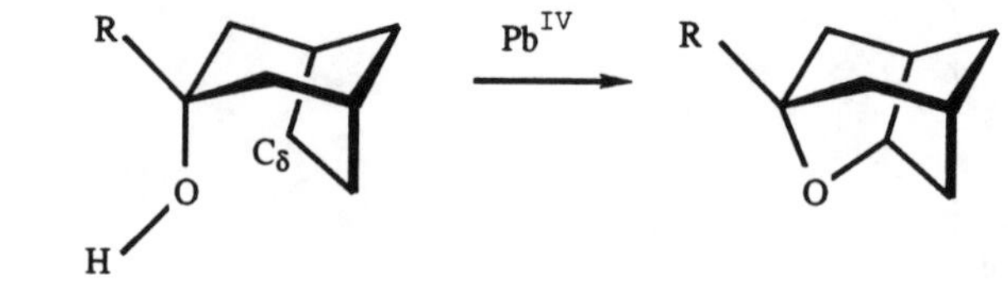

R, Pb^IV R, O, H, Cδ, O

	distance O.....Cδ	yield
28 R = H	244 pm	30 / 40 %
26 R = CH_3	237 pm	50%

Figure 11.49 ■ Influence of the nature of R group on the amount of ether formed by lead tetraacetate oxidation of alcohols featuring the bicyclo[3.2.1]octane skeleton.

The Δ value is 0.27° for bicyclo[3.2.1]octane and 0.66° for bicyclo[3.1.1] heptane. The bond populations, calculated by the Wiberg index[83] is lower for an axial C—H bond (0.959 in cyclohexane) than for the equatorial C—H bond (0.962). In agreement with anti reflex theory, the populations of the axial and equatorial C(2)—H bond become equal for the two bridged bicyclic derivatives (Figure 11.52).

The ketones corresponding to the previously mentioned hydrocarbons have been analyzed using the same semiempirical approach (Figure 11.53).[10b] The charge densities on the oxygen carbonyl decrease with the pinching increase when going from the ethano bridge of **2** to the etheno bridge of **44**. The calculated dipole moments decrease regularly from cyclohexanone **89** (2.95 D) to **2** (2.85 D) and then to **44** (2.68 D). For the same ketones there is a decrease of the σ bond population that parallels the infrared $\nu_{C=O}$ variation in carbon tetrachloride (1718 cm^{-1} for cyclohexanone, 1711 cm^{-1} for **2**, 1700 cm^{-1} for **44**). As mentioned earlier, the constraint is associated with Δ, that is, the difference between the interorbital angle of a cyclohexanone derivative under investigation and that of cyclohexanone itself, taken as a reference. The various Δ values for different angles of **2** and **44**, as well as the corresponding bonding populations of some bonds of these ketones and of the reference **89**, are reported in Table 11.2. It can be seen that, although the variation of the valence angle values is irregular, the constraint Δ increases with the pinching. The variation of the corresponding infrared $\nu_{C=O}$ follows the same trend. This supports the proposal[84] that has been made to use this infrared criterion as a constraint index.

As can be seen in Figure 11.53, the electronic density is higher for axial hydrogens than for equatorial ones (respectively, 1.0164 and 1.0075 in cyclohexane, for instance). Carbons are positively charged whereas the hydrogens bear a negative charge.[23] This charge distribution is reversed for the carbons and hydrogens α to the carbonyl in the corresponding ketones (**89** and **44**). It also can be noticed that the electron density is now higher for the related equatorial hydrogens as compared to the axial ones. In agreement with anti reflex theory, the electronic density differences decrease when the strain increases for the hydrocarbons and the corresponding ketones as well. This charge effect[23] also provides an explanation for the following experimental observa-

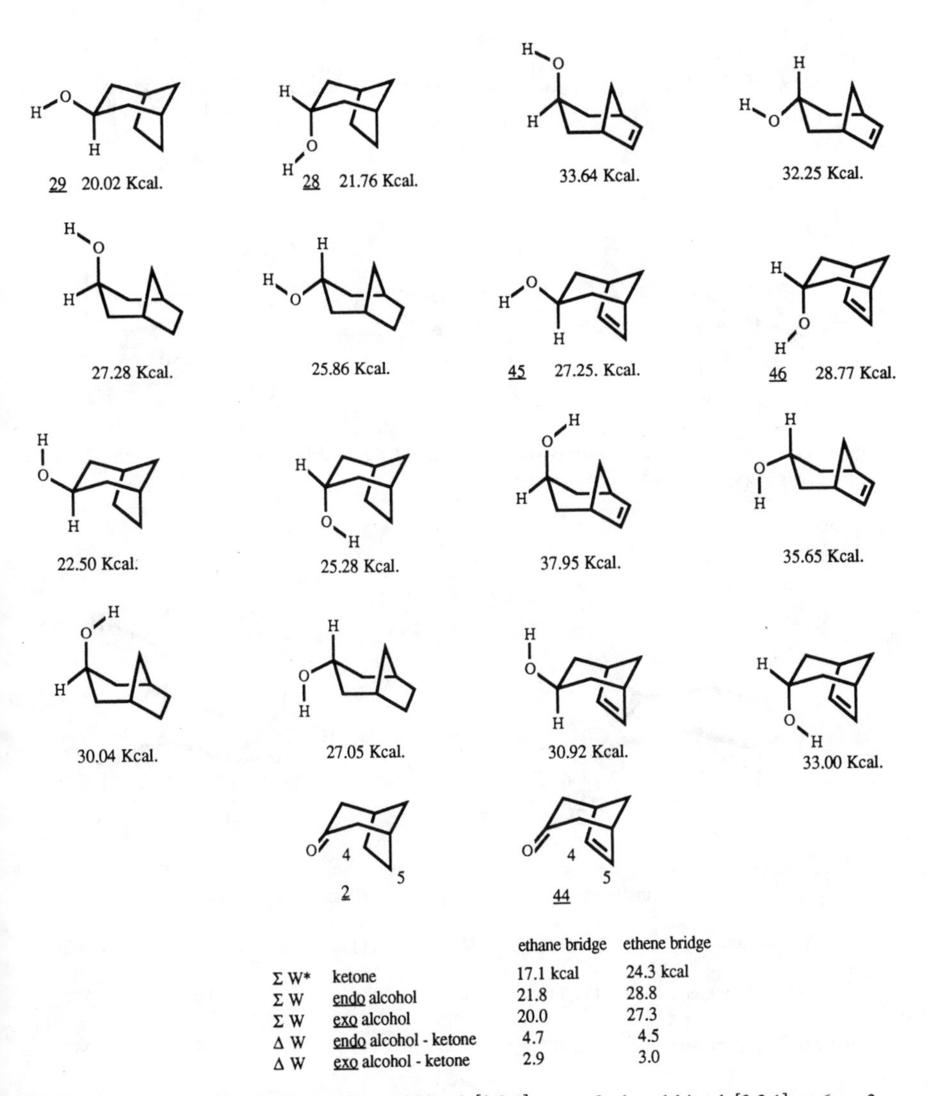

		ethane bridge	ethene bridge
Σ W*	ketone	17.1 kcal	24.3 kcal
Σ W	endo alcohol	21.8	28.8
Σ W	exo alcohol	20.0	27.3
Δ W	endo alcohol - ketone	4.7	4.5
Δ W	exo alcohol - ketone	2.9	3.0

Figure 11.50 ■ Various conformations of bicyclo[3.2.1]octane-3-ol and bicyclo[3.2.1]oct-6-en-3-ol; the energies given are steric (constraint energy). These energies are related to the thermodynamics, and it can be seen that exo derivatives (chair form for the cyclohexane ring) are the most stable for both alcohols.

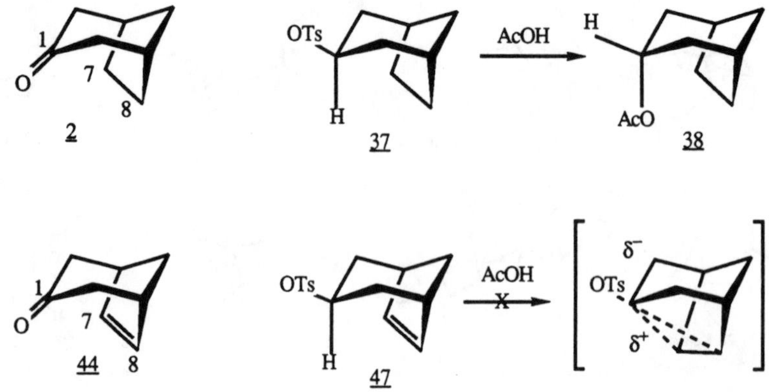

Figure 11.51 ▪ Acetolysis of *exo*-tosylates **37** and **47** derived from bicyclo[3.2.1]octan-3-one **2** and bicyclo[3.2.1]oct-6-en-3-one **44**, respectively.

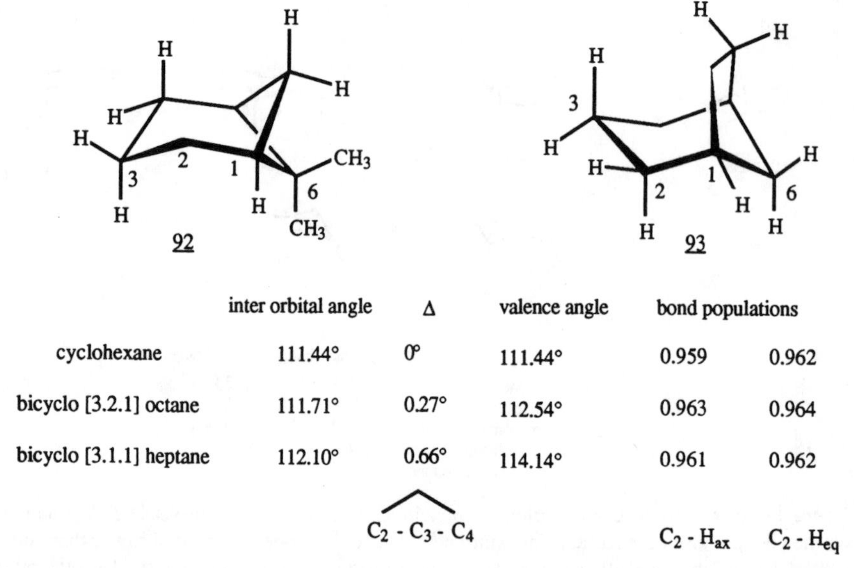

	inter orbital angle	Δ	valence angle	bond populations	
cyclohexane	111.44°	0°	111.44°	0.959	0.962
bicyclo [3.2.1] octane	111.71°	0.27°	112.54°	0.963	0.964
bicyclo [3.1.1] heptane	112.10°	0.66°	114.14°	0.961	0.962
		$\widehat{C_2 - C_3 - C_4}$		$C_2 - H_{ax}$	$C_2 - H_{eq}$

Figure 11.52 ▪ The anti reflex effect increases with Δ, which is the difference between the interorbital angle of the compound under investigation and the interorbital angle of cyclohexane taken as a reference. The bond populations of axial and equatorial groups become equal when they lose these axial and equatorial character.

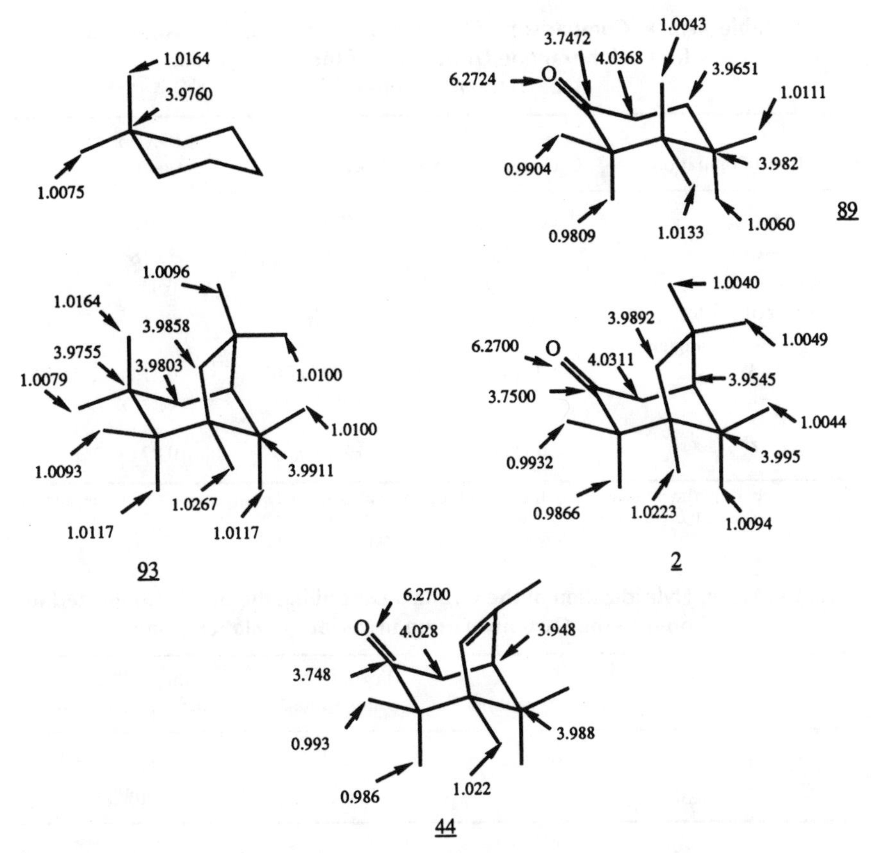

Figure 11.53 ▪ Electron densities of various hydrocarbons and related ketones (all bond termini are hydrogens).

tion: In cyclohexanones, the ^{1}H NMR signal due to α-equatorial protons is shielded relative to the corresponding α-axial proton, whereas this is reversed in hydrocarbons.

For cyclohexanones it is known that the s character of the α-carbon–hydrogen bonds is higher for axial bonds than it is for equatorial bonds, as can be seen in Table 11.3.[10b] It is also known that the $J^{13}{}_{C-H}$ coupling is proportional to this s character[85]; accordingly, the corresponding coupling for an axial C—H bond should have a higher value. Although this is not the case—and is even the reverse—in α-halogenoketones,[86] probably because other effects have to be taken into account, it is interesting to observe that this difference is lowered in bicyclo[3.2.1]octan-3-one, for instance, in agreement with the ring flattening that decreases the differentiation between axial and equatorial hydrogen.

We have previously discussed the possibility of either a cyclopropanic or a zwitterionic intermediate in the Favorskii mechanism. In order to appreciate the occurrence of one of these intermediate species, we have calculated the geometry of

Table 11.2 ■ Comparison of Bonding Population and Constraint Δ for Cyclohexanone Derivatives of Increasing Strain Due to the Anti Reflex Effect[a]

Bond Population	Cyclohexanone **89**	Bicyclo[3.2.1]-octan-3-one **2**	Bicyclo[3.2.1]-oct-6-ene-3-one **44**
C(1)—C(2)	1.008	0.991	0.982
C(2)—C(3)	1.016	1.019	1.018
C(1)—C(8)	1.020	0.993	0.983
Constraint Δ for			
C—C—C angle at:			
C-3	(115.69)	0.25 (117.07)	0.41 (116.23)
C-2		0.68	0.78
C-1		1.76	2.24
C-6	(111.17)	0.55 (99.68)	0.87 (99.26)

[a]The number in the brackets are the values of the valence angles in degrees (for the carbon numbering see Figure 11.51).

Table 11.3 ■ Hybridization of the Orbitals Describing the Bonds Connected to Carbon α to the Carbonyl Group in Various Cyclohexanones

	Cyclohexanone	Bicyclo[3.2.1]-octan-3-one **2**	Bicyclo[3.2.1]-oct-6-en-3-one **44**
$C_\alpha—H_{axial}$	$sp^{3.17}$	$sp^{3.38}$	$sp^{3.31}$
$C_\alpha—H_{equatorial}$	$sp^{3.32}$	$sp^{3.39}$	$sp^{3.47}$

the cyclopropanone derivative **94** (Figure 11.54) by the Westheimer method, and its electronic structure by the CNDO/2 method.

Because of the anti reflex effect of the *gem*-dimethylated methano bridge, the carbonyl angle has a value of 83.51° much larger than the value of 64.36° that would be observed without this additional strain.[8,87,88] As a consequence the C(2)–C(4) distance is increased. Furthermore, the whole system is flattened at the ring junction between the cyclopropanone and the remainder of the molecule. The C(1)—C(2) and

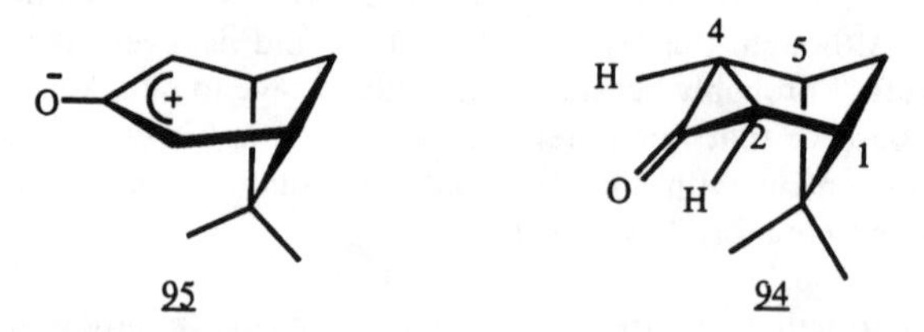

Figure 11.54 ■ Reactive intermediates involved in Favorskii rearrangement of α-bromo derivatives of isonopinone **53**.

C(2)—H bonds therefore tend to be in the plane defined by C(2)–C(3)–C(4). This is not at all the case in a nonsubstituted isolated cyclopropanonic intermediate, where the C—H bonds are almost perpendicular to the ring. All these constraints will obviously favor the zwitterionic intermediate **95** rather than the cyclopropanic intermediate **94**, and substitution products (rather than ring-contraction products) will be formed.

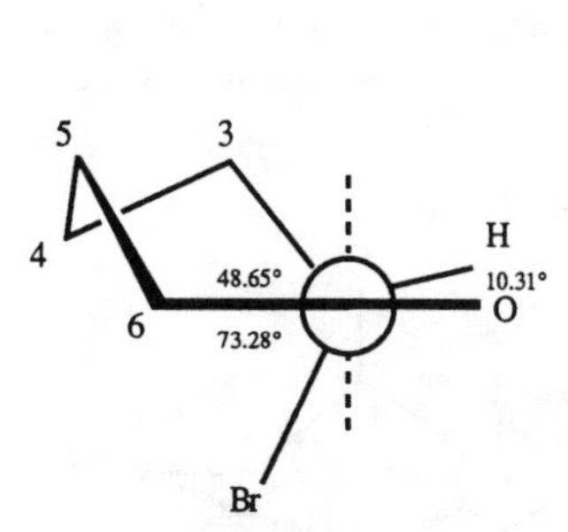

3 2-bromo cyclohexanone

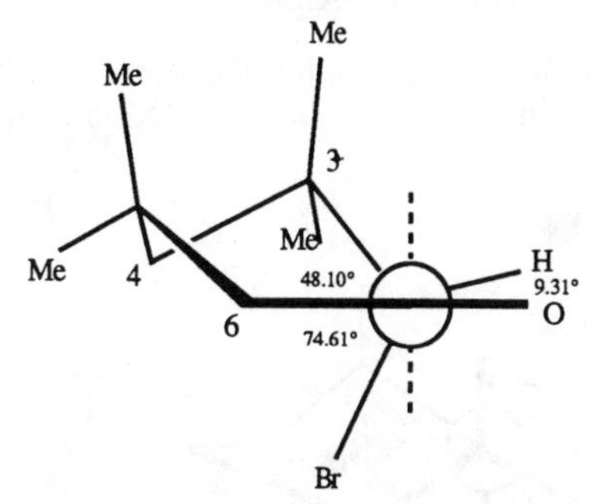

4 2-bromo-3,3,5,5-tetramethylcyclohexanone

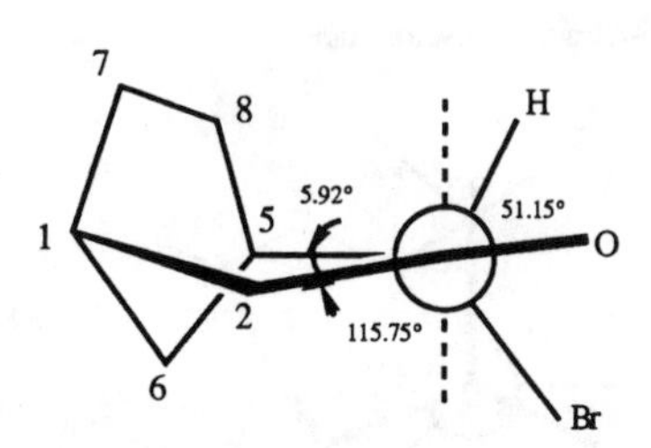

10 2-bromo bicyclo [3,2,1] octan-3-one

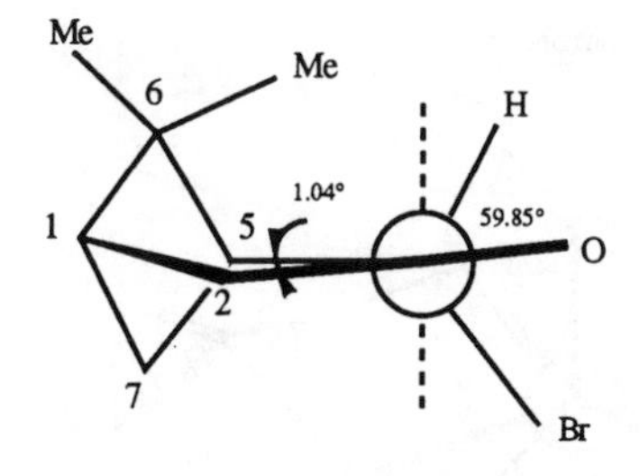

54 2-bromo-6,6-dimethyl bicyclo [3,1,1] heptan-3-one

	distance D (Å) between two α,α' diaxial hydrogen in the corresponding non brominated ketones	dihedral angle (φ) C - C(=O) - C - C of the corresponding non brominated ketones
3	2.768	52.83°
4	2.708	56.79°
10	3.044	40.26°
54	3.704	37.23°

Figure 11.55 ▪ Newman projections of the dihedral angles around C(1)—C(2) (for **3** and **4**) around C(3)—C(4) (for **10** and **54**) of α-brominated ketones exhibiting the reflex (**4**) and the anti reflex (**10** and **54**) effect, as compared to 2-bromo cyclohexanone. Typical data for D and ϕ are also reported for the corresponding nonhalogenated ketones.

11.4. Conclusion

The reflex effect due to the spreading apart of β,β'-axial substituents of a cyclohexanone and the anti reflex effect due to the pinching of the same substituents have a geometrical and energetical nature. These effects also occur in heterocycles and on the corresponding hydrocarbons and alcohols.

11.4A. Geometrical Aspects

The reflex and anti reflex effects are mainly due to the axial substituents (axial methyl groups in **1** and an ethano bridge in **2**, for instance), even so the equatorial methyl

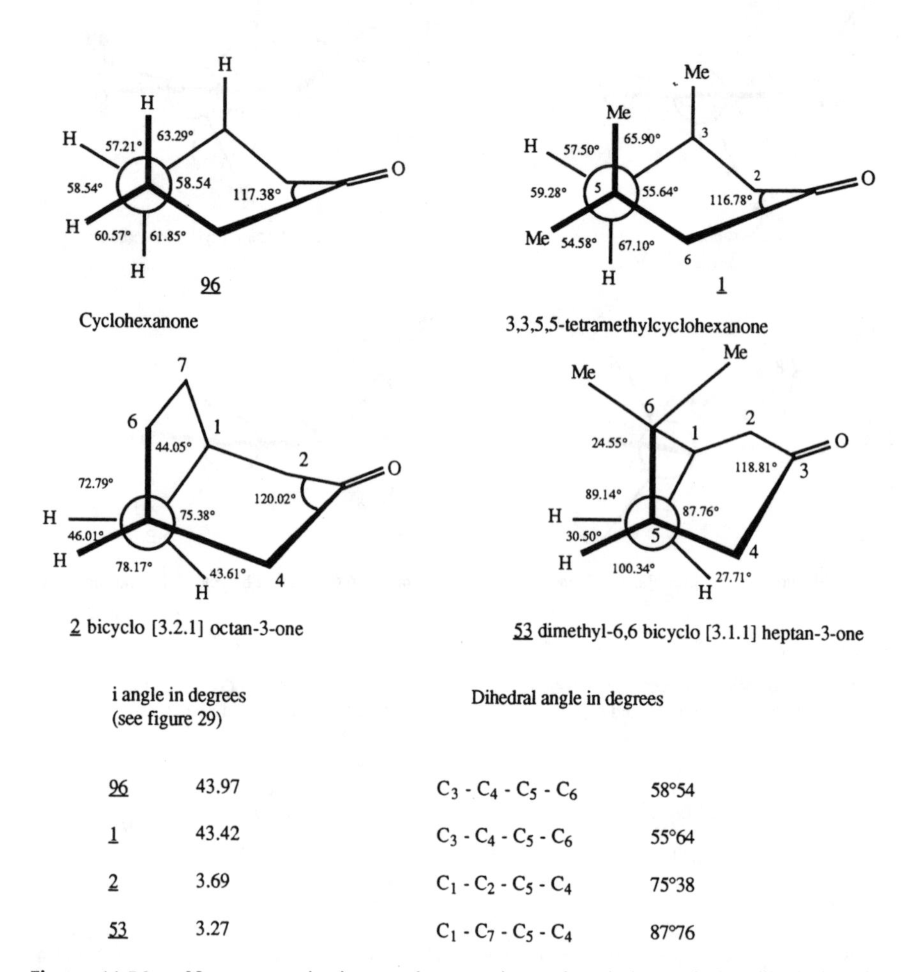

	i angle in degrees (see figure 29)		Dihedral angle in degrees
96	43.97	$C_3 - C_4 - C_5 - C_6$	58°54
1	43.42	$C_3 - C_4 - C_5 - C_6$	55°64
2	3.69	$C_1 - C_2 - C_5 - C_4$	75°38
53	3.27	$C_1 - C_7 - C_5 - C_4$	87°76

Figure 11.56 ■ Newman projections and comparison of variations of the dihedral angles around the C(5)—C(4) bond [C(5)—C(8) for **2**, C(5)—C(7) for isonopinone **53**].

group in **1** and the *gem*-dimethyl group in **53** superimpose their own effects. The initial idea of the reflex effect proved to be right: Separating two axial bonds in β,β' of a cyclohexanone carbonyl brings together the α,α'-axial substituents on the other side of the ring by simultaneously enlarging the dihedral angle around C(1)—C(2). The reverse, that is, the anti reflex effect, was also proven to be right: Bringing together (or pinching) two axial bonds in β,β' of a cyclohexanone carbonyl separates the α,α'-axial substituents on the other side of the ring by simultaneously decreasing the dihedral angle around C(3)—C(4); the latter effect results in the flattening of the ring (these values can be altered in α-bromoketones, as can be seen in Figure 11.55).

The consequences of the reflex and anti reflex effects can be clearly appreciated by examining the variations of the calculated distance between two axial α,α'-hydrogens in nonbrominated ketones (Figure 11.55).

The flattening of the cyclohexanone ring that is observed when the anti reflex effect operates can be seen clearly by examining the dihedral angles reported in Figure 11.55, but also by looking at the i angle variation (for the definition see Figure 11.29) reported in Figure 11.56. The consequences of the reflex and anti reflex effects also appear very nicely on the dihedral angle that includes the carbon bearing the *gem*-dimethyl group of **1** or the bridgehead carbons in **2** and **53** (Figure 11.56).

The valence angles around the carbonyl are not so informative. All these geometrical variations have an influence on the physical and chemical behavior of the corresponding molecules, as reported and described throughout this chapter. We have frequently emphasized the ring flattening at the level of the carbonyl for bridged bicyclic ketones **2** and **53** and their derivatives. Similarly, in 2,2,6,6-tetramethylcyclohexanone such a flattening is also observed.[89]

11.4B. Energetical Aspects

The most typical data are reported in Table 11.4 and present another approach to the reflex and anti reflex effects.

As a general conclusion it can be said that the initial hypotheses that had been made to explain physical data and the chemical behavior of 3,3,5,5-tetramethylcyclohexanone bromo derivatives were proven to be right and were confirmed by calcula-

Table 11.4 ▪ Energy Differences Assigned to the Reflex and Anti Reflex Effects; Cyclohexanone Might Be Taken as a Reference[a]

	ΔE_1[b]	ΔE_2[c]	ΔE_3[d]
Cyclohexanone	5.98	−0.88	1.30
3,3,5,5-Tetramethylcyclohexanone	14.12	+0.08	2.09
Bicyclo[3.2.1]octan-3-one	0.75	−1.32	−0.73
6,6-Dimethylbicyclo[3.1.1]heptan-3-one	−1.43	−1.58	−4.70

[a]All the energies are expressed in kilocalories per mole.
[b]ΔE_1: Energy difference between chair and boat conformation.
[c]ΔE_2: Energy difference between the α-brominated ketones with an axial bromine and with an equatorial bromine.
[d]ΔE_3: Energy difference between the α,α'-dibrominated ketones with two axial bromines and with two equatorial bromines.

tions. Steric constraints introduce a sort of mechanical deformation. This somewhat simple idea gives a fairly good approach to what is really happening, so that it is now possible to predict the chemical or physical behavior of molecules where such effects are occurring.

Acknowledgments. The authors wish to thank all those who participated in this work, whose names appear in the references. They thank particularly Professor Charles Williams Jefford of Temple University, Philadelphia, PA, presently at the University of Geneva Switzerland, and Professors André Baretta, Pierre Brun, and Jean-Pierre Zahra, University of Aix-Marseille, France. We also are grateful to the University of Aix-Marseille III, the University of Angers, and CNRS for providing financial support.

We also are very grateful to Mrs. Charlette Ruiz, who typed the manuscript, and Mr. Alain Uldry, who did the drawings.

References

1. Davis, M.; Hassel, O. *Acta Chem. Scand.* **1963**; *17*, 1181. Buys, H. R.; Geise, H. J. *Tetrahedron Lett.* **1970**, 2991. Wöhl, R. A. *Chimia* **1964**, *18*, 219. Atkinson, V. A. *Acta Chem. Scand.* **1961**, *15*, 599. Anet, F. A. L.; Ahmad, M.; Hall, L. D. *Proc. Chem. Soc.* **1964**, 145. Bovey, F. A.; Hood, F. P., III; Anderson, E. W.; Hornegay, R. L. *J. Chem. Phys.* **1964**, *41*, 2041.
2. Sandris, C.; Ourisson, G. *Bull. Soc. Chim. France* **1958**, 1524.
3. Corey, E. J. *J. Am. Chem. Soc.* **1953**, *75*, 2301, 3297, 4832.
4. Waegell, B.; Ourisson, G. *Bull. Soc. Chim. France* **1963**, 495, 496, 503.
5. Biellmann, J. F.; Hanna, R.; Ourisson, G.; Sandris, C.; Waegell, B. *Bull. Soc. Chim. France* **1960**, 1429.
6. Waegell, B. Thèse de Docteur-ès-Sciences, Université de Strasbourg, No. 287, 1964.
7. (a) Waegell, B.; Jefford, C. W. *Bull. Soc. Chim. France* **1964**, 844. (b) Durocher, J. G.; Fabre, H. *Can. J. Chem.* **1964**, *42*, 260.
8. Baretta, A. Thèse de Docteur-ès-Sciences, Université de Droit, d'Economie et des Sciences, Aix-Marseille, 1978. Baretta, A.; Waegell, B. *Tetrahedron Lett.* **1976**, 753. Baretta, A.; Waegell, B. A survey of Favorskii rearrangement mechanisms: Influence of the nature and strain of the skeleton; in *Reactive Intermediates*; Abramovitch, R. A., Ed.; Plenum Press: New York, 1982; vol. 2, p. 527.
9. (a) Waegell, B.; Ourisson, G. *Bull. Soc. Chim. France* **1961**, 2443. (b) Waegell, B.; Pouzet, P.; Ourisson, G. *Bull. Soc. Chim. France* **1963**, 1821. (c) Fournier, J. Thèse de Docteur-ès-Sciences, Université de Provence, Aix-Marseille, No. 7001, 1972.
10. (a) Fournier, J. *J. Mol. Struct.* **1974**, *23*, 337. (b) Fournier, J. *J. Mol. Struct.* **1974**, *23*, 177. (c) Fournier, J. *J. Chem. Res. (M)* 1977, 3440.
11. (a) Halford, J. O. *J. Chem. Phys.* **1956**, *24*, 830. (b) Brauman, J. I.; Lautie, Y. W. *Tetrahedron* **1968**, *24*, 2595.
12. (a) Jones, R. N.; Ramsay, D. A.; Herling, F.; Dobriner, K. *J. Am. Chem. Soc.* **1952**, *74*, 2828. (b) Jones, R. N. *J. Am. Chem. Soc.* **1953**, *75*, 4839.
13. Cantacuzène, J. J. *Chim. Phys.* **1962**, *59*, 186.
14. Cookson, R. C. *J. Chem. Soc.* **1954**, 282.
15. Julg, A. *J. Chim. Phys.* **1956**, *58*, 453.
16. Karplus, M. *J. Chem. Phys.* **1959**, *30*, 11.
17. Meinwald, J.; Lewis, A. *J. Am. Chem. Soc.* **1961**, *83*, 2769.

18. Shoolery, J. N. *J. Chem. Phys.* **1959**, *31*, 1427.
19. Corey, E. J. *J. Am. Chem. Soc.* **1954**, *76*, 175.
20. Waegell, B. *Bull. Soc. Chim. France* **1964**, 855.
21. Williamson, K. L.; Johnson, W. S. *J. Am. Chem. Soc.* **1961**, *83*, 4623.
22. Zahra, J. P. Thèse de Docteur-ès-Sciences, Université de Provence, Aix-Marseille, No. 8431, 1973.
23. Fournier, J. *J. Mol. Struct.* **1973**, *18*, 391.
24. Baretta, A.; Zahra, J. P.; Jefford, C. W.; Waegell, B. *Tetrahedron* **1970**, *26*, 15.
25. Fétizon, M.; Goré, J.; Laszlo, P.; Waegell, B. *J. Org. Chem.* **1966**, *31*, 4047.
26. Zahra, J. P.; Waegell, B.; Reisse, J.; Pouzard, G.; Fournier, J. *Bull. Soc. Chim. France* **1976**, 1896.
27. Piotrowska, H.; Wojnarowski, W.; Waegell, B. *Bull. Soc. Chim. France* **1965**, 3511.
28. Dürr, H.; Ourisson, G.; Waegell, B. *Chem. Ber.* **1965**, *98*, 1858.
29. Zahra, J. P.; Jefford, C. W.; Waegell, B. *Tetrahedron* **1969**, *25*, 5087.
30. Cherest, M.; Felkin, H.; Prudent, N. *Tetrahedron Lett.* **1968**, *18*, 2205.
31. Jefford, C. W.; Waegell, B.; Ramey, K. *J. Am. Chem. Soc.* **1965**, *87*, 2191.
32. Jefford, C. W.; Gunsher, J.; Hill, D. T.; Brun, P.; Le Gras, J.; Waegell, B. *Org. Synthesis* **1971**, *51*, 60.
33. Jefford, C. W.; Waegell, B. *Tetrahedron Lett.* **1963**, 1981.
34. Rassat, A.; Jefford, C. W.; Lehn, J. M.; Waegell, B. *Tetrahedron Lett.* **1964**, 233.
35. Jefford, C. W.; Mahajan, S.; Waslyn, J.; Waegell, *J. Am. Chem. Soc.* **1965**, *87*, 2183.
36. Jefford, C. W.; Mahajan, S.; Gunsher, J.; Waegell, B. *Tetrahedron Lett.* **1965**, 2333.
37. Crist, B. V.; Rodgers, S. L.; Lightner, D. A. *J. Am. Chem. Soc.* **1982**, *104*, 6040.
38. Cainelli, G.; Mihailovic, M. L.; Arigoni, D.; Jeger, O. *Helv. Chim. Acta* **1959**, *42*, 1124. Mihailovic, M. L.; Cekovic, Z.; Andrejevic, V.; Matic, R.; Jéremic, D. *Tetrahedron* **1968**, *24*, 4947.
39. Heusler, K.; Kalvoda, J. *Angew. Chem., Int. Ed. Engl.* **1964**, *3*, 525. Moriarty, R.; Walsh, H. G. *Tetrahedron Lett.* **1965**, 465.
40. (a) Brun, P.; Pally, M.; Waegell, B. *Tetrahedron Lett.* **1970**, 331. (b) Brun, P. Thèse de Docteur-ès-Sciences, Université de Droit, d'Economie et des Sciences, Aix-Marseille, 1975. (c) Brun, P.; Waegell, B. *Bull. Soc. Chim. France* **1972**, 1825. (d) Bensadoun, N.; Brun, P.; Casanova, J.; Waegell, B. *J. Chem. Res. (S)* **1981**, 236. (e) Brun, P.; Waegell, B. *Tetrahedron* **1976**, *32*, 1125. (f) Brun, P.; Waegell, B. Synthetic applications and reactivity of alkoxyl radicals; in *Reactive Intermediates*, Abramovitch, R. A., Ed.; Plenum Press: New York, 1983; Vol. 3, p. 367.
41. Arnaud, C.; Accary, A.; Huet, J. *C. R. Acad. Sci. Paris Série C* **1977**, *285*, 325.
42. Jefford, C. W.; Hill, D. T.; Gunsher, J. *J. Am. Chem. Soc.* **1967**, *89*, 6881.
43. Sneen, R. A.; Matheny, N. P. *J. Am. Chem. Soc.* **1964**, *86*, 5503.
44. Jefford, C. W.; Gunsher, J.; Waegell, B. *Tetrahedron Lett.* **1965**, 3405, 3421.
45. Berger, S. *J. Org. Chem.* **1978**, *43*, 209.
46. Lebel, N. A.; Maxwell, R. J. *J. Am. Chem. Soc.* **1969**, *91*, 2307.
47. Hoffmann, H. M. R.; Clemens, K. E.; Smithers, R. H. *J. Am. Chem. Soc.* **1972**, *94*, 3940.
48. Banks, R. M.; Maskill, H. *J. Chem. Soc., Perkin Trans. 2* **1976**, 1506.
49. Banks, R. M.; Maskill, H. *J. Chem. Soc., Perkin Trans. 2* **1977**, 1991.
50. Barfield, M.; Conn, S. A.; Marshall, J. L.; Miller, D. E. *J. Am. Chem. Soc.* **1976**, *98*, 6253.
51. Dallinga, G.; Toneman, L. H. *Rec. Trav. Chim. Pays-Bas* **1963**, *88*, 185.
52. Banthorpe, D. V.; Whittaker, D. *Chem. Rev.* **1966**, *66*, 643.
53. (a) Bessière-Chrétien, Y.; Grison, C. *Bull. Soc. Chim. France* **1971**, 1454. (b) Bessière-Chrétien, Y.; Meklati, B. *Bull. Soc. Chim. France* **1971**, 2591.
54. Abraham, R. J.; Bottom, F. H.; Cooper, M. A.; Salmon, J. R.; Whittaker, D. *Org. Magn. Res.* **1969**, *1*, 51.

55. Baretta, A. J.; Jefford, C. W.; Waegell, B. *Bull. Soc. Chim. France* **1970**, 3985.
56. Regan, A. F. *Tetrahedron* **1969**, *27*, 3801.
57. Mühlstädt, M.; Hermann, M.; Zschunke, A. *Tetrahedron* **1968**, *24*, 1611.
58. Baretta, A. J.; Jefford, C. W.; Waegell, B. *Bull. Soc. Chim. France* **1970**, 3899.
59. Reisse, J.; Piccinni-Leopardi, C.; Zahra, J. P.; Waegell, B.; Fournier, J. *Org. Magn. Reson.* **1977**, *9*, 512.
60. Zweifel, G.; Brown, H. C. *J. Am. Chem. Soc.* **1964**, *86*, 393.
61. Teisseire, P.; Galfré, A.; Plattier, M.; Cordier, B. *Recherches* **1966**, *15*, 52.
62. Hartshorn, M. P.; Wallis, A. F. A. *Tetrahedron* **1965**, *21*, 273.
63. Barthélémy, M.; Monthéard, J. P.; Bessière-Chrétien, Y. *Bull. Soc. Chim. France* **1969**, 2725.
64. Eaton, P. E.; Cole, T. W. *J. Am. Chem. Soc.* **1964**, *86*, 962.
65. Chenier, Ph. J.; Kao, J. C. *J. Org. Chem.* **1976**, *41*, 3730.
66. Fournier, J.; Waegell, B. *Tetrahedron* **1970**, *86*, 3195.
67. Allinger, N. L.; Da Rooge, M. *J. Am. Chem. Soc.* **1962**, *84*, 4561.
68. Lehn, J. M.; Levisalles, J.; Ourisson, G. *Bull. Soc. Chim. France* **1963**, 1096.
69. (a) Goaman, L. C. G.; Grant, D. F. *Acta Cryst.* **1964**, *17*, 1604. (b) Goaman, L. C. G.; Grant, D. F. *Tetrahedron* **1963**, *19*, 1531.
70. Westheimer, F. H. In *Steric Effects in Organic Chemistry*; Newman, M. S., Ed.; Wiley: New York, **1956**; p. 523.
71. (a) Melkanoff, M. A.; Swada, T.; Raynal, J. In *Methods in Computational Physics*; Academic Press: New York, **1966**; Vol. 6, p. 45. (b) Nelder, J. A.; Mead, R. *Computer J.* **1965**, *8*, 308. (c) Arndt, R. A.; McGregor, M. A. in *Methods in Computational Physics*; Academic Press: New York, **1966**; Vol. 6, p. 274.
72. Fournier, J.; Waegell, B. *Tetrahedron* **1972**, *28*, 3407.
73. Fournier, J.; Waegell, B. *Bull. Soc. Chim. France* **1973**, 436.
74. Fournier, J. *Bull. Soc. Chim. France* **1973**, 1954.
75. Burkert, U.; Allinger, N. L. *Molecular Mechanics*; A. C. S. Monograph 177; American Chemical Society: Washington, DC, **1982**.
76. Rickborn, B.; Wuesthoff, M. T. *J. Am. Chem. Soc.* **1970**, *92*, 6894.
77. Wigfield, D. C.; Phelps, D. J. *J. Chem. Soc., Chem. Commun.* **1970**, 1152.
78. Geneste, P.; Lamaty, G. *Bull. Soc. Chim. France* **1968**, 669; **1969**, 2027.
79. Allinger, N. L.; Yuh, Y. Y. *QCPE* #423; Indiana University: Bloomington, IN.
80. Jefford, C. W.; Baretta, A. J.; Fournier, J.; Waegell, B. *Helv. Chim. Acta* **1970**, *53*, 1180.
81. Fournier, J.; Waegell, B. *Bull. Soc. Chim. France* **1973**, 1599.
82. Dishler, B. *Z. Naturforsch.* **1964**, *19a*, 887.
83. Wiberg, K. B. *Tetrahedron* **1968**, *24*, 1083.
84. (a) Foote, C. S. *J. Am. Chem. Soc.* **1964**, *86*, 1853. (b) Schleyer, P. V. R. *J. Am. Chem. Soc.* **1964**, *86*, 1854.
85. (a) Grutzner, J. B.; Jautelat, M.; Dence, J. B.; Smith, R. A.; Roberts, J. D. *J. Am. Chem. Soc.* **1970**, *92*, 7107. (b) Maciel, G. E.; McIver, J. W.; Ostlund, N. S.; Pople, J. A. *J. Am. Chem. Soc.* **1970**, *92*, 1.
86. Cantacuzène, J.; Jantzen, R.; Tordeux, M. *Org. Magn. Reson.* **1975**, *7*, 407.
87. Fournier, J.; Baretta, A.; Waegell, B. Unpublished results.
88. Olsen, J. F.; Kang, S.; Burnelle, L. *J. Mol. Struct.* **1971**, *9*, 305.
89. Bory, S.; Fétizon, M.; Laszlo, P.; Williams, D. H. *Bull. Soc. Chim. France* **1965**, 2541.
90. Anet, F. A. L. *J. Am. Chem. Soc.* **1962**, *84*, 1053.

12

Cage Molecules in Photochemistry

Tsutomu Miyashi, Yoshiro Yamashita, and Toshio Mukai

Tohoku University, Sendai, Japan

Contents

12.1. Introduction
12.2. Photochemical Syntheses of Cage Molecules
12.3. Photochemical Reactions of Cage Molecules
References

12.1. Introduction

According to the Woodward–Hoffmann rule, [2 + 2] cycloaddition reactions are thermally forbidden, although photochemically allowed.[1] In fact, a cyclobutane ring can be constructed by a photochemical [2 + 2] cycloaddition reaction. The intramolecular reaction is useful for preparing a strained cyclobutane ring. Thus, such reactions have provided convenient syntheses of cage molecules like cubane **1**, the preparation of which seems impossible without the use of photochemical reactions. The photochemical syntheses of cage molecules are summarized in the first half of this chapter, in which structural factors affecting the photochemical [2 + 2] cycloadditions are also described.

Cage molecules possess high strain energy, which often causes unusual reactions such as transition-metal-promoted valence isomerizations.[2] The reactions of strained systems have also been studied in the field of photochemistry. In particular, photoinduced electron-transfer reactions have attracted much attention due to low ionization potential as well as the high reactivities of the cage molecules. The photochemical reactions are described in the latter half of this chapter.

12.2. Photochemical Syntheses of Cage Molecules

Cubane **1** was expected to be prepared by a photochemical [2 + 2] cyclization of *syn*-tricyclo[4.2.0.0^{2,5}]octa-3,7-dienes **2**. However, the attempts did not succeed in the case of R = H or Me.[3] One explanation of this failure in the cyclization is that an effective through-bond coupling of the two π orbitals, brought by ideally oriented high-lying σ orbitals, converts a symmetry-allowed photochemical [2 + 2] cycloaddition to a symmetry-forbidden reaction.[3a,4] An extremely large strain increase involved in cyclization is the alternative explanation.[3b] However, introduction of substituents makes the photoreaction possible by controlling the electronic states or the steric conditions. Thus, the octakis(trifluoromethyl) derivative **2** (R = CF_3) affords the corresponding cubane.[5] The sterically hindered diene **3** affords cubane **4**.[6] In the latter case, the nonbonded distance between unsaturated centers (2.6 Å) is shorter than that of the parent diene **2** (3.05 Å).

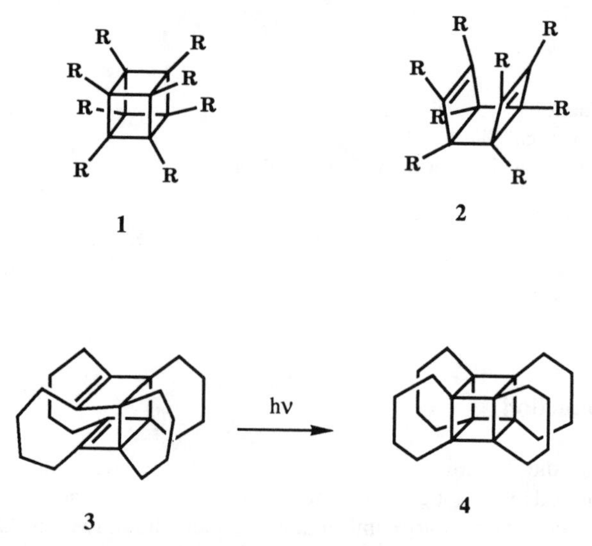

The photochemical [2 + 2] cycloaddition of diene **5**, a homolog of **2**, smoothly proceeds to afford homocubane **6** by acetone-sensitized irradiation.[7] A derivative **7** affords homocubane **8**, from which homocubanone **9** is derived.[8] The heterocyclic analog of homocubane can be prepared photochemically also. For example, phosphin oxide **10** undergoes photochemical [2 + 2] cyclization to give cage compound **11**.[9] The *anti*-tricyclic dienes **12** also afford homocubane derivatives **13** by irradiation.[10] In this case, a 1,3 shift is proposed to take place prior to cyclization. Similarly, the *anti*-tricyclic dienone **14** affords homocubanone **9**, in which a 1,3-acyl shift occurs before the [2 + 2] cyclization.[11]

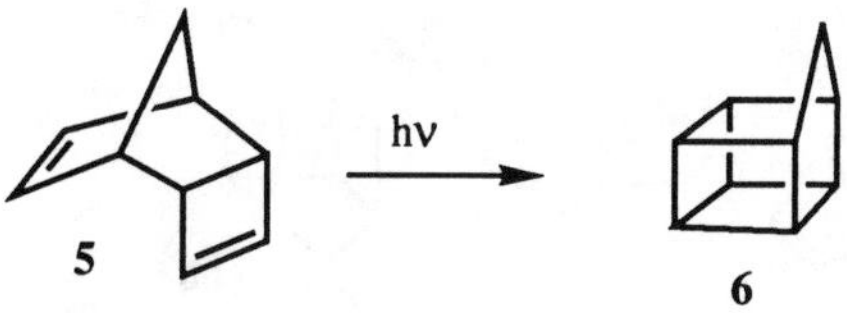

5 hν → 6

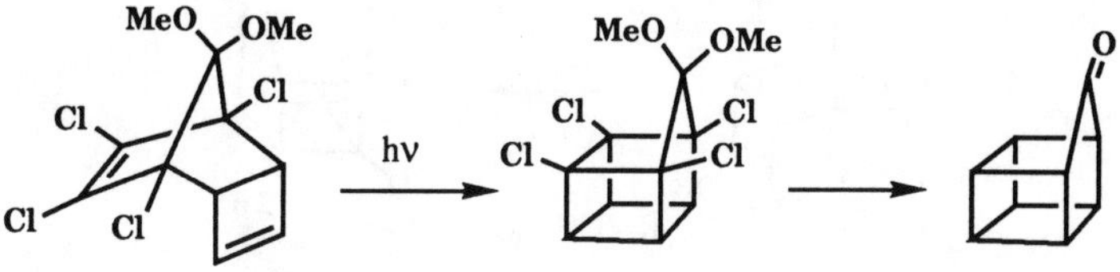

7 hν → 8 → 9

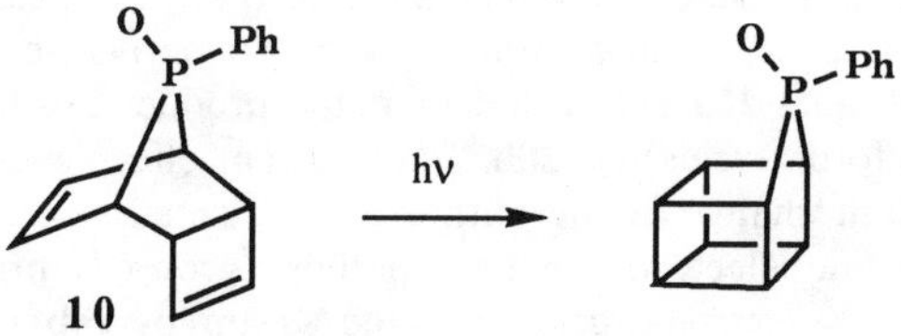

10 hν → 11

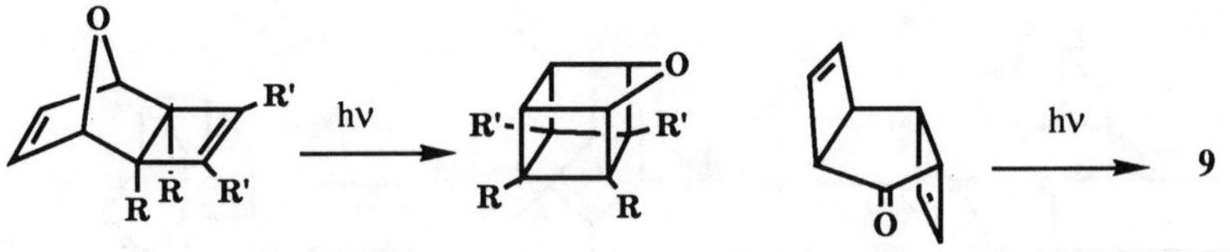

12 hν → 13 a; R=H, R'=CO_2Me b;R=R'=CF_3

14 hν → 9

Diene **15** affords bishomocubane **16** in 80% yield by acetone-sensitized irradiation.[12] On the other hand, the yield of the photocyclization of diene **17** containing an ethano bridge to trishomocubane **18** is only 15%.[3b] The effect of the bridge is also observed in the photoreaction of trimethylsilyloxy derivatives **19**, which can be excited by direct irradiation.[13] The yields of the cyclization to **20** are dependent on the length of the bridge, that is, n = 1 (80%), n = 2 (5%), n = 3 (0%). Osawa explained this effect in terms of the difference in strain increase by cyclization.[3b]

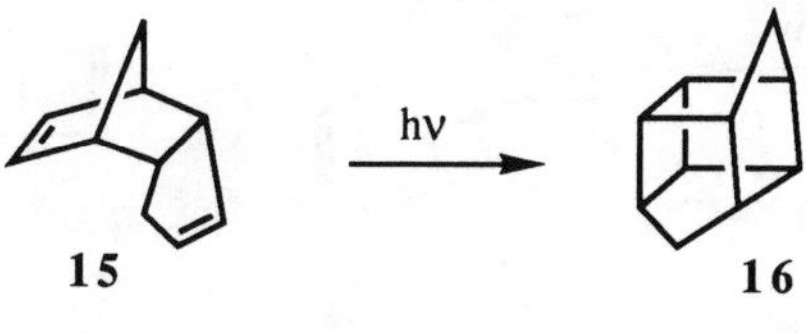

15 hν → 16

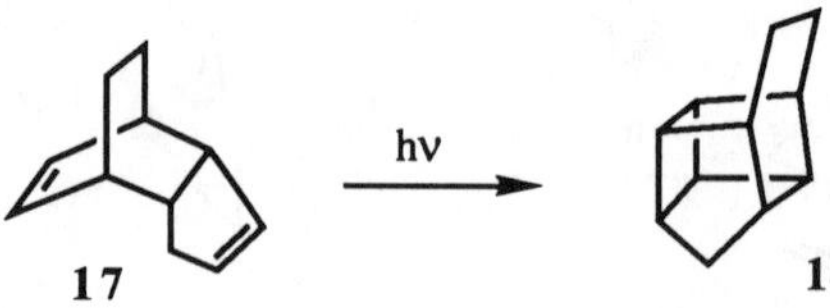

17 hν 18

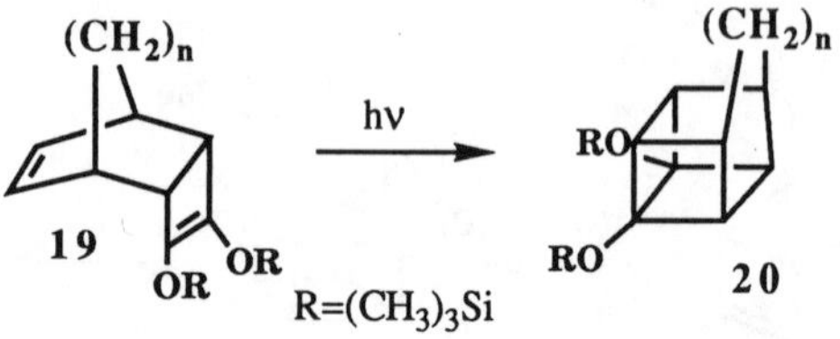

$(CH_2)_n$ hν $(CH_2)_n$

RO

19 OR OR RO 20

R=$(CH_3)_3$Si

Diene **21**, an adduct of cyclooctatetraene with maleic anhydride, affords a cage compound **22** in high yield, from which basketene **23** is derived.[14] The homolog **24** is prepared by irradiation of diene **25** in acetone.[15] A benzo derivative **26a** gives a mixture of a cyclization product **27a** and a di-π-methane product **28a** by sensitized irradiation, whereas **26b** affords exclusively **28b**.[16] Substituent effects were explained by the substituent-dependent change in the nature of an excited state. The lowest excited state of **26a** with the electron-donating methoxy group is predominantly ethylene $\pi\pi^*$, whereas the electron-accepting cyano group of **26b** localizes the excited state on the aromatic moiety.

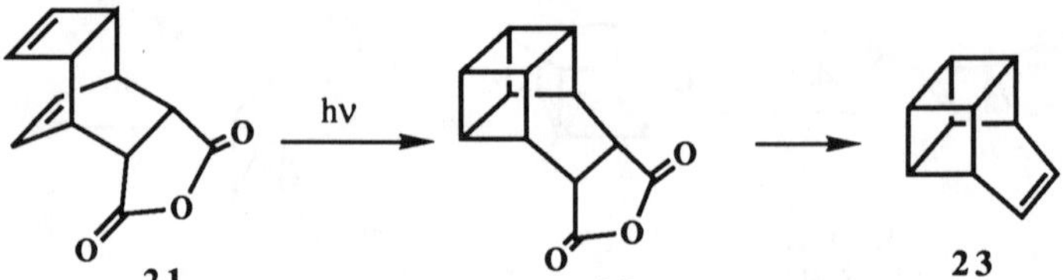

21 hν 22 23

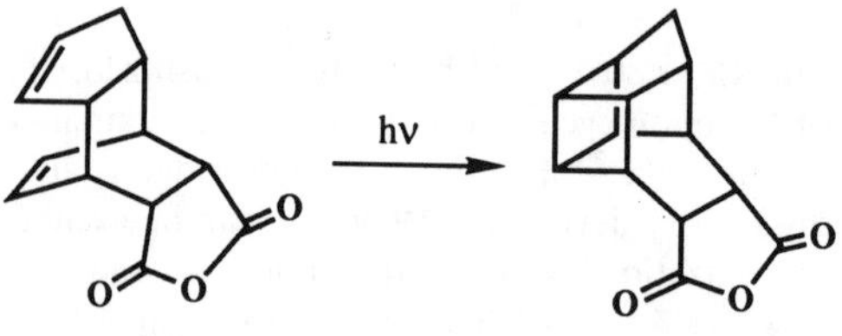

25 hν 24

26 hν a; X=OMe b; X=CN 27 + 28

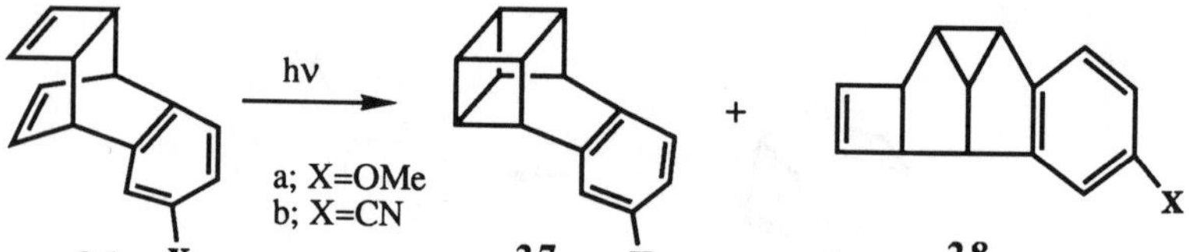

The photocyclization of **29** to **30** efficiently takes place.[17] The reaction of **30** with *t*-BuLi leads to the formation of an interesting cage compound **31**, which is expected to afford hexaprismane **32** by [2 + 2] cyclization. However, the photoreaction has not been successful. The cyclization of hypostrophene **33** to pentaprismane **34** also has failed.[18] The reason is attributed to a through-bond interaction of the two π orbitals,[19] as discussed in the case of cubanes **1**. On the other hand, homohypostrophene **35** affords homopentaprismane **36** by irradiation in acetone.[20]

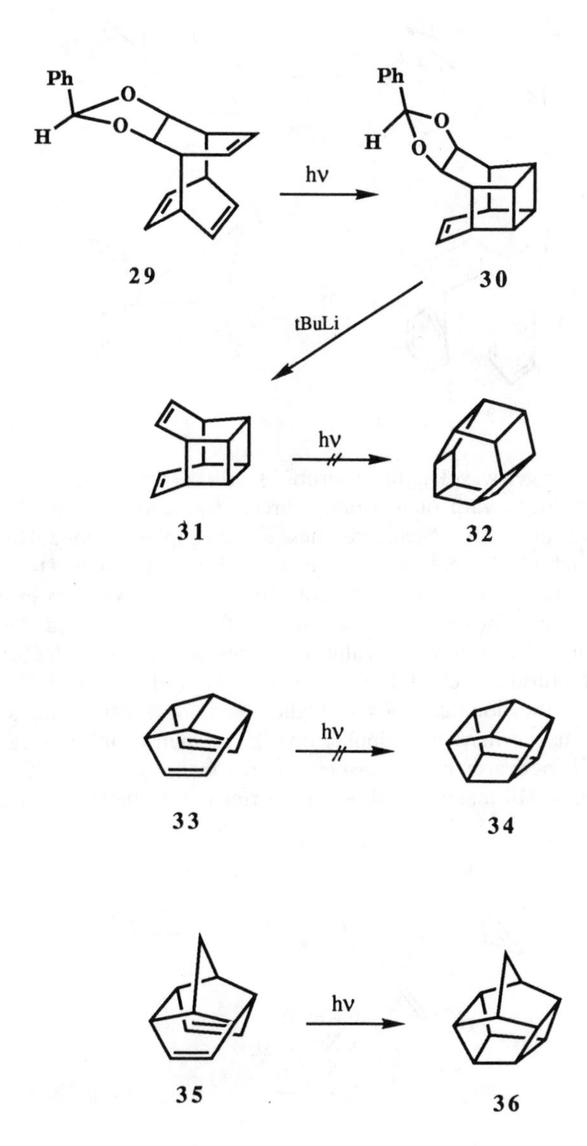

The photocyclization of diene **37** to cage compound **38** under the triplet sensitization efficiently takes place because of the good π overlap in the diene.[21] Even the dibenzo derivative **39** undergoes cyclization to give **40** with destruction of aromatic rings.[22] Diene **37** has a low ionization potential (IP = 8.1 eV).[23] Interestingly, the cyclization of **37** to **38** can be induced by irradiation with electron-acceptor sensitizers such as 9,10-dicyanoanthracene (DCA). The quantum yield in a polar solvent exceeds 1.0, indicating a chain mechanism through radical cation intermediates.[21]

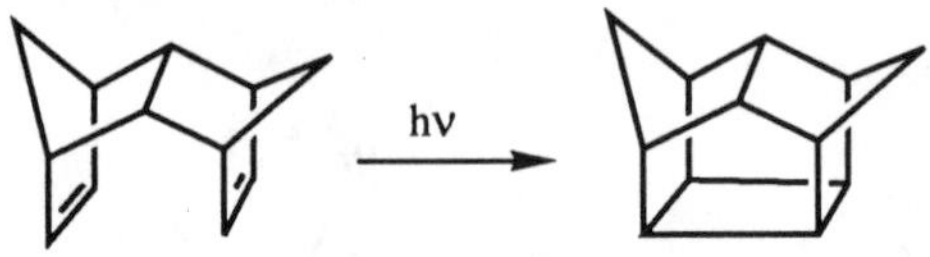

hν

37 **38**

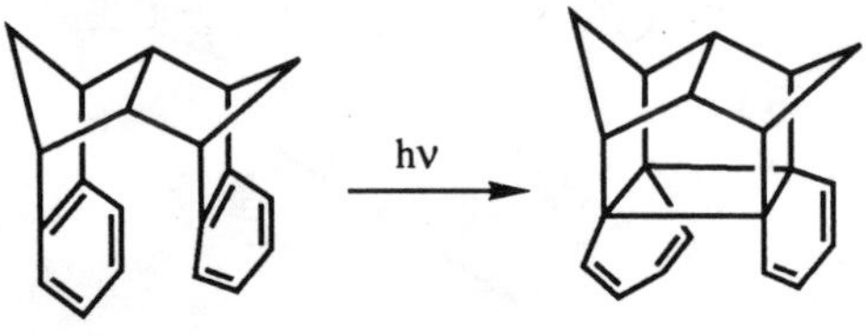

hν

39 **40**

Enones with longer-wavelength absorptions are known to undergo facile [2 + 2] cycloaddition reactions with olefins upon direct irradiation. Intramolecular cyclizations lead to the formation of cage ketones. For example, dienone **41a** affords cage ketone **42a** in high yield.[24] Similarly, irradiation of dienones **41b–41d** in acetonitrile with a high-pressure Hg lamp gives the corresponding cage ketones in good yields.[25] An interesting feature in this cyclization is that the quantum yield depends on the length of the bridge X, as shown in Table 12.1, thus decreasing as the bridge becomes longer. The structural effect of the bridge may be rationalized by the increasing distance between the double bonds in higher homologs in addition to the greater reactivity of the norbornene-type double bond. The high quantum yield observed in **42d** can be explained also by the distance between the reaction sites because the cyclobutene ring in **41d** makes the distance shorter due to the steric effect.

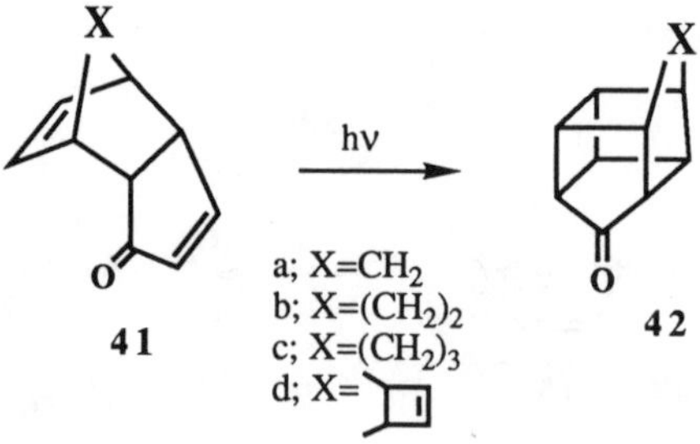

X X

hν

O O

a; X=CH_2
b; X=$(CH_2)_2$
c; X=$(CH_2)_3$
d; X=

41 **42**

Table 12.1 ▪ Photoreaction of Dienones 41a–41d

Compound	Bridge X	Isolated Yield (%)	Quantum Yield
41a	CH_2	70	0.37
41b	$(CH_2)_2$	62	0.17
41c	$(CH_2)_3$	70	0.08
41d	[cyclobutene]	69	0.37

A similar type of photocyclization of dienone **43** to cage ketone **44** is used as a key step in the synthesis of nonsubstituted cubane **1** (R = H).[56] Irradiation of enones **45** and **46** leads to the formation of trishomocubane derivatives **47** and **48**, respectively, by [2 + 2] cyclization.[27] Dienones **49** afford trishomocubanones **50** in quantitative yields.[28] The isomerization is considered as a potential solar-energy storage system because the cage compounds **50** undergo a quantitative acid-catalyzed cycloreversion to **49**. Cage ketones **51** containing a sulfone group are also obtained by irradiation of dienones **52**.[29]

O O Br O Br **43** hν O O Br Br O **44** → **1**

R H H O **45** hν R=H,Me,Ph O R **47**

O H H H H Me **46** O hν O O Me **48**

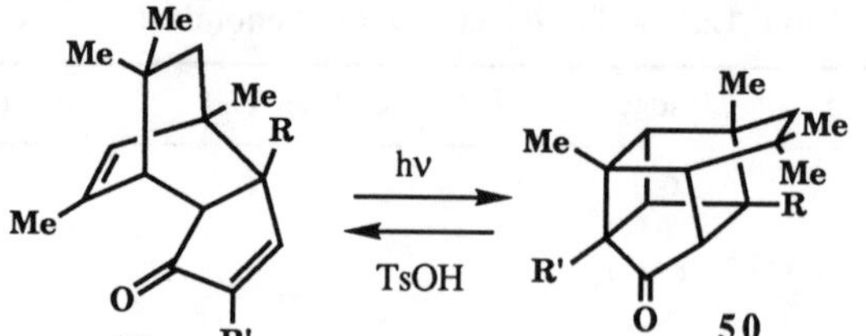

49 hν ⇌ TsOH 50

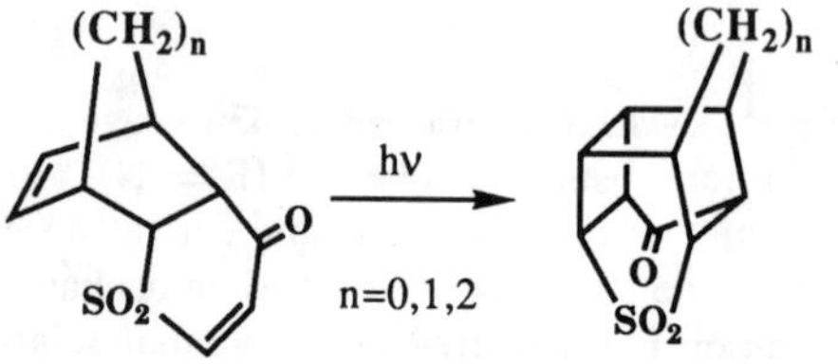

52 hν n=0,1,2 → 51

The substituted cage ketones **53** with various bridges X are prepared by irradiation of dienones **54**,[30] which are synthesized by Diels–Alder reaction of 1,3-dienes with cyclopentadienone derivatives followed by Cope rearrangement as shown in Scheme 12.1.[31] Cage ketones **53**, excepting one containing a methano bridge, undergo a novel thermally induced decarbonylation to give dienes **55**.[30] Irradiation of the dienes **55** leads to cage compounds **56**. Because dienes **55** possess styrene chromophores, they can be directly photoexcited by long-wavelength light.[32] The quantum yields are fairly

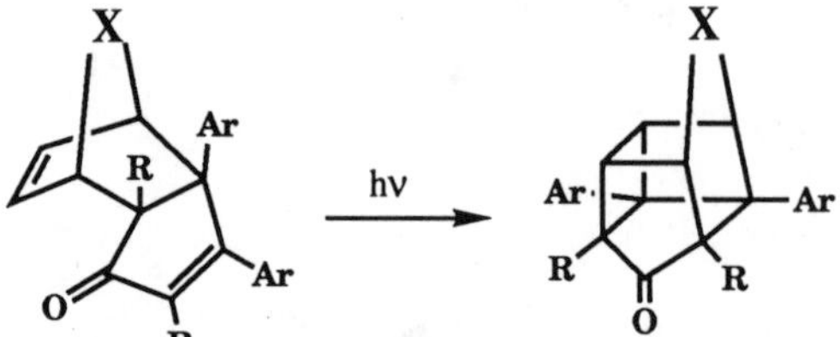

54 hν → 53

X=$(CH_2)_{1\text{-}3}$
CH_2=$CHCH_2$
no link
R=Me, CO_2Me

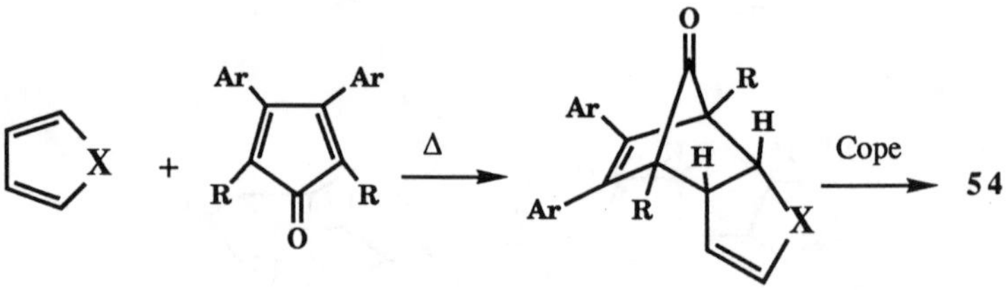

Δ Cope → 54

Scheme 12.1 ■

high (0.40–0.56). Cage compounds **56** readily cyclorevert to dienes **55** upon heating. This reversible valence isomerization was proposed as a model for solar-energy storing systems.[32] Cage ketones **57** containing heteroatoms in the bridge are also prepared starting from Diels–Alder reaction of a cyclopentadienone derivative with heteropines such as 1,2-diazepine.[33] Ketones **57** similarly undergo the thermal decarbonylation to ketones **53**.[34]

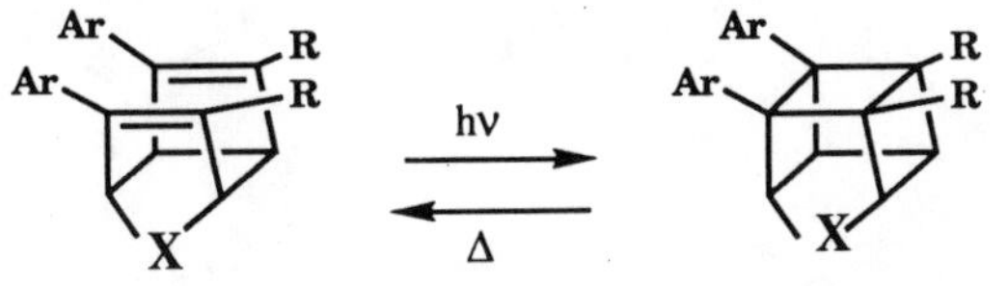

55 **56**

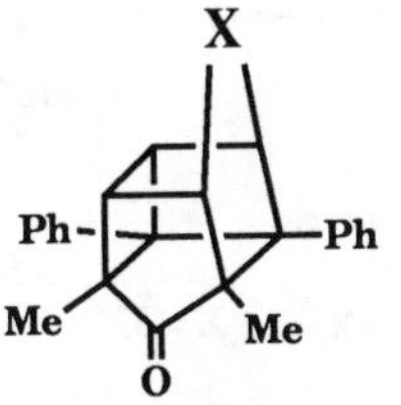

X=N(CO_2Et)CH=CH
N(CO_2Et)N=CH
N(COPh)N=CH
N=C(Ph)O

57

The excited singlet state of cyclopentadienone dimer **58** undergoes a [1,3]sigmatropic shift to give **59**, whereas the triplet state of **58** forms a cage compound **60** by [2 + 2] cyclization.[35] Irradiation of 2,5-dimethyl-3,4-diphenylcyclopentadienone dimer **61** leads to the formation of two types of cage compound, **62** and **63**.[36] The symmetric molecule **62** is considered to be formed by a 1,3 shift followed by [2 + 2] cyclization. The ratio of the two products is dependent on the reaction conditions and the symmetric ketone **62** is a main product upon irradiation in solution.[37] On the other hand, the replacement of the phenyl group with the thienyl or furyl group changes the reactivity. Irradiation of dienones **64** in solution exclusively affords the asymmetric cage ketones **65** in high yields by a normal [2 + 2] cyclization.[38]

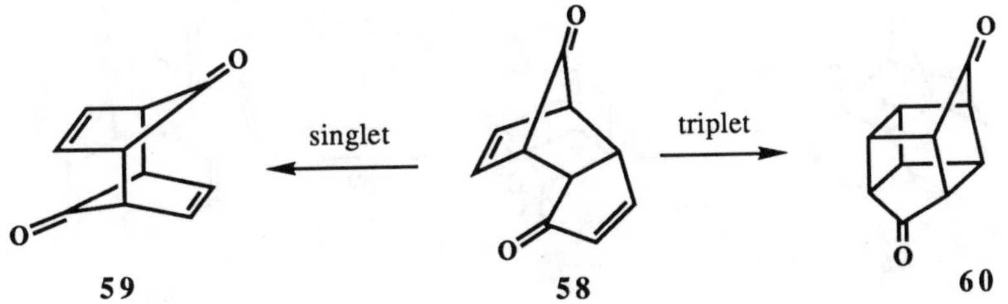

59 **58** **60**

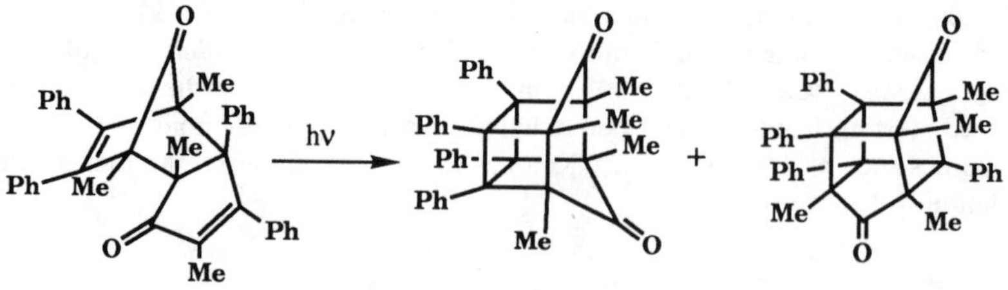

61 62 63

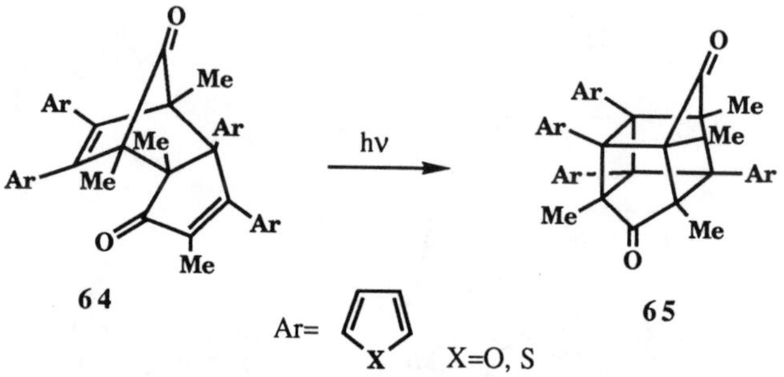

64 65

Ar= X=O, S

Dienedione **66**, a Diels–Alder adduct of *p*-benzoquinone and cyclopentadiene, affords a cage compound **67** by photochemical cyclization.[39] Flash thermolysis of **67** affords bis-enone **68**, which recyclizes to **67** by photolysis.[40] The cyclizations of derivatives **69** to **70** are also reported.[41] Interestingly, the benzo derivative **71** undergoes photochemical cyclization to give **72**.[42] The tricyclic endo adduct **73** of *p*-benzoquinone with cyclopentadienone derivatives affords cage triones **74**. The methyl derivative **74** (R = Me) undergoes thermal [2 + 2] cycloreversion to give bis-enone **75**, which reverts to **74** by photolysis.[40] However, the carbomethoxy derivative **74** (R = CO_2Me) undergoes thermal decarbonylation to give bis-enone **76**.[43]

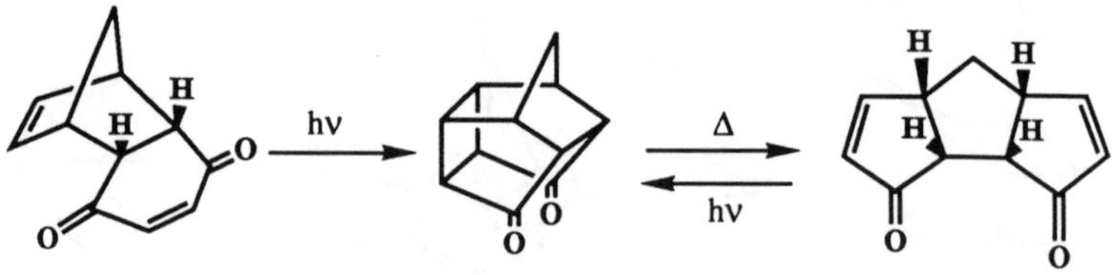

66 67 68

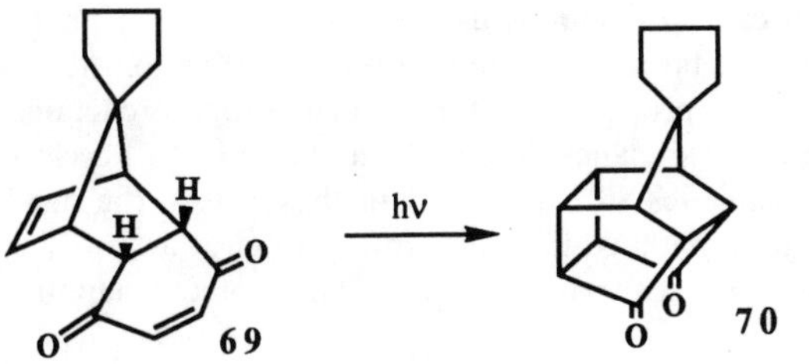

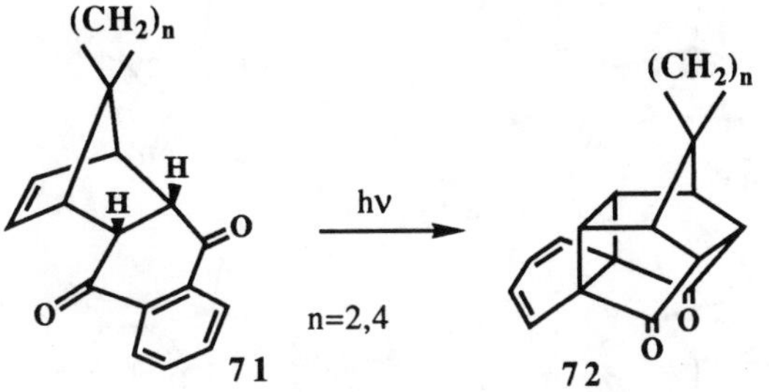

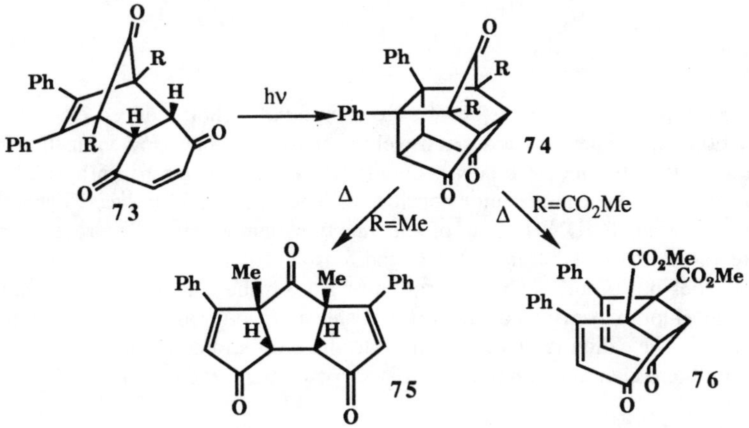

12.3. Photochemical Reactions of Cage Molecules

Cage compounds containing C=C double bonds are usually photochemically labile. For example, upon excitation of the C=C bond, basketene **23** undergoes dimerization to give **77** by irradiation in acetone.[44]

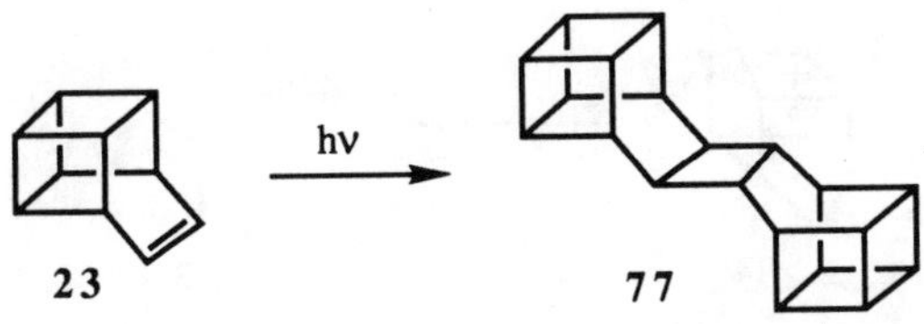

In contrast, saturated cage compounds usually do not undergo photochemical reactions because they cannot be photoexcited by direct irradiation nor by sensitization. However, introduction of proper substituents makes photoreactions possible. For example, substituted quadricyclanes **78** and **79** undergo [2 + 2] cycloreversion to the corresponding norbornadienes **80**[45] and **81**.[46] In these cases, the quadricyclanes and the norbornadienes are in photoequilibrium. Interestingly, the [2 + 2] cycloaddition of **81** occurs upon irradiation with visible light because of the naphthoquinone chromophore.

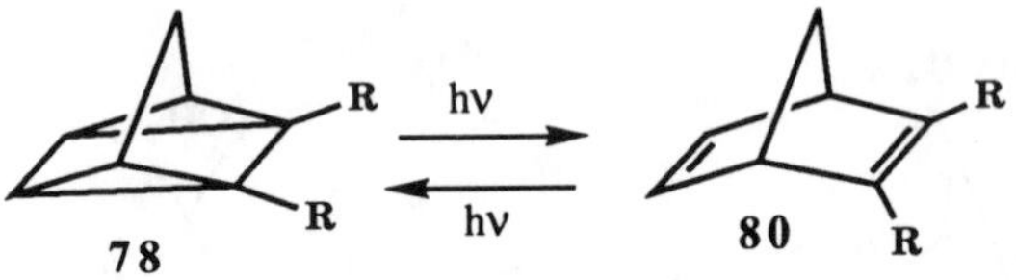

R=Ph, 1-Methylpyridino

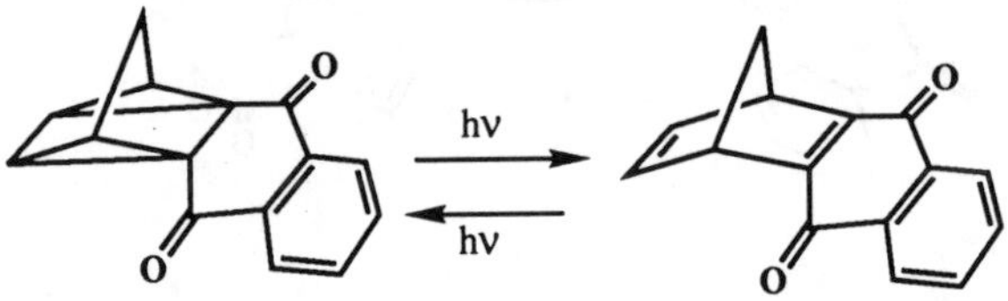

As to the [2 + 2] cycloreversion of cage compounds, photoinduced electron-transfer reactions have been studied extensively because they have low ionization potentials due to the strained σ bonds.[47] Quadricyclane isomerizes to norbornadiene by irradiation with electron-acceptor sensitizers. The chemically induced dynamic nuclear polarization (CIDNP) study of the reaction demostrates the presence of two discrete radical cations **82** and **83**; **82** readily isomerizes to **83**, whereas the reverse reaction does not occur.[48] The generation of a 1,3-bishomocubane radical cation **84** under the chloranil-sensitized electron-transfer reaction was studied by a CIDNP technique.[49] The cation radical is localized to the most strained bond, but **84** does not undergo a bond cleavage reaction to a dicyclopentadiene radical cation **85**.

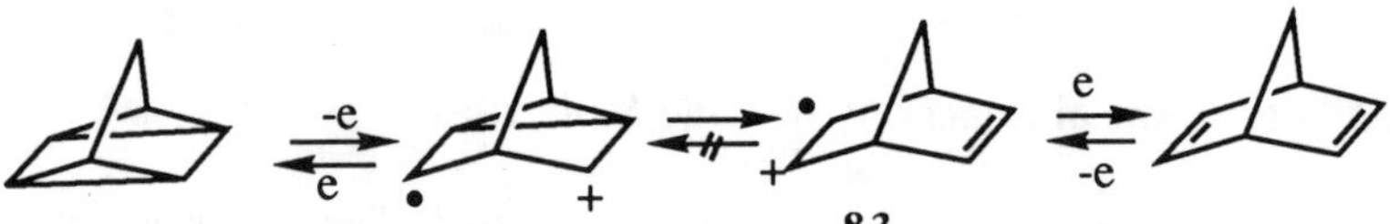

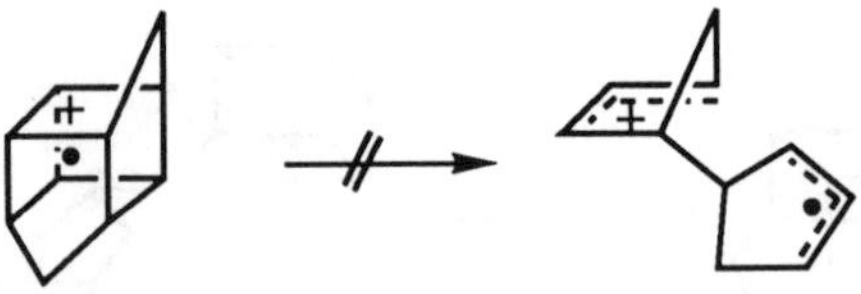

The [2 + 2] cycloreversion of 1,2-diaryl-substituted cyclobutanes has been studied in detail.[50] Bishomocubanes **86** containing this skeleton are noteworthy compounds. Irradiation (> 400 nm) of **86** (Ar = *p*-$MeOC_6H_4$) in the presence of DCA or *p*-benzoquinone in acetonitrile affords diene **87** in high yield.[51] The quantum yields are dependent on the concentration of the diene and the limiting quantum yield is found to be infinite. This result shows the occurrence of a chain reaction. The CIDNP study of this photoreaction shows that the cage radical cations **88** have shallow energy minima and suffer fast ring opening to the diolefin cation radicals **89**.[52] Photochemical cycloreversion also occurs when cage compounds are irradiated with cationic sensitizers such as triarylpyrylium salts **90** or trityl salts **91**.[53] The mechanism involves a radical cation chain process, which is confirmed by a quenching study using 1,2,4,5-tetramethoxybenzene.[54]

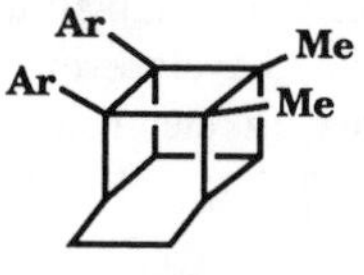

86 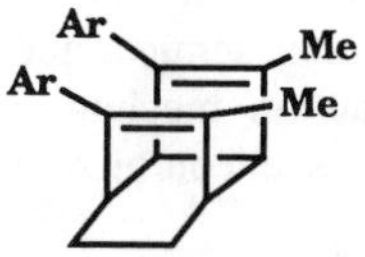 **87**

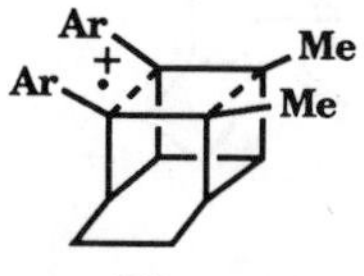

88 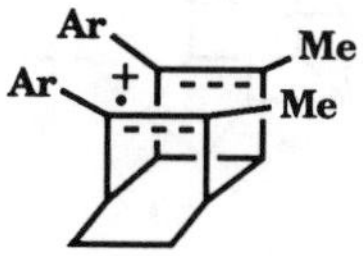 **89**

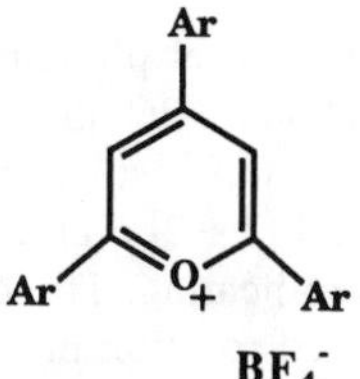

BF_4^- **90**

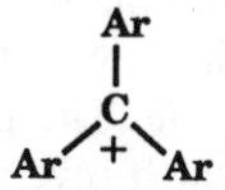

X^- ($X=BF_4$, Cl) **91**

Irradiation (> 350 nm) of bishomocubane derivatives **86** in the presence of 3,9-dicyanophenanthrene (DCP) or DCA in nonpolar solvents such as benzene and dichloromethane affords dienes **87** quantitatively. This reaction has been proved to be an adiabatic exciplex isomerization.[55] The mechanism is shown in Scheme 12.2. The efficiency of the adiabatic process is very high (> 90% in dichlorobenzene). This unusual behavior is attributed to a high reactivity of cycloreversion of cage compounds **86**.[55] The cycloreversion reaction can be electrochemically induced by constant-potential anodic oxidation.[56] Treatment of **86** with triarylaminium radical salts also results in the same electron-transfer isomerization.[57]

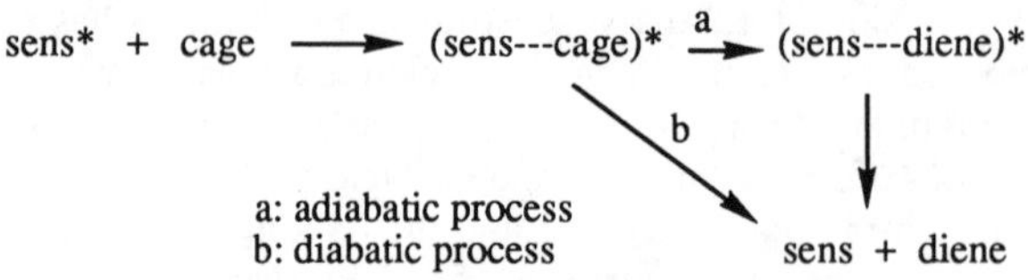

sens* + cage ⟶ (sens---cage)* $\xrightarrow{a}$ (sens---diene)*

b

a: adiabatic process
b: diabatic process

sens + diene

Scheme 12.2 ■

Photocatalysts of powdered semiconductors have attracted considerable attention. The cycloreversion of cage compounds **86** to dienes **87** is found to take place upon irradiation in the presence of ZnO or CdS powder.[58] Nonsubstituted 1,8-bishomocubane **92** undergoes a similar semiconductor-mediated cycloreversion to afford diene **93**.[59] The proposed reaction mechanism involves an electron transfer from cage molecules to holes of the semiconductors.

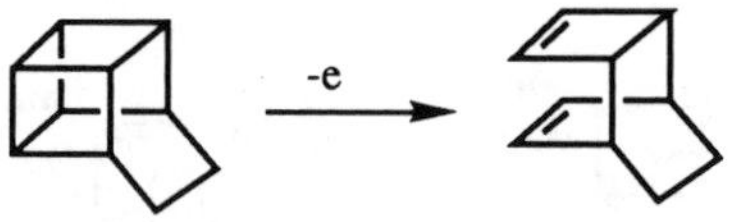

-e

92 **93**

Cage ketone **94** (Ar = *p*-$MeOC_6H_4$) undergoes a novel photoinduced rearrangement.[60] Irradiation of a dichloromethane solution of **94** (> 360 nm) in the presence of 2,4,6-triphenylpyrylium perchlorate (TPP^+) or chloranil results in the quantitative formation of a novel product **95**. The photochemical [2 + 2] cyclization of **95** affords another cage molecule **96**, which reverts to **95** upon heating. The quantum yield for the formation of **95** is more than unity. The proposed reaction mechanism involves a cation-radical mechanism via an intermediate **97** that is isolated when the photoreaction is carried out in the presence of tetramethoxybenzene (TMB), which acts as a quencher of the electron transfer. The cage molecule **97** is considered to be formed by a $[\sigma 2 + \sigma 1]$ pericyclic reaction of **94**$^{\cdot +}$ or via an intermediate **98**, which leads to **97**$^{\cdot +}$ by a 1,2-vinyl shift. The rearrangement of **97**$^{\cdot +}$ to **95**$^{\cdot +}$ is a formal $[\pi 2 + \sigma 2 + \sigma 1]$ pericyclic reaction. This photoisomerization is not sensitized by DCA, showing that the back electron transfer from $DCA^{\cdot -}$ takes place more easily than that from $TPP^{\cdot}$. When $NaClO_4$ is added to a solution of **94** and DCA, the isomerization proceeds quantitatively. The result shows that the isomerization of cage molecule **94** to **95** is a good model for investigating salt effects to suppress the back electron transfer.

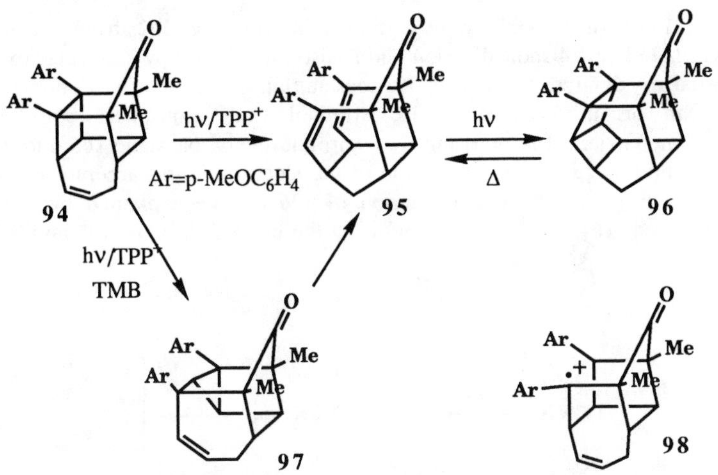

Cage ketone **99** is unchanged under photosensitized electron-transfer conditions using chloranil, DCA, or TPP^+ as sensitizer.[60] This fact indicates that back electron transfer takes place faster than chemical reactions such as cycloreversion. However, the radical cation **100** is successfully intercepted with TCNE to afford an unusual adduct **101** by irradiation of the electron donor–acceptor complex of **99** with TCNE.[61] The reaction mechanism is proposed, as shown in Scheme 12.3, that the radical

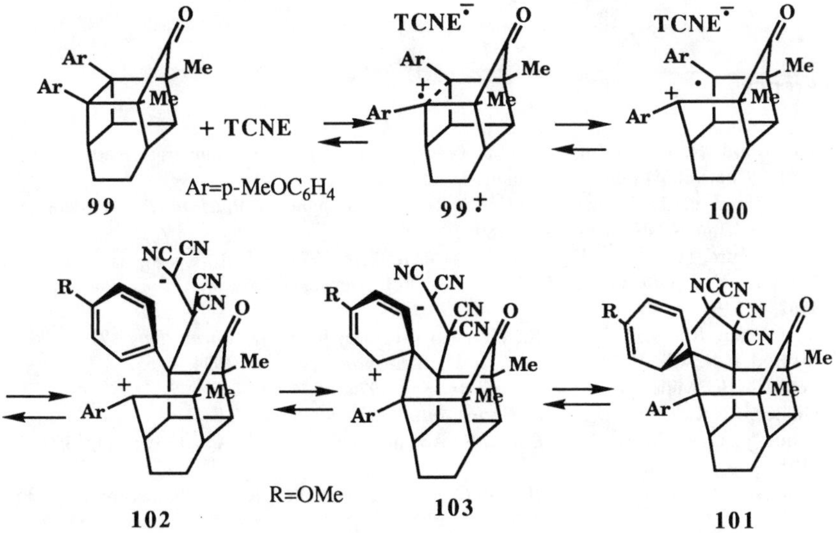

Scheme 12.3 ■

coupling of **100** with TCNE$^{\cdot -}$ gives a dipolar intermediate **102**, from which **101** is formed via **102**. The 1,4-radical cation **100** is also intercepted by molecular oxygen to give unusual oxygenated products.[62] Photoexcitation of an electron donor–acceptor complex can be used to synthesize strained cage molecules. Spirofluorene-bicyclo[6.1.0]nonatriene **104** rearranges to spirofluorenebarbaralane **105** under DCA-sensitized conditions, whereas irradiation of the electron donor–acceptor complex of **104** and TCNE gave **106**. The formation of **106** can be explained by the initial rearrangement of **104**$^{\cdot +}$ to **105**$^{\cdot +}$ followed by the cycloaddition of **105** and TCNE.[63]

104 + TCNE $\xrightarrow{h\nu}$ [104$^{+\cdot}$/TCNE$^{-\cdot}$] → [$^{+\cdot}$ TCNE$^{-\cdot}$] → 106

R, R =

References

1. Woodward, R. B.; Hoffman, R. *The Conservation of Orbital Symmetry*; Academic Press, Verlag Chemie: Weinheim, Germany, 1970.
2. For reviews see the following: (a) Griffin, G. W.; Marchand, A. P. *Chem. Rev.* **1989**, *89*, 997. (b) Marchand, A. P. *Chem. Rev.* **1989**, *89*, 1011.
3. (a) Iwamura, H.; Morio, K.; Kihara, H. *Chem. Lett.* **1973**, 457. (b) Ōsawa, E.; Aigami, K.; Inamoto, Y. *J. Org. Chem.*, **1977**, *42*, 2621. (c) Criegee, R. *Angew. Chem., Int. Ed. Engl.* **1962**, 1, 519.
4. (a) Gleiter, R.; Heilbronner, E.; Hekman, M.; Martin, H.-D. *Chem. Ber.* **1973**, *106*, 28. (b) Boder, N.; Chen, B. H.; Worley, S. D. *J. Electron Spectrosc.* **1974**, *4*, 65.
5. Pelosi, L. F.; Miller, W. T. *J. Am. Chem. Soc.* **1976**, *98*, 4311.
6. Gleiter, R.; Karcher, M. *Angew. Chem., Int. Ed. Engl.* **1988**, *27*, 840.
7. Paquette, L. A.; Ward, J. S.; Boggs, R. A. Farnham, W. B. *J. Am. Chem. Soc.* **1975**, *97*, 1101.
8. Anderson, C. M.; Bremner, J. B.; McCay, I. W.; Warrener, R. N. *Tetrahedron Lett.* **1968**, 1255.
9. Katz, T. J.; Carnahan, J. C., Jr.; Clarke, G. M.; Acton, N. *J. Am. Chem. Soc.* **1970**, *92*, 734.
10. (a) Eberbach, W.; Perroud-Arguelles, M. *Chem. Ber.* **1972**, *105*, 3078. (b) Kobayashi, Y.; Kumadaki, I.; Ohsawa, A.; Sekine, Y. *Tetrahedron Lett.* **1974**, 2841.

11. Cargill, R. L.; King, T. Y. *Tetrahedron Lett.* **1970**, 409.
12. Dilling, W. L. *J. Org. Chem.* **1975**, *40*, 2380.
13. Miller, R. D.; Dolce, D. *Tetrahedron Lett.* **1972**, *44*, 4541.
14. Masamune, S.; Cuts, H.; Hogben, M. G. *Tetrahedron Lett.* **1966**, 1017.
15. Mauer, W.; Grimme, W. *Tetrahedron Lett.* **1976**, 1835.
16. (a) Paquette, L. A.; Cottrell, D. M.; Snow, R. A.; Gifkins, K. B.; Clardy, J. *J. Am. Chem. Soc.* **1975**, *97*, 3275. (b) Santiago, C.; Houk, K. N. *J. Am. Chem. Soc.* **1976**, *98*, 3380.
17. Yang, N. C.; Horner, M. G. *Tetrahedron Lett.* **1986**, *27*, 543.
18. Paquette, L. A.; James, D. R.; Klein, G. *J. Org. Chem.* **1978**, *43*, 1287.
19. Schmidt, W.; Wilkins, B. T. *Tetrahedron Lett.* **1972**, *28*, 5649.
20. Eaton, P. E.; Cassar, L.; Hudson, R. A.; Hwang, D. R. *J. Org. Chem.* **1976**, *41*, 1445.
21. Jones, G. II; Becker, W. G.; Chiang, S.-H. *J. Photochem.* **1982**, *19*, 245.
22. Martin, H.-D.; Schwesinger, R. *Chem. Ber.* **1974**, *107*, 3143.
23. Fessner, W.-D.; Sedelmeier, G.; Spurr, R. P.; Rihs, G.; Printzbach, H. *J. Am. Chem. Soc.* **1987**, *109*, 4626.
24. Jones, G. II; Ramachandran, B. R. *J. Org. Chem.* **1976**, *41*, 798.
25. Yamashita, Y.; Mukai, T. *Chem. Lett.* **1984**, 1741.
26. Eaton, P. E.; Cole, T. W., Jr. *J. Am. Chem. Soc.* **1964**, *86*, 3157.
27. (a) Ogino, T.; Awano, K. *Chem. Lett.* **1982**, 891. (b) Ogino, T.; Awano, K.; Ogihara, T.; Isogai, K. *Tetrahedron Lett.* **1983**, *24*, 2781. (c) Mehta, G.; Srikrishna, A. *J. Chem. Soc., Chem. Commun.* **1982**, 218.
28. Hamada, T.; Iijima, H.; Yamamoto, T.; Numao, N.; Hirao, K.; Yonemitsu, O. *J. Chem. Soc., Chem. Commun.* **1980**, 696.
29. Paquette, L. A.; Lawrence, D. W. *J. Am. Chem. Soc.* **1967**, *89*, 6659.
30. Tezuka, T.; Yamashita, Y.; Mukai, T. *J. Am. Chem. Soc.* **1976**, *98*, 6051.
31. Yamashita, Y.; Mukai, T. *Chem. Lett.* **1978**, 919.
32. Mukai, T.; Yamashita, Y. *Tetrahedron Lett.* **1978**, 357.
33. Mukai, T.; Yamashita, Y.; Sukawa, H.; Tezuka, T. *Chem. Lett.* **1975**, 423.
34. (a) Yamashita, Y.; Mukai, T.; Tezuka, T. *J. Chem. Soc., Chem. Commun.* **1977**, 532. (b) Harano, K.; Ban, T.; Yasuda, M.; Ōsawa, E.; Kanematsu, K. *J. Am. Chem. Soc.* **1981**, *103*, 2310.
35. Baggiolini, E.; Herzog, E. G.; Iwasaki, S.; Schorta, R.; Schaffner, K. *Helv. Chim. Acta* **1967**, *50*, 297.
36. Fuchs, B. *J. Am. Chem. Soc.* **1971**, *93*, 2544.
37. Fuchs, B.; Pasternak, M.; Pazhenchevsky, B. *Tetrahedron* **1980**, *36*, 3443.
38. Yamashita, Y.; Masumura, M. *Heterocycles* **1980**, *14*, 29.
39. Cookson, R. C.; Crundwell, E.; Hill, R. R.; Hudec, J. *J. Chem. Soc.* **1964**, 3062.
40. Mehta, G.; Srikrishna, A.; Reddy, V.; Nair, M. S. *Tetrahedron* **1981**, *37*, 4543.
41. Singh, V. K.; Deota, P. T.; Raju, B. N. S. *Synth. Commun.* **1987**, *17*, 593.
42. Singh, V. K.; Raju, B. N. S.; Deota, P. T. *Synth. Commun.* **1987**, *17*, 1103.
43. Okamoto, Y.; Harano, K.; Yasuda, M.; Osawa, E.; Kanematsu, K. *Chem. Pharm. Bull.* **1983**, *31*, 2526.
44. Jones, N. J.; Deadman, W. D.; Legoff, E. *Tetrahedron Lett.* **1973**, 2087.
45. (a) Kaupp, G. *Liebigs Ann. Chem.* **1973**, 844. (b) Yamashita, Y.; Hanaoka, T.; Takeda, Y.; Mukai, T.; Miyashi, T. *Bull. Chem. Soc. Jpn.* **1988**, *61*, 2451.
46. Suzuki, T.; Yamashita, Y.; Mukai, T.; Miyashi, T. *Tetrahedron Lett.* **1988**, *29*, 1405.
47. Bordor, N.; Dewar, M. J. S.; Worley, S. D. *J. Am. Chem. Soc.* **1970**, *92*, 19.
48. (a) Roth, H. D.; Schilling, M. L. M. *J. Am. Chem. Soc.* **1981**, *103*, 7210. (b) Roth, H. D.; Schilling, M. L. M.; Jones, G., II. *J. Am. Chem. Soc.* **1981**, *103*, 1246.
49. Roth, H. D.; Schilling, M. L. M.; Abelt, C. J. *Tetrahedron* **1986**, *42*, 6157.
50. Pac, C. *Pure Appl. Chem.* **1986**, *58*, 1249.

51. Mukai, T.; Sato, K.; Yamashita, Y. *J. Am. Chem. Soc.* **1981**, *103*, 670.
52. Roth, H. D.; Schilling, M. L. M.; Mukai, T.; Miyashi, T. *Tetrahedron Lett.* **1983**, *24*, 5815.
53. Okada, K.; Hisamitsu, K.; Mukai, T. *Tetrahedron Lett.* **1981**, *22*, 1251.
54. Okada, K.; Hisamitsu, K.; Miyashi, T.; Mukai, T. *J. Chem. Soc., Chem. Commun.* **1982**, 974.
55. Hasegawa, E.; Okada, K.; Mukai, T. *J. Am. Chem. Soc.* **1984**, *106*, 6852.
56. Takahashi, Y.; Sato, K.; Miyashi, T.; Mukai, T. *Chem. Lett.* **1984**, 1553.
57. Hasegawa, E.; Mukai, T. *Bull. Chem. Soc. Jpn.* **1985**, *58*, 3391
58. Okada, K.; Hisamitsu, K.; Mukai, T. *J. Chem. Soc., Chem. Commun.* **1980**, 941.
59. Baird, N. C.; Draper, A. M.; De Mayo, P. *Can. J. Chem.* **1988**, *66*, 1579.
60. Yamashita, Y.; Ikeda, H.; Mukai, T. *J. Am. Chem. Soc.* **1987**, *109*, 6682.
61. Ikeda, H.; Yamashita, Y.; Kabuto, C.; Miyashi, T. *Tetrahedron Lett.* **1988**, *29*, 5779.
62. Ikeda, H.; Yamashita, Y.; Kabuto, C.; Miyashi, T. *Chem. Lett.* **1988**, 1333.
63. Roth, H. D.; Schilling, M. L.; Abelt, C. J.; Miyashi, T.; Takahashi, Y.; Konno, A.; Mukai, T. *J. Am. Chem. Soc.* **1988**, *110*, 5130. Miyashi, T.; Takahashi, Y.; Konno, A.; Mukai, T.; Roth, H. D.; Schilling, M. L.; Abelt, C. J. *J. Org. Chem.* **1989**, *54*, 1445.

13

Recent Studies on Valence Isomerization between Norbornadiene and Quadricyclane

Ken-ichi Hirao,[a] Asami Yamashita,[b] and Osamu Yonemitsu[a]

[a]Hokkaido University, Sapporo, Japan
[b]Hokkaido Institute of Pharmaceutical Sciences, Otaru, Japan

Contents

13.1. Introduction
13.2. Acylnorbornadiene and Acylquadricyclane System
13.3. The Photochemical Isomerization of Norbornadienes into Quadricyclanes
    13.3A. Photosensitizers
    13.3B. Modified Norbornadienes
13.4. The Reversion of Quadricyclanes into Norbornadienes
    13.4A. Catalysts
        13.4A(1). Homogeneous Catalysts
        13.4A(2). Heterogeneous Catalysts
    13.4B. Mechanistic Aspects of the Isomerization of Quadricyclane to Norbornadiene by Electron-Transfer Process
13.5. Miscellaneous Studies
Acknowledgments
References

13.1. Introduction

In 1976 Jones and Ramachandran[1] assessed photovalence isomerism forming *strained* polycyclic cage compounds in the conversion of solar energy, and a fair number of systems were examined through 1980. Because the experimental results and the essential characteristics for designing the system have been discussed repeatedly by many reviewers,[2] redescribing them here is unnecessary. Among the valence isomerism systems studied so far, interconversion between norbornadiene **1(N)** and quadricyclane **2(Q)** derivatives (**N/Q** system) has several preferable characteristics and has been considered to be the most prospective candidate. The difficulty that has arisen in these studies is that many requirements must be wholly satisfied simultaneously. Therefore, in actuality the **N/Q** system presently has little practical value, and much further development needs to be done.

hν

-ΔE

1 (N) 2 (Q)

Some improvements were made in the **N/Q** system from 1980 through 1984; for example, a bathochromic shift to the visible region in the electronic absorption by introducing both electron-donating and -accepting substituents into **N** olefins, an increase in quantum yield, water-soluble **N/Q** systems, solid catalysts for the reversion of **Q** to **N**, and so on. These were discussed extensively in the literature.[3] Although the **N/Q** cycle has been thought to be the best candidate for solar energy conversion and has been studied intensively, much remains to be solved before attachment of practical importance to the work. Disadvantages of the **N/Q** system for photochemical energy storage were pointed out from economical viewpoints;[4] isomerization of **Q** under UV to ring-cleaved substances other than **N** was also noted.[5] As the remaining problems seem to be almost insurmountable, introduction of fundamental changes of the basic model for the system may be required. However, attempts to store solar energy via valence isomerization of **N** derivatives have continued despite the rather limited progress. Under these circumstances, it is our interest in this review to introduce and discuss our own studies on this subject and the representative results that have appeared in the literature since 1985.

13.2. Acylnorbornadiene and Acylquadricyclane System

In the course of our studies of the reaction of strained polycyclic cage compounds,[6] some pentacyclic cage molecules (**6**, **7**, and **8**) having a substituent R that stabilizes an intermediate cation (**9** for **6**, for example) have been found to revert to the corresponding dienes (**3**, **4**, and **5**) under very mild acidic conditions.[7] The mechanism that accounts for the acid-catalyzed reaction is shown in Scheme 13.1. Direct photolysis of the dienes (**3–5**), the reverse valence isomerization, has also been found to proceed quantitatively to give the corresponding cage compounds. Over 30 compounds related to the **3** ⇄ **6** and analogous systems (**4/7** and **5/8**) were synthesized in order to study whether these can be applicable to the conversion of solar energy.

The results from these studies can be summarized as follows[7]: Both the reversion by acid catalysis developed by the authors[8] and photolysis are generally clean reactions.

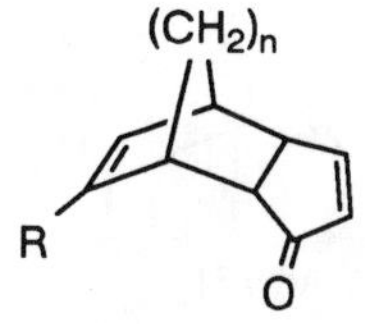

When the concept of Yoshida's DONAC system[3c] is extended to our dienes **3–5**, bathochromic shift of end of absorption (EA)[9] can be observed. Some such dienes actually absorb visible light. Diene **10**, having a cyano group conjugated with an enone (acceptor) and two anisyl groups (donors), actually shows an EA at 540 nm.[10] No notable difference in the storage enthalpies among C_{12}-**6**, C_{11}-**7**, and C_{10}-**8** cage systems have been observed (18–23 kcal mol^{-1}).

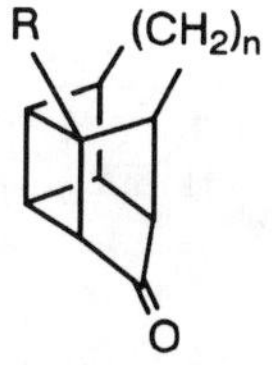

Because the quantum yields of these systems are not sufficient (0.2–0.4) for storage of light energy and because they require inordinate amounts of time and effort to synthesize various derivatives for further studies, the principle of our acid-catalyzed reversion was next applied to the well-studied **N**/**Q** system.[11,12,15] To this end we synthesized over 25 acylnorbornadiene derivatives (acyl **N**) having electron-donating substituents [such as H-, Me-, Ph-, and Ar-(*p*-methoxyphenyl-)] and electron-accepting groups (such as —CO_2Me, —CONHPh, —COEt, —COBu, —COMe, —CN, and fluoroalkyl groups). The characteristics of these acyl **N** and corresponding acylquadricyclane (acyl **Q**) pairs can be summarized as follows: The acid-catalyzed reversion of the keto quadricyclane, among the rest of the acyl **Q**, is rapid and clean. This fact shows that the same mechanism also operates in this reversion reaction of the acyl **Q**s as does in the reaction of our original pentacyclic cage compounds (Schemes 13.1 and 13.2). As expected, all the **N**s having both electron-withdrawing and electron-donating groups (DONAC[3c] acyl **N**) show a distinct red shift with an EA > 430 nm (visible-light region). In particular, EA for compound **12** shows 581 nm.

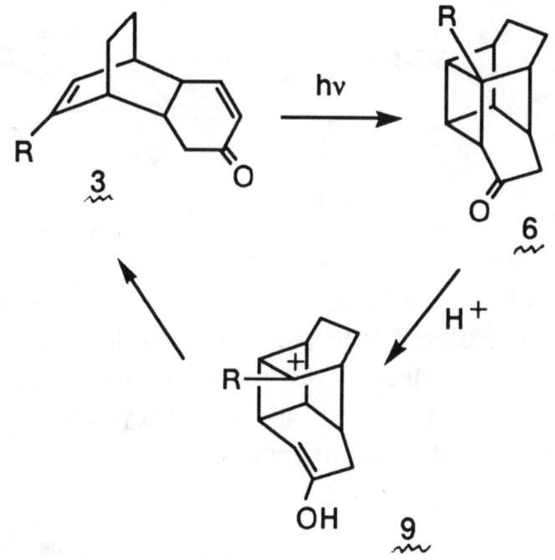

Scheme 13.1 Formation of pentacyclic cage compounds and their acid-catalyzed reversion reaction.

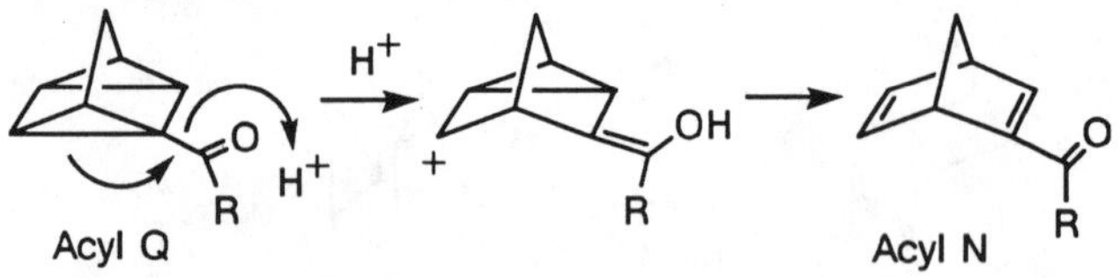

Scheme 13.2 Acid-catalyzed reversion of acyl quadricyclanes.

Photolysis of these acyl **N**s generally proceeds well, giving the corresponding acyl **Q**s with high quantum yields, around 0.6–0.8, at 383 nm. Introduction of a methoxyphenyl group to **N** nucleus as a donor has made the photocyclization of **12** less efficient again,[10] because of the thermal instability of the corresponding acyl **Q** as shown in the case of pentacyclic cage compound **11**.

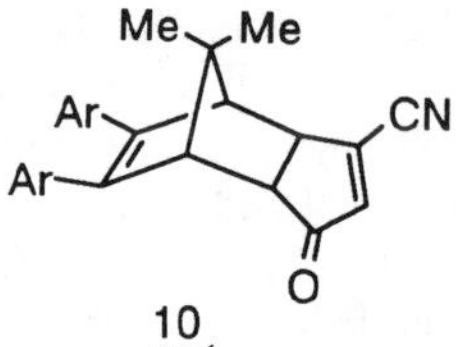

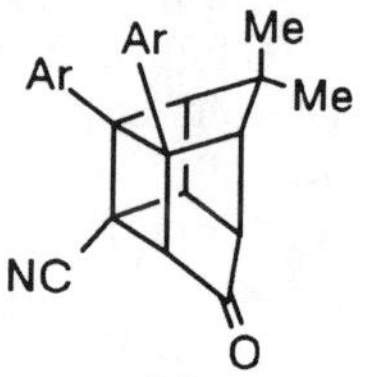

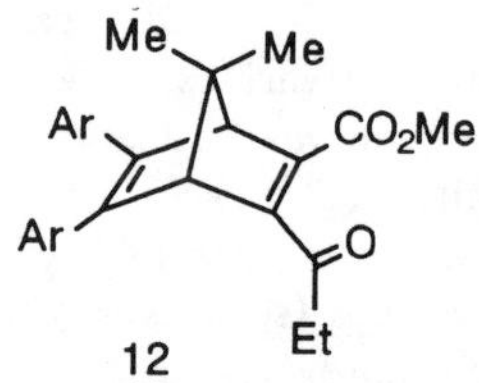

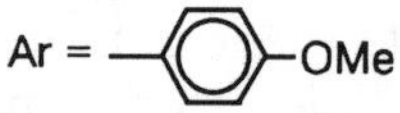

Among the acyl **N**/acyl **Q** systems that we have synthesized, **13**/**14** is the most favorable for the present purpose of storing light energy. The advantages of this system are as follows: absorption of visible light (EA = 558 nm); good quantum yield ($\Phi = 0.75$ at 383 nm) for the photocycloaddition; and clean and instant reversion ($k_m = 1.3\ s^{-1}\ [CF_3CO_2H]^{-1}$) on acid treatment.

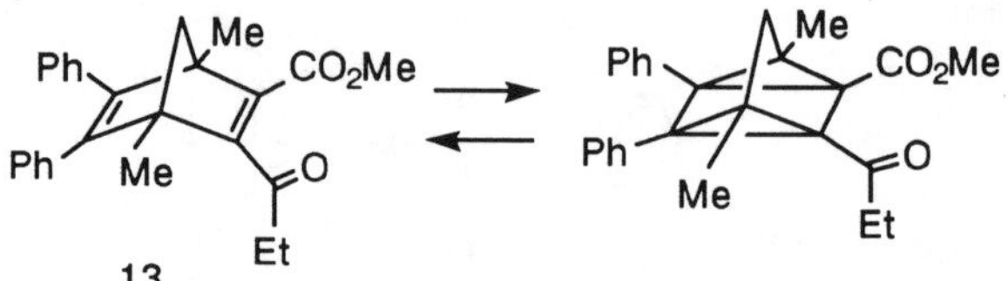

Some fluoroalkyl **N**s, **15** and **16**, have been prepared.[12] The absorption of a wide range of visible light by these dienes is in accordance with our expectation. Contrarily, the anticipated "perfluoroalkyl effect"[13] did not contribute to stabilizing the strained molecules. Thus the corresponding fluoroalkylquadricyclanes gradually reverted to the starting dienes, **15** and **16**, at ambient temperature.

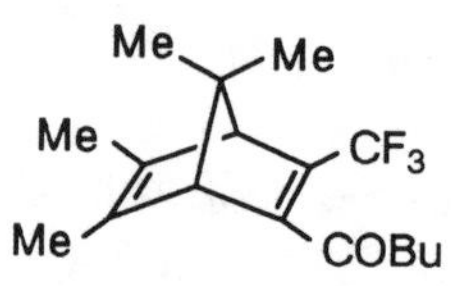

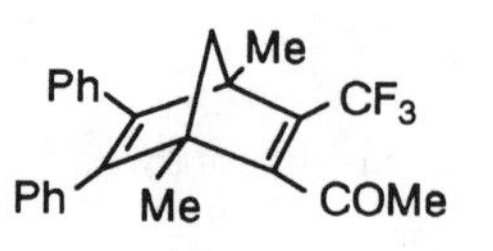

One of the most crucial problems for the removal of the inherent disadvantages of the **N/Q** system is to find the most compatible catalyst in every respect for the reversion reaction. Many studies on the catalysts have been reported. However, they all have at least one of the following disadvantages: use of expensive noble metals or intricate ligands; deactivation by side reactions on repeated runs, etc.

The most mild, stable, immobilized, and inexpensive metal oxides are known to be the typical solid acid catalysts.[14] These were applied to the acid-catalyzed reversion of acyl **Q 14**.[15] In spite of uncertain factors in shape of the catalysts such as particle size, crystal or surface structure, moisture on the surface, etc., the rates of the reaction can be approximately correlated to the electronegativities of metal ions.[18] Thus, the abundant and tractable metal oxides such as MoO_3, WO_3, V_2O_5, and so on, have been found to be practical catalysts suitable for the reversion of **14**. Furthermore, acidic mixed metal oxides, for example TiO_2–ZnO (both components are very weak acid, thus show no catalytic activity on the reaction),[19] exhibited a great catalytic activity. More preferable combinations of metal oxides are anticipated to be found following further investigations.

13.3. The Photochemical Isomerization of Norbornadienes into Quadricyclanes

It has been recognized that the interconversion between **N** and **Q** is the most preferential system in the field of energy storage models utilizing organic compounds. However, the **N** skeleton itself does not absorb in the visible region of available solar radiation (> 300 nm). Accordingly there has been much interest in the use of sensitizers and in the design of modified norbornadienes in order to shift the selected wavelength region of the solar light to longer wavelengths.

13.3A. Photosensitizers

Since effective sensitization of inorganic Cu(I) salt was pointed out by Kutal et al.,[20] many studies on the photochemistry of Cu(I)-complex–**N** pairs have appeared. In particular, the feature of the more-detailed mechanism of the sensitization of Cu(I) complexes and also findings of more-efficient complexes have been reported in recent years.

Morse and co-workers[21] reported the details of their own $(MePh_2P)_3CuCl$ **17**,[22] whose efficiency as a sensitizer for the **N** → **Q** conversion is unprecedented ($\Phi = 1.0$) compared to other inorganic complexes presented to date. Complex **17** has an absorption up to 365 nm and shows energy transfer to **N** from the triplet state that is stabilized by donation of electron from Cu(I) metal to phosphorus atom. Unfortunately, however, **17** has a prime disadvantage that originates from extensive ligand dissociation, especially in dilute solutions, giving less-efficient copper complexes. Kutal and co-workers[23] have revealed that formation of the chelate rings in the complexes such as **18** and **19** using bidentate phosphine causes change in C—P—C angles along with lowering the triplet state energy, and they concluded that triplet–triplet energy transfer is the principal mechanism by which **18** and **19** sensitize the **N** → **Q** photoreaction.

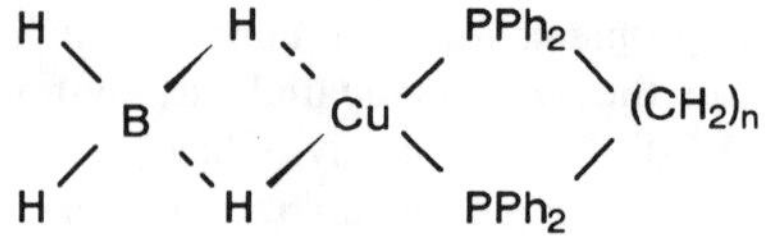

18 (n = 2)
19 (n = 3)

Thus it has become apparent that the complex **19** with the six-membered ring has a much more suitable energy level for the purpose of sensitization than **18**. Copper(I) complexes such as $[Cu(PPh_3)_2(dpk)]NO_3$ **20**, $[Cu_2I_2(Ph_3)_2(dpk)]$ **21**, and $[Cu_4I_4(dpk)_3]$ **22** with both di-2-pyridylketone(dpk) and phosphine ligands were prepared for providing an absorption in the visible region.[24] Only moderate quantum yields ($\Phi =$ 0.17–0.36) for the sensitization of **N** $\rightarrow$ **Q** isomerization were obtained with these complexes, which, again decompose gradually in the course of the photolysis. In addition, Cu(I) complex with ferrocenyl phosphine ligands were prepared. Unfortunately, however, they do not sensitize the **N** $\rightarrow$ **Q** conversion satisfactorily.[25]

The sensitization of **N** using metallic compounds other than Cu(I) complex has been studied. Thus, the sensitization with orthometalated iridium(III) complex **23** having lower triplet-state energy than that of **N** has been explained by unusual partial electron transfer from **N** to excited Ir complex **23**.[26] Conversion of **N** to **Q** on semiconductors such as ZnO, ZnS, CdS, or Ge as sensitizers has been reported in the presence of air under solar light.[27] Quite variant results have been presented by other research groups and are described in Section 13.4.[61,62]

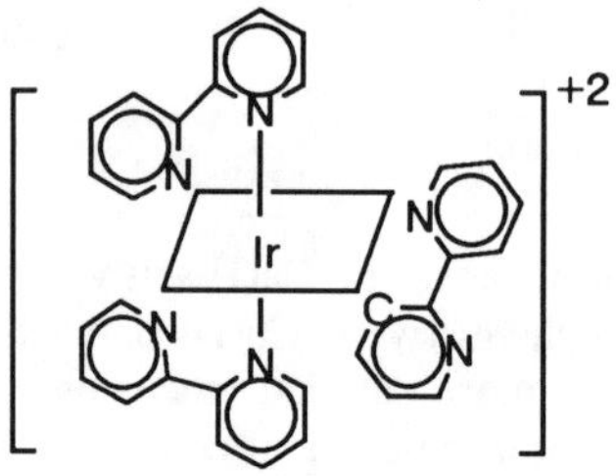

23

Mechanistic studies on rather equivocal triplet sensitization of **N** $\rightarrow$ **Q** photoisomerization by aromatic ketones are still in progress. Using several benzophenones substituted with electron-donating or -accepting groups as triplet sensitizers, an addition–elimination process through the adducts such as **24** was proposed for the **N** $\rightarrow$ **Q** conversion, and an electron-transfer process via norbornadiene-like radical cations for the **Q** $\rightarrow$ **N** photoisomerization.[28] Norbornadiene naphthyl carboxylates such as **25**, whose absorption spectra shift to the visible region, do not isomerize to the corresponding **Q**s on direct photolysis. Contrarily, sensitized reaction with biacetyl, a sensitizer with low triplet energy ($E_T = 55.5$ kcal mol^{-1}), has been found to work ($\Phi = 0.7$) and a mechanism of reversible energy transfer was proposed by quenching measurements of biacetyl phosphorescence.[29]

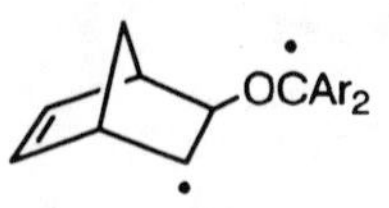

OCAr$_2$

24

R

R

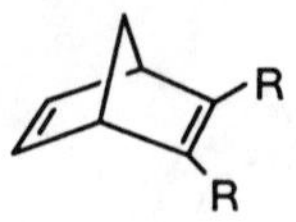

25 R = CO$_2$naphthyl

To prevent the formation of polymeric material during the repeated usage of triplet organic sensitizers, phenolic radical scavengers, such as 2,6-di-*tert*-butyl-*p*-cresol, have been found to be useful for the practical purpose of suppressing the extraneous reactions effectively.[30]

13.3B. Modified Norbornadienes

Norbornadiene itself does not undergo fully efficient valence isomerization upon direct or sensitized irradiation with visible light. Regardless of several disadvantages, such as lavish expenditure on preparations, small energy-storage capacity, and so on, the research on various modifications of the **N** structure have continued to receive attention, and **N**s bearing appropriate chromophores have been synthesized.

The direct photolysis of 1,2,3-trimethyl-5,6-dicyano-norbornadiene **26**, typical DONAC norbornadiene,[3c] has been found to proceed with much higher efficiency ($\Phi = 0.68$ at 366 nm) than the triplet sensitized reaction ($\Phi = 0.06$ at 546 nm with Ru(2,2′-bipyridine)$_3^{2+}$ and $\Phi = 0.11$ at 436 nm with biacetyl).[31] This opposite reactivity of **26** to that of parent **N** has been interpreted in the following elegant manner using the framework of the Turro model.[32]

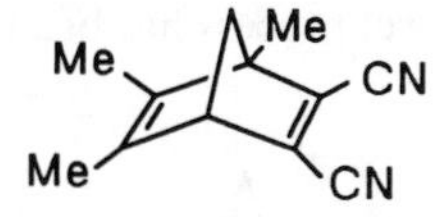

Me Me Me CN CN

26

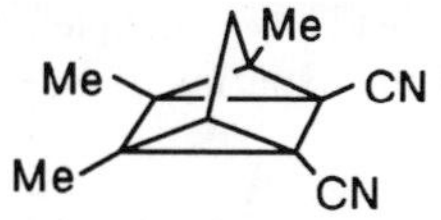

Me Me Me CN CN

27

Me Me Me + − CN CN

Me + Me Me − CN CN

28

The singlet excited state of **26** decays to **27** structure much more efficiently than the triplet state does, because the minimum in the surface of singlet excited state of **26** shifts closer to the quadricyclane. This shift of the minimum was accounted for in terms of ionic contribution of **28** stabilized by the cyano groups to the 1,3-diradicaloidal geometry.[32] The similar contribution of ionic structure **29** originated by the presence of phenylsulfonyl substituent has been recognized in the course of photoisomerization of *syn*-sesquinorbornatriene **30**.[33]

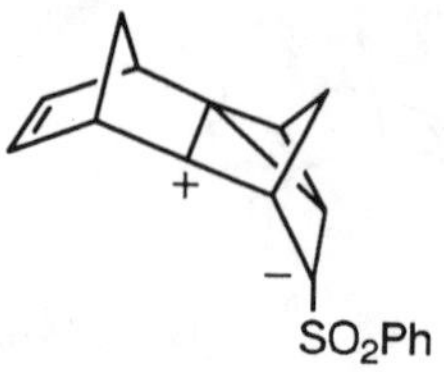

SO2Ph SO2Ph

29 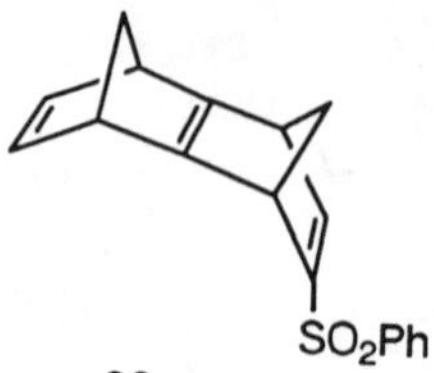 30

Dipyridylnorbornadienes **31** and corresponding ammonium salts result in the bathochromic shifts of their electronic absorptions, and the photochemical behavior (cycloaddition, photostationary state, addition of protic solvents) of these **N**s has been found to be dependent upon the structure.[34] Furthermore, the photolysis of norbornadienes **32** fused with methyl or methoxy naphthoquinones with visible light results in the formation of corresponding **Q**s ($\Phi = 0.1$–0.2 at 460 nm).[35]

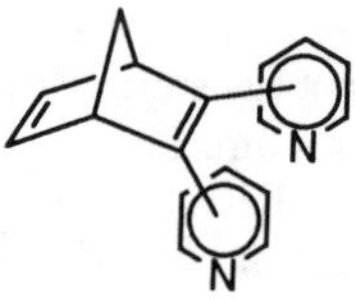

31 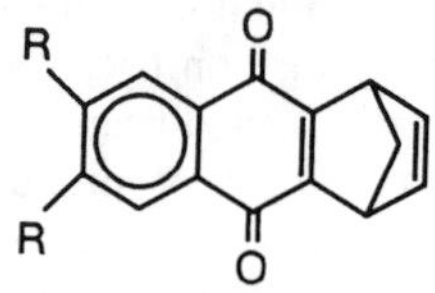 32

Although it has been concluded that **N** isomerizes by triplet sensitizers to **Q** through a biradicaloidal intermediate (diabatic photoreaction),[32] Takamuku and co-workers have found that naphthyl **Q** **33** photoreverts adiabatically to the corresponding naphthyl **N** **34** via triplet state. This novel process has been substantiated by various convincing experimental results.[36]

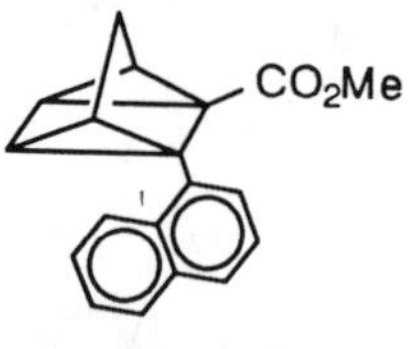

CO2Me CO2Me

33 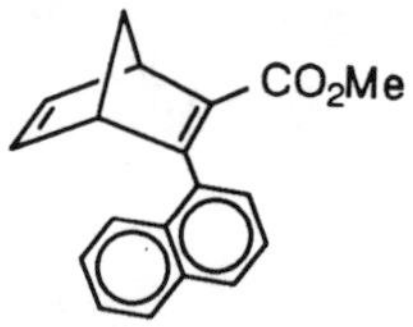 34

13.4. The Reversion of Quadricyclanes into Norbornadienes

For using valence isomerization between **Q** and **N** as a means of storing solar energy, the process of reversion must be freely controlled. Although the highly strained **Q**[37] is kinetically inert at ambient temperature because of orbital symmetry restrictions on the back reaction,[37,38] a number of catalysts have been presented to bring about the isomerization of **Q**. This chapter deals with the recently proposed or modified catalysts for the reversion and is broken down into two specific areas, that is, (1) homogeneous and heterogeneous catalysts and (2) mechanistic aspects of the reaction by electron-transfer process.

13.4A. Catalysts

13.4A(1). Homogeneous Catalysts. The literature through 1985 in the field of the homogeneous catalyst has been summarized previously.[39]

Using various methylated quadricyclane dicarboxylic acids **35** as substrates, it was proved that an apparent carbocation mechanism[40] of the reversion catalyzed by silver(I) salts, which have been extensively studied for the isomerization of strained hydrocarbons,[41] also plays a role both in aqueous ammonia and in benzene.[42] The attack of silver ion to the quadricyclane system has been ascribed to a good σ acceptor of the ion and has been suggested to be described with the aid of structure **36**, an alternative approach proposed in the cobalt(II)-catalyzed isomerization.[43]

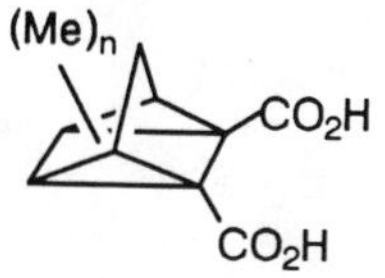

(Me)n CO2H CO2H

35

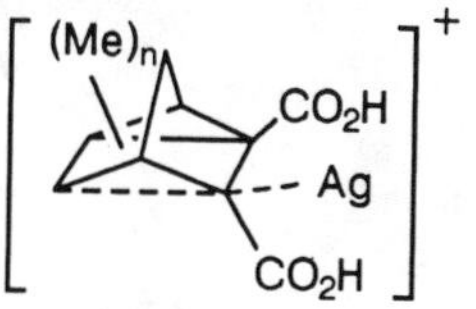

[(Me)n CO2H Ag CO2H]+

36

New cyclopropenylidene palladium(II) complexes **37** and **38** have been found to be efficient and stable catalysts for the cycloreversion of **Q** → **N**.[44] Di-μ-chloro species **37** was shown to have higher catalytic activities than the corresponding **38**, because the former dissociates more rapidly than the latter into the reaction species **39**, which is a coordinatively unsaturated molecule and is the predominant catalyst for the reversion.

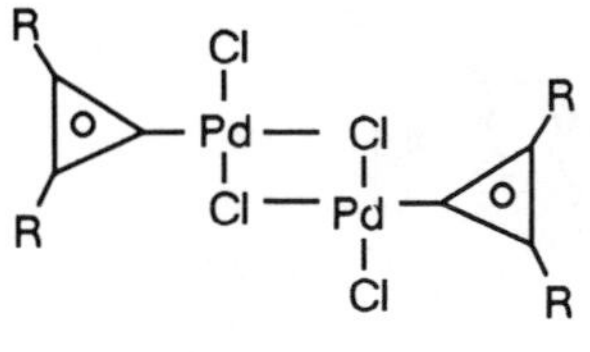

37

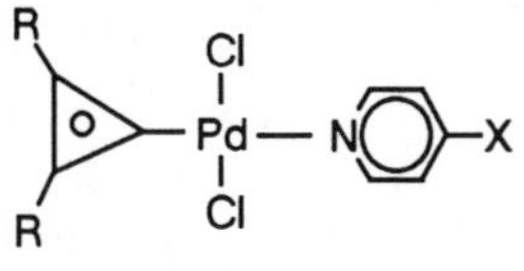

38

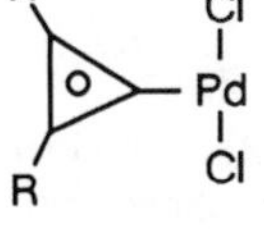

39

R = *t*-Bu, *i*-Pr_2N
X = H, Me, CN

Homogeneous catalysis of various cobalt porphyrins on the isomerization of **Q**-2-carboxylic acid[45] was enhanced, a k_a value approximately 10^4 times larger than that of Rh(I) catalysis,[46] by introducing a pyridyl or anisyl group to the ligand; the results have been interpreted by polar effects between basic catalysts and acidic **N** derivatives. The catalytic actions of the cobalt complexes with porphyrins and their related analogs has been studied in more detail under heterogeneous conditions; these will be described in Section 13.4A(2).

Strongly electrophilic lanthanoide $[Eu(CH_3CN)_3(BF_4)_3]_x$, which catalyzes oligomerization, polymerization, and isomerization of olefins, has been shown to initiate the spontaneous reaction of **Q** forming **N** in CH_3NO_2.[47]

Dimethyldioxirane **40**, whose reactivities have been investigated in detail as a new class of organic oxidant,[48] acts as a catalyst for the isomerization of **Q** with a turnover as high as 60.[49] Because **40** is regarded as an efficient oxygen transfer agent,[48] the feature of the reaction mechanism needs further clarification.

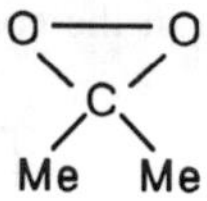

O—O
C
Me Me

40

13.4A(2). Heterogeneous Catalysts. In an actual energy plan based on the **N/Q** system, the use of heterogeneous catalysts is favorable in many respects. For this purpose immobilization of cobalt complex, which is one of the most active metal complexes for the isomerization of **Q**, has been attempted,[39] a typical example being bonding to macroreticular polystyrene.[50] Some investigations along this line have appeared since 1985. To simplify the method of preparation of the polymer-anchored catalyst, a Co(II) sulfonatophenyl porphine complex **41** has been bound ionically to an anion exchange resin.[51]

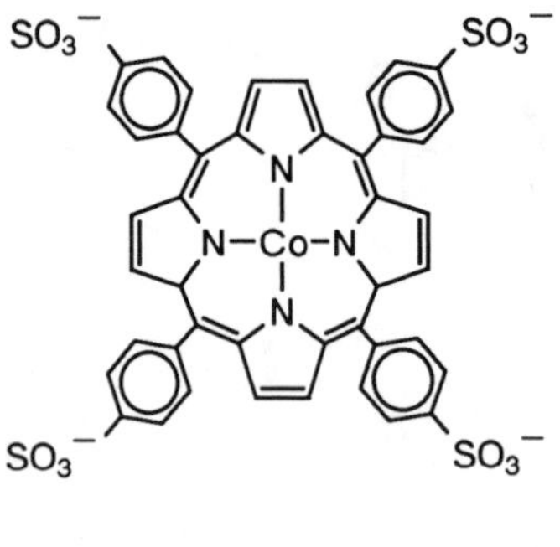

SO_3^- SO_3^-
N
N—Co—N
N
SO_3^- SO_3^-

41

This readily preparable agent shows catalytic activity to the isomerization of **Q** with selectivity similar to that of the Co(II) 5,10,15,20-tetraphenylporphyrin (homogeneous). Co(II)-complex **42** covalently bound on polystyrene with 30% crosslinking was prepared by Wöhrle and Buttner.[52] It has been claimed that catalyst **42** was up to that time the most active heterogeneous agent for valence isomerization of **Q**. Some acidic Co(II) complexes including **41** and cobalt(II) Schiff base complex **43** are anchored on alumina beads coated with polyamine sulfone **44**.[53] The catalysts show excellent properties: high activity, a small deactivation rate, and, moreover, the ability of regenerating their activities simply by heating at 200°C in vacuo.

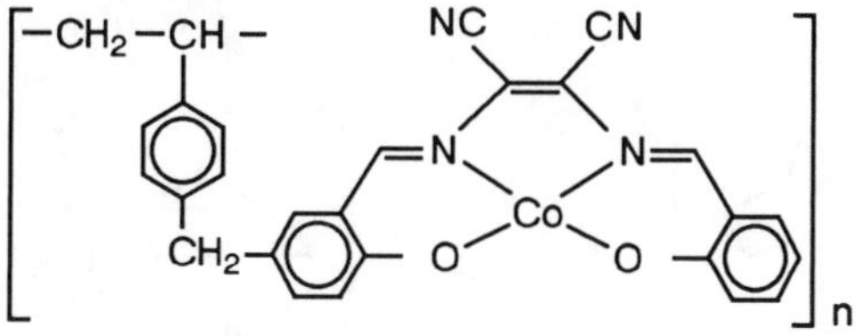

42

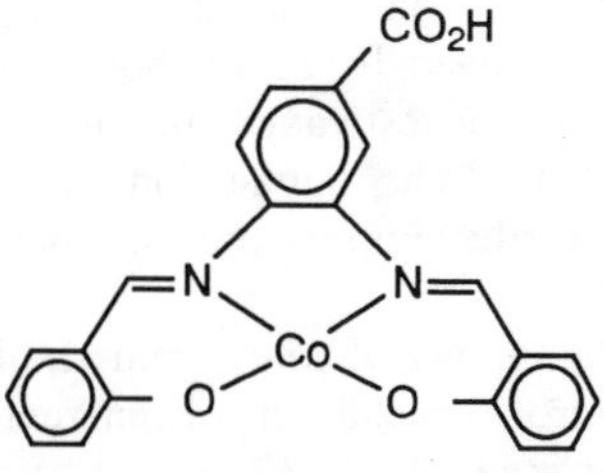

43

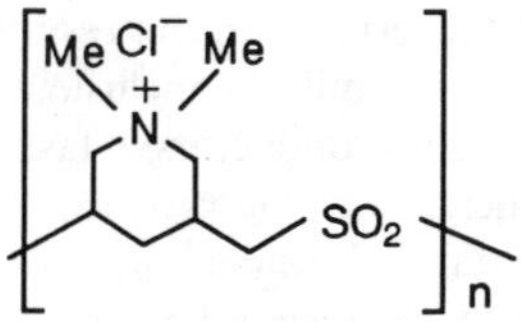

44

13.4B. Mechanistic Aspects of the Isomerization of Quadricyclane to Norbornadiene by Electron-Transfer Process

Applications of electron-transfer processes to the reversion reaction of **Q** to **N** emerged in the early 1980s. These came from the following findings: (a) the valuable data by Gassman and Yamaguchi[54] of half-wave oxidation potentials of **Q** (+0.91 V vs. SCE), **N** (+1.56 V vs. SCE), and related strained compounds; (b) photoelectron data of **Q** and **N**,[55] preparation of **N** radical cation **45** from both **Q** and **N** by γ-irradiation,[55] and observations of **N** and **Q** radical cations, **45** and **46**, using a chemically induced dynamic nuclear polarization (CIDNP) technique[56] that has experimentally made clear the relative stabilities of these radical cations; and (c) the theoretical studies[55,57] on both radical cations being in good agreement with the experimental data. These studies have stimulated a wide variety of applications of this process to the isomerization of **Q** to **N**.

Splitting water into its elements, available hydrogen and oxygen, by photoreaction on powdered semiconductors has been the focus of many intensive studies.[58] Semiconductors have also been used as heterogeneous photocatalysts[59] for diverse organic transformations.[60] Contrary to the variant results by Lahiry and Haldar,[27] although the reaction was conducted under different conditions, de Mayo and co-workers[61]

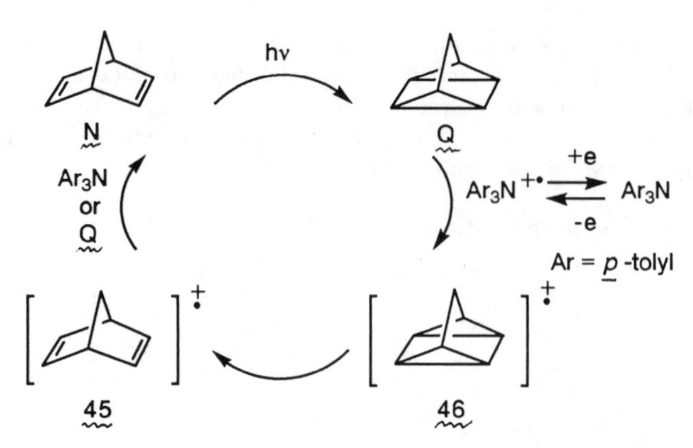

Scheme 13.3 Electrochemical transformation of quadricyclanes to norbornadienes.

reported the photoinduced reversion of **Q** to **N** on illuminated CdS and ZnO powders, proposing the electron-transfer mechanism based on both their various experimental results and the results of the MNDO investigation. Ikezawa and Kutal[62] also obtained similar conclusions independently. The application of catalysis on illuminated semiconductors is fascinating, but further improvements, especially of the overall efficiency of the process, are required.

ECE-catalyzed isomerization of **Q** to **N**[63] by the use of triarylamine radical cation, derived from corresponding triarylamine electrochemically, has been reported. This may enable the system to be suitable for recycling (Scheme 13.3).[64]

By a correlation of the quantum yields of **Q**s to **N**s with the half-wave oxidation potentials of **Q**s,[65] a redox-chain mechanism, which has been proposed as an electron transfer from **Q** to photoexcited **47**,[66] has achieved further confirmation.

Cl Pd Cl

47

Novel studies on the reaction of **Q** to **N** via radical cation **46** have recently been published. Thus, rapid conversion of **Q** "salted" in CsI or KBr into **N** under conditions that induce color center formation in the alkali halide has been reported recently. The reaction proceeds at a temperature where the color centers are mobile. A mechanism is suggested by the analogy of the known electron-transfer process forming the radical cations as intermediates.[67] Another example is that cytochrome P-450 has been found to catalyze the oxidation of **Q**, giving no **N** but rather nortricyclanol **48** and rearranged aldehyde as the principal metabolites; nortricyclyl radical cation has been suggested as a possible intermediate.[68]

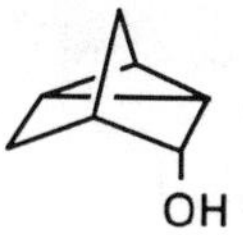

CHO
OH

48 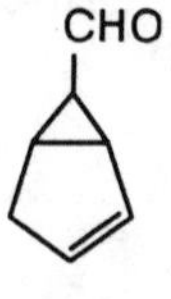 49

The studies of transient spectra of **Q** and **N**,[55] which were obtained after irradiation with high-energy electron pulses (0.6 kGy) at 96 K, proved directly the presence of distinct **Q** and **N** radical cations.[69] The radical cation **45** in $CF_2ClCFCl_2$ and CF_3CCl_3 matrices has been studied in detail by electron spin resonance (ESR) and external nuclear double resonance (ENDOR) spectroscopy.[70]

A unique method of the isomerization of **Q** to **N** has appeared. Thus, a hot monoenergetic ground state azulene, which is internally converted from its photoexcited S_2 state, has been shown to activate **Q** by collisional energy-transfer process. It has been estimated that the probability of collision is 10^{-3} but a large quantity of energy, ~35 kcal mol^{-1}, is transferred in a single event.[71]

13.5. Miscellaneous Studies

A series of Maruyama's studies of water-soluble norbornadienes has appeared successively.[39,42a,43,72] The concept was extended to the preparation of a novel amino acid derivative **50**, whose photoreactivity and stability in water were found to be most suitable.[73]

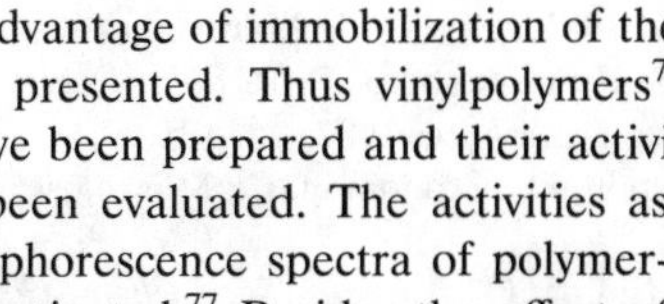

$CONHCH_2CH_2\overset{+}{N}H(CH_3)CH_3$
COO^-

50

In order to study the advantage of immobilization of the system, some photosensitive polymers have been presented. Thus vinylpolymers[74] and Merrifield's[75] polystyrene[76] with **N** units have been prepared and their activities to the interconversion between **N** and **Q** have been evaluated. The activities as triplet sensitizers for the **N** → **Q** reaction and phosphorescence spectra of polymer-bound benzyloxybenzaldehydes also have been investigated.[77] Besides the effect of these solid polymers, the photoreactivities of **N** in host–guest inclusion complexes have been tested.[78]

To investigate the properties of **Q**s bearing appropriate substituents, the thermal reversion of various 2-aryl-hexachloroquadricyclanes **51** has been studied.[79] By Arnold's treatment,[80] the ratio $\rho^{\cdot}/\rho$, which expresses the relative importance of the radical nature at the transition state, is estimated to be 1.11. Although **Q** itself has been reported to isomerize thermally via a diradical mechanism,[81] the value indicates that both radical and ionic characters contribute to stabilizing the transition state of the thermal reversion of **51**.

51

Adamson and co-workers[82] determined directly the enthalpies of the isomerization of **N** derivatives to corresponding **Q**s and also determined percentage ratios of stored to absorbed light energy of the photoprocesses by the method of his own photocalorimetry.[83] In contrast to the conventional methods utilizing combustion or hydrogenation process or differential thermal analysis, the experimental conditions in this case are the same as those of an actual photochemical process. The results show that bathochromic shift of the electronic absorption of **N** does not impair the energy storage ability of the system.

Acknowledgments. We are very grateful to Professor Keiichi Ito, Hokkaido Institute of Pharmaceutical Sciences, for his helpful advice and encouragement. This work was financially supported by Grant-in-Aid for Scientific Research from the Ministry of Education, Science and Culture of Japan.

References

1. Jones, G., II; Ramachandram, B. R. *J. Org. Chem.* **1976**, *41*, 798.
2. Representative reviews: (a) Hautala, R. R.; Little J.; Sweet, E. *Solar Energy* **1977**, *19*, 503. (b) Laird, T. *Chem. Ind.* **1978**, 186. (c) Kutal, C. *Adv. Chem. Ser.* **1978**, *168*, 158. (d) Scharf, H.-D.; Fleischhauer, J.; Leismann, H.; Ressler, I.; Schleker, W.; Weitz, R. *Angew. Chem., Int. Ed. Eng.* **1979**, *18*, 652. (e) Jones, G., II; Chiang, S.-H.; Thanhxuan, P. *J. Photochem.* **1979**, *10*, 1.
3. (a) Kutal, C. *J. Chem. Ed.* **1983**, *60*, 882. (b) Kutal, C. *Sci. Papers I.P.C.R.* **1984**, *78*, 186. (c) Yoshida, Z. *J. Photochem.* **1985**, *29*, 27. (d) Maruyama, K.; Tamiaki, H. *Kagaku to Kogyo* **1984**, *58*, 319. (e) Canas, L. R.; Greenberg, D. B. *Solar Energy* **1985**, *34*, 93. (f) Vicente, M.; Esplugas, S. *J. Chem. Tech. Biotechnol.* **1987**, *40*, 101.
4. Philippopoulos, C.; Economou, D.; Economou, C.; Marangzis, J. *Ind. Eng. Chem. Prod. Res. Dev.* **1983**, *22*, 627.
5. Srinivasan, R.; Baum, T.; Epling, G. *J. Chem. Soc., Chem. Commun.* **1982**, 437.
6. (a) Hirao, K.; Yonemitsu, O. *Farumashia* **1976**, *12*, 595. (b) Hirao, K. *Yakugaku Zasshi* **1980**, *100*, 473.
7. (a) Hamada, T.; Iijima, H.; Yamamoto, T.; Numao, N.; Hirao, K.; Yonemitsu, O. *J. Chem. Soc., Chem. Commun.* **1980**, 696. (b) Hirao, K.; Yamashita, A.; Ando, A.; Iijima, H.; Yamamoto, T.; Hamada, T.; Yonemitsu, O. *J. Chem. Res. (S)* **1987**, 162.
8. Iwakuma, T.; Hirao, K.; Yonemitsu, O. *J. Am. Chem. Soc.* **1974**, *96*, 2570.

9. We define a provisional threshold wavelength (end of absorption, abbreviated as EA) as that at which the optical density of a solution having a concentration of ca. 1.3×10^{-2} mol dm^{-3} is not perceived by a spectrophotometer.[7b] A widely acceptable definition of this special term should be given.
10. The diene **10**, however, cannot be converted to the corresponding cage compound **11** by photolysis, probably because of its instability being responsible for the contribution of the anisyl group to the stabilization of intermediate cation **i** for the reversion.

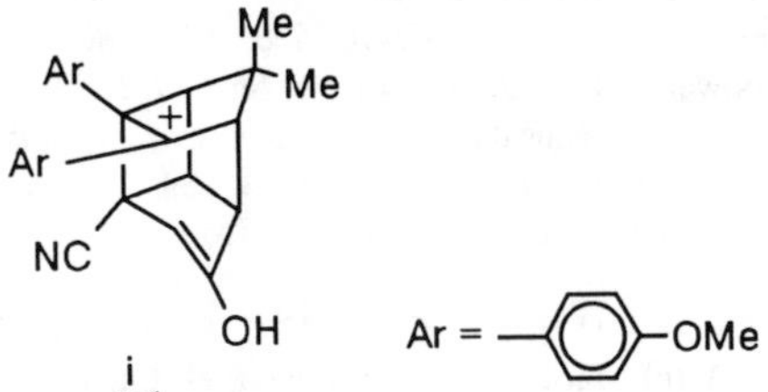

11. (a) Hirao, K.; Ando, A.; Hamada, T.; Yonemitsu, O. *J. Chem. Soc., Chem. Commun.* **1984**, 300. (b) Hirao, K.; Yamashita, A.; Ando, A.; Hamada, T.; Yonemitsu, O. *J. Chem. Soc., Perkin Trans. 1* **1988**, 2913.
12. Hirao, K.; Yamashita, A.; Yonemitsu, O. *J. Fluorine Chem.* **1987**, *36*, 293.
13. Compare Greenberg, A.; Liebman, J. F. *Strained Organic Molecules*; Academic: New York, 1978; pp. 333.
14. (a) Tanabe, K. *Solid Acids and Bases*; Kodansha: Tokyo, 1970. (b) Seiyama, T. *Metal Oxides and Their Catalytic Activities*; Kodansha: Tokyo, 1978. (c) Tanabe, K.; Seiyama, T.; Fueki, K. *Metal Oxides and Mixed Metal Oxides*; Kodansha: Tokyo, 1978. (d) Laszlo, P. *Acc. Chem. Res.* **1986**, *19*, 121.
15. Hirao, K.; Yamashita, A.; Yonemitsu, O. *Tetrahedron Lett.* **1988**, *29*, 4109. For the reversion of unstable **Q**[16] and 2-aroyl **Q**,[17] Al_2O_3 and SiO_2, respectively, have been used as the catalysts.
16. Butler, D. N.; Gupta, I. *Can. J. Chem.* **1982**, *60*, 415.
17. Toda, T.; Hasegawa, E.; Mukai, T.; Tsuruta, H.; Hagiwara, T.; Yoshida, T. *Chem. Lett.* **1982**, 1551.
18. Tanaka, J.; Ozaki, A. *J. Catal.* **1967**, *8*, 1.
19. Tanabe, K.; Ishiya, C.; Matsuzaki, I.; Ichikawa, I.; Hattori, H. *Bull. Chem. Soc. Jpn.* **1972**, *45*, 47.
20. (a) Kutal, C.; Schwendiman, D. P.; Grutsch, P. A. *Sol. Energy* **1977**, *19*, 651. (b) Kutal, C. *Adv. Chem. Ser.* **1978**, *168*, 158.
21. Fife, D. J.; Moore, W. M.; Morse, K. W. *J. Am. Chem. Soc.* **1985**, *107*, 7077.
22. (a) Fife, D. J.; Moore, W. M.; Morse, K. W. *Inorg. Chem.* **1984**, *23*, 1545. (b) Fife, D. J.; Moore, W. M.; Morse, K. W. *Inorg. Chem.* **1984**, *23*, 1684.
23. Liaw, B.; Orchard, S. W.; Kutal, C. *Inorg. Chem.* **1988**, *27*, 1311.
24. Basu, A.; Saple, A. R.; Sapre, N. Y. *J. Chem. Soc., Dalton Trans.* **1987**, 1797.
25. Onishi, M.; Hiraki, K.; Itoh, H.; Eguchi, H.; Abe, S.; Kawato, T. *Inorg. Chim. Acta* **1988**, *145*, 105.
26. Grutsch, P. A.; Kutal, C. *J. Am. Chem. Soc.* **1986**, *108*, 3108.
27. Lahiry, S.; Haldar, C. *Sol. Energy* **1986**, *37*, 71.
28. Arai, T.; Oguchi, T.; Wakabayashi, T.; Tsuchiya, M.; Nishimura, Y.; Oishi, S.; Sakuragi, H.; Tokumaru, K. *Bull. Chem. Soc. Jpn.* **1987**, *60*, 2937.
29. Favaro, G.; Aloisi, G. G. *Z. Phys. Chem. (Munich)* **1988**, *159*, 11.
30. Taoda, H.; Hayakawa, K.; Kawase, K. *J. Chem. Eng. Jpn.* **1987**, *20*, 335.
31. Ikezawa, H.; Kutal, C.; Yasufuku, K.; Yamazaki, H. *J. Am. Chem. Soc.* **1986**, *108*, 1589.

32. Turro, N. J.; Cherry, W. R.; Mirbach, M. F.; Mirbach, M. J. *J. Am. Chem. Soc.* **1977**, *99*, 7388.
33. Paquette, L. A.; Künzer, H.; Kesselmayer, M. A. *J. Am. Chem. Soc.* **1988**, *110*, 6521.
34. (a) Yamashita, Y.; Hanaoka, T.; Takeda, Y.; Mukai, T. *Chem. Lett.* **1986**, 1279. (b) Yamashita, Y.; Hanaoka, T.; Takeda, Y.; Mukai, T.; Miyashi, T. *Bull. Chem. Soc. Jpn.* **1988**, *61*, 2451.
35. Suzuki, T.; Yamashita, Y.; Mukai, T.; Miyashi, T. *Tetrahedron Lett.* **1988**, *29*, 1405.
36. Nishino, H.; Toki, S.; Takamuku, S. *J. Am. Chem. Soc.* **1986**, *108*, 5030.
37. Wiberg, K. B.; Connon, H. A. *J. Am. Chem. Soc.* **1976**, *98*, 5411.
38. Hoffmann, R.; Woodward, R. B. *J. Am. Chem. Soc.* **1965**, *87*, 2046.
39. Maruyama, K.; Tamiaki, H.; Kawabata, S. *J. Chem. Soc., Perkin Trans. 2* **1986**, 543.
40. Hogeveen, H.; Nusse, B. J. *Tetrahedron Lett.* **1974**, 159.
41. (a) Bishop, K. C., III. *Chem. Rev.* **1976**, *76*, 461. (b) Crabtree, R. H. *Chem. Rev.* **1985**, *85*, 245.
42. (a) Maruyama, K.; Tamiaki, H. *J. Org. Chem.* **1987**, *52*, 3967. (b) Maruyama, K.; Tamiaki, H. *Chem. Lett.* **1987**, 683. (c) Maruyama, K.; Tamiaki, H. *Chem. Lett.* **1987**, 485.
43. Maruyama, K.; Tamiaki, H. *J. Org. Chem.* **1986**, *51*, 602.
44. Miki, S.; Ohno, T.; Iwasaki, H.; Yoshida, Z. *Tetrahedron* **1988**, *44*, 55.
45. Maruyama, K.; Tamiaki, H. *Chem. Lett.* **1986**, 819.
46. Maruyama, K.; Terada, K.; Yamamoto, Y. *J. Org. Chem.* **1981**, *46*, 5294.
47. Thomas, R. R.; Chebolu, V.; Sen, A. *J. Am. Chem. Soc.* **1986**, *108*, 4096.
48. (a) Adam, W.; Curci, R.; Edwards, J. O. *Acc. Chem. Res.* **1989**, *22*, 205. (b) Murray, R. W. *Chem. Rev.* **1989**, *89*, 1187. (c) Compare Adam, W.; Hadjiarapoglou, L.; Nestler, B. *Tetrahedron Lett.* **1990**, *31*, 331.
49. (a) Murray, R. W.; Pillay, M. K. *Tetrahedron Lett.* **1988**, *29*, 15. (b) Murray, R. W.; Pillay, M. K.; Jeyaraman, R. *J. Org. Chem.* **1988**, *53*, 3007.
50. Rollmann, L. D. *J. Am. Chem. Soc.* **1975**, *97*, 2132.
51. Smierciak, R. C.; Giordano, P. J. *Appl. Catal.* **1985**, *18*, 353.
52. Wöhrle, D.; Buttner, P. *Polymer Bull.* **1985**, *13*, 57.
53. (a) Miki, S.; Asako, Y.; Morimoto, M.; Ohno, T.; Yoshida, Z.; Maruyama, T.; Fukuoka, M.; Takada, T. *Bull. Chem. Soc. Jpn.* **1988**, *61*, 973. (b) Miki, S.; Maruyama, T.; Ohno, T.; Tohma, T.; Toyama, S.; Yoshida, Z. *Chem. Lett.* **1988**, 861.
54. (a) Gassman, P. G.; Yamaguchi, R.; Koser, G. F. *J. Org. Chem.* **1978**, *43*, 4392. (b) Gassman, P. G.; Yamaguchi, R. *Tetrahedron* **1982**, *38*, 1113.
55. Haselbach, E.; Bally, T.; Lanyiova, Z.; Baertschi, P. *Helv. Chim. Acta* **1979**, *62*, 583, and references (4–6) cited therein.
56. Roth, H. D.; Schilling, M. L. M.; Jones, G., II. *J. Am. Chem. Soc.* **1981**, *103*, 1246.
57. Raghavachari, K.; Haddon, R. C.; Roth, H. D. *J. Am. Chem. Soc.* **1983**, *105*, 3110.
58. (a) Wait, T. D. In ACS Symposium Series 323; American Chemical Society: Washington, DC, 1986; p. 426. (b) Ollis, D. F. *NATO ASI Series, Series C*, 1986, 174, 651. (c) Cooper, W. J.; Herr, F. L. In ACS Symposium Series 327; American Chemical Society; Washington, DC, 1987; p. 1.
59. Kohl, P. A.; Bard, A. J. *J. Am. Chem. Soc.* **1977**, *99*, 7531.
60. Typical review: Fox, M. A. *Acc. Chem. Res.* **1983**, *16*, 314.
61. (a) Draper, A. M.; de Mayo, P. *Tetrahedron Lett.* **1986**, *27*, 6157. (b) Baird, N. C.; Draper, A. M.; de Mayo, P. *Can. J. Chem.* **1988**, *66*, 1579.
62. Ikezawa, H.; Kutal, C. *J. Org. Chem.* **1987**, *52*, 3299.
63. Hoffmann, R. W.; Barth, W. *J. Chem. Soc., Chem. Commun.* **1983**, 345.
64. Gassman, P. G.; Hershberger, J. W. *J. Org. Chem.* **1987**, *52*, 1337.
65. Kelley, C. K.; Kutal, C. *Organometallics* **1985**, *4*, 1351.
66. Borsub, N.; Kutal, C. *J. Am. Chem. Soc.* **1984**, *106*, 4826.

67. Kirkor, E. S.; Maloney, V. M.; Michl, J. *J. Am. Chem. Soc.* **1990**, *112*, 148.
68. Stearns, R. A.; Ortiz de Montellano, P. R. *J. Am. Chem. Soc.* **1985**, *107*, 4081.
69. Gebicki, J. L.; Gebicki, J.; Mayer, J. *Radiat. Phys. Chem.* **1987**, *30*, 165.
70. Gerson, F.; Qin, X.-Z. *Helv. Chim. Acta* **1989**, *72*, 383.
71. Hassoon, S.; Oref, I.; Steel, C. *J. Chem. Phys.* **1988**, *89*, 1743.
72. Maruyama, K.; Tamiaki, H.; Yanai, T. *Bull. Chem. Soc. Jpn.* **1985**, *58*, 781.
73. Tamiaki, H.; Maruyama, K. *Chem. Lett.* **1988**, 1875.
74. Kamogawa, H.; Yamada, M. *Bull. Chem. Soc. Jpn.* **1986**, *59*, 1501.
75. Merrifield, R. B. *J. Am. Chem. Soc.* **1963**, *85*, 2149.
76. Nishikubo, T.; Shimokawa, T.; Sahara, A. *Macromolecules* **1989**, *22*, 8.
77. Martinez-Utrilla, R.; Catalina, F.; Sastre, R. *J. Photochem. Photobiol., A* **1988**, *44*, 187.
78. (a) Yumoto, T.; Hayakawa, K.; Kawase, K.; Yamakita, H.; Taoda, H. *Chem. Lett.* **1985**, 1021. (b) Guarino, A.; Possagno, E.; Bassanelli, R. *Bull. Soc. Chim. France* **1988**, 253.
79. Tomioka, H.; Hamano, Y.; Izawa, Y. *Bull. Chem. Soc. Jpn.* **1987**, *60*, 821.
80. Dust, J. M.; Arnold, D. R. *J. Am. Chem. Soc.* **1983**, *105*, 1221.
81. Kabakoff, D. S.; Bünzli, J.-C. G.; Oth, J. F. M.; Hammond, W. B.; Berson, J. A. *J. Am. Chem. Soc.* **1975**, *97*, 1510.
82. Harel, Y.; Adamson, A. W.; Kutal, C.; Grutsch, P. A.; Yasufuku, K. *J. Phys. Chem.* **1987**, *91*, 901.
83. Adamson, A. W.; Vogler, A.; Kunkely, H.; Wachter, R. *J. Am. Chem. Soc.* **1978**, *100*, 1298.

Index

ab initio calculations 109, 252
ab initio calculations 3-21G and 6-31G* basis sets 46
ab initio transition structure 110
Absolute rotations 63, 67
Achiral [2.2](2,6)naphthalenophane 163
Achiral [3.3](2,6)naphthalenophane 163
Acid catalysts 174
Acid-catalyzed reaction 384
Acid-catalyzed reversion 385
Acid-promoted ring opening 11
Acylnorbornadiene 384
Acyloin condensation 156
Acylquadricyclane 384
Adamantane 11, 78, 294, 296–300, 311, 315
Adamantaneland 294, 299
Adamantane rearrangement 294, 298, 299, 302, 315
Adamantane system 53
Adamantyladamantane 234
Africanol 104, 106
Ag(I) 14
Alcohols 300, 302–305, 310, 313, 315
AM1 251
Amino acid derivative 395
Androstene-3,17-dione 110
6,(5β)-androstene-3,17-dione 110
Anionic C–C bond formation 156
Anionic cubane functionalization 50
anti-[4.4]metacyclophane 170
anti-[2.2](1,4)naphthalenophane 163
anti-[3.3](1,4)naphthalenophane 163
anti-naphthalenophanes 164, 166
Anti reflex effect 320, 321, 326, 328, 329, 342, 343, 344, 345, 348, 351, 360, 361
Anti reflex theory 354
anti-tricyclo[4.2.1.1^{2,5}]decane 298, 313
Arnold's treatment 395
Aromatization 11
Arylcubanes 46, 235
[6]-asterane 200
Asymmetrical substitution 172
Azahomocubene 230, 231
Azaprismanes 189

Barton, D. H. R. 225
Base-induced homoketonization 15
Base-promoted ring fragmentation 21
Base-promoted semibenzilic acid rearrangement of cage α-haloketones 21
Basketane 6, 70, 250
Basketanone 14
Basketene 69, 193, 250, 375
Benzyne 278
Bicyclohumulenone 104, 106
Biocatalyst 86
Biogenetic and biomimetic transannular cyclization 108
Biogenetic-like transannular reactions 108
Biological transformation 62
Biomimetic cyclization 108
Biomimetic synthesis 103
Biosynthesis 101, 103, 104
Biphenylophanes 172
Birch reduction 164, 165, 167, 172
α,ω-bis(ethylphenyl)alkane 165
α,ω-bis(m-ethylphenyl)alkane 172
α,ω-bis(m-vinylphenyl)alkane 161
α,ω-bis(p-vinylphenyl)alkane 161

α,ω-bis(vinylaryl)alkane 160
bis-seco-[6]-prismane tetraone 202
3,4:10,11-bis(2′,3′-quinolino)-tricyclo[6.3.0.0^{2,6}]undecane 32
2,3:6,7-bis(2′,3′-quinolino)pentacyclo-[6.5.0.0^{4,12}.0^{5,10}.0^{9,13}]tridecane 32
1,4-bishomo-6-seco-[7]-prismane dione 209
1,3′-bishomocubadiene 251
1,8-bishomocubane 6
bishomocubane 73, 227, 229, 242, 367, 377
1,8-bishomocubane (i.e., basketane) 14
1,3-bishomocubane 14, 24, 28
1,3-bishomocubanone 11
1,3-bishomopentaprismanedione 5
1,3-bishomopentaprismane 5, 38
Bisnorditwistane 6
1,5-bis(*p*-ethylphenyl)pentane 165
1,2-bis(*p*-vinylphenyl)ethane 161
1,5-bis(*p*-vinylphenyl)pentane 165
1,3-bis(*p*-vinylphenyl)propane 156, 158
Bissecocubane 63
Bissecododecahedradiene 39
Borden, W. T. 233
Boron trifluoride 300
Brendane 6, 64
Brendane derivative 11
Brexane 69
Bridgehead alkene 141
Bridgehead olefin 129, 145
Bromination 281

C_1-homobasketane 74
C_2-bishomocubane 73
C_2-bissecocubane 63
C_2-dehydroditwistane 74
C_2-ketone 87
C_2 symmetry 63
C_3 symmetry 79
C_4 unit cycloaddition 96
C_{16}-hexaquinacene 38
C_{60} formation in flames 94
Cage-compound 159
Cage fragmentation 30
Cage ketone 370, 372, 373, 378, 379
C_4 unit cycloaddition 96
Cage propellane 50
Cage-shaped molecule 62
Cage trione 374
Caged divalent carbon species (carbenes) 50
Caged divalent carbon species (carbenoids) 50
Capnellene 30
Carbocations 294, 296, 298, 300–302, 304, 305, 308–316
Carbon migrations 298, 309, 312, 315
Cation binding studies 31
Cationic C–C bond formation 172, 173
Cationic cyclization 179
Cationic cyclocodimerization 156, 158, 172, 174, 178
Cationic rearrangement 14
Cationic rearrangements of substituted homocubanone 14
CD spectra 83
Central core 63
Chiral [2.2](2,6)naphthalenophane 163
Chiral [3.3](2,6)naphthalenophane 163
Chiral polyhedron (*I* symmetry) 78
Chiral polyhedron (*O* symmetry) 78
Chiral polyhedron (*T* symmetry) 78
Chiroptical Properties 83
Chlorinated pentacyclo-[5.4.0.0^{2,6}.0^{3,10}.0^{5,9}]undecane-8,11-dione 21
CIDNP 275, 376, 377, 393
Cieplak's theoretical model 53
cis-(1,2)-ethano-*syn*-[2.5]metacyclophane 169
cis-(1,2)ethano[2.5]paracyclophane 169
cis-2,7-dimethylcyclooctanone 117
cis-bicyclo[6.3.0]undecane 104
cis,cisoid,cis linear triquinanes 5
cis,cisoid,cis-tricyclo[5.3.0.0^{2,6}]deca-3,9-diene 28
cis,exo-1,2-ethano-*syn*-[2.3](1,4)naphthalenophane 167
cis,syn,cis linear triquinane 3, 5, 28
Cleavage 21
CNDO 346
CNDO/2 352, 358
Cobalt porphyrin 392

Co(II) sulfonatophenyl porphine complex 392
Collisionally activated dissociation 287
Conformational biasing 101, 111, 115
Conformationally selective transannular cyclization 106, 108
Corannulene 94
Coriolin 30
Cotton effect 83
Crich, D. 225
Crown compound 171
Crown ether 30, 171, 172
Crowned *para*-benzoquinone 31
1,4-cubadiyl 46
Cubane (O_h symmetry) 78
Cubane 6, 24, 40, 63, 188, 250, 366
Cubene 39, 196
Cubene (dehydrocubane) 233
Cubyl anion 46, 50, 220, 223
Cubyl bromide 46
Cubyl cation 40, 43, 196, 225, 227, 229, 236
Cubyl chloride 46
Cubyl radical 196
Cubyl radical by halodecarboxylation 46
Cubyl radical by reductive decarboxylation 46
Cubyl triflate 43
Cubylcubane 196, 233
Cu(I) complexes 387
Cuneane 6, 14
Cunkle, G. T. 225
Cyclobutadiene 192
Cyclobutane ring 161
Cyclobutyl–cyclopropylcarbinyl 43
Cyclocodimerization 174, 176, 177
Cyclocodimers 173, 174
1,4- and 1,5-cyclodecadienes 111
Cyclodimerization of 7-substituted norbornadienes 10
Cyclohumulanoids 104, 108
Cyclohumulanoids biosynthesis 106
Cyclooctatetraene 204, 270
Cyclooctatetraenophanes 148, 149
Cyclopentadiene 209
Cyclophane syntheses 155
Cyclophane 155, 156, 159, 160, 161, 172
[3.3]cyclophane 158, 173
Cyclopropenylidene palladium(II) complex 391
[2 + 2] cycloreversion 11
Cytochrome P-450 394

D_2 symmetry 65
D_2-twist-boat cyclohexane 77
D_{2d}-bisnoradamantane 80
D_{2d} symmetry 80
D_3 symmetry 74
D_3-Trishomocubane 11, 74
D_3-tritwistane 77
D_3-twisted bicyclo[2.2.2]octane 62
D_{3h} symmetry 62
Daedalus 91
Decarbonylation 372, 373, 374
Decarboxylation 192
1,2-dehydrocubane 39
1,4-dehydrocubane 46
Dehydroisocalamendiol 108
Dehydrotwist-brendane 71
Delocalization energy 175
$\Delta\delta$ value 162
3-deoxyrosaranolide 119
Dewar benzene 189
Diacetylene carbinol 242
Diamantane 78
Diazomethane ring expansion 73
2,3-dichloro-5,6-dicyano-1,4-benzoquinone (DDQ) 166
9,10-dicyanoanthracene 370
Dicyclopentadiene derivative 14
Dicyclopentadienone ketal 192
Diels–Adler cycloaddition 38
Diels–Alder cycloaddition of 2,3-dimethylbutadiene to 5-fluoroadamantane-2-thione 53
Diels–Alder reactions with cyclopentadiene 31
Differential scanning calorimetry 279
1,9-dihalo-PCUD-8,11-diones 11
Dihydrohomocubane 6
9,10-dihydrohumulene-9-yl-cation 104
4,5-dihydroxyhomocubane 15
1,4-diiodocubane 40
Dimethyl 4,4′-bicubyldicarboxylate 46

Dimethyl bicyclo[2.2.2]octane-1,4-dicarboxylate 6
Dimethyl cubane-1,4-dicarboxylate 6
Dimethyldioxirane 392
Dimethyldioxolane 241
1,2-diphenylcyclobutane 158
Direct radical iodination of the cubane skeleton 46
Dissymmetric 62
2,5-disubstituted adamantane 53
Dithiacyclophane 157
Ditwist-brendane 71
Ditwistane 73
Ditwistane-5,11-dione, 6
Divalent carbon species 50
D_{nh} symmetry 184
Dodecahedradiene 39
Dodecahedrane 36, 139
Dodecahedrane (I_h symmetry) 78
Dodecahedrene 39
DONAC norbornadiene 389
DONAC system 385
Donor–acceptor complex 156
Donor–acceptor interaction 157, 158
DP (detonation pressure) 240
DPIBF(9,10-diphenylisobenzofuran) 233
Driving force for cationic rearrangement 11

Eaton, P. E. 218, 220, 225, 230, 231, 233
ECE-catalyzed isomerization 394
Electron donor–acceptor complex 379, 380
Electron transfer 275, 378, 379
Electron-transfer isomerization 377
Electron-transfer process 393
Electronic effect 51
Electronic substituent effect 50
Electrostatic repulsion 156
Empirical rule of Djerassi and Klyne 67
End of absorption 385
endo-dicyclopentadiene 258
endo-tetrahydrodicyclopentadiene (*endo*-THDCP) 294, 298, 299, 309–313
1,2-*endo*-trimethylene-8,9,10-trinorbornane 296, 301, 305
2-*endo*,6-*endo*-trimethylene-8,9,10-trinorbornane 296, 302, 304, 305
ENDOR spectroscopy 395
Energetic material 38
Energy minimization 103
Entropy advantage 157, 161
10-*epi*-periplanol B benzoate 114
ESR 395
(1,2)ethano[2.2]-metacyclophane 172
(1,2)ethano[2.3]-metacyclophane 172
Ethanoditwist-brendane 76
9-ethoxy-1-phenyl[1-^{13}C] 39
9-ethoxy-9-phenyl[9-^{13}C]pentacyclo-[4.3.0.0^{2,5}.0^{3,8}.0^{4,7}]nonane 39
exo-tetrahydrodicyclopentadiene (*endo*-THDCP) 294, 297, 298, 309–311, 313
1,2-*exo*-trimethylene-8,9,10-trinorbornane 296, 301, 305

Face-to-face arrangement 156
Farnesyl pyrophosphate 103, 104
Favorskii rearrangement 80, 322, 325, 343, 344, 345, 358
Favorskii ring contraction 188
Ferrahexaphospha-1,3′-bishomocubane 257
Ferratetraphospha-1,2-bishomocubane 255
Flash pyrolysis 265
5-fluoro-2-methyleneadamantane 53
Fluoroalkylquadricyclanes 386
Friedländer condensation of tetracyclo[6.3.0.0^{4,11}.0^{5,9}]undecane-3,8-dione 32
Full geometry optimization 109

Garudane 207
Gas-phase pyrolysis 5
Gauche conformation 168
Germacranoid 111
Germacrene 102, 103, 108, 111
Gleiter, R. 239

Half-wave oxidation potential 393
Haller–Bauer 21, 284
Hasebe, M. 227
Hasegawa, T. 228

HCTD 10
Helical model 67
Helvetane 187
Heptacyclo[6.6.0.0^{2,6}.0^{3,13}.0^{4,11}.0^{5,9}.0^{10,14}]-tetradecane 10
Hexacyclo[5.4.1.0^{2,6}.0^{3,10}.0^{5,9}.0^{8,11}]dodecane-4,12-dione (i.e., 1,3-bishomopentaprismanedione 38
Hexaphospha-1,2-bishomocubane 257
Hexaphospha-1,3-bishomocubane 261
Hexaphospha-1,3′-bishomocubane 256
Hexaprismane 199, 369
High-dilution technique 173
High-molecular-weight vinylarene 156
High-symmetry chiral 62
Higuchi, H. 223, 236, 242
Hirsutane 104
Hirsutene 30
Hirsutanoids 104
1,6-Hoffman degradation 163
Homoazulene 141
Homoconjugation 82
Homoconjugative interaction 38
Homocub-1(9)-ene 39
Homocub-4(5)-ene 39
Homocubane 6, 14, 70, 251, 366
Homocubanone 366
Homocubene (dehydrohomocubane) 233
Homocuneane 6
Homohypostrophene 369
Homoketonization 20, 21
Homopentaprismane 28, 198, 265, 369
Homoprismane nitrile 230
Hormann, R. E. 230
Host–guest chemistry 178
Hrovat, D. A. 233
Humulene 102, 103, 104, 108, 111
Humulene 9,10-epoxide 106, 108
Humulenic and germacrenic cation 103
Hunsdicker reaction 64
Hydride shifts 298, 302, 304, 305, 308–315, 317
Hydrocarbons 298–300, 304, 308, 309, 314, 315
Hydrogenolyses of strained carbon–carbon σ bond 5
Hydrogenolysis of homocubane 6
Hydrophobicity 172
σ hyperconjugation 53
Hypostrophene 196, 253, 369

Icospiral PAH embryos 93
Illudoid 103, 104
Immobilization of cobalt complex 392
Intramolecular [2 + 2] photocyclization 31
Intramolecular [2 + 2] photocycloaddition 163
Intramolecular reaction 156
Iodinated cubanes 221, 222, 223, 225, 233
4-iodocubyl cation 40
Ionic hydrogenation 300, 301, 310, 315
Ionization potential 174
Ionophore 31
Isogarudane 207
Isomerization 314
Israelane 187
Iterative maximum overlap calculation 250

Karcher, M. 239
Kepone 289
Kimura, Y. 229

Lanthanoide 392
Large-membered paracyclophane 172
Lead tetraacetate in the presence of trifluoroacetic acid 10
Least-motion pathway 108, 110, 111, 115, 119
Least-motion reaction 116
Lewis acid catalyzed isomerization 294, 298, 299, 308
Lewis acid promoted [2 + 2] cycloreversion 11
Lithiated cubane 220, 221, 222, 223
Lithium metal 172

M-C_2 ketone rule 87
M helicity 83
Macrocyclic Diels–Alder reaction 109
Macrocyclic diterpene 108
Macrocyclic epoxidation 114

Macrocyclic stereocontrolled epoxidation 114
Macrocyclic stereocontrolled epoxidation 115
Macrocyclic stereocontrolled reaction 102, 116, 119
Macrocyclic terpene 101, 102, 103, 108
Macrocyclic terpenoid 101
Macrocyclically stereocontrolled transannular cyclization 109
Macrolide 101, 119
Maggini, M. 233
Malonate synthesis 163
Mass spectrometry 275
2-mercaptopyridine-*N*-oxide sodium salt 227
[2.2]metacyclophane 172
Metacyclophane 162, 165
Metal oxides 387
Metallocubane 220, 222, 235
Methanoditwistane 76
1-methoxy-PCUD-8,11-dione 11
7-methyl-2-methylenecyclooctanone 118
Methyl 4-bicubylcarboxylate 46
Methyl 4-*tert*-butylcubanecarboxylate 46
9-methylcyclooct-2-enone 118
2-methylcyclooctanone 117
3-methylcyclooctanone 117
Mirex 287
Mixed metal oxide 387
MM2 114, 115, 117, 118, 170, 187, 251
MM2 energy minimizations 109, 110, 114
MM2 method 169
MM2 transition structure 109
MMI calculation 104, 106, 108
MMI energy minimization 108
MMI program 103
MMRS program 109, 110, 114
Modhephene 126, 149
Molecular cleft 30
Molecular mechanics calculations 102, 103, 108, 109, 114, 116, 119, 321, 346
Molecular mechanics calculations of bridgehead carbonium ion reactivities 42, 43
Molecular modeling 109, 116, 117
Moriarty, R. M. 225
Motherwell, W. B. 225
Multibridged cyclophane 167, 170

[*n*]-prismane 184
[4.4.*n*](1,3,5)cyclophane 168
[2.*n*](2,6)naphthalenophane 163
[2.2](2,6)(2,7)naphthalenophane 163
[2.2](2,7)naphthalenophane 163
(2,6)naphthalenophane 163
[3.4](1,4)naphthalenophane 167
Naphthalenophane 163, 166, 172
(1,4)naphthalenophane 178
(1,5)naphthalenophane 167
(2,6)naphthalenophane 178
[3.4](2,6)naphthalenophane 166
Natural product chemistry 102, 119
[4.*n*]biphenylophane 165
[4.*n*]cyclophane 164, 172
[3.*n*]cyclophane 156
Nickel(0) 24
Nitration 282
[*n*]metacyclophane 141
[*n.m*]paracyclophane 157, 167
NOE interaction 170
NOESY 170
Nondissymmetric 62
Nonplanar sp^2 network 95
Noradamantane 69
Norbornadiene 207, 376
Norbornenobenzoquinone 209
1-norbornyl cation 43
1-norbornyl triflate 43
Norsnoutane 6
[*n*]paracyclophane 146, 147
[4.*n*]paracyclophane 165
[*n*]staffane 140
Nuclear fusion 172
Nuclear magnetic resonance 50

Octant rule 83
Okahara cyclization 171
Oligoethylene glycol 171
[2.3]-one 166
[4.4]-one 166
ortho-aminobenzaldehyde 32
Ortho-directing ability of the amide group 50

Ortho-metalation 50
Ortho-metalation reaction 46
[1.2]Orthophane 172
Osawa, E. 238
Outer ring 83
Oxacyclophane 168
Oxidation 10, 281
Oxidative deiodination of cubyl iodide 40

P-C_2 ketone rule 87
P-helicity 87
Paddlane 168, 171
Pagodane 36, 138
Pagodane dication 43
Palladium(II) 28
[2.2]paracyclo(1,4) 163
[2.2]paracyclophane 157, 158, 172
[3.3]paracyclophane 156, 157
[2.3]paracyclophane 172
[3.4]paracyclophane 167
[4.4]paracyclophane 167
Pentacyclic cage alcohol and acetate 15
Pentacyclo[4.4.0.$0^{2,5}$.$0^{3,8}$.$0^{4,7}$]decane, 1,1-bishomocubane 250
Pentacyclo[4.4.0.$0^{2,5}$.$0^{3,9}$.$0^{4,7}$]decane, 1,2-bishomocubane 250
Pentacyclo[4.4.0.$0^{2,5}$.$0^{3,9}$.$0^{4,8}$]decane, 1,3′-bishomocubane 250
Pentacyclo[5.3.0.$0^{2,5}$.$0^{3,9}$.$0^{4,8}$]decane, 1,3-bishomocubane 250
Pentacyclo[5.3.0.$0^{2,6}$.$0^{3,9}$.$0^{4,8}$]decane, 1,4-bishomocubane 250
Pentacyclo[5.4.0.$0^{2,6}$.$0^{3,10}$.$0^{5,9}$]undecane 3, 30, 51
Pentagons among the hexagons 94
Pentaphospha-1,2-bishomocubane 256
Pentaprismane 196, 252, 369
Peripheral attack 111, 113, 114, 115, 116
Peripheral epoxidation 113, 115
Peripheral face 111
Peripheral stereocontrol 111, 114
Periplanol B 114
Periplanone B 111, 113, 114, 115
[4]peristylane 38
9-phenyl-1(9)-homocubene 39
Phenylcubane 50
Phosphacubane 239
Photocalorimetry 396
Photocyclization 166
[2 + 2] photocyclization 159
Photocycloaddition 158, 171, 258
[2 + 2] photocycloaddition 189
Photocyclodimerization 159
Photodimerization 158
Photoelectron data 393
Photoelectron (PE) spectroscopy 51
Photoinduced electron-transfer reaction 376
Photoinduced electron transfer 275
Photooxidation 171
Photosensitive polymer 395
Photothermal metathesis 3, 30, 36
Pinacolic coupling 197
Platonic solid 78
Polyaza cavity-shaped molecule 32
Polycyclic sesquiterpene 103
Polycyclic terpene 101
Polycyclic terpenoid 103
Polyhedrane 183
Polynitro[4]peristylane 38
Polynitrocubane 240, 241
Polyquinane natural product 30
Postfullerene organic chemistry 91
Preisocalamendiol 108
[3]-prismane 184
[4]-prismane 187
[5]-prismane 188
[6]-prismane 188
[7]-prismane 210
[8]-prismane 209
propellacubane 196
propellane 125
Protoadamantane 82, 296
Protoilludane 104
Protonation 174
Pyramidalized alkene 39
Pyramidalized olefin 232, 233

Q "salted" in CsI or KBr 394
Quadrant rule 87
Quadricycl-1(7)-ene 39
Quadricyclane 376
Quadrone 144

Quasi-Favorskii 263
Quinone methide cubane 236

Reactivity parameter 174
Rearrangements 294, 295, 299, 304
Reddy, D. S. 225
Reduction 282
Reflex effect 320, 321, 322, 323, 324, 325, 332, 346, 348, 351, 360, 361
Regular polyhedron 78
Remote asymmetric induction 115, 119
Resonance interaction 43
Reverse transmetalation 46
$[Rh(CO)_2Cl]_2$ 24
$[Rh(diene)Cl]_2$ 24
Rhodium(I) 24

SCF 252
Schlegel diagram 98
Secocubane 6, 69
Secocuneane 6
Secohexaprismane 205
Secohomocubane 15
Selectivity 163
Semiconductor 388, 393
Semiempirical quantum mechanical calculation 321
Sesquiterpene biosynthesis 103
Sesquiterpene 102, 108
Silver(I) 28
Silver(I) salt 391
Sodium borohydride 325, 331
Sodium borohydride reduction 334, 342
Solar energy 383
Solar energy conversion and storage 28
Solar-energy storage system 371, 373
Solid acid 387
Solvolysis 163
Solvolysis of cubyl triflate 43
Sooting system 93
Spiro[2.2.2.2]deca-4,9-diene 157
Spirocyclization 97
Spirodimerization 98
Spontaneous fullerene formation 92
Stabilomer 11, 74
Stereoselectivity 178
Stilbene 159
Strain 156, 161, 165
Strain energy 172
Styrene 156, 159, 173
Styrene photodimerization 160
Substituted 1,3-bishomocubanone 11
5-substituted adamantanone 53
4-substituted cubane-1-carboxylic acid 50
Substituted cubane 14
Substituted heptacyclo-6.6.0.0^{2,6}.0^{3,13}.0^{4,11}.0^{5,9}.0^{10,14}]tetradecane 51
Substituted secopentaprismanone 28
Substituted trishomocubanes by protic acid and/or Lewis acid 11
Sulfur extrusion 158
syn-[2.2](1,4)naphthalenophane 163
syn-[2.2]metacyclophane 158
syn-[2.4]-naphthalenophane 167
syn-[2.*n*](1,4)naphthalenophane 163
syn-[2.*n*](1,5)naphthalenophane 163
syn-[2.*n*]paracyclonaphthalenophane 163
syn-[3.3]-naphthalenophane 167
syn-[3.3](1,4)naphthalenophane 163
Syn-diene 237, 239
syn-orthocyclophane 32
syn-tricyclo[4.2.1.1^{2,5}]decane 298, 313

T_d conformation 78
Terpene 101, 102
tert-butyl cubanepercarboxylate 46
Tetra-*tert*-butyl-tetrahedrane 78
Tetrabora-tetraphospha-1,4-bishomocubane 267
Tetracyanoquinodimethane 157
Tetracyclo[5.2.1.0^{2,6}.0^{4,8}]decanes 6
Tetracyclo[6.2.2.0^{2,7}.0^{4,9}]dodecane-5,11-dione, 6
Tetrahedrane (T_d symmetry) 78
Tetrahydrohomocubane 6
Tetramethylene unit 164
Through-bond 50
Through-bond (T–B) interaction 238
Through-bond versus through-space interaction 51
Through-space 50
Tosmic synthesis 163
trans-2,7-dimethylcyclooctanone 118

trans-2,8-dimethylcyclooctanone 117
Transannular cyclization 106, 108
Transannular cyclization 101, 102, 103, 104
Transannular Diels–Alder reaction 109, 110
Transannular nonbonded repulsion 102, 111
Transannular nonbonding interaction 102
Transannular reaction 104, 106, 110
Transannular repulsion 118
Transition-metal-promoted valence isomerization of highly strained cage compound 24
Transmetalation 46, 220, 221, 222
Transmission of electronic substituent effect 50
Tricyclo[5.2.1.$0^{2,6}$]deca-2,5,8-triene 38
Tricyclo[5.3.0.$0^{4,8}$]decane 297, 308
Tricyclodecane 297, 298, 313
Triethano[2_3](1,3,5)cyclophane 159
Triethyl silane 300
1,7-trimethylene-8,9,10-trinorbornane 297, 305, 308, 309
Trimethylsilyl iodide 169
Triplet sensitization of **N** → **Q** photoisomerization by aromatic ketone 388
Triprismane 189
Triquinane natural product 36
Trishomocubane 367, 371
1,2,4-trishomocubane 5, 28
Trishomocubanone 371
Twistane 6, 65
Twistbrendane 6, 65
Twisted olefin 230, 231
Two-electron reduction of iodocubane 50

Ueda, I. 242

Vertical delocalization 234
Via cationic rearrangement of suitably substituted PCUD 11
Vinylarene 155, 156
Vinylnaphthalene 163

Wagner-Meerwein shift 230, 231
Westheimer approach 347
Westheimer method 352, 358
Wiberg index 354

X-ray 36
X-ray crystal structure of the novel molecular cleft 35